ELECTRICITY ELECTRONICS|

PRINCIPLES

AND

APPLICATIONS

DEDICATION

I dedicate this book to my wife Jo whose understanding and patience assisted me in writing this text; and to my daughter Joanne who gave up many a weekend to type the manuscript.

DELMAR PUBLISHERS INC. • ALBANY, NEW YORK 12205

ELECTRICITY ELECTRONICS

PRINCIPLES AND APPLICATIONS

Joseph M. De Guilmo P.E.

COPYRIGHT © 1982
BY DELMAR PUBLISHERS INC.

10 9 8 7 6 5 4 3 2 1

LIBRARY OF CONGRESS CATALOG CARD NUMBER: 79-54909
ISBN: 0-8273-1686-0

Printed in the United States of America
Published simultaneously in Canada by
Delmar Publishers, A Division of
Van Nostrand Reinhold, Ltd.

Preface

The advances made in electricity and electronics in the past several decades have brought profound changes in modes of modern living, education, technology, and directions for the future. The study of electricity/electronics opens challenging career opportunities to the student armed with a thorough knowledge of basic concepts and an inquisitive mind.

ELECTRICITY/ELECTRONICS: PRINCIPLES AND APPLICATIONS is an introductory text that provides a thorough explanation of the major topics in electrical technology. Extensive coverage of solid-state electronics is also included, ranging from simple measurement and control to microprocessors. The text covers the basic concepts, circuitry, and components involved in the manufacturing of machinery, instruments, and consumer products. The fundamentals of control are also examined.

The text is divided into five sections, proceeding from basic concepts to industrial applications. Section I covers electrical fundamentals, starting with units and notation, through basic concepts, and ending with resonance and filters.

Section II covers applied electrical technology by discussing the following topics: polyphase circuits, transformers, generators, and motors.

Section III introduces the student to electronics fundamentals, including the study of rectifier diodes, bipolar junction transistors, and field effect transistors. Electronic amplifiers are covered in Chapter 25.

Section IV on applied electronics technology presents an in-depth explanation of topics not generally seen in texts of this type. Chapter 27 covers modulation and communication circuits. Operational amplifiers are extensively described in

Chapter 28. Regulating semiconductors such as Zener diodes, SCRs, TRIACs, and DIACs are studied in detail in Chapter 29. This chapter also describes regulated power supplies. Chapter 30 is unique in its presentation of logic circuits and applications. Audio systems are covered in Chapter 31.

Section V on instrumentation technology is an innovative addition to a basic text. Chapter 32 analyzes negative feedback, instrumentation systems, devices, and control. Chemical and industrial technicians will find this chapter particularly appropriate. Although this unit has material seldom seen in an introductory text, the use of many drawings, photographs, and solved problems makes this chapter easy to understand.

Chapters 33 and 34 discuss the measuring and testing instruments used by all technicians. The oscilloscope is discussed in great detail. Detailed explanations are given of a number of new instruments such as logic analyzers and spectrum analyzers. Students are advised to use Chapters 33 and 34 as references while completing the various laboratory exercises given in the text.

The material presented in this text has been classroom tested and revised over a number of years of teaching. As a result of this refining process, the text is designed to be used by a diverse student population.

The level of presentation is suitable for students in nonelectrical technology programs who require some background in electrical and electronic components, circuits, and applications. Persons engaged in two-year programs in chemical, mechanical, civil, environmental, power, plant, maintenance, quality control, and other nonelectrical fields will find the text invaluable. Students in trade schools and adult education programs will find the text suited to their specialized needs.

In addition to the programs listed, this text is ideal for an introductory course for electrical/electronics technicians. More advanced electrical/electronics students will find it useful for further study for selected topics, as a reference, or for a review of basic principles.

Practicing technicians and engineers will find the text useful because of its up-to-date range of topics. Those technicians and engineers trained in other fields will be able to use the text as a handy guide to the understanding and use of electricity/electronics concepts.

About the Author:

In writing this text, Joseph M. De Guilmo has been able to call upon forty-three years of experience as an engineer in industry and as a teacher of electrical/electronics technology in both two-year and four-year post-secondary programs.

Mr. De Guilmo earned his undergraduate engineering degree from Stevens Institute of Technology. His Master of Science degree was awarded by the New Jersey Institute of Technology (formerly the Newark College of Engineering).

For many years Mr. De Guilmo taught at the Industries Training School of Stevens Institute of Technology. As a Professor of Engineering Technology for twelve years, he taught at the Bronx Community College of the City University of New York.

At present, Mr. De Guilmo is the Program Coordinator of the Electronics Technology program of the Hudson County Community College at Stevens Institute of Technology. He is a professional engineer registered in the state of New Jersey. In addition, he is a member of the Board of Visitors of ABET (Accreditation Board for Engineering and Technology), for the purpose of accrediting engineering technology programs in two-year and four-year colleges.

Mr. De Guilmo's professional affiliations include the following:

- Institute of Electrical and Electronics Engineers, Senior Member and member of The Power Engineering Society.

- Instrument Society of America, Senior Member and District I Education Chairman.

- American Society for Engineering Education, Member affiliated with the Relations with Industry Division, the Continuing Engineering Studies Division, and the Engineering Technology Division.

- Member of Tau Alpha Pi, National Engineering Technology Honor society.

- Faculty Advisor to Omicron Delta Chapter of Tau Alpha Pi of Hudson County Community College at Stevens Institute of Technology.

Contents

The digital computer — The microprocessor (μP) — The
microcomputer (μC) — The calculator — μP applications

Section 1
Electrical Fundamentals

| Chapter 1 | Units and Notation |

OBJECTIVES

After studying this chapter, the student will be able to

- apply the principles of scientific notation and powers of ten notation.
- discuss the SI system of units and the metric system of units.
- analyze problems by dimensional analysis.
- use mathematical symbols in problems.

POWERS OF TEN NOTATION

Powers of ten notation is a mathematical tool based on the properties of the powers of ten. This type of notation expresses numbers in a form that simplifies mathematical operations. Table 1-1 shows how large numbers given in ordinary, or decimal, notation are converted to powers of ten notation. Large numbers are numbers greater than one.

TABLE 1-1 Conversion of Large Numbers

Ordinary or Decimal Notation	Powers of Ten Notation
1	10^0
10	10^1
100	10^2
1 000	10^3
10 000	10^4
100 000	10^5
1 000 000	10^6
10 000 000	10^7
100 000 000	10^8
1 000 000 000	10^9

TABLE 1-2 Conversion of Small Numbers

Ordinary or Decimal Notation	Powers of Ten Notation
0.1	10^{-1}
0.01	10^{-2}
0.001	10^{-3}
0.000 1	10^{-4}
0.000 01	10^{-5}
0.000 001	10^{-6}
0.000 000 1	10^{-7}
0.000 000 01	10^{-8}
0.000 000 001	10^{-9}

Problem 1 Express 100 000 000 000 in powers of ten notation.
Solution $100\ 000\ 000\ 000 = 10^{11} = 1.0 \times 10^{11}$

(R1-1) Express 10 000 000 000 in powers of ten notation.

Table 1-2 shows how small numbers are converted from ordinary, or decimal, notation to powers of ten notation. Small numbers are numbers less than one.

Problem 2 Express 0.000 000 000 1 in powers of ten notation.
Solution $0.000\ 000\ 000\ 1 = 10^{-10} = 1.0 \times 10^{-10}$

(R1-2) Express 0.000 000 000 001 in powers of ten notation.

SCIENTIFIC NOTATION Scientific notation is a method of expressing both very large and very small numbers as a product of a number less than ten and a power of ten.

Large Numbers. To express a number greater than ten in scientific notation, the following procedure is used.
1. The decimal point is moved to the *left* until one digit not equal to zero remains to the *left* of the decimal point.
2. The remaining number is called the coefficient of the power of ten.
3. The number of places the decimal point is moved to the *left* is the *positive* exponent of the power of ten.
4. The product of the coefficient and the power of ten is the original number expressed in scientific notation.

Problem 3 Express 14 650 000 in scientific notation.
Solution $14\ 650\ 000 = 1.465 \times 10^{7}$

Problem 4 Express 34 800 in scientific notation.
Solution $34\ 800 = 3.48 \times 10^{4}$

(R1-3) Express 29 000 in scientific notation.
(R1-4) Express 1 000 000 000 000 in scientific notation.

To convert a number expressed in scientific notation to ordinary (decimal) notation, the decimal point is moved to the right if the exponent is positive. The exponent indicates the number of places that the decimal point must be moved.

Problem 5 Convert 3.02×10^7 to ordinary (decimal) notation.
Solution $3.02 \times 10^7 = 30\ 200\ 000$

Problem 6 Convert 10^3 to ordinary (decimal) notation.
Solution $10^3 = 1.0 \times 10^3$
$ = 1\ 000$

The student must understand that $10^3 = 1.0 \times 10^3$. The exponent of 10^3 is a positive three. This means that the decimal point in 1.0 is to be moved three places to the right to obtain the value of 1.0×10^3 in ordinary notation.

(R1-5) Convert 100×10^4 to ordinary notation.

(R1-6) Convert 1.458×10^7 to ordinary notation.

Small Numbers. To express a number less than 1.0 in scientific notation, the following procedure is used.
1. The decimal point is moved to the *right* until one digit remains to the left of the decimal point.
2. The remaining number is called the coefficient of the power of ten.
3. The number of places the decimal point is moved to the *right* is the *negative* exponent of the power of ten.
4. The product of the coefficient and the power of ten is the original number expressed in scientific notation.

Problem 7 Convert 0.000 25 to scientific notation.
Solution $0.000\ 25 = 2.5 \times 10^{-4}$

A number with a negative exponent can also be expressed as the reciprocal of the number with the exponent changed to a positive value. As a proof:

$$2.5 \times 10^{-4} = 2.5 \times \frac{1}{10^4}$$
$$= 0.000\ 25$$

Problem 8 Convert 0.000 054 to scientific notation.
Solution $0.000\ 054 = 5.4 \times 10^{-5}$

Problem 9 Convert 0.000 1 to scientific notation.
Solution $0.000\ 1 = 1.0 \times 10^{-4}$
$ = 10^{-4}$

(R1-7) Convert 0.002 22 to scientific notation.

(R1-8) Convert 0.000 000 000 05 to scientific notation.

To convert a number less than 1.0 written in powers of ten notation to ordinary or decimal notation, the decimal point is moved to the *left* because the exponent is *negative*. The exponent indicates both the direction in which the decimal point is moved and the number of places it is moved.

Problem 10 Convert 8.55×10^{-4} to ordinary or decimal notation.
Solution $8.55 \times 10^{-4} = 0.000\ 855$

Problem 11 Convert 9.85×10^{-6} to ordinary or decimal notation.
Solution $9.85 \times 10^{-6} = 0.000\ 009\ 85$

Problem 12 Convert 10^{-3} to ordinary or decimal notation.
Solution $10^{-3} = 1.0 \times 10^{-3}$
$= 0.001$

(R1-9) Convert 9.22×10^{-6} to ordinary or decimal notation.

(R1-10) Convert 10^{-5} to ordinary or decimal notation.

ADDITION AND SUBTRACTION USING SCIENTIFIC NOTATION

The powers of ten must be the same before adding or subtracting quantities expressed in scientific notation.

Problem 13 $(5.3 \times 10^3) + (1.4 \times 10^3) = ?$
Solution $(5.3 \times 10^3) + (1.4 \times 10^3) = (5.3 + 1.4) \times 10^3$
$= 6.7 \times 10^3$

Problem 14 $(7.7 \times 10^{-4}) - (4.3 \times 10^{-4}) = ?$
Solution $(7.7 \times 10^{-4}) - (4.3 \times 10^{-4}) = (7.7 - 4.3) \times 10^{-4}$
$= 3.4 \times 10^{-4}$

Problem 15 $(2.48 \times 10^5) + (5.6 \times 10^5) = ?$
Solution $(2.48 \times 10^5) + (5.6 \times 10^5) = (2.48 + 5.60) \times 10^5$
$= 8.08 \times 10^5$

(R1-11) $(8.5 \times 10^{-1}) + (1.2 \times 10^{-1}) = ?$

MULTIPLICATION AND DIVISION USING SCIENTIFIC NOTATION

When multiplying or dividing quantities expressed in scientific notation, the powers of ten may be different. Exponents are added in multiplication and subtracted in division.

Problem 16 $(5.5 \times 10^2) \times (2.0 \times 10^4) = ?$
Solution $(5.5 \times 10^2) \times (2.0 \times 10^4) = (5.5 \times 2.0) \times (10^2 \times 10^4)$
$= 11.0 \times 10^{(4 + 2)} = 11.0 \times 10^6$
$= 1.1 \times 10^7$

Problem 17 $\dfrac{4.8 \times 10^6}{1.6 \times 10^2} = ?$

Solution $\dfrac{4.8 \times 10^6}{1.6 \times 10^2} = \left(\dfrac{4.8}{1.6}\right) \times \left(\dfrac{10^6}{10^2}\right) = \left(\dfrac{4.8}{1.6}\right) \times (10^{6-2})$

$\qquad\qquad\qquad = 3.0 \times 10^4$

(R1-12) $(-1.5 \times 10^5) \times (1.5 \times 10^{-3}) = ?$

(R1-13) $\dfrac{-7.5 \times 10^{-2}}{1.5 \times 10^3} = ?$

FINDING ROOTS IN POWERS OF TEN NOTATION The square root of a quantity expressed as a power of ten is determined by (1) finding the square root of the coefficient, and (2) then finding the square root of the power of ten. To arrive at the square root of the power of ten, divide the exponent by two.

Problem 18 $\sqrt{625 \times 10^4} = ?$

Solution $\sqrt{625 \times 10^4} = \sqrt{625} \times \sqrt{10^4}$

$\qquad\qquad\qquad = 25 \times 10^2$

$\qquad\qquad\qquad = 2.5 \times 10^3$

Problem 19 $\sqrt[3]{27 \times 10^9} = ?$

(This expression means that the student is to find the cube root of 27×10^9. The same general procedure is followed in finding roots other than square roots. The desired root is determined for both the coefficient and the power of ten.)

Solution $\sqrt[3]{27 \times 10^9} = \sqrt[3]{27} \times \sqrt[3]{10^9}$

$\qquad\qquad\qquad = 3 \times 10^{\left(\frac{9}{3}\right)}$

$\qquad\qquad\qquad = 3 \times 10^3$

(R1-14) $\sqrt[4]{16 \times 10^{12}} = ?$

Problem 20 $\sqrt{4.9 \times 10^5} = ?$

Solution $\sqrt{4.9 \times 10^5} = \sqrt{49 \times 10^4}$

$\qquad\qquad\qquad = \sqrt{49} \times \sqrt{10^4} = 7.0 \times 10^{\left(\frac{4}{2}\right)}$

$\qquad\qquad\qquad = 7.0 \times 10^2$

In Problem 20, it was necessary to change the number to obtain an even power of ten. The decimal point of the coefficient was shifted one place to the right. Then, the exponent of the power of ten was decreased by one. As a result, the exponent of the final quantity was a whole number.

Problem 21 $\sqrt[3]{21.6 \times 10^7} = ?$

Solution $\sqrt[3]{21.6 \times 10^7} = \sqrt[3]{216 \times 10^6}$

$\qquad\qquad\qquad = \sqrt[3]{216} \times \sqrt[3]{10^6} = \sqrt[3]{216} \times 10^{\left(\frac{6}{3}\right)}$

$\qquad\qquad\qquad = 6.0 \times 10^2$

Problem 22 $\sqrt{25\ 000}\ =\ ?$

Solution $\sqrt{25\ 000}\ =\ \sqrt{2.5 \times 10^4}$

$$=\ \sqrt{2.5} \times \sqrt{10^4}\ =\ \sqrt{2.5} \times 10^{(\frac{4}{2})}$$

$$=\ 1.58 \times 10^2$$

(R1-15) $\sqrt[4]{62.5 \times 10^9}\ =\ ?$

SI SYSTEM   In October 1965, the Institute of Electrical and Electronics Engineers adopted the International System of Units as a way of expressing physical quantities. This SI system is based on the metric MKS system. The basic MKS units are meters, kilograms (one kilogram equals 1 000 grams), and seconds.

Four of the basic or fundamental units used in the SI system are listed in Table 1-3.

Additional SI units are required for electronics measurements. These units are derived from the four fundamental units. Some of the derived units used in this text are listed in Table 1-4.

Problem 23   Express a current of 10 amperes using the correct SI unit symbol.

Solution $I\ =\ 10\ A$

Problem 24   Express a resistance of 1 000 ohms using the correct SI unit symbol.

Solution $R\ =\ 1\ 000\ \Omega$

Problem 25   Express an electric charge of 10^{-6} coulombs using the correct SI unit symbol.

Solution $Q\ =\ 10^{-6}\ C$

(R1-16)   Express a frequency of 710 000 Hz (cycles per second) using the correct SI unit symbol.

(R1-17)   A toaster has a power rating of 1 000 watts. Express this value using the correct SI unit symbol.

TABLE 1-3 Basic SI Units

Physical Quantity	SI Unit	SI Unit Symbol
Length (l)	meter	m
Mass (M)	kilogram	kg
Time (t)	second	s
Current (I)	ampere	A

TABLE 1-4 Derived SI Units

Physical Quantity	SI Unit	SI Unit Symbol
Capacitance (C)	farad	F
Charge (Q)	coulomb	C
Conductance (G)	siemens	S
Energy (W)	joule	J
Frequency (f)	hertz	Hz
Force (F)	newton	N
Inductance (L)	henry	H
Power (P)	watt	W
Resistance (R)	ohm	Ω
Voltage (V or E)	volt	V

TABLE 1-5 Metric System Prefixes

Prefix	Symbol	Multiplier
atto	a	10^{-18}
femto	f	10^{-15}
pico	p	10^{-12}
nano	n	10^{-9}
micro	μ	10^{-6}
milli	m	10^{-3}
centi	c	10^{-2}
deci	d	10^{-1}
deka	da	10^{1}
hecto	h	10^{2}
kilo	k	10^{3}
mega	M	10^{6}
giga	G	10^{9}
tera	T	10^{12}

PREFIXED UNITS The physical quantities a technician works with are often much smaller or much larger than the basic units. Prefixed units are used to express these quantities. A prefix is added to the name of the basic unit to modify it to obtain the appropriate unit. Each of the prefixed units, such as a kilometer, consists of a prefix (kilo-) and the basic unit (meter). The prefix (kilo-) is a *multiplier* which represents the quantity (10^{3}). Table 1-5 lists the common prefixes, symbols, and multipliers used in the SI metric system.

Problem 26 1 kilometer = ?

Solution 1 kilometer = 1 km $\times$ 10^{3} $\dfrac{m}{km}$

= 1 000 m

Problem 27 Express 85 millivolts in SI units, using the correct prefix.

Solution 85 millivolts = 85 mV

Problem 28 Express 0.08 microampere in SI units, using the correct prefix.

Solution 0.08 microampere = 0.08 μA

(R1-18) Express 900 picocoulombs in SI units, using the correct prefix.

THE METRIC SYSTEM OF UNITS The metric system of units, also known as the MKS system, uses the meter as its basic unit of length. The meter is equal to 39.37 inches. Quantities in the metric system cover a wide range of values. Thus, various prefixes are used to designate larger and smaller values of the base unit of the meter. Tables 1-6 and 1-7 show how measurements are expressed using prefixes for multiples and submultiples of the meter.

TABLE 1-6 Metric Unit Relationships

1 meter (m) = 100 centimeters = 1 000 millimeters
1 centimeter (cm) = 10 millimeters = 0.01 meter
1 millimeter (mm) = 0.1 centimeter = 0.001 meter
1 kilometer (km) = 1 000 meters

TABLE 1-7 Metric Conversions

Given Unit	X Multiplier =	Resultant Unit
m	10^3	mm
m	10^2	cm
m	10^{-3}	km
cm	10^{-2}	m
cm	10^1	mm
mm	10^{-1}	cm
mm	10^{-3}	m
km	10^3	m
m^2	10^4	cm^2
cm^2	10^{-4}	m^2

Problem 29 Convert 25 cm to mm.

Solution The conversion multiplier (10^1) is obtained from Table 1-7.

$$25 \text{ cm } = 25 \text{ cm } \times 10 \frac{\text{mm}}{\text{cm}}$$
$$= 250 \text{ mm}$$

Problem 30 Convert 30 mm to cm.

Solution
$$30 \text{ mm } = 30 \text{ mm } \times 10^{-1} \frac{\text{cm}}{\text{mm}}$$
$$= 3.0 \text{ cm}$$

Problem 31 Convert 0.055 m to cm.

Solution
$$0.055 \text{ m } = 0.055 \text{ m } \times 10^2 \frac{\text{cm}}{\text{m}}$$
$$= 5.5 \text{ cm}$$

In addition to the meter, the other basic units in the MKS system are kilograms and seconds. A comparison of the MKS system of units and the English system of units is made in Table 1-8.

Problem 32 Convert 10 m to feet (ft).

Solution
$$10 \text{ m } = 10 \text{ m } \times 39.37 \frac{\text{in}}{\text{m}}$$
$$= 393.7 \text{ in}$$
$$= 393.7 \text{ in } \times \frac{1 \text{ ft}}{12 \text{ in}}$$
$$= \frac{393.7}{12} \text{ ft } = 32.8 \text{ ft}$$

TABLE 1-8 Comparison of Metric and English Systems of Units

Physical Quantity	MKS System	English System
Length	Meter (m) (39.37 in)	Yard (yd) (0.914 m)
Mass	Kilogram (kg)	Slug (14.6 kg)
Force	Newton (N)	Pound (lb) (4.45 N)
Energy	Joule (J) or Newton-meter (0.737 8 ft-lb)	Foot-pound (ft-lb) (1.356 joules)
Temperature	Celsius ($^\circ$C) $^\circ C = \dfrac{5}{9}(^\circ F - 32^\circ)$	Fahrenheit ($^\circ$F) $^\circ F = \dfrac{9}{5}{}^\circ C + 32^\circ$
Time	Second (s)	Second (s)

Problem 33 Convert 200 ft to m.

Solution

$$200 \text{ ft} = 200 \text{ ft} \times 12 \frac{\text{in}}{\text{ft}}$$

$$= 2\ 400 \text{ in}$$

$$= 2\ 400 \text{ in} \times \frac{1}{39.37} \frac{\text{m}}{\text{in}}$$

$$= \frac{2\ 400}{39.37} \text{ m} = 61 \text{ m}$$

Problem 34 The boiling point of water is 212°F. What is the boiling point of water in $^\circ$C?

Solution

$$^\circ C = \frac{5}{9}(^\circ F - 32^\circ)$$

$$= \frac{5}{9}(212^\circ - 32^\circ)$$

$$= \frac{5}{9}(180^\circ)$$

$$= 100^\circ$$

(R1-19) Which is the larger unit: the kilometer or the mile?

INTERCONVER-SION OF UNITS Very often it is necessary to change from one unit to another to solve problems in electronics. This type of conversion is done easily using scientific notation and the prefixes given in Table 1-7.

Problem 35   Convert 150 microhenries (μH) to henries (H).

Solution $150\ \mu\text{H} = 150 \times 10^{-6}\ \text{H}$

To convert from a unit with a prefix to a fundamental unit, the prefixed unit is multiplied by a conversion factor. When the conversion is being made from a smaller unit to a larger unit, the exponent of the conversion factor is negative. In this problem, the multiplier 10^{-6} replaces the prefix μ.

Problem 36   Convert 0.004 5 H to mH.

Solution

$$0.004\ 5\ \text{H} = 0.004\ 5\ \text{H} \times 10^3\ \frac{\text{mH}}{\text{H}}$$

$$= 4.5\ \text{mH}$$

In this problem, the conversion is being made from a larger unit to a smaller unit and the exponent of the conversion factor is positive.

(R1-20)   Change 30.2 μH to H.

(R1-21)   Change 0.003 72 H to μH.

Problem 37   Convert 20.5 mH to μH.

Solution

$$20.5\ \text{mH} = 20.5\ \text{mH} \times 10^{-3}\ \frac{\text{H}}{\text{mH}} = 20.5 \times 10^{-3}\ \text{H}$$

$$= 20.5 \times 10^{-3}\ \text{H} \times 10^6\ \frac{\mu\text{H}}{\text{H}}$$

$$= 20.5 \times 10^3\ \mu\text{H}$$

To convert from one prefixed unit to another prefixed unit, the given unit is first converted to a fundamental unit. The fundamental unit is then converted to a unit having the desired prefix.

(R1-22)   Change 0.055 mH to μH.

NUMERICAL ACCURACY   The answers to most practical problems in electronics can be expressed to an accuracy of three significant figures. The significant figures in a number are the digits of that number. Zeros placed at the beginning or end of the number are *excluded* if their only purpose is to locate the decimal point.

Problem 38   How many significant figures are contained in the following numbers: 1.55, 502, 62.5, 1.555, 1 555, 1.8?

Solution

NUMBER	SIGNIFICANT FIGURES
1.55	3
502	3
62.5	3
1.555	4
1 555	4
1.8	2

Problem 39 The current in a circuit is measured at 2.55 mA. Express the current in amperes. How many significant figures does the answer contain?

Solution $2.55 \text{ mA} = 2.55 \text{ mA} \times 10^{-3} \dfrac{A}{mA}$

$$= 2.55 \times 10^{-3} \text{ A}$$
$$= 0.002\ 55 \text{ A}$$

The answer (0.002 55 A) contains three significant figures. The two zeros in front of the digits 255 are added to locate the decimal point. Since the zeros are not involved in determining the accuracy of the reading, they are not significant figures.

Problem 40 What is the difference between a voltage reading of 50 V and 50.0 V?
Solution The 50-V reading is accurate to two significant figures. This means that the voltage reading is closer to 50 V than it is to 49 V or 51 V. The voltmeter must be calibrated in one-volt intervals to give a reading to the nearest whole number of volts.

A 50.0-V reading is accurate to three significant figures. This means that the voltage reading is closer to 50.0 V than it is to 49.9 V or 50.1 V. The voltmeter must be calibrated in one-tenth volt intervals to give a reading to the nearest tenth of a volt.

Problem 41 When multiplying 5 555 by 55, how many significant figures should the answer contain?

Solution Since one of the original numbers (55) is accurate to two significant figures, the product of the two numbers should contain two significant figures.

$$5\ 555 \times 55 = 305\ 525$$

The answer is closer to 310 000 than it is 300 000. Therefore, the correct answer to two significant figures is 310 000.

(R1-23) Round off the following numbers to three significant figures:
a. 56.358 4
b. 0.222 22
c. 254 977

DIMENSIONAL ANALYSIS The units of measurement in all terms of an equation must be consistent. In other words, only one *system* of units can be used in any one equation. If the SI system of units is used on one side of an equation, it must also be used on the other side. *Dimensional analysis* is a technique that treats units as algebraic quantities. Mathematical operations are expressed by words representing the units and figures representing the number of the units.

Problem 42   Convert 32 ounces (oz) to pounds (lb).

Solution   Conversion relation: 1 lb = 16 oz

For example, 1 lb per 16 oz or $\dfrac{1 \text{ lb}}{16 \text{ oz}}$

The expression "per" represents "divided by".

$$32 \text{ oz} = 32 \text{ oz} \times \frac{1 \text{ lb}}{16 \text{ oz}}$$

$$= \left(\frac{32 \times 1}{16}\right) \left(\frac{\text{oz} \times \text{lb}}{\text{oz}}\right)$$

$$= 2 \text{ lb}$$

There are two steps in the solution. The first step is to find the numerical ratio by dividing 32 by 16 to obtain 2. In the second step, the units are treated as algebraic quantities. The ounce units cancel and the resulting unit is pounds. The correct use of the conversion relation is very important.

Problem 43   An automobile travels a distance of 320 miles in 480 minutes (min). What is the speed in miles per hour?

Solution   Conversion relation: 60 min = 1 h

$$\frac{320 \text{ miles}}{480 \text{ min}} = \frac{320 \text{ miles}}{480 \text{ min}} \times \frac{60 \text{ min}}{1 \text{ h}}$$

$$= \left(\frac{320 \times 60}{480 \times 1}\right) \left(\frac{\text{miles} \times \text{min}}{\text{min} \times \text{h}}\right)$$

$$= 40 \frac{\text{miles}}{\text{h}}$$

$$= 40 \text{ miles per hour (m/h)}$$

Problem 44   Convert 162 square inches (in^2) to square feet (ft^2).

Solution   Conversion relations: 1 ft = 12 in

$$1 \text{ ft}^2 = (12 \text{ in})^2 = 144 \text{ in}^2$$

$$162 \text{ in}^2 = (162 \text{ in}^2) \left(\frac{1 \text{ ft}^2}{144 \text{ in}^2}\right)$$

$$= \left(\frac{162 \times 1}{144}\right) \left(\frac{\text{in}^2 \times \text{ft}^2}{\text{in}^2}\right)$$

$$= 1.125 \text{ ft}^2$$

(R1-24)   How many minutes are there in one week?

THE RIGHT TRIANGLE   Figure 1-1 is used to show how the trigonometric functions of a right triangle are obtained. The functions of the angle θ (theta) are defined as follows:

1. The sine of an angle is the ratio of the length of the opposite side to the length of the hypotenuse.

$$\sin \theta = \frac{\text{opposite side}}{\text{hypotenuse}}$$

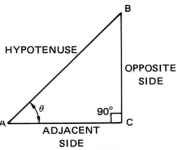

Fig. 1-1 The right triangle

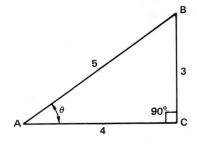

Fig. 1-2

2. The cosine of an angle is the ratio of the length of the adjacent side to the length of the hypotenuse.

$$\cos \theta = \frac{\text{adjacent side}}{\text{hypotenuse}}$$

3. The tangent of an angle is the ratio of the length of the opposite side to the length of the adjacent side.

$$\tan \theta = \frac{\text{opposite side}}{\text{adjacent side}}$$

Problem 45   Find the sine, cosine, and tangent of the angle θ in figure 1-2.

Solution

$$\sin \theta = \frac{\text{opposite side}}{\text{hypotenuse}}$$

$$= \frac{3}{5}$$

$$= 0.6$$

$$\cos \theta = \frac{\text{adjacent side}}{\text{hypotenuse}}$$

$$= \frac{4}{5}$$

$$= 0.8$$

$$\tan \theta = \frac{\text{opposite side}}{\text{adjacent side}}$$

$$= \frac{3}{4}$$

$$= 0.75$$

Problem 46   Find the sine of the angle θ in figure 1-3.

Solution

$$\sin \theta = \frac{6}{10}$$

$$= 0.6$$

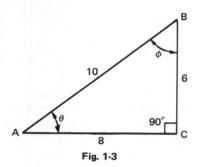

Fig. 1-3

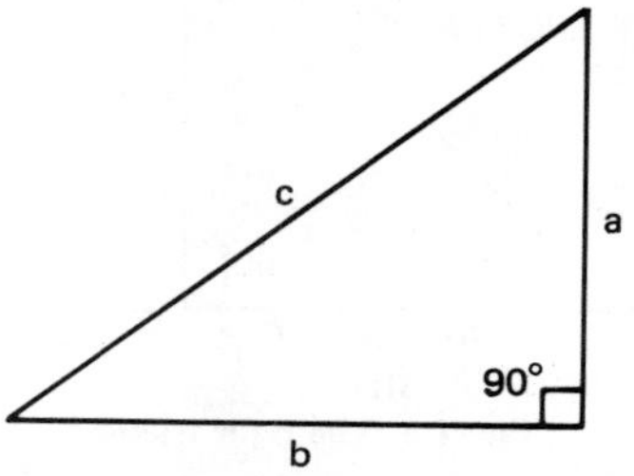

Fig. 1-4 Pythagorean Theorem, the right angle triangle

The sine of the angle θ is the same as the sine of the angle θ in figure 1-2. This means that the angle θ has the same value in both triangles. In other words, the sine of an angle depends upon the ratio of the sides involved and not on the size of the triangle.

$\theta = \sin^{-1} 0.6$ or arc sin $\theta = 0.6$

This equation states that θ is an angle whose sine is equal to 0.6. In this case: $\theta = 36.9°$.

(R1-25)   What is the tangent of the angle ϕ in figure 1-3?

PYTHAGOREAN THEOREM   The Pythagorean Theorem expresses the relationship between the lengths of the three sides of a right triangle. The theorem states that *the square of the hypotenuse is equal to the sum of the squares of the other two sides.* Figure 1-4 illustrates this theorem.

The Pythagorean Theorem is expressed in mathematical form by:

$$c^2 = a^2 + b^2$$

This equation can be rearranged to obtain the following expressions:

$$c^2 = a^2 + b^2$$
$$b^2 = c^2 - a^2$$
$$a^2 = c^2 - b^2$$

(R1-26)   The sides of the right triangle shown in figure 1-4 have the following values:

$a = \sqrt{3} = 1.732$

$b = 1.0$
Find c.

ANGLES IN FOUR QUADRANTS The four quadrants are shown in figure 1-5. The signs of the trigonometric functions in the four quadrants are listed in Table 1-9. Functions of selected angles are shown in Table 1-10. Variations in the values of the functions in the four quadrants are listed in Table 1-11.

(R1-27) What angles are included in the first quadrant?

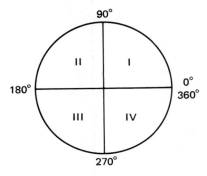

Fig. 1-5 Angles in four quadrants

TABLE 1-9 Signs of the Functions

Quadrant	Sine	Cosine	Tangent
I	+	+	+
II	+	−	−
III	−	−	+
IV	−	+	−

TABLE 1-10 Functions of Selected Angles

	0°	30°	45°	60°	90°	180°	270°	360°
Sine	0.000	0.500	0.707	0.866	1.000	0.000	−1.000	0.000
Cosine	1.000	0.866	0.707	0.500	0.000	−1.000	0.000	1.000
Tangent	0.000	0.577	1.000	1.732	+∞	0.000	+∞	0.000

TABLE 1-11 Variations of Functions

Quadrant	Sine	Cosine	Tangent
I	0 to +1	+1 to 0	0 to +∞
II	+1 to 0	0 to −1	−∞ to 0
III	0 to −1	−1 to 0	0 to +∞
IV	−1 to 0	0 to +1	−∞ to 0

TABLE 1-12 Mathematical Symbols and Meanings

Symbol	Meaning
+	plus or positive
−	minus or negative
±	plus or minus positive or negative
∓	minus or plus negative or positive
× or ·	multiplied by
÷	divided by
=	equals
≠	not equal to
≡	identical with
≢	not identical with
≅	approximately equal to
>	greater than
≫	much greater than
<	less than
≪	much less than
⩾	greater than or equal to
⩽	less than or equal to
∴	therefore
∢	angle
$\sin^{-1} A$ or arc sin A	angle whose sine is A
Σ	sum of
∞	infinity
|x|	absolute magnitude of x

MATHEMATICAL SYMBOLS   The mathematical symbols listed in Table 1-12 will be used throughout the text. Additional symbols will be defined at the appropriate places in the text.

Problem 47   What does $0° < \theta < 360°$ mean?

Solution θ is an angle between $0°$ and $360°$, but is not equal to $0°$ or $360°$.
θ is greater than $0°$ and less than $360°$.

(R1-28)   What does $0° \leqslant \theta < 360°$ represent?

(R1-29)   What does $0° < \theta \leqslant 360°$ represent?

EXTENDED STUDY TOPICS

1. Convert 71 000 000 to scientific notation.
2. Convert 4 500 000 000 to scientific notation.
3. Convert 100 000 000 000 to scientific notation.
4. Convert 3.75×10^4 to ordinary or decimal notation.
5. Converr 10^{10} to ordinary or decimal notation.
6. Convert 0.000 94 to scientific notation.
7. Convert 0.000 000 000 1 to scientific notation.
8. Convert 0.000 000 000 17 to scientific notation.
9. Convert 5.36×10^{-11} to ordinary or decimal notation.
10. Convert 10^{-7} to ordinary or decimal notation.
11. Solve and express the answer in scientific notation.
    $(31.2 \times 10^3) + (4.8 \times 10^4)$

12. Solve and express the answer in scientific notation.
$(8.2 \times 10^{-3}) (5 \times 10^8)$
13. Solve and express the answer in scientific notation.
$$\frac{8.1 \times 10^{-5}}{9.0 \times 10^6}$$
14. Solve and express the answer in scientific notation.
$$\sqrt{64 \times 10^{15}}$$
15. Solve and express the answer in scientific notation.
$$\sqrt[3]{34.3 \times 10^{13}}$$
16. Express a conductance of 10^{-3} siemens (S) in SI units.
17. Express an inductance of $\dfrac{5}{1\ 000}$ H in SI units.
18. Express 5.7 megavolts in SI units, using the correct metric system prefixes.
19. Convert 20°C to F°.
20. Change 0.000 25 μF to F.
21. Change 3.5×10^{-12} F to pF.
22. Convert 3.33 μV to mV.
23. Convert 15 kV to mV.
24. Convert 60 miles per hour to meters per second (m/s).
25. If $\theta = \cos^{-1} 0.5$, find θ.
26. The sides of the right triangle shown in figure 1-4 have the following values:
a = 1.0   c = 1.414
Find b.
27. What does $0° \leqslant \theta \leqslant 360°$ represent?

Chapter 2	Basic Electrical Concepts

OBJECTIVES   After studying this chapter, the student will be able to

- state the uses of electricity.

- discuss the concept of electric charges.

- explain the basic concepts of the atom.

- define free electrons.

- describe the differences between conductors, insulators, and semiconductors.

THE USES OF ELECTRICITY   One way in which energy is moved readily from one location to another is by means of an electric current. The processes of generating, transmitting, and using electrical energy require a number of energy conversions, or changes in the form of the energy, as shown in figure 2-1.

(R2-1)   List the energy conversions that occur in figure 2-1.

(R2-2)   List some of the electrical energy conversions that occur in the home. Include the name of the device that converts the energy.

Electricity can be controlled easily and accurately. This factor is primarily responsible for the many scientific advances of the twentieth century. Electricity made it possible for landings to take place on the moon and for information to be transmitted back to earth. The use of electricity and electronics enabled the Viking Landers to transmit photographs of the surface of the

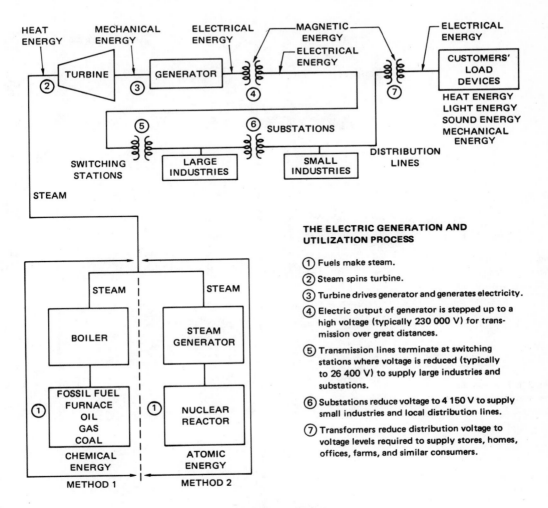

Fig. 2-1 Energy conversions

THE ELECTRIC GENERATION AND UTILIZATION PROCESS

① Fuels make steam.

② Steam spins turbine.

③ Turbine drives generator and generates electricity.

④ Electric output of generator is stepped up to a high voltage (typically 230 000 V) for transmission over great distances.

⑤ Transmission lines terminate at switching stations where voltage is reduced (typically to 26 400 V) to supply large industries and substations.

⑥ Substations reduce voltage to 4 150 V to supply small industries and local distribution lines.

⑦ Transformers reduce distribution voltage to voltage levels required to supply stores, homes, offices, farms, and similar consumers.

planet Mars, figures 2-2, 2-3, and 2-4. To understand the nature of electricity, it is necessary to review some of the basic concepts of physics, including the structure of the atom.

PHYSICS FOR ELECTRICITY

Mass. *Mass* is a fundamental physical quantity. It is the property of a material body that causes it to resist any change in its state of motion. Mass is independent of any gravitational force. Therefore, a body has a constant mass regardless of its location.

(R2-3) Is there any difference in the mass of an astronaut when on the earth, in space, or on the moon?

Fig. 2-2 Panoramic view of Mars from Viking 1

Fig. 2-3 The first picture of the surface of Mars from Viking 2

Fig. 2-4 Viking's soil sampler collector arm at work on Mars

Fig. 2-5 Magnetic attraction

Force. *Force* is the physical quantity that is required to change the state of motion of a mass. A *contact force* exists when the object exerting the force makes physical contact with the mass. A *field force* acts on a mass at a distance from the mass exerting the force. In this case, there is no physical contact.

For example, a magnet exerts force that attracts magnetic materials without contacting them. Figure 2-5 illustrates an iron nail being attracted by a magnet.

(R2-4) What type of force is illustrated in figure 2-5?

As another example, an electric field is that region in which an electric force acts on an electric charge.

The earth exerts a force or pull on objects which are on its surface. In addition, the earth exerts a gravitational (field) force on objects which are not on its surface. This gravitational force is always a force of attraction.

Weight. *Weight* is a gravitational force. It is the force of gravity acting on an object. This force (weight) is directly proportional to the size of the masses and inversely proportional to the distance between the masses.

(R2-5) An astronaut weighs 180 lb on earth before launching. What is the weight of this astronaut when he is far enough away from earth (or any other body) to cause the effect of gravity to be zero?

(R2-6) Has any change taken place in the mass of the astronaut?

(R2-7) The mass of the moon is smaller than that of the earth. Does a 180-lb astronaut on the moon weigh more or less than 180 lb?

ELECTRIC CHARGES When a comb is run through dry hair, the comb becomes electrically charged. The friction between the comb and the hair produces *static electricity*. The act of combing removes some small atomic particles of matter, or *electrons*, from the hair. These electrons are deposited on the comb. Electrons, by nature, have a charge of electricity that is said to be negative. The comb becomes negatively charged because it now has an excess of electrons.

Before it is combed, the hair is neutral. That is, it has no charge. When it loses some of its electrons, it becomes positively charged. The action of combing causes both the hair and the comb to become charged, but with two distinct and different types of electricity.

Experiments show that two bodies charged with the same type of electricity will repel each other when they are brought close together. But, when a positively charged body is brought close to a negatively charged body, the two bodies will attract each other. Such observations led to the statement of the fundamental law of electrostatic charges: "like charges repel, and unlike charges attract."

POSITIVE CHARGES

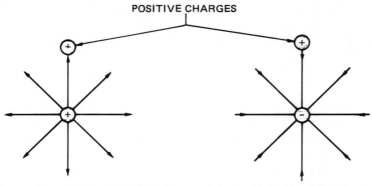

(A) FIELD FOR POSITIVE CHARGE   (B) FIELD FOR NEGATIVE CHARGE

Fig. 2-6 Electrostatic field (dielectric field) surrounding positively and negatively charged bodies.

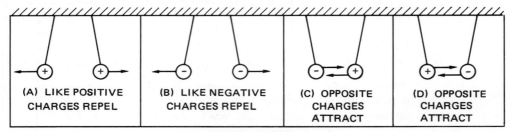

(A) LIKE POSITIVE   (B) LIKE NEGATIVE   (C) OPPOSITE   (D) OPPOSITE
CHARGES REPEL   CHARGES REPEL   CHARGES   CHARGES
                                                 ATTRACT   ATTRACT

Fig. 2-7 Effects of static charges

An electrostatic field surrounds a charged body. The magnitude of the field is a function of the strength of the electric charge on the body. Figure 2-6 shows that electrostatic fields surround a positively charged body and a negatively charged body. The arrows show the direction of the invisible lines of force. This is the direction in which a postitively charged body will move in the presence of an electrostatic field.

Forces between charged bodies are called *electrostatic forces* and are field forces. Electrostatic forces increase as the size of the charge is increased, or as the distance between the charges is decreased.

The effects of static electricity are as follows:

- A positive charge repels another positive charge.
- A negative charge repels another negative charge.
- A negative charge attracts a positive charge.
- A positive charge attracts a negative charge.

Figure 2-7 shows the effects of static charges on lightweight balls hanging from a support and free to move.

(R2-8)   What type of force is responsible for the action of the balls in figure 2-7?

The electrostatic field is also called the *dielectric field of force*. The electrostatic field between two unlike charges is shown in figure 2-8. Figure 2-9 shows

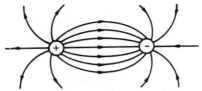

Fig. 2-8 The dielectric field of force between two opposite charges represents a field of attraction.

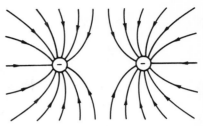

Fig. 2-9 The dielectric field of force between two negative charges represents a field of repulsion.

the electrostatic field between two negative charges. The field between two positive charges is shown in figure 2-10.

(R2-9) Two glass rods are rubbed with silk. Both rods become positively charged. When they are brought together, figure 2-11, will the rods attract or repel each other?

(R2-10) Explain the reason for the answer to (R2-9).

(R2-11) If an ebonite rod is rubbed with fur, it becomes negatively charged. A positively charged glass rod is brought close to the ebonite rod, figure 2-12. What is the effect?

COULOMB'S LAW Coulomb's Law of electrostatic force is used to find a value for the force of attraction or repulsion between electric charges:

$$F = \frac{k\,Q_1\,Q_2}{d^2}$$

Eq. 2.1

where

Q_1 = charge of body one, in coulombs (C)
Q_2 = charge of body two, in coulombs (C)
d = distance between Q_1 and Q_2, in meters
k = constant of proportionality having a numerical value of 9.0×10^9
F = electrostatic force, in newtons (a negative value for F represents an attractive force; a positive value represents a repulsive force)

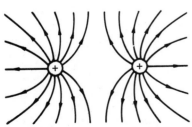

Fig. 2-10 The dielectric field of force between two positive charges represents a field of repulsion.

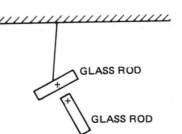

Fig. 2-11 Two glass rods with positive charges (R2-9)

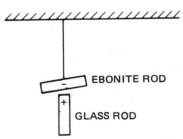

Fig. 2-12 Ebonite rod and glass rod with opposite charges (R2-11)

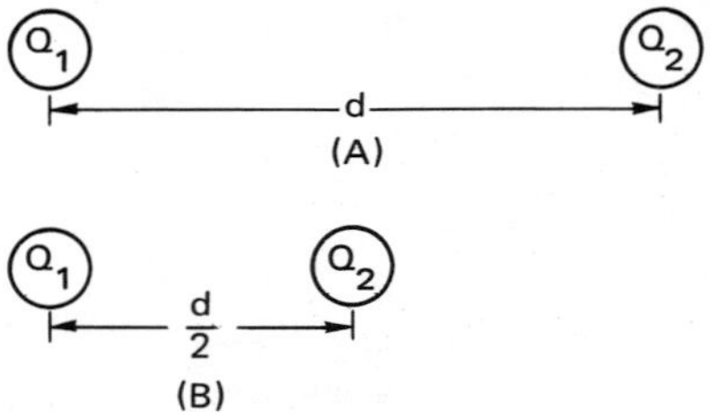

Fig. 2-13

A coulomb (C) is the charge of 6.24×10^{18} electrons. Coulomb's Law of electrostatic force states that:

The electrostatic force between two electric charges is directly proportional to the product of the two charges, and inversely proportional to the square of the distance between the charges.

Problem 1   Figure 2-13A shows two equal positive charges which are d meters apart. The resulting force of repulsion between these charges is equal to F_1 newtons. In figure 2-13B, the distance between the charges is reduced to $\dfrac{d}{2}$ meters. Determine the change in F_1.

Solution   1. Figure 2-13A:

$$F_1 = \frac{k\, Q_1\, Q_2}{d^2}$$

2. Figure 2-13B:

$$F_2 = \frac{k\, Q_1\, Q_2}{\left(\dfrac{d}{2}\right)^2}$$

$$F_2 = \frac{k\, Q_1\, Q_2}{\dfrac{d^2}{4}}$$

$$F_2 = 4\left(\frac{k\, Q_1\, Q_2}{d^2}\right) = 4F_1$$

Therefore, the force of repulsion (F) is increased to four times the value for the case in figure 2-13A.

Problem 2   If $Q_1 = +2$ C and $Q_2 = -3$ C, calculate the force of attraction for a separation of two meters.

Solution

$$F = \frac{k\, Q_1\, Q_2}{d^2}$$

$$= \frac{(9 \times 10^9)\,(2)\,(-3)}{(2)^2}$$

$$= -13.5 \times 10^9 \text{ N}$$

(The minus sign indicates that the force is one of attraction.)

Refer to figure 2-13A for questions (R2-12), (R2-13), and (R2-14).

(R2-12) If the charges on Q_1 and Q_2 are both positive, what type of electrostatic force exists?

(R2-13) If the charge on Q_1 is positive and the charge on Q_2 is negative, what type of electrostatic force exists?

(R2-14) If the positive charge on Q_1 is doubled, and the negative charge on Q_2 remains constant, what change takes place in the electrostatic force?

THE ATOM All matter consists of basic materials called *elements*. To date, 105 elements have been identified. Common elements are copper, silver, and aluminum. An *atom* is the smallest part of an element that can exist and still retain the chemical properties of the element. All matter consists of atoms of one or more elements.

The atom is composed of a nucleus and electrons. The nucleus is the center of the atom and has most of the mass of the atom. Electrons are small particles which move about the nucleus. The simple diagram in figure 2-14 represents a hydrogen atom.

Hydrogen Atom. The hydrogen atom has a nucleus consisting of one proton. It also has one electron which moves around this nucleus. The *proton* is a particle of mass which has an electrical charge defined as positive. The protons in all elements, and thus of all materials, are identical. The electron is a particle whose mass is much less than that of a proton. (The proton is approximately 1 837 times more massive than the electron.) The electrical charge of an electron is equal in magnitude, but opposite in character, to the charge of a proton. The charge of an electron is designated as negative. The electrons of all elements and all materials are identical.

Lithium Atom. Figure 2-15 is a simple diagram of a lithium atom. The nucleus consists of three protons and four neutrons. *Neutrons* are particles of mass that do *not* have an electrical charge. Neutrons are about the same size as protons and have about the same mass. The neutrons of all elements, and thus of all materials, are identical. Three electrons move around the nucleus of the lithium atom.

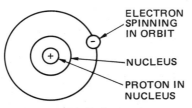

Fig. 2-14 Hydrogen atom

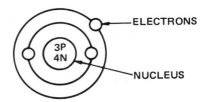

Fig. 2-15 Lithium atom

(R2-15)   What is the difference between a proton and a neutron? In what ways are these particles similar?

(R2-16)   What is the difference between a proton and an electron? In what ways are these particles similar?

All material substances consist of the same basic components: protons, neutrons, and electrons. The only substance that contains a single proton and electron is hydrogen. The atoms of the elements differ in the number of particles in the nucleus, and in the number of electrons. An atom is electrically neutral because the number of electrons is equal to the number of protons.

(R2-17)   The helium atom has two neutrons and two protons. How many electrons does a helium atom have?

(R2-18)   Uranium has 146 neutrons and 92 electrons. How many protons does it have?

ENERGY LEVELS   Electrons move around the nucleus of an atom in orbits. The type of electron orbit depends upon the energy of the electron. The greater the energy of an electron, the greater is the distance of its orbit from the nucleus. Electron energies fall into energy bands or shells which are defined by the maximum number of electrons that may occupy each shell.

The energy band closest to the nucleus contains no more than two electrons. The maximum number of electrons in a shell is equal to $2n^2$, where n is the shell number starting at the nucleus and moving outward. The lithium atom has three electrons, figure 2-16. For the first, or K, shell, n is one and the shell has $2(1)^2$ or two electrons. The remaining electron is in the second shell from the nucleus, or the L shell. As shown in Table 2-1, the electron shells in an atom are designated by the letters K, L, M, N, O, P, and Q. The K shell is the one nearest the nucleus.

An analysis of Table 2-1 shows that the $2n^2$ rule has the following limitations:
1. The outermost shell is filled when it contains eight electrons.
2. The next to the last shell cannot contain more than 18 electrons regardless of its quota.
3. No shell can contain more than 32 electrons.

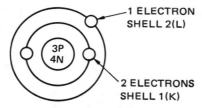

Fig. 2-16   Arrangement of electrons in shells of lithium atom

TABLE 2-1   Capacities of Electron Shells

Shell	Maximum Number of Possible Electrons
K	2
L	8
M	18
N	32
O	32
P	18
Q	8

These limitations are demonstrated by the following examples.

1. Argon contains 18 protons. The atom also has 18 electrons arranged in groups of 2, 8 and 8. The last shell is filled even though its maximum quota is 18.

2. Cesium contains 55 protons. The atom also has 55 electrons arranged in groups of 2, 8, 18, 18, 8 and 1. The last electron cannot go into shell 4 (N) since it would violate the limit of 18 electrons in the next to the last shell. This electron cannot go into shell 5 (O) since it would violate the limit of 8 electrons in the last shell.

3. Lead contains 82 protons. The atom also has 82 electrons arranged in groups of 2, 8, 18, 32, 18 and 4. This arrangement meets all of the given limitations.

(R2-19) Nitrogen has seven protons. How many electrons does nitrogen have in the second shell?

(R2-20) What is the maximum number of electrons that can be contained in the fourth shell of any atom?

Carbon Atom. The carbon atom, figure 2-17, contains six protons and six electrons. The neutrons are not shown in the figure because they do not affect the electrical conductivity of a material. The carbon atom has two electrons, the maximum number permitted, in the first shell. The remaining four electrons are in the second shell.

Copper Atom. The copper atom, figure 2-18, contains 29 protons. The atom also has 29 electrons arranged in shells according to the $2 n^2$ rule (up to shell No. 3):

Shell No. 1 (K)   2 electrons
Shell No. 2 (L)   8 electrons
Shell No. 3 (M)   18 electrons
Shell No. 4 (N)   1 electron

The first three shells of the copper atom are complete. The fourth shell contains only one electron and is incomplete. The maximum number of electrons that may occupy this shell is 32, according to Table 2-1. Because copper atoms have

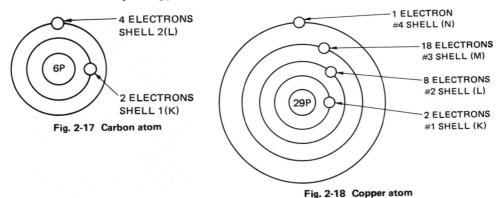

Fig. 2-17 Carbon atom

Fig. 2-18 Copper atom

only one electron in the last shell, this material is a much better conductor of electricity than carbon.

FREE ELECTRONS   The single electron that occupies the outer shell of the copper atom is bound loosely to the atom because of its distance from the nucleus. If this electron can gain enough energy from its surroundings, it will leave its atom and become a free electron. With no external force applied, the thermal (heat) energy at room temperature creates 1.4×10^{24} free electrons in one cubic inch of copper.

When the copper atom loses this free electron in the outer shell, the atom is no longer electrically neutral. The atom now has 29 positive charges (protons) and 28 negative charges (electrons). The atom has a net positive charge of one. This structure is called a *positive ion*.

(R2-21)   Since 1 in^3 of copper at room temperature has 1.4×10^{24} free electrons, how many positive ions does it have?

When there is no external force on a copper conductor, the free electrons in the conductor can move in a random motion. However, the positive ions do not move. The ions oscillate, or vibrate, in a mean fixed position. The free electron is the charge carrier in the copper conductor.

Three factors cause the random motion of free electrons:
1. Collisions with other electrons and with positive ions.
2. The attractive forces of the positive ions for the negative electrons.
3. The force of repulsion that exists between electrons (since they all have the same negative charge).

Figure 2-19 shows that the number of free electrons moving to the right of a fixed reference point over a period of time equals the number of free electrons moving to the left of the reference point. This means that if an external force is *not* applied, the net flow of charge in any one direction is zero.

The free electrons do *not* leave the surface of the copper wire. At room temperature, the copper wire contains neutral copper atoms, free negative electrons, and positive ions.

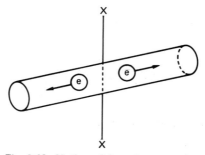

Fig. 2-19 Motion of free electrons: the number of electrons moving to the left of X – X equals the number of electrons moving to the right of X – X (over a period of time)

(R2-22)   What is the net charge of a copper wire at room temperature?

The electrons in the outer shell of an atom are called *valence electrons.* Copper has only one electron in this shell. Therefore, the copper atom is said to have a valence of one. The valence electrons of copper atoms leave each atom at room temperature and become free electrons. The remaining 28 electrons of each atom are tightly bound in their shells. The positive copper ions repel each other, but they do not move.

(R2-23)   What is the valence of carbon?

CONDUCTORS, INSULATORS, AND SEMI-CONDUCTORS

Conductors. Conductors are materials in which the free electrons move easily. All metals are conductors. Table 2-2 compares the relative conductivity of various metals, with silver as the best conductor.

(R2-24)   Which material is the better conductor of electricity, iron or tungsten?

Insulators. An insulator is a material in which the atoms contain electrons that tend to stay in the outer shell. Because it is difficult to free these electrons from the outer shell, insulators will not conduct electricity easily. For example, these materials are used to cover current-carrying wires to protect personnel from injury and to prevent short circuits. Rubber gloves are used for protection by public utility employees when working on high-voltage power lines. Table 2-3 lists insulators in the order of their insulating ability. Practical insulating materials have a very small number of free electrons as compared to metallic conductors. Insulators oppose the flow of electrons.

Various types of insulating materials are used to cover electric conductors. The conditions in which the wires are to be used determine the thickness and type of insulation applied. Table 2-4 lists the insulating materials commonly used on conductors. Typical applications and/or special properties of the insulating materials are also given.

TABLE 2-2 Relative Conductivity of Various Metals

Metal	Relative Conductivity
Silver Copper Gold Aluminum Tungsten Nickel Iron Constantan Nichrome	High ↓ Decreasing Conductivity ↓ Low

TABLE 2-3 Insulators

Material	Relative Insulating Ability
Mica Glass Teflon Paper (paraffin coated) Rubber Bakelite Oils Porcelain Air	High ↓ Decreasing Insulating Ability ↓ Low

TABLE 2-4 Insulation for Conductors

Insulation	Use or Special Property
Asbestos	High-temperature circuits
Cambric	High-voltage circuits
Cellulose acetate	Flame resistant
Cotton	Single layer Double layer General hookup wire
Enamel	Winding coils Antenna installations
Plastic	High-voltage wiring Moistureproof Space saving
Rubber	General service cables Outdoor leads
Silk	Single layer Double layer Coil winding
Spun glass	Space saving Moistureproof

Fig. 2-20 Volt-ohm-milliammeter (VOM)

Semiconductors. Semiconductors are elements whose ability to conduct electricity is between that of conductors and insulators. Semiconductors have fewer free electrons than conductors, but more free electrons than insulators. The most frequently used semiconductors are germanium and silicon. Diodes and transistors are manufactured from these materials.

LABORATORY EXERCISE 2-1 VOM FAMILIARI-ZATION

PURPOSE

After completing this Laboratory Exercise, the student will be able to

• operate a commercial VOM.

• read the various voltage, current, and resistance scales of the VOM.

• set the ZERO OHMS adjustment on the meter.

EQUIPMENT AND MATERIALS

1 Commercial VOM, similar to the instrument shown in figure 2-20.

PROCEDURE

NOTE: The commercial volt-ohm-milliammeter (VOM) is a collection of instruments housed in a single case. The following measurements can be made with this one device: dc voltage, ac voltage, dc current, and resistance. Additional functions, such as decibel (dB) measurements, may be included also. In general, a VOM is a dc instrument that operates on batteries installed in the case. Ac powered models are available. The VOM is explained in more detail in Chapter 35 of this text.

TABLE 2-5 VOM Characteristics

Controls	
Dc voltage ranges	
Dc current ranges	
Resistance ranges	
Center scale reading on R X 1	
Ac voltage ranges	

1. Study the controls and the scales of the VOM. Read the instruction manual.
2. Draw a panel view of the meter, showing the operating controls and the functions and ranges indicated for the switches.
3. Use Table 2-5 to tabulate the information given for the following controls: dc voltage ranges, dc current ranges, resistance ranges, center scale reading on the R X 1 range, and ac voltage ranges.
4. The voltmeter scale is a linear scale. The meter pointer travels equal distances along the scale arc for equal voltage changes. The arc distance between 0 and 10 volts is the same as the distance between 20 and 30 volts, or between 90 and 100 volts.
   a. Draw the dc voltage scales.
   b. What scale is used to read 125 V dc?
   c. Draw a meter pointer at the 125-V dc reading.
5. Draw the ohms scale.
   a. Is the ohms scale linear? (This subject is covered in detail in Chapter 35 of this text.)
   b. At which end of the ohms scale are the readings more accurate? Explain.
   c. Draw a meter pointer at the 12.5-ohm reading.
6. a. Draw a linear scale with equal divisions marked "one" through "ten".
   b. Subdivide each major division into ten minor divisions.
   c. Draw a meter pointer at the 9.3 reading.
7. Where is the normal position of the VOM pointer?
8. Set the function switch to +DC.
9. a. Set the range selector switch to R X 1. Connect leads to the + and – (common) terminals. Short the leads by touching the metal tips of the leads to each other. The meter pointer should swing to zero ohms. If the pointer does not come to rest on zero when the leads are shorted, use the ZERO OHMS adjustment to set the pointer on zero.

b. Repeat step 9.a. with the range selector switch set on R X 100.

c. Repeat step 9.a. with the range selector switch set on R X 10 000.

CAUTION. Do not leave the meter leads shorted together while the range selector switch is on one of the R positions. An internal battery is connected in the circuit of the meter for measuring resistance. If the leads remain shorted together for any length of time, the battery will be discharged.

EXTENDED STUDY TOPICS

1. A rubber comb is rubbed on a sheet of paper. Why does the comb attract the paper?

2. Hair may stand on end after it is combed on a very dry day. Explain why this happens.

3. After dry hair is combed, is the charge of static electricity on the hair positive or negative? Explain the answer.

4. A germanium atom has 41 neutrons and 32 protons. How many electrons does the germanium atom have?

5. Is there any difference between the protons, neutrons, and electrons of lead and gold?

6. If the distance between two positively charged bodies is doubled, what change takes place in the electrostatic force?

7. If Q_1 = +1 C and Q_2 = +1 C, calculate the electrostatic force between the bodies for a separation of two meters. What type of force does this value represent?

8. Find the force of attraction between charges Q_1 and Q_2 in figure 2-21.

9. Silicon has 14 protons, figure 2-22. How many electrons does it have in its outer shell? Draw the shell arrangement.

10. Explain why silicon is or is not as good a conductor of electricity as copper.

11. What are the physical and electrical properties of an electron?

12. What are the physical and electrical properties of a proton?

13. What are the physical and electrical properties of a neutron?

14. Define a coulomb.

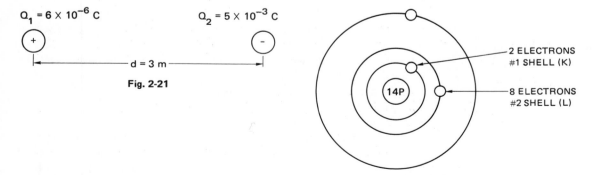

$Q_1 = 6 \times 10^{-6}$ C $Q_2 = 5 \times 10^{-3}$ C

d = 3 m

Fig. 2-21

2 ELECTRONS
#1 SHELL (K)

14P

8 ELECTRONS
#2 SHELL (L)

Fig. 2-22

<table>
<tr><td>Chapter
3</td><td>Current and Voltage</td></tr>
</table>

OBJECTIVES After studying this chapter, the student will be able to

- discuss the basic principles of current and voltage.
- apply the double subscript voltage notation.
- use ammeters and voltmeters in electric circuits.

THE ELECTRIC CURRENT An electric current is the directed motion, or flow, of charges through the material of a conductor.

(R3-1) What is the polarity (positive or negative) of the charges that make up an electric current in a copper wire?

According to the definition of current, a force is required to direct the motion or flow of charges. If there is no force, the charged particles move in a random direction only. In addition, there must be a material, or conductor, in which the charged particles can move readily with little opposition or resistance. The complete conducting path through which electrons move is called an *electric circuit*. A complete electric circuit is shown in figure 3-1.

(R3-2) List the parts that make up the complete electric circuit in figure 3-1.

(R3-3) What is the polarity of terminal B in figure 3-1?

(R3-4) What is the polarity of terminal A in figure 3-1?

THE ELECTRIC CIRCUIT A schematic diagram of figure 3-1 is shown in figure 3-2. Standardized symbols are used to represent the various electrical or electronic devices in the circuit.

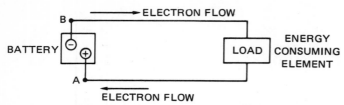

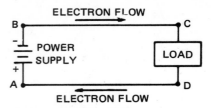

(THE BATTERY IS A FORCE WHICH DIRECTS FLOW OF
ELECTRONS, THAT IS, ELECTROMOTIVE FORCE OR EMF)

Fig. 3-1 The complete electric circuit

Fig. 3-2 Schematic diagram for the circuit
of figure 3-1

The use of symbols makes it easier to draw electric circuit diagrams. The symbol
for the power supply consists of a series of long and short lines. The outer long
line of the symbol indicates the positive terminal of the power supply. The outer
short line of the symbol represents the negative terminal.

A switch may be inserted at any point in the circuit of figure 3-2. If this switch
is opened, the flow of electricity (electron flow) stops. (In industry, the flow of
charges is called a current flow. This concept will be followed in this text.) The
circuit must be closed before current can flow. The current must be able to leave
the power supply, travel through the conducting wire to the load, and return to
its starting point. If the circuit is open at *any* point, current will not flow. The
circuit may be opened intentionally by opening a switch, or it may be opened
unintentionally because of a broken wire or a loose connection. Regardless of
how the circuit is opened, this condition is called an *open circuit.*

If an accidental connection is made between points E and F, as shown in figure
3-3, the resulting condition is known as a *short circuit.* The electrons flow from
terminal B to point E, through the short circuit to point F, and then return to terminal A of the power supply. This short circuit prevents the flow of current through
the useful or main path. Note that the term "short circuit" does not refer to the
length of the path. Rather, it indicates that the electrons do not follow the main
circuit path. Electrons always follow the path of least resistance or opposition.

(R3-5)   What is the useful, or main, path of the current in figure 3-3?

THE COULOMB   The quantity of electric charge carried by a single electron is too small for practical use. Instead, electric charge is expressed by the SI unit known as the coulomb. One *coulomb* is defined as the quantity of electric charge carried by 6.24
$\times$ 10^{18} electrons.

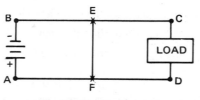

Fig. 3-3 The short circuit

The letter symbol for the quantity of electric charge is Q. The unit symbol for the coulomb is C. For example, the statement Q = 3.4 C means that the amount of charge (Q) is equal to 3.4 coulombs (C).

Problem 1 If there are 4.80×10^{19} free electrons moving in one direction in a copper conductor, what is the value of the electric charge expressed in coulombs?

Solution
$$Q = \frac{4.80 \times 10^{19} \text{ electrons}}{6.24 \times 10^{18} \text{ (electrons/coulombs)}}$$
$$= 7.69 \text{ C}$$

THE AMPERE An electric current is the rate of charge flow in coulombs per second. The SI unit of electric current is the ampere. Thus, one ampere is equal to one coulomb of electric charge carriers passing a certain point in an electric circuit in one second.

Amperes and coulombs per second express the same rate. The letter symbol for electric current is I. The unit symbol for the ampere is A. The relationship between amperes and coulombs per second is expressed as follows:

$$I = \frac{Q}{t}$$

Eq. 3.1

where

    I = current, in amperes
    Q = quantity of electric charge, in coulombs
    t = time, in seconds

Problem 2 What is the current in amperes if 150 coulombs of electric charge pass a certain point in an electric circuit in half a minute?

Solution
$$I = \frac{Q}{t}$$
$$= \frac{150 \text{ coulombs}}{30 \text{ seconds}}$$
$$= 5.0 \text{ A}$$

Problem 3 A circuit has a current of 10 A. How long will it take for 0.1 C to pass through a certain point in the circuit?

Solution
$$I = \frac{Q}{t}$$
$$It = Q$$
$$t = \frac{Q}{I}$$
$$= \frac{0.1 \text{ C}}{10 \text{ A}}$$
$$= 0.01 \text{ s}$$

Problem 4 A 6-A current flows through an electric toaster for 30 seconds. What quantity of electric charge (Q) passes through the toaster in that time?

Solution

$$I = \frac{Q}{t}$$

$$It = Q$$

$$Q = It$$

$$= \left(6 \, \frac{C}{s}\right)(30 \text{ s})$$

$$= 180 \text{ C}$$

(R3-6)   The current in a circuit is 5 A. What quantity of charge will pass a certain point in the circuit in 0.2 s?

(R3-7)   In problem 4, how many electrons flow through the toaster?

(R3-8)   How many electrons flow through the toaster of problem 4 in one minute, if the current is constant at 6 A?

CURRENT DIRECTION   The charge carriers in an electric circuit are electrons. Electrons have a negative charge. Therefore, an electric current is the flow of negative charges. Benjamin Franklin and other experimenters working soon after the discovery of electricity said that an electric current was the movement of positive charges. Therefore, the *conventional direction* of electric current flow was established. This convention is stated as follows: electric current flows from the positive terminal of the power supply, travels through the external circuit, and returns to the negative terminal of the power supply.

This way of expressing the direction of current flow can be thought of as a mathematical convention. There is extensive acceptance of this convention by the electronics industry. For this reason, the conventional direction of current flow will be used in this text. Figure 3-4 shows both the conventional direction of current flow and the actual physical direction of electron flow.

The use of conventional current flow does not affect the solution of electric circuit problems. The mathematical solution of a problem gives the same answer for both electron current flow and conventional current flow. Occasionally, the electron flow direction is used when studying the physical process of electrical conduction in electronic components.

(R3-9)   What are the polarities of terminals A and B in figure 3-5? (The current I represents conventional current flow.)

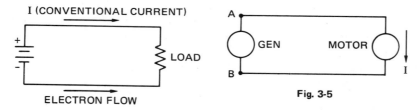

I (CONVENTIONAL CURRENT)

LOAD

ELECTRON FLOW

Fig. 3-4   Conventional current flow and electron flow directions

GEN   MOTOR

Fig. 3-5

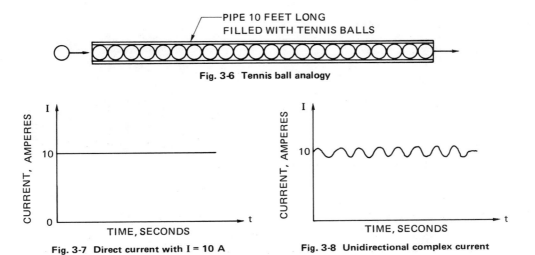

PIPE 10 FEET LONG
FILLED WITH TENNIS BALLS

Fig. 3-6 Tennis ball analogy

Fig. 3-7 Direct current with I = 10 A

Fig. 3-8 Unidirectional complex current

SPEED OF ELECTRICITY The effect of an electric current is felt at the speed of light, or 186 000 miles per second. This means that a current in a wire exists almost instantaneously. Individual electrons move very slowly. When the electric current in a conductor measures one ampere, the velocity of each electron (electron drift) is approximately 1/1 000 inch per second. However, the combined movement of these electrons causes an almost instant spread of energy throughout the circuit. This situation is somewhat like filling a 10 ft long pipe with tennis balls, figure 3-6. If a tennis ball is pushed into one end of the pipe, another tennis ball is instantly forced out at the other end of the pipe. As each additional tennis ball is inserted into the pipe, individual balls advance slowly through the pipe. Although the velocity of each tennis ball is very small, the effect of inserting a tennis ball at one end is instantaneously felt at the other end of the pipe.

DIRECT CURRENT A direct current of 10 A is shown in figure 3-7. Two characteristics of this type can be stated. First, the current is unidirectional; that is, it travels in one direction only. Second, the current is constant in value and does not change with time.

A current is shown in figure 3-8. This current is unidirectional, but it is not a direct current by the definition given because it is not constant in amplitude at all times. This current is a complex current and consists of both direct-current and alternating-current components. The subject of complex waves is studied in chapter 10.

Problem 5 The current shown in figure 3-9 is called a pulsating unidirectional current. Does this type of current fit into the previous definition of a direct current? Explain why or why not.

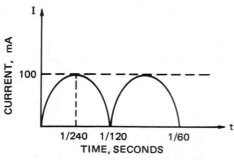

Fig. 3-9 Pulsating unidirectional current

Fig. 3-10 Dc ammeter

Solution   The current is a unidirectional current. However, it is a complex current consisting of a direct-current component and many alternating-current components. It is not a direct current because its value is not constant over a period of time.

(R3-10)   What is the value of the current shown in figure 3-9 when t = 1/120 s?

(R3-11)   What is the value of the current shown in figure 3-9 when t = 1/240 s?

MEASURING CURRENT   An instrument called an *ammeter* is used to measure current. The ammeter shown in figure 3-10 can measure direct currents ranging from 0 to 50 amperes.

(R3-12)   What is the maximum number of coulombs that the ammeter in figure 3-10 can measure in 30 seconds?

An ammeter is placed in a circuit so that the same current passes through the load and through the ammeter. This type of circuit is called a *series circuit*, figure 3-11. Each terminal of an ammeter is marked with its polarity (+ or –). The positive terminal of the ammeter must be connected facing the positive terminal of the power supply. This means that current will enter the ammeter at the positive terminal and will leave it at the negative terminal. As a result, the pointer of the ammeter will swing upscale.

When meters are connected in a circuit, the currents in the circuit are modified. The technician must insure that the effect of adding a meter to a circuit can be accounted for or can be neglected. This characteristic is called the loading effect of a meter and is described in more detail in Chapter 34.

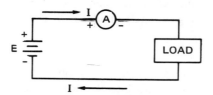

Fig. 3-11 Ammeter connected in a series circuit

Fig. 3-12 Multirange ammeter with scales marked from 0 to 2.5 A and 0 to 5.0 A

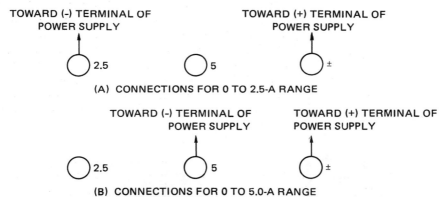

(A) CONNECTIONS FOR 0 TO 2.5-A RANGE

(B) CONNECTIONS FOR 0 TO 5.0-A RANGE

NOTE: THIS AMMETER CAN BE USED WITH THE
SAME CONNECTIONS TO ALTERNATING CURRENTS.

CAUTION. All connections must have the proper polarities. Failure
to observe the polarities may cause serious damage to the pointer. If
the polarity is reversed, the pointer will try to move counterclockwise from zero and will be forced against the resting stop.

Fig. 3-13 Connections for a multirange ammeter

The student should estimate the maximum current in a circuit, so that an
ammeter of the proper range is used.

Figure 3-12 shows a multirange ammeter. One scale has a maximum value of
2.5 amperes. The second scale has a maximum value of 5.0 amperes. Figure
3-13 shows how the ammeter is connected to obtain the maximum values.

**POTENTIAL
DIFFERENCE**

There must be a force to direct the motion of free electrons in conductors before current can flow in an electric circuit. This force can be supplied by a battery, a power supply, or a generator. The force is known as an electron-moving force, or an electromotive force (emf).

(R3-13)   Name two sources of emf in an automobile.

The source of emf has one terminal with a positive polarity and one terminal with a negative polarity. In other words, the negative terminal has an excess of electrons, and the positive terminal has a deficiency of electrons.

The difference in force between the positive and negative terminals of the emf source is called a *potential difference*, or *voltage*. This difference in force is the potential of the source for electrical work. The source has this potential whether it is lying on a shelf or is connected in an electrical circuit. Potential energy is the same as electric pressure.

A battery is a device in which chemical energy is converted to an energy potential difference. Figure 3-14 is a drawing of a battery which can be used in a calculator. A selection of several types of batteries is shown in figure 3-15.

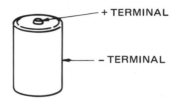

Fig. 3-14 1.5–V AA battery for a calculator

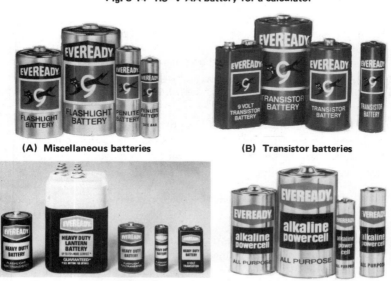

(A) Miscellaneous batteries   (B) Transistor batteries

(C) Heavy duty batteries   (D) Alkaline powercells

Fig. 3-15 Batteries

An automobile uses a wet battery and a flashlight uses a dry battery. Batteries consist of cells which are connected in series to obtain the total desired potential energy. Figure 3-16 shows the standard electrical symbols for a single-cell battery and one with two or more cells.

POWER SUPPLIES　School and industrial laboratories commonly use dc voltage sources for experimental work. These electronic soruces of emf are variable regulated power supplies. Typical units are shown in figure 3-17. This type of power supply can be adjusted by hand to deliver any voltage within its range of operation. The output voltage remains constant as the load current changes. (The power supply must be operated within its specified limits.)

An electronic power supply receives its energy from the ac line supply. Generally, a typical supply has an ON-OFF switch, a VOLTAGE control to set the output voltage level, and a meter switch which is set to measure either the dc voltage output or the direct current delivered to the load.

Some power supplies can provide two or more independent dc voltages. For this type of instrument, the front panel contains separate controls and separate output terminals.

The dc terminals of the power supply are usually marked +, -, and gnd (ground). By convention, a red jack is used for the positive terminal of the power supply, and a black jack is used for the negative terminal.

CAUTION. To prevent damage to the power supply, the output terminals must not be shorted together. The test leads from the output terminals must not contact each other.

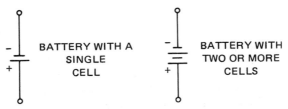

Fig. 3-16　Battery symbols

Fig. 3-17A　High-voltage power supply

Fig. 3-17B　High-voltage bipolar power supply

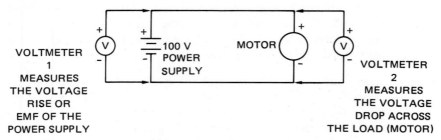

Fig. 3-18 Emf voltage and load voltage drop

THE VOLT   The potential difference between two points in an electric circuit is the rise or fall of the energy required to move a unit quantity of electric charge from one point to the other. Work and energy are numerically equal. Both work and energy are expressed by the same letter symbol, W. The *joule* (J) is the SI unit of work and energy.

The *volt* is the SI unit of *potential difference.* One volt is the potential difference between two points in an electric circuit, when one *joule* of energy is required to move one coulomb of electric charge from one point to the other.

The letter symbol for the potential difference provided by a source of emf is E. The letter symbol for the potential difference, or voltage drop, across a load or energy consuming device is V. The unit symbol for voltage is V. A volt is one joule per coulomb. The relationship of voltage, energy, and charge is expressed by the following equation:

$$E \text{ (or } V) = \frac{W}{Q}$$

Eq. 3.2

where
$$E \text{ (or } V) = \text{ the potential difference, in volts}$$
$$W = \text{ energy, in joules}$$
$$Q = \text{ quantity of electric charge, in coulombs.}$$

Figure 3-18 distinguishes between the voltage rise of an emf and the voltage drop across the load. Voltmeter 1 reads the voltage rise of the power supply, or E = 100 V. Voltmeter 2 reads the voltage of the motor (the voltage drop across the load), or V = 100 V.

Problem 6   A 6-volt automobile battery delivers 50 A for one second. How much chemical energy, in joules, is required to produce this amount of electrical energy?

Solution
$$I = \frac{Q}{t}$$
$$Q = It$$
$$= (50 \text{ A}) (1 \text{ s})$$
$$= 50 \text{ C}$$
$$E = \frac{W}{Q}$$
$$EQ = W$$
$$W = EQ$$
$$= (6 \text{ V}) (50 \text{ C}) = 300 \text{ J}$$

Fig. 3-19A Dc voltmeter

Fig. 3-19B Voltmeter for measuring dc voltage or ac voltage

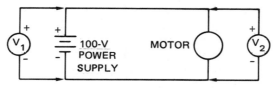

Fig. 3-20 Voltmeter polarity connections

Problem 7 A 0.5-A current flows through a resistor for 10 seconds. The amount of heat generated is 12 joules. What is the voltage drop across the resistor?

Solution
$$Q = It$$
$$= (0.5 \text{ A}) (10 \text{ s}) = 5.0 \text{ C}$$
$$V = \frac{W}{Q}$$
$$= \frac{12 \text{ J}}{5 \text{ C}}$$
$$= 2.4 \text{ V}$$

(R3-14) It takes 10 J of chemical energy to move 20 C of electric charge between the positive and negative terminals of a battery. What is the emf between the battery terminals?

MEASURING VOLTAGE Voltage is measured with an instrument called a *voltmeter*, figure 3-19. A voltmeter must be placed across the circuit device whose voltage is to be measured. The proper meter polarity connections must be used, figure 3-20.

(R3-15) What will happen if the leads of voltmeter 1 in figure 3-20 are reversed?

DOUBLE SUB-SCRIPT VOLT-AGE NOTATION A voltage reading has both magnitude and polarity. Problems 8–11 illustrate the use of double subscripts.

Problem 8 $V_{AB} = +5$ V. What does this value expressed in double subscript notation mean?

Solution The magnitude of the voltage drop across terminals A and B is five volts. The second letter (B) of the double subscript is the reference terminal. Therefore, terminal A has a positive polarity potential of +5 V above reference terminal B. Figure 3-21 shows both a graphical and a schematic representation of this voltage.

Problem 9   What is the voltage V_{BA} in figure 3-21?

Solution   Terminal A is the reference. Since terminal B is five volts below terminal A, V_{BA} = -5 V. An analysis of the results of problems 8 and 9 proves the following rule: reversing the subscripts is the same as multiplying the voltage value by -1.

Check: V_{AB} = +5 V
$$V_{BA} = (-1)(+5 \text{ V})$$
$$= -5 \text{ V}$$

Problem 10   What is the voltage V_{AD} in figure 3-22?

Solution   For this type of arrangement, the voltages of the batteries are added. Thus, this connection is said to be *series aiding*. Terminal D is the reference terminal. Starting at terminal D, the voltages are added algebraically to obtain V_{AD}.

$$V_{AD} = +6 \text{ V} + (+12 \text{ V})$$
$$= +18 \text{ V}$$

Terminals B and C are at the same potential. They are connected by a highly conductive wire with negligible opposition or resistance. This connection has a double polarity. Terminal B is -12 V below terminal A and +6 V above terminal D.

(R3-16)   Draw a graphical representation of V_{AD} in figure 3-22.

Problem 11   Determine the voltage V_{AD} in figure 3-23.

Solution   In this type of arrangement, the connection is *series opposing* because the voltages of the batteries oppose each other.
$$V_{AD} = +3 \text{ V} + (-9 \text{ V})$$
$$= -6 \text{ V}$$

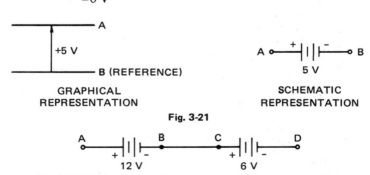

GRAPHICAL
REPRESENTATION

SCHEMATIC
REPRESENTATION

Fig. 3-21

Fig. 3-22   The batteries in this problem are said to be series aiding

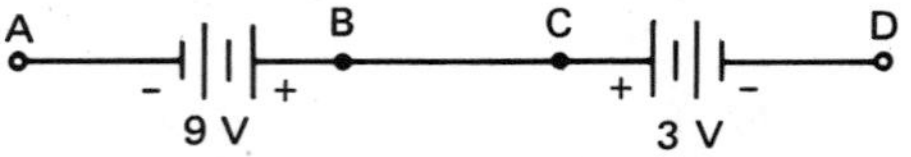

Fig. 3-23   The batteries in this problem are said to be series opposing

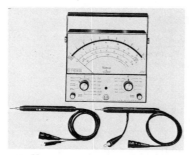

Fig. 3-25 Digital multimeter

Fig. 3-24 Volt-ohm-milliammeter
(VOM) analog multimeter

(R3-17) Draw a graphical representation of V_{AD} in figure 3-23.

MULTIMETERS Multimeters are instruments used to measure voltage, current, and resistance. They are operated as three instruments in one case. The volt-ohm-milliammeter (VOM) shown in figure 3-24 is an analog instrument. It has a pointer that moves in an arc to the appropriate reading. The instrument has one meter movement and one calibrated meter face.

The digital multimeter shown in figure 3-25 displays the numerical magnitude of a reading in a digital form and also indicates the polarity.

LABORATORY EXERCISE 3-1 CURRENT MEASUREMENTS

PURPOSE After completing this exercise, the student will be able to make current measurements.

EQUIPMENT AND MATERIALS
1 Dc power supply
1 Milliammeter, 0–10 mA
1 Analog VOM
1 Resistor, 1 kΩ, 2 W
1 Resistor, 2 kΩ, 2 W
1 Resistor, 3 kΩ, 2 W

PROCEDURE
1. Connect the circuit shown in figure 3-26.
   a. Adjust the power supply voltage to 10 V, as measured with the VOM.

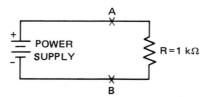

Fig. 3-26 Laboratory Exercise 3-1, step 1

TABLE 3-1

Voltage Volts	Resistance Ohms	Current (mA)	
		Point A	Point B
10	1 000		
10	2 000		
10	3 000		

b. Connect the milliammeter at point A. Measure the current, and record the value in Table 3-1.

c. Remove the milliammeter from point A and connect it at point B. Measure the current and record the value in Table 3-1.

d. Repeat steps a, b, and c with R = 2 kΩ. .

e. Repeat steps a, b, and c with R = 3 kΩ. .

ANALYSIS

1. What conclusion can be reached after taking current readings at points A and B?

2. Under what conditions will there be current in a circuit?

3. Explain how a milliammeter is connected to measure current.

4. What precautions must be observed when measuring current, in order to prevent damage to the meter?

LABORATORY EXERCISE 3-2 VOLTAGE MEASUREMENTS

PURPOSE

After completing this exercise, the student will be able to take voltage measurements.

EQUIPMENT AND MATERIALS

1 Analog VOM
1 Digital multimeter
4 Dry cells, 1.5 V
1 Dry cell battery, 9 V

PROCEDURE

1. Measure the voltage of each of the five batteries.

2. Connect four 1.5-V dry cells as shown in figure 3-27.

a. Calculate E_{AB}.

b. Measure E_{AB} using the VOM.

CAUTION. Connect the + lead of the VOM to point A and the common lead to point B.

c. Measure E_{BA} using the digital multimeter. Connect the + lead to point B and the common lead to point A.

d. What conclusion can be reached as a result of the readings obtained in steps 2.b. and 2.c?

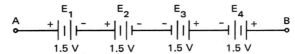

Fig. 3-27 Laboratory Exercise 3-2, step 2

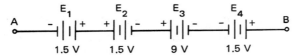

Fig. 3-28 Laboratory Exercise 3-2, step 3

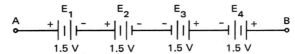

Fig. 3-29 Laboratory Exercise 3-2, step 4

3. Connect the four dry cells as shown in figure 3-28.
   a. Measure E_{AB} using the VOM.
   b. Explain the reading obtained in step 3.a.
4. Connect the circuit shown in figure 3-29.
   a. Measure E_{BA} using the digital multimeter.
   b. Explain the reading obtained in step 4.a.

EXTENDED STUDY TOPICS

1. If 30 C of charge passes through a circuit in three seconds, what is the current in the circuit?
2. The rate of flow of electron drift is 10^{15} electrons per second. What is the current in amperes?
3. Express a current of four coulombs per minute in SI units.
4. How long will it take for 16 C of electric charge to flow through a load whose current measures 250 mA?
5. Ten joules of chemical energy are required to move 20 C of electric charge between the terminals of a battery. What is the emf of the battery?
6. An incandescent lamp releases 12 J of total energy as 0.1 C flows through the filament. What is the voltage drop across the lamp?
7. What is the electric charge movement from one terminal of a load to the other terminal, if 100 J of energy is released? The voltage drop across the load is 120 V.
8. What energy is required to move 10 mC of electric charge through a potential rise of 120 V?
9. How much energy is required to maintain a current of 0.5 A through a battery for 1.0 minute if the battery has an emf of 1.5 V?
10. A 4.5-V flashlight lamp releases 9 J of total energy as a current of 500 mA flows through it. How long will it take to release this energy?

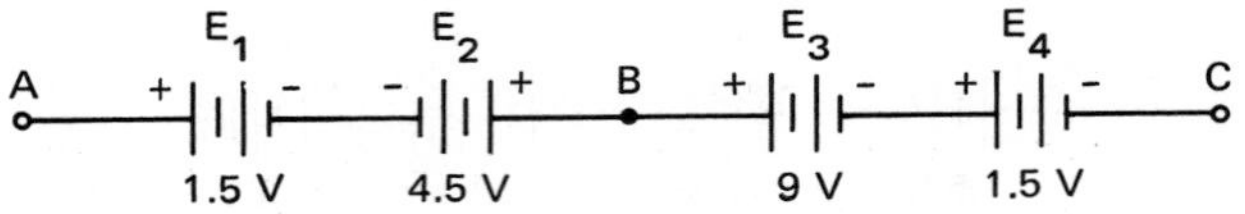

Fig. 3-30

11. If a lamp is connected to a 12-V battery and draws a current of 250 mA, at what rate is the chemical energy of the battery being converted?

12. What current is flowing in a circuit if 24 joules per second is delivered to a load which has a voltage drop equal to 120 V?

13. Calculate E_{AC} in figure 3-30.

14. Calculate E_{CA} in figure 3-30.

15. Calculate E_{AB} in figure 3-30.

16. Calculate E_{BA} in figure 3-30.

17. Show the voltmeter connections with the proper polarities for reading E_{AC} and E_{CA} in figure 3-30.

Chapter 4

Resistance and Ohm's Law

OBJECTIVES

After studying this chapter, the student will be able to:

- explain the nature of resistance.
- apply Ohm's Law to circuit problems.
- identify resistors by means of the color code.
- describe the various types of resistors available commercially.
- discuss the concept of conductance.

NATURE OF RESISTANCE

Resistance is a physical property of a material. This property is the opposition to the flow of current offered by an electric circuit or a specific element. This opposition is caused by collisions between electrons. Resistance is the property that causes electrical energy to be converted into heat and also causes a power loss.

The SI unit of measurement for resistance is the ohm. The unit symbol is the Greek capital letter omega, Ω. The letter symbol for resistance is R. These symbols and the schematic symbol for resistance are shown in figure 4-1.

Resistors are commercially produced devices that provide specific amounts of resistance when added to circuits. Manufacturers generally mark each resistor

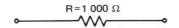

R = 1 000 Ω

Fig. 4-1 Schematic and unit symbols for resistance

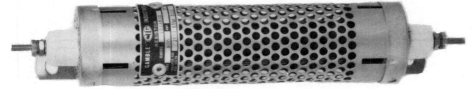

Fig. 4-2 A cage-mounted commercial resistor

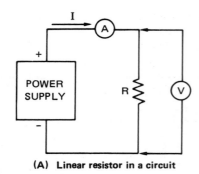

(A) Linear resistor in a circuit

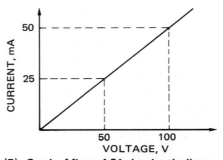

(B) Graph of figure 4-3A showing the linear relationship between voltage and current

Fig. 4-3

with its resistance value using a color code. The value of each resistance is guaranteed by the manufacturer to a specified accuracy under normal use. A cage-mounted commercial resistor is shown in figure 4-2. A linear resistor is a device for which the resistance value is independent of the current through it and the voltage across it. The current in a linear resistor varies directly with the applied voltage. In other words, the ratio between the voltage and the current is a constant. This constant is the resistance of the circuit or element and is expressed as follows: K = R = voltage/current. This relationship is the same as Equation 4.1 which is presented later in this chapter. The circuit shown in figure 4-3 contains a linear resistor. A graph of the relationship between the current and the voltage is also given.

Problem 1   What is the value of R in the circuit in figure 4-3 if the voltage V = 50 V?

Solution $R = \dfrac{\text{Voltage}}{\text{Current}}$

$= \dfrac{50 \text{ V}}{25 \times 10^{-3} \text{ A}}$

$= 2\ 000\ \Omega$

$= 2\ k\Omega$

(R4-1)   Refer to figure 4-3B and calcualte the value of R when V = 100 V.

(R4-2)   What conclusion can be reached when the answers to question (R4-1) and Problem 1 are analyzed?

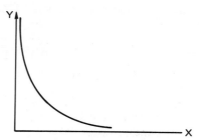

Fig. 4-4 A nonlinear relationship

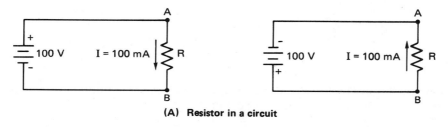

(A) Resistor in a circuit

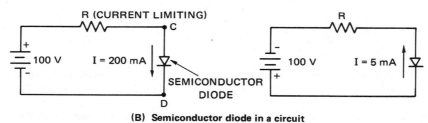

(B) Semiconductor diode in a circuit

Fig. 4-5

Figure 4-4 is a graph of a nonlinear relationship between Y and X. In this relationship, Y is inversely proportional to X. This means that increasing values of X give decreasing values of Y.

A bilateral device offers the same opposition to a current for any direction of current flow. A unilateral device offers a low resistance to current flow in one direction, and a very high, or infinite, resistance to current flow in the opposite direction.

(R4-3) Study the resistor in the circuit of figure 4-5A and the semiconductor diode in the circuit of figure 4-5B. State whether each device is unilateral or bilateral.

OHM'S LAW The relationship between voltage, current, and resistance is expressed by Ohm's Law:

$$R = \frac{E}{I} \ (or, \frac{V}{I})$$

Eq. 4.1

where

$$R = \text{resistance, in ohms}$$
$$E \ (or \ V) = \text{voltage, in volts}$$
$$I = \text{current, in amperes}$$

By rearranging the terms of Equation 4.1, two other forms of Ohm's Law can be derived:

$$I = \frac{E}{R} \left(\text{or, } \frac{V}{R} \right) \qquad \text{Eq. 4.2}$$

$$E \text{ (or, V)} = IR \qquad \text{Eq. 4.3}$$

Problem 2 A resistor has a value of 100 Ω. What is the value of the current if the voltage across this resistor measures 200 V?

Solution
$$I = \frac{V}{R}$$
$$= \frac{200 \text{ V}}{100 \ \Omega}$$
$$= 2 \text{ A}$$

Problem 3 In Problem 2, the resistance remains constant but the voltage increases to 400 V. What is the value of the current?

Solution
$$I = \frac{V}{R}$$
$$= \frac{400 \text{ V}}{100 \ \Omega}$$
$$= 4 \text{ A}$$

(R4-4) What conclusion can be reached by studying the results of Problems 2 and 3?

CIRCULAR MILS NOTATION The area of a circular conductor is measured in circular mils (CM).

$$\text{Area of a circle} = \pi \, r^2$$
$$= \frac{\pi \, D^2}{4} \qquad \text{Eq. 4.4}$$

By definition, a wire with a diameter of one mil has an area of one circular mil (CM). The area in circular mils is equal to the diameter (in mils) squared. (1 mil = $\frac{1}{1\,000}$ inch and 1 inch = 1 000 mils)

$$A \text{ (CM)} = D^2 \qquad \text{Eq. 4.5}$$

Problem 4 The diameter of a circular conductor is $\frac{25}{1\,000}$ inch. What is the area of the conductor in CM?

Solution
$$\frac{25}{1\,000} \text{ in} = 0.025 \text{ in}$$
$$= 0.025 \text{ in} \times \frac{1\,000 \text{ mils}}{\text{in}}$$
$$= 25 \text{ mils}$$

Note that inches are converted to mils by moving the decimal point three places to the right. The same result is obtained by multiplying inches by 1 000, or 10^3.

$$A = D^2$$
$$= (25 \text{ mils})^2$$
$$= 625 \text{ CM}$$

Problem 5 The diameter of a circular conductor is 1/4 in. What is the cross-sectional area in CM?

Solution
$$D = 1/4 \text{ in}$$
$$= 0.250 \text{ in}$$
$$= 250 \text{ mils}$$
$$A = D^2$$
$$= (250 \text{ mils})^2$$
$$= 62\ 500 \text{ CM}$$

Problem 6 A wire has a cross-sectional area of 1 600 CM. What is its diameter, in inches?

Solution
$$D^2 = A$$
$$D = \sqrt{A}$$
$$= \sqrt{1\ 600 \text{ CM}}$$
$$= 40 \text{ mils} = 40 \text{ mils} \times \frac{1 \text{ inch}}{1\ 000 \text{ mils}}$$
$$= 0.040 \text{ in}$$

Note that $(\text{mils} \times \frac{1 \text{ inch}}{1\ 000 \text{ mils}})$ results in an answer in inches.

FACTORS AFFECTING RESISTANCE

A wire has a uniform cross-sectional area. The resistance of this wire is determined by four factors:

1. the type of material used.
2. the length of the wire.
3. the cross-sectional area of the wire.
4. the temperature of the wire.

The relationship of the first three factors is expressed by the following equation:

$$R = \rho \frac{L}{A}$$

or

$$R = \rho \frac{L}{D^2}$$

Eq. 4.6

where

R = resistance, in ohms
ρ = resistivity; this property is a characteristic of the material at $20°C$. The symbol for resistivity is the Greek letter rho.
L = length of wire, in feet
A = cross-sectional area of the wire, in circular mils (CM)
D = diameter of wire, in mils

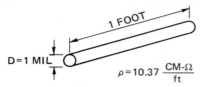

$$D = 1 \text{ MIL}$$
$$\rho = 10.37 \ \frac{\text{CM-}\Omega}{\text{ft}}$$

Fig. 4-6   Specific resistance of copper wire at 20°C

TABLE 4-1 Resistivity

Material	$\rho(20°C) \ \dfrac{\text{CM-ohms}}{\text{foot}}$
Silver	9.9
Copper	10.37
Gold	14.7
Aluminum	17.0
Tungsten	33.0
Nickel	47.0
Iron	74.0
Constantan	295.0
Nichrome	600.0
Calorite	720.0
Carbon	21 000.0

Resistivity is also called the *specific resistance* of a conductor. For a specific material, this resistance is determined at 20°C for one foot of wire having a diameter of one mil, figure 4-6.

The units of resistivity (ρ) are $\dfrac{\text{CM} - \text{ohms}}{\text{foot}}$. The resistivities of various materials are listed in Table 4-1.

The following conclusions can be reached by analyzing Equation 4.6.

- R varies directly with L (the length of the wire).

- R varies inversely with the square of the diameter of the wire.

- R is inversely proportional to the cross-sectional area of the wire.

Problem 7 What is the resistance of 1 000 feet of copper wire having a diameter of $\dfrac{20}{1\ 000}$ in at 20°C?

Solution
$$D = \frac{20}{1\ 000} \text{ in}$$

$$= 0.020 \text{ in} = 0.020 \text{ in} \times 1\ 000 \ \frac{\text{mils}}{\text{in}}$$

$$= 20 \text{ mils}$$

$$\rho = 10.37 \text{ (from Table 4-1 for copper)}$$

$$R = \rho \frac{L}{D^2}$$

$$= \frac{(10.37 \ \frac{\text{CM -}\Omega}{\text{ft}}) (1\ 000 \text{ ft})}{(20)^2 \text{ CM}}$$

$$= 25.9 \ \Omega$$

Problem 8 A reel of copper wire has a total resistance of 2.0 ohms. If the wire has a diameter of 1/16 in, what is the length of wire on the reel?

Solution

$D \quad = \dfrac{1}{16}$ in

$\quad\quad = 0.062\ 5$ in

$\quad\quad = 62.5$ mils

$R \quad = \rho \dfrac{L}{D^2}$

$RD^2 = \rho\,L$

$\rho\,L \quad = RD^2$

$L \quad = \dfrac{RD^2}{\rho}$

$L \quad = \dfrac{(2\ \text{ohms})\ [(62.5)^2\ \text{CM}]}{10.37\dfrac{(\text{CM-ohms})}{\text{ft}}}$

$\quad\quad = 753$ feet

Problem 9 The resistance of 200 yards of copper wire is 10 ohms at 20°C. What is the diameter of the wire in inches?

Solution

$L \quad = 200\ \text{yd} = 200\ \text{yd} \times \dfrac{3\ \text{ft}}{\text{yd}}$

$\quad\quad = 600$ ft

$R \quad = \rho \dfrac{L}{D^2}$

$RD^2 = \rho\,L$

$D^2 \quad = \dfrac{\rho\,L}{R}$

$D \quad = \sqrt{\dfrac{\rho\,L}{R}}$

$\quad = \sqrt{\dfrac{(10.37\dfrac{\text{CM-ohms}}{\text{ft}})\,(600\ \text{ft})}{10\ \text{ohms}}}$

$\quad = \sqrt{622\ \text{CM}}$

$\quad = 24.94$ mils

$\quad = 0.024\ 94$ in

EFFECT OF TEMPERATURE The change of resistance with temperature is expressed by the temperature coefficient of resistance (TCR). Most metals have a positive temperature coefficient (PTC). Thus, the resistance of these metals increases as the temperature increases. Figure 4-7 shows the effect of temperatures on the resistance of copper.

Some materials, such as carbon, porcelain, glass, and semiconductors, have a negative temperature coefficient (NTC). The resistance of these materials de-

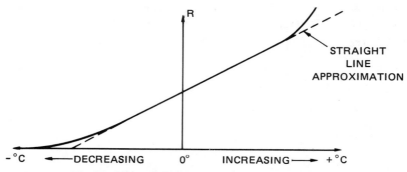

Fig. 4-7 Effect of temperature on the resistance of copper

creases with an increase in the temperature. The resistance of some metal alloys changes very little as the temperature changes. An example of such an alloy is constantan, which consists of 55% copper and 45% nickel. Table 4-2 lists the temperature coefficients of resistance for various materials at 20°C. The symbol for this characteristic is the Greek letter alpha, α.

The resistance of a material can be calculated for a temperature other than 20°C. The following equation can be used:

$$R_2 = R_1 \left[1 + \alpha(T_2 - T_1) \right]$$ Eq. 4.7

where

R_1 = resistance at temperature T_1 (20°C)

R_2 = resistance at temperature T_2

α = temperature coefficient of resistance (See Table 4-2 for specific material); the units of α are $°C^{-1}$ or $\dfrac{1}{°C}$

T_2 = specified temperature in °C

T_1 = 20°C

TABLE 4-2 Temperature Coefficient of Resistance (TCR) for Various Materials

Material	TCR (α) ($°C^{-1}$ or $\dfrac{1}{°C}$)
Silver	0.003 8
Copper	0.003 93
Gold	0.003 4
Aluminum	0.003 91
Tungsten	0.005
Nickel	0.006
Iron	0.005 5
Constantan	0.000 008
Nichrome	0.000 44
Carbon	–0.000 5

Problem 10 A length of copper wire has a resistance of 50 ohms at $20°C$. Determine the resistance of this wire at $100°C$.

Solution
$T_2 = 100°C$
$T_1 = 20°C$
$\alpha = 0.003\ 93°C^{-1}$ (Table 4-2)
$R_1 = 50\ \Omega$
$R_2 = R_1\ [1 + \alpha(T_2 - T_1)]$

$$= 50\ \Omega[1 + \frac{0.003\ 93}{°C}\ (100°C - 20°C)]$$

$$= 50\ \Omega[1 + \frac{0.003\ 93}{°C}\ (80°C)]$$

$$= 50\ \Omega(1 + 0.314)$$

$$= 50\ \Omega(1.314)$$

$$= 65.7\ \Omega$$

WIRE TABLES Table 4-3 lists the standardized wire sizes that form the American Wire Gauge (AWG); The following statements can be made, based on an analysis of Table 4-3:

1. A lower AWG number indicates a greater cross-sectional area in CM.

TABLE 4-3 American Wire Gauge

AWG No.	Area (CM)	Ω/1 000 ft at $20°C$	AWG No.	Area (CM)	Ω/1 000 ft at $20°C$
0000	211 600	0.049 0	19	1 288.1	8.051
000	167 810	0.061 8	20	1 021.5	10.15
00	133 080	0.078 0	21	810.1	12.80
0	105 530	0.098 3	22	642.4	16.14
1	83 694	0.124 0	23	509.45	20.36
2	66 373	0.156 3	24	404.01	25.67
3	52 634	0.197 0	25	320.40	32.37
4	41 742	0.248 5	26	254.10	40.81
5	33 102	0.313 3	27	201.50	51.47
6	26 250	0.395 1	28	159.79	64.90
7	20 816	0.498 2	29	126.72	81.83
8	16 509	0.628 2	30	100.50	103.2
9	13 094	0.792 1	31	79.70	130.1
10	10 381	0.998 9	32	63.21	164.1
11	8 234	1.260	33	50.13	206.9
12	6 529.9	1.588	34	39.75	260.9
13	5 178.4	2.003	35	31.52	329.0
14	4 106.8	2.525	36	25.00	414.8
15	3 256.7	3.184	37	19.83	523.1
16	2 582.9	4.016	38	15.72	659.6
17	2 048.2	5.064	39	12.47	831.8
18	1 624.3	6.385	40	9.89	1 049.0

2. A decrease of one gauge number represents an increase in the cross-sectional area of approximately 25%.
3. If the gauge numbers decrease by three, the cross-sectional area increases two to one.
4. A change of ten wire gauge numbers represents a ten to one change in the cross-sectional area.

(R4-5)   What is the cross-sectional area of No. 18 wire?

(R4-6)   What is the resistance of 1 000 feet of No. 18 wire at 20°C?

Problem 11   Calculate the resistance of 500 feet of No. 13 copper wire at 20°C.
Solution   According to Table 4-3, the resistance of 1 000 feet of No. 13 wire is 2.003 Ω.

$$R(500 \text{ ft}) = 500 \text{ ft} \times \frac{2.003 \ \Omega}{1 \ 000 \text{ ft}}$$

$$= 1.001 \ 5 \ \Omega$$

(R4-7)   What is the resistance of 2 000 feet of No. 13 copper wire at 20°C?

Problem 12   Find the resistance of 2 000 feet of No. 20 copper wire at 40°C.
Solution $R_2 = R_1[1 + \alpha(T_2 - T_1)]$
But,

$$R_1 = \rho\frac{L}{A}$$

Therefore,

$$R_2 = \frac{\rho L}{A}[1 + \alpha(T_2 - T_1)]$$

ρ = 10.37 (Table 4-1)
α = 0.003 93 (Table 4-2) ,
A   = 1 021.5 CM (TAble 4-3)
L   = 2 000 ft
T_1 = 20°C
T_2 = 40°C
R_1 = resistance at 20°C
R_2 = resistance at 40°C

$$R_2 = \frac{\left(10.37 \ \frac{\text{CM-ohms}}{\text{ft}}\right)(2 \ 000 \text{ ft})}{1 \ 021.5 \text{ CM}} [1 + (0.003 \ 93 °\text{C}^{-1})(40°\text{C} - 20°\text{C})]$$

$$= 20.3 \ \Omega \ (1 + 0.078 \ 6)$$

$$= 21.896 \ \Omega$$

FIXED AND VARIABLE RESISTORS

Resistors are manufactured in both fixed value types and variable value types.

Fixed Resistors. The power, or wattage, rating of the resistor determines its size and shape. As the power rating increases, the size of the resistor also increases.

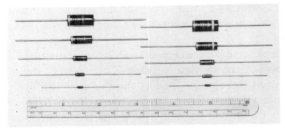

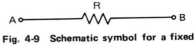

Fig. 4-9 Schematic symbol for a fixed resistor

Fig. 4-8 Variation in size of hot-molded fixed resistors; the power ratings of these resistors, from the smallest size to the largest size, are: 1/8 W, 1/4 W, 1/2 W, 1 W, and 2 W

(Power is covered in detail in Chapter 5.) The increased size is required to dissipate more heat energy. Figure 4-8 shows the size range for hot-molded fixed resistors having power ratings of 1/8 W, 1/4 W, 1/2 W, 1 W, and 2 W. The schematic symbol for a fixed resistor is shown in figure 4-9. Figure 4-10 shows a schematic representation of a tapped resistor.

(R4-8)   Refer to figure 4-10. If the required resistor is 500 Ω, what terminals of this resistor must be connected?

Variable Resistors. A decade resistance box is shown in figure 4-11. This device provides a range of resistance values from 1 Ω to 9 999 Ω for general use in circuits.

The *three-point variable resistor* is called a rheostat or a potentiometer, depending upon how it is used in a circuit. Two commercial variable resistors are

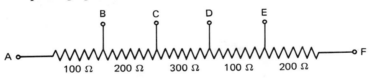

Fig. 4-10 Tapped resistor

Fig. 4-11 Decade resistance box

shown in figure 4-12. Figure 4-13 is a schematic drawing of a variable resistor. In this figure, the arrow represents a contact that slides or moves over a continuous resistive element.

A three-point variable resistor connected as shown in figure 4-14 is called a *rheostat*. The position of the moving, or sliding, contact (C) determines the magnitude of the current through the load of the circuit shown in figure 4-15. When the sliding contact is at point A, the resistance R has a minimum value of zero ohms. When the sliding contact is at point B, the resistance R has a maximum value of R_{AB}.

When the three-point variable resistor is connected as shown in the circuit of figure 4-16, it is called a *potentiometer*. Note that all of the terminals are connected into the circuit. In this case, the moving contact, or wiper arm, of the · variable resistor controls the voltage or potential difference impressed across the load. Figure 4-17 shows several commercial variable resistors which are used as potentiometers.

Refer to figure 4-16 to answer questions (R4-9), (R4-10), and (R4-11).

(R4-9)   What voltage is supplied to the load when the wiper arm C is at terminal A?

Fig. 4-12A Screw-drive rheostat   Fig. 4-12B Hot-molded variable resistor

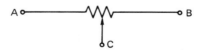

Fig. 4-13 Schematic representation of a three-point variable resistor

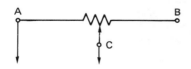

(A) Three-point resistor connected as a rheostat.   Terminal B is not used.   Terminals A and C are connected into a circuit.

(B) Schematic drawing of a variable rheostat

Fig. 4-14

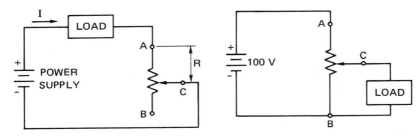

Fig. 4-15 Rheostat controls current   **Fig. 4-16 Potentiometer in a circuit**

(A) Hot-molded potentiometer

(B) Trimming potentiometers

(C) Trimmer potentiometers

Fig. 4-17

(R4-10)   What voltage is supplied to the load when the wiper arm C is at terminal B?

(R4-11)   Is the voltage between wiper arm C and point B greater than, less than, or equal to 50 V when the wiper arm is midway between terminals A and B?

**RESISTOR TERMS
AND
PARAMETERS**   The nominal, stated value of a resistor is based on operation at a standard room temperature of 20°C. The *tolerance* of a resistor is an indication of the maximum permitted variation from the nominal value.

Problem 13   What is the range of values for a resistor whose nominal value is 1 000 Ω ±10%?

The tolerance of the resistor is ±10%
(10%) (1 000 Ω) = 100 Ω
Minimum value of R   = 1 000 Ω − 100 Ω = 900 Ω
Maximum value of R = 1 000 Ω + 100 Ω = 1 100 Ω

Therefore, the range is 900 Ω to 1 100 Ω
This resistor is within its stated tolerance if its actual value is between 900 Ω and 1 100 Ω.

Resistors are given power ratings in watts. This rating is the maximum continuous power that a resistor can dissipate at a temperature up to 70°C. Above 70°C, the power rating of the resistor is reduced, or *derated*. Figure 4-18 shows a typical derating curve.

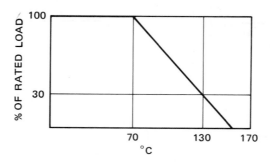

Fig. 4-18 Derating curve

Problem 14 A resistor has a power rating of 60 watts. Determine the power rating of the resistor at 130°C.

Solution According to figure 4-18, a resistor operating at 130°C can dissipate 30% of its nominal power rating.

$$(0.30)\,(60\ W)\ =\ 18\ W$$

Therefore, at a temperature of 130°C, the resistor dissipates only 18 W.

As a precaution, it is a common practice to increase the calculated dissipated power by a safety factor of two. For example, if a calculation indicates that a 1-W resistor is required, a 2-W resistor is normally specified.

$$1\ W\ \times\ 2\ (\text{safety factor})\ =\ 2\ W$$

A resistor rating is complete if it states the nominal value of the resistor, the tolerance, and the wattage rating. An example of a complete rating is:

$$R = 100\ \Omega \pm 5\%,\ 2\ W$$

CHARACTERIS-TIC RANGES OF FIXED RESISTORS Figure 4-19 shows several types of fixed resistors. Table 4-4 lists these resistors with standard ranges of values for resistances, tolerances, and power.

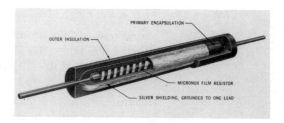

(B) Ultra-stable, high-voltage resistor

(A) Radial lug-type, vitreous, power wirewound resistors

Fig. 4-19

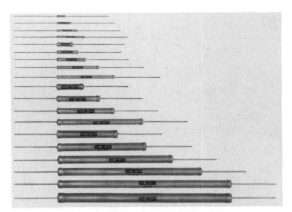

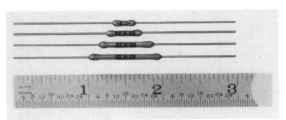

(D) Miniature resistors; resistance values range from 1 MΩ to 220 MΩ; power ratings range from 0.5 W to 5.0 W at 600 V to 4 000 V

(C) High-voltage resistors with long term stability; values range as high as 2 000 MΩ

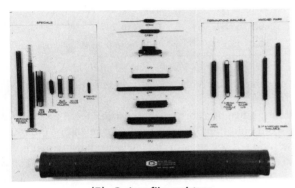

(E) Carbon film resistors

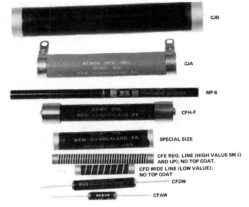

(F) Carbon film resistors

(G) Discrete resistors

(H) Cutaway view of precision wirewound resistor

(I) Cermet film fixed resistors

Fig. 4-19 Cont.

TABLE 4-4

Resistor Type	Resistance Range (Ohms)	Tolerance Range (%)	Power Range (Watts)
Wirewound power	0.1 Ω to 1 M Ω	±1 to ±10%	5 W to 1 500 W
Wirewound precision	0.001 Ω to 60 M Ω	±0.001 to ±1%	0.04 W to 250 W
Carbon composition (standard)	1 Ω to 22 M Ω	±5 to ±20%	1/8 W to 2 W
Carbon composition (hot-molded)	1 Ω to 100 M Ω	±5 to ±20%	1/8 W to 5 W
Carbon film	1 Ω to 200 M Ω	±0.5 to ±10%	0.1 W to 100 W
Metal film	0.27 Ω to 100 M Ω	±0.01 to ±2%	1/20 W to 20 W
Glazed metal	1 Ω to 500 M Ω	±2 to ±5%	0.1 W to 10 W
Metal oxide	10 Ω to 15 M Ω	±1 to ±5%	1/4 W to 115 W
Cermet film	10 Ω to 500 M Ω	±0.5 to ±5%	0.1 W to 2 W
Bulk property film	30 Ω to 10 M Ω	±0.01 to ±1%	0.05 W to 0.75 W

(R4-12)   What is the tolerance range of standard carbon composition resistors?

COLOR CODING OF CARBON COMPOSITION RESISTORS   The tolerance and the resistance range of a carbon composition fixed resistor are identified by a color code. Four color bands are printed on one end of the resistor body. To determine the rating of the resistor, it is held so that the bands are on the left, as shown in figure 4-20. The colors are read from left to right. Table 4-5 lists the values assigned to the colors in the bands.

Problem 15   Determine the rating of a resistor having four color bands as follows:

Band 1   yellow
Band 2   violet
Band 3   red
Band 4   silver

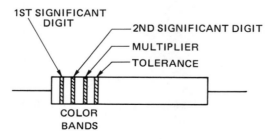

Fig. 4-20 Resistor color bands

TABLE 4-5 Color Code for Carbon Composition Resistors

Color	First Band	Second Band	Third Band	Fourth Band
	First Significant Digit	Second Significant Digit	Multiplier	Tolerance
Black	–	0	10^0	–
Brown	1	1	10^1	–
Red	2	2	10^2	–
Orange	3	3	10^3	–
Yellow	4	4	10^4	–
Green	5	5	10^5	–
Blue	6	6	10^6	–
Violet	7	7	10^7	–
Gray	8	8	10^8	–
White	9	9	10^9	–
Gold	–	–	0.1	±5%
Silver	–	–	0.01	±10%
No Color	–	–	–	±20%

Solution   yellow = 4 (first significant digit)
  violet = 7 (second significant digit)
  red = 10^2 (multiplier)
  silver = ±10% (tolerance)
  Resistor = 47 × 10^2 Ω ±10%
   = 4 700 Ω ±10%

(R4-13)   What is the resistance range of the resistor in Problem 15?

Problem 16   A resistor has the following color bands:

  Band 1   red
  Band 2   red
  Band 3   orange
  Band 4   no color

  Determine the nominal resistance value and the resistance range for this resistor.

Solution   Band 1, red = 2
  Band 2, red = 2
  Band 3, orange = 10^3
  Band 4, no color = ±20% tolerance
  Nominal value   = 22 × 10^3 Ω ±20%
    = 22 000 Ω ±20%
  ±(20%)(22 000 Ω) = ±4 400 Ω tolerance
  22 000 Ω – 4 400 Ω = 17 600 Ω
  22 000 Ω + 4 400 Ω = 26 400 Ω
    Range = 17 600 Ω to 26 400 Ω

Problem 17   A resistor has the following color bands:

Band 1   white
Band 2   brown
Band 3   gold
Band 4   gold

Determine the rating of this resistor.

Solution   Band 1, white = 9
Band 2, brown = 1
Band 3, gold = 0.1 (multiplier in 3rd band)
Band 4, gold = ±5% tolerance
Resistor rating = (91 × 0.1) Ω ±5%
$$= 9.1 \ \Omega \ \pm 5\%$$

(R4-14)   Calculate the rating of a resistor with the following sequence of color bands: gray, white, gold, and silver.

PREFERRED RESISTANCE VALUES   Certain values of carbon composition fixed resistors have been standardized by the electronics industry to reduce manufacturing costs. Table 4-6 lists the standard resistance values that are available commercially. Any value which is a multiple of ten of one of these values, up to 100 MΩ, may be obtained from suppliers.

TABLE 4-6 Preferred Resistance Values for Carbon Composition Fixed Resistors

±5% Tolerance	±10% Tolerance	±20% Tolerance
1.0	1.0	1.0
1.1		
1.2	1.2	
1.3		
1.5	1.5	1.5
1.6		
1.8	1.8	
2.0		
2.2	2.2	2.2
2.4		
2.7	2.7	
3.0		
3.3	3.3	3.3
3.6		
3.9	3.9	
4.3		
4.7	4.7	4.7
5.1		
5.6	5.6	
6.2		
6.8	6.8	6.8
7.5		
8.2	8.2	
9.1		

THERMISTORS A thermistor is a nonlinear resistor. It is made of a semiconductor material. It is extremely sensitive to changes in temperature. Therefore, it can have either a positive temperature coefficient (PTC), or a negative temperature coefficient (NTC). The resistance versus temperature characteristics are shown in figure 4-21 for both types of thermistors.

Figure 4-22 is a comparison of a conductor and a thermistor with a negative temperature coefficient. The resistance of the thermistor (NTC) decreases rapidly as the temperature increases. On the other hand, the conductor (PTC) shows a small increase in resistance with a temperature increase.

The electrical symbol for a thermistor is shown in figure 4-23. Thermistors have many industrial applications, including measurement of temperature, control of temperature, time delay, temperature compensation, and liquid-level indication. Thermistor configurations are shown in figure 4-24, and thermistor

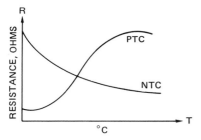

Fig. 4-21 Thermistor characteristics

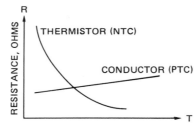

Fig. 4-22 Effect of temperature variations on the resistance of a conductor (PTC) and a thermistor (NTC)

Fig. 4-23 Electrical symbol for a thermistor

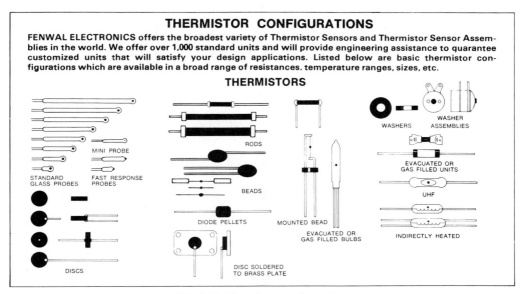

THERMISTOR CONFIGURATIONS

FENWAL ELECTRONICS offers the broadest variety of Thermistor Sensors and Thermistor Sensor Assemblies in the world. We offer over 1,000 standard units and will provide engineering assistance to guarantee customized units that will satisfy your design applications. Listed below are basic thermistor configurations which are available in a broad range of resistances, temperature ranges, sizes, etc.

THERMISTORS

STANDARD GLASS PROBES
FAST RESPONSE PROBES
MINI PROBE
DISCS
RODS
BEADS
DIODE PELLETS
DISC SOLDERED TO BRASS PLATE
MOUNTED BEAD
EVACUATED OR GAS FILLED BULBS
WASHERS
WASHER ASSEMBLIES
EVACUATED OR GAS FILLED UNITS
UHF
INDIRECTLY HEATED

Fig. 4-24 Thermistor configurations

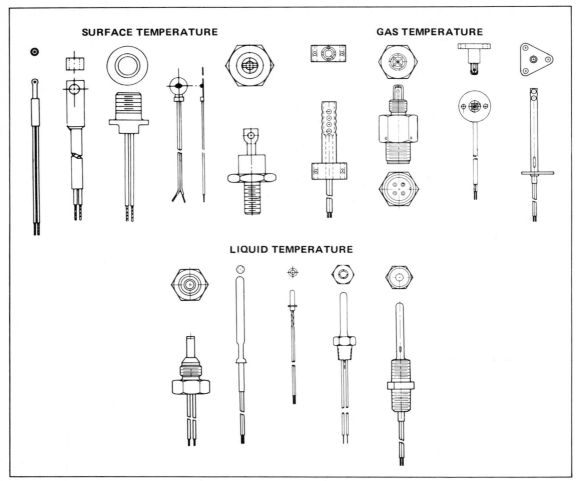

Fig. 4-25 Thermistor probe assemblies

Fig. 4-26 Thermistor thermometer

probe assemblies are shown in figure 4-25. Figure 4-26 shows the application of a thermistor in a digital thermometer.

CONDUCTANCE   *Conductance* is a measure of the ability of an electric circuit to pass current. The letter symbol for conductance is G. The SI unit of conductance is the siemens. (In the English system of measurement, the unit of conductance was the mho.) The unit symbol for siemens is S.

Conductance is the reciprocal of resistance, as shown by the following equation:

$$G = \frac{1}{R}$$

Eq. 4.8

where

G = conductance, in siemens

R = resistance, in ohms

The following equations are derived from Equation 4.8 and Ohm's Law:

$$R = \frac{1}{G}$$

Eq. 4.9

$$E \text{ (or V)} = \frac{I}{G}$$

Eq. 4.10

$$I = EG \text{ (or VG)}$$

Eq. 4.11

Problem 18   A circuit has a conductance of 0.005 S. Determine the resistance of this circuit.

Solution

$$R = \frac{1}{G}$$

$$= \frac{1}{0.005 \text{ S}}$$

$$= 200 \ \Omega$$

Problem 19   A circuit has a resistance of 20Ω. Calculate the conductance of this circuit.

Solution

$$G = \frac{1}{R}$$

$$= \frac{1}{20 \ \Omega}$$

$$= 0.05 \text{ S}$$

Table 4-7 summarizes the four electrical characteristics that were studied in this chapter.

TABLE 4-7 Summary of Electrical Characteristics

Characteristic	Unit	Letter Symbol	Unit Symbol
Current	ampere	I	A
Voltage	volt	E or V	V
Resistance	ohm	R	Ω
Conductance	siemens	G	S

LABORATORY EXERCISE 4-1 RESISTANCE MEASUREMENTS

PURPOSE   After completing this exercise, the student will be able to

- make resistance measurements.

- use the resistor color code.

EQUIPMENT AND MATERIALS

1 Digital multimeter
1 Resistor, 1 k Ω ±5%, 2 W
1 Resistor, 2 k Ω ±5%, 2 W
1 Resistor, 33 k Ω ±10%, 2 W
1 Resistor, 470 k Ω ±10%, 2 W
1 Resistor, 10 M Ω ±20%, 2 W
1 Resistor, 22 MΩ ±20%, 2 W

PROCEDURE

1. Applying the color code, identify the six carbon resistors listed and fill in the data requested in Table 4-8.
2. Is the 10-M Ω resistor or the 22-M Ω resistor physically larger? Explain the reason for the size difference.
3. Measure the resistance of each of the six resistors using the digital multimeter. Record the values in Table 4-9.
   a. Calculate the percentage difference between the measured value and the nominal value for each resistor. Use the following formula:

   $$\% \text{ difference} = \frac{\text{measured value} - \text{nominal value}}{\text{nominal value}} \times 100$$

   b. In the "Remarks" column of Table 4-9, indicate if the resistance values are within the acceptable range determined from the color code.
   c. Explain any variations in the values from the range of resistances listed in Table 4-8.

TABLE 4-8

Resistor	Color Sequence				Nominal Value	Tolerance %	Resistance Range
	Band 1	Band 2	Band 3	Band 4			
R_1							
R_2							
R_3							
R_4							
R_5							
R_6							

TABLE 4-9

Resistor	Nominal Value	Measured Value	% Difference	Remarks
R_1				
R_2				
R_3				
R_4				
R_5				
R_6				

LABORATORY EXERCISE 4-2 OHM'S LAW

PURPOSE   After completing this exercise, the student will be able to prove Ohm's Law experimentally.

EQUIPMENT AND MATERIALS
1 Power supply, dc
1 Digital multimeter
1 Milliammeter, dc, 0–30-mA range
1 Milliammeter, dc, 0–100-mA range
1 Resistor, 220 Ω ±5%, 2 W
1 Resistor, 330 Ω ±5%, 2 W
1 Resistor, 470 Ω ±5%, 2 W
1 Resistor, 680 Ω ±5%, 2 W
1 Resistor, 1 000 Ω ±5%, 2 W

PROCEDURE
1. Measure the value of each resistor using the digital multimeter. Make sure that each resistor is within the manufacturer's tolerance range of ±5%.
2. Connect the circuit shown in figure 4-27. This part of the exercise shows how the current varies as the voltage is changed across a fixed resistor.

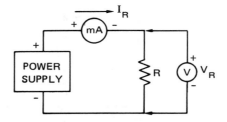

Fig. 4-27 Circuit for confirmation of Ohm's Law

    a. Use R = 220 Ω

    b. Adjust the voltage of the power supply in the following steps: V_R = 5 V, 10 V, 15 V, and 20 V. Measure the current I_R for each voltage level.

    c. Calculate the expected value of I_R. Use the measured values of resistance for this calculation.

    d. Calculate the % error of I_R.

    e. List the data required in Table 4-10.

    f. Plot a curve of I_R versus V_R.

    g. What conclusion can be reached by analyzing this curve?

3. This part of the experiment shows how the current varies with changes in resistance in a circuit having a constant voltage source.

    a. Adjust the voltage of the power supply so that V_R = 20 V.

    b. Measure I_R through each individual resistor.

    c. Calculate I_R.

    d. Calculate the % error of I_R.

    e. List the data required in Table 4-11.

    f. Plot a curve of I_R versus R.

    g. What conclusion can be reached by analyzing this curve?

4. This part of the experiment shows how the voltage varies with changes in resistance in a circuit having a constant current source.

    a. Adjust the voltage of the power supply so that V_R = 20 V across the 1 000-Ω resistor.

    b. Measure I_R and note the value.

    c. Calculate V_R using the measured values of I_R and R.

    d. Calculate the % error of V_R.

    e. Repeat steps 4.a.–4.d. for each of the four remaining resistors. The value of I_R is held constant at the value measured in step 4.b. by adjusting the power supply.

    f. For each resistance, measure and record V_R.

    g. List the data required in Table 4-12.

    h. Plot a curve of V_R versus R.

    i. What conclusion can be reached by analyzing this curve?

TABLE 4-10

R Nominal	R Measured	V_R	I_R Measured	I_R Calculated	I_R % Error
220 Ω		5			
		10			
		15			
		20			

TABLE 4-11

R	V_R	I_R Measured	I_R Calculated	I_R % Error
220 Ω	20			
330 Ω				
470 Ω				
680 Ω				
1 000 Ω				

TABLE 4-12

R	V_R	I_R Measured	V_R Calculated	V_R % Error
1 000 Ω	20			
680 Ω				
470 Ω				
330 Ω				
220 Ω				

EXTENDED STUDY TOPICS

1. A circuit draws a current of 2 A from a 100-V source. What is the resistance of the circuit?
2. If the applied voltage is 40 V, what resistance is required to limit the current to 2 mA?
3. What is the potential drop across a 75-ohm resistor when the current is 3A?
4. What voltage must be applied across a 240-kΩ resistor to obtain a current of 20 μA?
5. Find the resulting current when a 120-V source is connected to a 3-MΩ resistor.
6. Determine the current when there is a voltage drop of 150 V across a 75-Ω resistor.
7. What is the voltage across a 56-Ω resistor when 2 A flows through it?
8. A copper wire has a cross-sectional area of 1 624 CM. Determine its diameter in inches.
9. What is the gauge number of the wire in topic 8?
10. A reel of copper wire is 1 000 ft long. The total resistance of the wire is 2.525 Ω. Determine the cross-sectional area of the wire in CM.
11. What is the gauge number of the wire in topic 10?

12. A 2 000-ft length of copper wire has a cross-sectional area of 2 583 CM. Determine its resistance.

13. A 3 000-ft reel of wire has a total resistance of 3 Ω. The cross-sectional area of the wire is 10 381 CM. Is this a reel of copper wire?

14. A reel contains 1 000 ft of wire. The wire has a diameter of 1/4 in and a total resistance of 100 Ω. A second reel of the same type of wire contains 2 000 ft. The diameter of the wire is 1/2 in. Find the resistance of the second reel of wire.

15. A 1 000-ft reel of wire has a resistance of 160 Ω. The cross-sectional area of the wire is 25 CM. A second reel of the same type of wire contains 500 ft of wire. The cross-sectional area of the wire is 100 CM. Find the resistance of the second reel of wire.

16. Find the resistance of 1 000 ft of No. 20 copper wire at 30°C.

17. A length of copper wire has a resistance of 100 Ω at 20°C. Find the resistance of the wire at 10°C.

18. Find the resistance of 1 000 ft of No. 20 copper wire at the freezing point of water.

19. Four carbon resistors have the following color code sequence:
    a. red-red-orange-gold
    b. red-violet-blue-silver
    c. green-blue-yellow-silver
    d. orange-orange-orange
    Determine the resistance value and the tolerance for each.

20. List the color code sequence for each of the following carbon resistors:
    a. 100 000 Ω ±20%
    b. 22 MΩ ±5%
    c. 8.2 Ω ±10%
    d. 36 kΩ ±5%
    e. 2 700 Ω ±10%

21. Find the conductance of a resistor which has a voltage drop of 120 V when 6 A flows through it.

22. Calculate the resistance of an element having a conductance of 0.033 S.

23. A circuit has a conductance of 0.05 S and a current of 2 A. Calculate the applied voltage.

24. Determine the current in a circuit if the impressed voltage is 120 V and the conductance is 0.008 S.

25. A circuit has a total resistance of 750 Ω. Calculate the conductance of this circuit.

Chapter 5	Work, Energy, and Power

OBJECTIVES   After studying this chapter, the student will be able to

- state the basic concepts of work, energy, and power.

- explain power dissipation.

- define horsepower and kilowatt.

- explain the relationship between horsepower and kilowatt.

- discuss the internal resistance of a power supply and maximum power transfer.

WORK   The definition of mechanical work is stated as follows:

Work is motion that takes place against the action of a force that opposes the desired motion.

Work (W) is independent of time. The SI unit for work is the joule (J). One joule is the work accomplished when an object is moved a distance of one meter against an opposing force of one newton. The mechanical formula for work is:

$$W = FD \qquad\qquad \text{Eq. 5.1}$$

where
W = work, in joules
F = force, in newtons
D = distance, in meters

Problem 1   Determine the work that is done when an object moves 8 m against a force of 2 N.

Solution $W = FD$
$$= (2 \text{ N}) (8 \text{ m})$$
$$= 16 \text{ J}$$

Problem 2   The work accomplished in moving an object 4 m is 8 J. Determine the opposing force.

Solution $W = FD$

$$\frac{W}{D} = F$$

$$F = \frac{W}{D}$$

$$= \frac{8 \text{ J}}{4 \text{ m}}$$

$$= 2 \text{ N}$$

Problem 3   An object is moved against an opposing force of 3 N. The work done is equal to 21 J. How far is the object moved?

Solution $W = FD$

$$\frac{W}{F} = D$$

$$D = \frac{W}{F}$$

$$= \frac{21 \text{ J}}{3 \text{ N}}$$

$$= 7 \text{ m}$$

(R5-1)   What is the relationship between pounds and newtons? Suggestion: refer to Table 1-8.

ENERGY   When an object is moved against an opposing force, energy is used. The law of conservation of energy states that energy cannot be created or destroyed. However, energy may be converted from one form to some other form. Energy and work are numerically equal and both are expressed by Equation 5.1. The capacity to do work is a form of energy. Thus, work is performed whenever energy is changed from one form to another.

Figure 5-1 shows the electrical circuit of a flashlight with the switch in the *off* position. The flashlight cells contain chemical energy which has the capacity, or potential, to do work. This energy is known as *potential energy*. Figure 5-2 shows the various forms of energy that exist in the circuit when the flashlight is in the *on* position. The chemical energy in the flashlight cells is converted into electrical energy. This energy, in turn, is converted into light energy.

(R5-2)   What form of energy conversion is not shown in figure 5-2?

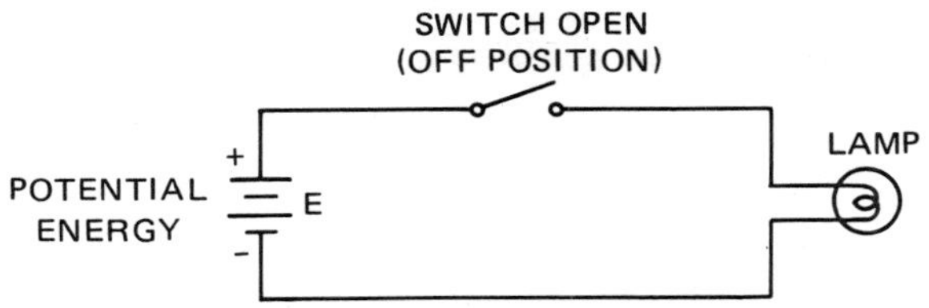

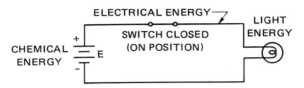

**SWITCH OPEN
(OFF POSITION)**

LAMP

POTENTIAL ENERGY

Fig. 5-1 Electrical circuit of a flashlight when the switch is in the *off* position

ELECTRICAL ENERGY

LIGHT ENERGY

SWITCH CLOSED
(ON POSITION)

CHEMICAL ENERGY

Fig. 5-2 Types of energy existing in a flashlight when the switch is in the *on* position

The joule is the basic unit of both electrical work and energy. Equation 5.2 shows that one joule of electric energy is required to raise one coulomb of electric charge through a potential difference of one volt.

$$W = EQ \qquad\qquad \text{Eq. 5.2}$$

where
W = energy, in joules
E = potential difference, in volts
Q = quantity of electric charge, in coulombs

Problem 4 A 150-mA current flows through an electronic device for a period of 20 seconds. The value of heat energy released is 18.9 J. Determine the voltage across the terminals of the device.

Solution
$$Q = It$$
$$= (0.150 \text{ A}) (20 \text{ s})$$
$$= 3 \text{ C}$$
$$W = VQ$$
$$\frac{W}{Q} = V$$
$$V = \frac{W}{Q}$$
$$= 18.9 \text{ J}/3 \text{ C}$$
$$= 6.3 \text{ V}$$

(R5-3) A 150-mA current through the electronic device described in Problem 4 releases 18.9 J of heat energy in 10 s. Determine the effect on the voltage across the terminals of the device.

POWER   Power (P) is the rate of doing work. Power is also the rate at which energy is used. The formula for power is:

$$P = \frac{W}{t}$$

Eq. 5.3

where

W = work or energy, in joules
t = time, in seconds
P = power, in joules per second

Equation 5.3 clearly shows that power is the *rate* of doing work. The unit of electric power is called the *watt* (W). The student must remember that (1) *watts are equivalent to joules per second*, and (2) power is always the *rate* of doing work. In addition, power is the *rate* of energy consumption.

Problem 5   Determine the power which must be supplied to an electric motor if it does 900 000 joules of work in five minutes.

Solution   W = 900 000 J

t = 5 × 60 = 300 s

$$P = \frac{W}{t}$$

$$= \frac{900\ 000\ J}{300\ s}$$

= 3 000 W

$$= 3\ 000\ W \times \frac{1\ kW}{1\ 000\ W}$$

= 3.0 kW

(R5-4)   What is the rate at which electric energy is used by the motor described in problem 5?

The following equations are also used to calculate electric power:

P = EI **or** VI   Eq. 5.4

P = I²R   Eq. 5.5

$$P = \frac{E^2}{R} \ \text{or} \ \frac{V^2}{R}$$ Eq. 5.6

where

P = power, in watts
E = emf, in volts
V = voltage drop, in volts
I = current, in amperes

Problem 6   What is the power rating if an incandescent lamp draws 625 mA when connected to a 120-V source?

Solution   P = EI

= (120 V) (0.625 A)

= 75 W

Problem 7   A 150-W light bulb draws 1.25 A. What is the operating voltage of the bulb?

Solution $P = EI$

$$\frac{P}{I} = E$$

$$E = \frac{P}{I}$$

$$= \frac{150 \text{ W}}{1.25 \text{ A}}$$

$$= 120 \text{ V}$$

Problem 8   An electronic device operates at 12 V and is supplied with 0.18 W of power. Determine the current drawn by the device.

Solution $P = EI$

$$\frac{P}{E} = I$$

$$I = \frac{P}{E}$$

$$= \frac{0.18 \text{ W}}{12 \text{ V}}$$

$$= 0.015 \text{ A} \times \frac{1 \ 000 \text{ mA}}{1 \text{ A}}$$

$$= 15 \text{ mA}$$

Problem 9   Calculate the power delivered to the resistor shown in figure 5-3.

Solution $I = \dfrac{E}{R}$

$$= \frac{100 \text{ V}}{2 \ \Omega}$$

$$= 50 \text{ A}$$

$$P = EI$$

$$= (100 \text{ V}) (50 \text{ A})$$

$$= 5 \ 000 \text{ W} \times \frac{1 \text{ kW}}{1 \ 000 \text{ W}}$$

$$= 5 \text{ kW}$$

(R5-5)   What is the current drawn by a 100-W incandescent lamp operating at 120 V?

(R5-6)   Is it correct to say: How much power is consumed by an electric heater drawing 4 A from a 120-V power supply? Explain the answer.

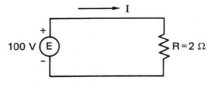

Fig. 5-3

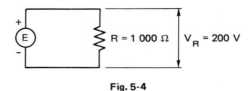

Fig. 5-4

Problem 10   Determine the power delivered to the resistor shown in figure 5-4.
Solution $P = \dfrac{V^2}{R}$

$= \dfrac{(200\ V)\ (200\ V)}{1\ 000\ \Omega}$

$= 40\ W$

Problem 11   Calculate the current drawn by the resistor described in Problem 10.
Solution $I = \dfrac{V_R}{R}$

$= \dfrac{200\ V}{1\ 000\ \Omega}$

$= 0.20\ A = (0.20\ A)\left(\dfrac{1\ 000\ mA}{1\ A}\right)$

$= 200\ mA$

Alternate Solution $P = I^2 R$

$\dfrac{P}{R} = I^2$

$I^2 = \dfrac{P}{R}$

$I = \sqrt{\dfrac{P}{R}}$

$I = \sqrt{\dfrac{40\ W}{1\ 000\ \Omega}}$

$= \sqrt{0.04\ A^2}$

$= 0.2\ A = (0.2\ A)\ \dfrac{1\ 000\ mA}{1\ A}$

$= 200\ mA$

POWER   Voltage sources supply the power to produce a current flow through resistive
DISSIPATION   devices. Heat is produced as the current flows through the resistive material. As
the free electrons move through the material, they collide with each other and
produce heat. Since heat is a form of energy, the fact that it is present indicates
that power is being supplied to the resistive device. This power is the rate at
which the energy is being used, or the rate of doing work.

Any device which offers resistance to the flow of current causes some electrical energy to be converted to heat energy. The heat energy is radiated, or dissipated, into the surrounding medium. Therefore, it cannot be returned to the circuit as electrical energy. The rate at which energy is dissipated is known as the *power dissipation* of the device. The total energy dissipated in a circuit must be provided by the power supply, or input source.

Problem 12 The following items are in use at one time in a home:

Refrigerator	P_1 = 300 W
Lamp	P_2 = 100 W
Toaster	P_3 = 1 200 W
Electric heater	P_4 = 1 500 W
Television	P_5 = 350 W

What is the total power dissipation?

Solution Power dissipation is the arithmetic sum of the individual power values.

$$P_T = P_1 + P_2 + P_3 + P_4 + P_5$$
$$= 300 \text{ W} + 100 \text{ W} + 1\ 200 \text{ W} + 1\ 500 \text{ W} + 350 \text{ W}$$
$$= 3\ 450 \text{ W}$$

Problem 13 A current of 40 mA flows through an electronic device. The voltage drop across the terminals of the device is 0.8 V. How much power does the device dissipate?

Solution
$$P = VI$$
$$= (0.8 \text{ V}) (40 \times 10^{-3} \text{ A})$$
$$= (32 \times 10^{-3} \text{ W}) \left(\frac{10^3 \text{ mW}}{1 \text{ W}} \right)$$
$$= 32 \text{ mW}$$

Problem 14 How much power is dissipated by a 5-k Ω resistor which passes 4 mA of current?

Solution
$$P = I^2 R$$
$$= (4 \times 10^{-3} \text{ A})^2 (5 \times 10^3 \ \Omega)$$
$$= (16 \times 10^{-6} \text{ A}^2)(5 \times 10^3 \ \Omega)$$
$$= (80 \times 10^{-3} \text{ W}) \left(\frac{10^3 \text{ mW}}{1 \text{ W}} \right)$$
$$= 80 \text{ mW}$$

Problem 15 How much power is dissipated by a 300-ohm resistor if the voltage drop across it is 10 V?

Solution
$$P = \frac{V^2}{R}$$
$$= \frac{(10 \text{ V})^2}{300 \ \Omega}$$
$$= \frac{100 \text{ V}}{300 \ \Omega}$$
$$= (0.333 \text{ W}) \ \frac{10^3 \text{ mW}}{1 \text{ W}}$$
$$= 333 \text{ mW}$$

Heat dissipation is used to good advantage in a number of devices, including an electric stove, a toaster, a soldering iron, and an electric heater. In an incandescent lamp, for example, the power dissipation causes the filament to heat until it becomes white-hot and glows, giving off light.

If a device overheats, it generally stops operating. If it continues to overheat, it may burn up. Therefore, every resistive device is given a power dissipation rating. This rating indicates the amount of power (watts) in the form of heat that the device can dissipate safely in its normal ambient environment without being damaged. For example, a 2-W resistor can dissipate up to 2 W of power. This is the rate at which the resistor can convert electrical energy into heat energy. If a resistor has a larger power rating, it is physically larger. This means that the resistor can radiate a greater amount of heat.

When selecting a resistor for an application, the power dissipation must be determined. The resistor should be large enough that it will not overheat and burn out. A design safety factor of 100% is often applied when the resistor size is selected. For example, if the calculated power dissipation of a resistor is 1.0 W, a resistor with a 2.0-W rating is specified.

Problem 16 Determine the power rating for the resistor shown in the circuit of figure 5-5. Use a design safety factor of 100%.

Solution
$$I = \frac{E}{R}$$
$$= \frac{10\,V}{100\,\Omega}$$
$$= 0.1\,A$$
$$P = I^2 R$$
$$= (0.1\,A)^2\,(100\,\Omega)$$
$$= 1\,W \text{ (calculated power dissipation)}$$
$$P = (2)\,(1\,W)$$
$$= 2\,W \text{ (design value)}$$

MEASUREMENT OF POWER IN A DC ELECTRIC CIRCUIT

According to Equation 5.4:
$$P = EI$$
$$\text{or}$$
$$\text{watts} = \text{(volts)(amperes)}$$

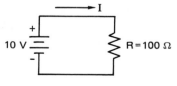

Fig. 5-5

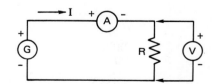

Fig. 5-6 Voltmeter-ammeter method for measuring power

By studying this equation, it can be seen that a voltmeter and an ammeter can be used to obtain measurements which will determine the power dissipated by a device. Figure 5-6 shows this method of measuring power. The *ammeter* must be connected in *series* with the resistor. The reading of the ammeter is the current drawn by the resistor. The *voltmeter* must be connected in *parallel* with the resistor. The resulting reading is the voltage across the resistor.

Problem 17   The readings on the meters shown in figure 5-6 are as follows:

Ammeter   20 A
Voltmeter   100 V

Calculate the power dissipated by the resistor R.

Solution $P = VI$

$$= (100 \text{ V}) (20 \text{ A})$$

$$= 2\,000 \text{ W} \times \frac{1 \text{ kW}}{1\,000 \text{ W}}$$

$$= 2 \text{ kW}$$

A single instrument known as a *wattmeter* can also be used to measure power, figure 5-7. This instrument has two coils, a voltage coil and a current coil. The voltage coil is connected across the resistor just as the voltmeter is connected in figure 5-6. The current coil is connected in series with the resistor, in the same manner as the ammeter in the circuit of figure 5-6. The pointer of the wattmeter is subjected to a turning force. This force is proportional to the product of the current and the voltage supplied to the load. The wattmeter does the multiplication that is called for in the power equation: $P = VI$. Figure 5-8 shows the connections of a wattmeter used to measure the power dissipated by a load.

Fig. 5-7 Portable wattmeter for measuring power

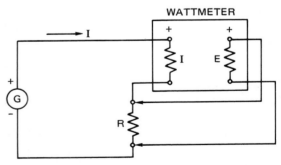

Fig. 5-8 Wattmeter connections for measuring power

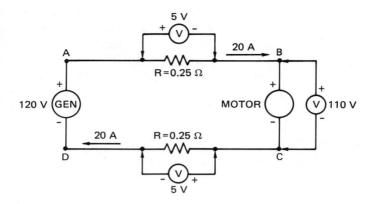

Fig. 5-9 Line loss analysis (Problem 18)

LINE LOSS   In the analysis of electric power distribution circuits, it is necessary to account for the resistance of the connecting line wires. In figure 5-9, the resistance of each wire between the generator and the motor is added. This resistance is treated as if it were a physical resistor with a value of 0.25 Ω. The *line loss* is the power lost in heat in the line wire supplying a load.

Problem 18   For the circuit of figure 5-9, determine: a. the voltage across the motor, b. the motor input power, c. the total line losses, and d. the generator output power.

Solution   a. Voltage across the motor:

A 20-A current flows through the circuit.

There is a 5-V drop in the line between terminals A and B.

$$V = IR$$
$$= (20 \text{ A}) (0.25 \text{ Ω})$$
$$= 5 \text{ V}$$

There is also a 5-V drop in the line between terminals C and D. The total line drop = 5 V + 5 V = 10V. The generator has a voltage output of 120 V. Therefore, the motor receives a voltage of:

120 V (generator output) – 10 V (total line drop) = 110 V

b. Motor input power:

$$P = VI$$
$$= (110 \text{ V}) (20 \text{ A})$$
$$= 2\,200 \text{ W} \times \frac{1 \text{ kW}}{1\,000 \text{ W}}$$
$$= 2.2 \text{ kW}$$

c. Total line losses:

Line loss (line A to B)

$P = I^2 R$

$= (20 \text{ A})^2 (0.25 \, \Omega)$

$= 100 \text{ W}$

Line loss (line C to D)

$P = I^2 R$

$= (20 \text{ A})^2 (0.25 \, \Omega)$

$= 100 \text{ W}$

Total line loss $= 100 \text{ W} + 100 \text{ W} = 200 \text{ W}$

d. Generator output power:

$P = EI$

$= (120 \text{ V}) (20 \text{ A})$

$= 2\,400 \text{ W} \times \dfrac{1 \text{ kW}}{1\,000 \text{ W}}$

$= 2.4 \text{ kW}$

The solution is checked as follows:

Generator output power − Line losses = Motor input power

$2\,400 \text{ W} - 200 \text{ W} = 2\,200 \text{ W}$

$2\,200 \text{ W} \qquad\quad = 2\,200 \text{ W}$

(R5-7) What are the units for (a) line drop and (b) line loss?

(R5-8) What formula is used to calculate (a) line drop, and (b) line loss?

HORSEPOWER The watt is the SI unit for both electrical and mechanical power. However, the horsepower (hp), a unit in the British system of measurement, is still commonly used to express mechanical power. Conversions can be made between mechanical power and electrical power by the use of the following relationships:

$$1 \text{ hp} = 746 \text{ W}$$

$$= 0.746 \text{ kW} \qquad\qquad \text{Eq. 5.7}$$

$$1 \text{ kW} = 1.34 \text{ hp} \qquad\qquad \text{Eq. 5.8}$$

Rotating machinery is usually rated in terms of its expected power output. Generator ratings normally are given in kW; motors are usually rated in hp, but will be rated in kW when the SI system is universally adopted.

Problem 19 Determine the mechanical power output of a motor that draws 20 A at 220 V Assume there are no losses.

Solution $P = EI$

$= (220 \text{ V}) (20 \text{ A})$

$= 4\,400 \text{ W}$

Since 1 hp $= 746$ W

Mechanical power output $= \dfrac{4\,400 \text{ W}}{746 \text{ W/hp}}$

$= 5.9 \text{ hp}$

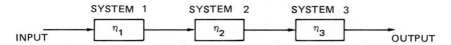

Fig. 5-10 A cascaded system

EFFICIENCY   Efficiency is the ratio of useful output to total input. This ratio is represented by the Greek letter eta, η. Efficiency can be expressed as an energy ratio or as a power ratio:

$$\eta = \frac{W_o}{W_i} \times 100\%$$
Eq. 5.9

$$\eta = \frac{P_o}{P_i} \times 100\%$$
Eq. 5.10

The input and the output must be expressed in the same units. The efficiency of any device is always less than 100%.

   The overall efficiency of a system consisting of several devices or systems is the product of the individual efficiencies, when η is expressed as a ratio and not as a percentage. The system shown in figure 5-10 is known as a cascaded system. The overall efficiency of this system is:

$$\eta_T = (\eta_1)(\eta_2)(\eta_3)$$
Eq. 5.11

Problem 20   A 5-hp motor draws 4.8 kW from a power source. What is the efficiency of the motor?

Solution P_o = 5 hp

   = (5 hp) (746 W/hp)

   = 3 730 W $\times \dfrac{1 \text{ kW}}{1\,000 \text{ W}}$

   = 3.73 kW (output power)

P_i = 4.8 kW (input power)

$\eta = \dfrac{P_o}{P_i} \times 100\%$

   = $\dfrac{3.73 \text{ kW}}{4.8 \text{ kW}} \times 100$

   = 77.7%

Problem 21   A motor has an efficiency of 80%. What current does it draw from a 110-V source when its output is 5 hp?

Solution P_o = 5 hp $\times$ 746 W/hp

 = 3 730 W (output power)

$$\eta = \frac{P_o}{P_i}$$

$$P_i = \frac{P_o}{\eta}$$

$$= \frac{3\ 730\ W}{0.8}$$

 = 4 663 W (input power)

P = EI

$$I = \frac{P}{E}$$

$$= \frac{4\ 663\ W}{110\ V}$$

 = 42.4 A

Problem 22   Find the overall efficiency of the system in figure 5-10 if η_1 = 90%, η_2 = 85%, and η_3 = 80%.

Solution $\eta_T = (\eta_1)(\eta_2)(\eta_3)$

$$= \left(\frac{90}{100}\right)\left(\frac{85}{100}\right)\left(\frac{80}{100}\right) = 0.612$$

 = (0.612)(100%) = 61.2%

(R5-9)   If the efficiency of system 3 in Problem 22 drops to 60%, calculate the new overall efficiency.

THE KILO-WATTHOUR   The basic unit of electric energy is the joule. Because this unit is too small for typical problems, a more practical unit of electric energy is required. This unit, the *kilowatthour*, is obtained by transposing the terms of Equation 5.3:

$$P = \frac{W}{t}$$

$$W = Pt$$

When P is the power in watts, and t is the time in seconds, then W is the work or energy in wattseconds, or joules. When P is the power in kW, and t is the time in hours, then W is the work or energy in kilowatthours (kWh). The unit of kilowatthours is kilowatts $\times$ hours.

Problem 23   A generator delivers 2 kW to a customer for 40 hours. How much electric energy does the customer receive in this time?

Solution   W = Pt
        = (2 kW) (40 h)
        = 80 kWh

Problem 24   At 6¢ ($.06) per kWh, how much will it cost to use a 100-W lamp for 30 days?

Solution
$$30 \text{ days} = 30 \text{ days} \times \frac{24 \text{ h}}{\text{day}}$$
            = 720 hours
$$100 \text{ W} = \frac{100 \text{ W}}{1\,000 \text{ W/kW}}$$
            = 0.1 kW
W   = Pt
        = (0.1 kW) (720 h)
        = 72 kWh
Cost   = (72 kWh) ($.06/kWh)
        = $4.32

(R5-10)   How much electric energy is used by a 1 000-W heater in 24 hours?

Figure 5-11 shows a watthour meter. This meter is typical of the type used to record energy consumption in residences. It records the kilowatthours used in each billing period. Figure 5-12 shows the type of meter used by small industrial consumers. This meter records both the energy consumption in kilowatthours and the maximum number of kilowatts used anytime during the billing period. This maximum value is called the *maximum*, or *peak, demand*. Both values are shown on the bill to the customer. The peak demand amount must be known because the utility company must install the generating capacity to meet the peak demand (kW load) of its customers. A panel mounted meter which is used for industrial applications is shown in figure 5-13. This meter displays watt and watthour readings. It is used by the utility company customer to monitor plant operations.

Fig. 5-11 Watthour meter for residential applications

Fig. 5-12 Combination watthour meter with maximum demand indication

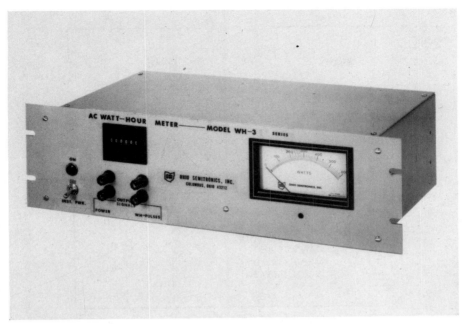

Fig. 5-13 Watthour meter mounted in a rack panel and used for industrial applications

INTERNAL RESISTANCE An *ideal* source of emf is shown in figure 5-14. However, a practical power supply has a resistance that must be considered in problems. The total effect of this resistance, which includes internal resistances of the power supply and the resistances of the terminals, is combined into a single internal resistance, as shown in figure 5-15. The circuit in this figure represents the equivalent circuit of a practical source of emf.

The voltage V_{AB} is equal to the ideal voltage E when the power supply is not supplying current to a load. The voltage V_{AB} is called the *terminal voltage* of the power supply when supplying a load. When the power supply is connected, figure 5-16, the following equations apply:

$$V_{int} = I_L R_{int} \qquad \text{Eq. 5.12}$$

Equation 5.12 is used to calculate the internal voltage drop caused by the load current passing through the internal resistance of the power supply. The terminal voltage, V_{AB} or V_T, is equal to V_L which is the voltage across the load. This relationship is shown in the following equation:

$$V_{AB} = V_T = V_L \qquad \text{Eq. 5.13}$$

The student must remember that:

$$V_L = I_L R_L \text{ (from Ohm's Law)}$$

$$V_T = E - V_{int} \qquad \text{Eq. 5.14}$$

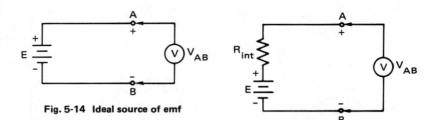

Fig. 5-14 Ideal source of emf

Fig. 5-15 Equivalent circuit of a practical source of emf

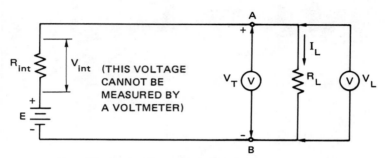

Fig. 5-16 Practical source of emf supplying load current

Equation 5.14 shows that the terminal voltage, V_T, is less than the ideal voltage, E, by an amount equal to the internal voltage drop.

Problem 25   A power supply has an open circuit voltage of 6 V. The load resistance is 5 ohms. Calculate the equivalent internal resistance of the power supply if the voltage across the load measures 5 V.

Solution

$$V_L = E - V_{int}$$
$$V_{int} = E - V_L$$
$$= 6\,V - 5\,V$$
$$= 1\,V$$

$$I_L = \frac{V_L}{R_L}$$
$$= \frac{5\,V}{5\,\Omega}$$
$$= 1\,A$$

$$V_{int} = I_L R_{int}$$
$$R_{int} = \frac{V_{int}}{I_L}$$
$$= \frac{1\,V}{1\,A}$$
$$= 1\,\Omega$$

MAXIMUM POWER TRANSFER According to the maximum power transfer theorem:

> A practical source delivers the maximum power to a load when the load resistance equals the internal resistance of the network, as seen by the load.

When applied to the circuit shown in figure 5-16, the theorem means:

$$R_L = R_{int} \qquad \text{Eq. 5.15}$$

That is, maximum power transfer occurs if the total resistance to the left of terminals A and B equals R_L.

(R5-11) Write the equation for maximum power transfer for the circuit shown in figure 5-17.

If R_L is varied, and power is plotted versus R_L, the resulting graph is similar to the one shown in figure 5-18. According to the figure, the maximum value of power occurs when $R_L = R_{int}$. A lower value of output power is obtained for any other value of R_L.

When the maximum power transfer takes place in a circuit, the following conditions are true:

1. $\eta = 50\%$; that is, one-half of the power is lost internally in the power supply.

2. $V_T = V_L = \dfrac{E}{2}$; the load voltage is reduced to one-half of the open circuit voltage of the power supply.

3. $R_T = R_L + R_{int} = 2R_L$, where R_T is the total resistance seen by the power supply.

4. The load is matched to the power supply. This condition is very important in electronic engineering technology. For example, the antenna of a transmitting station is matched to the output circuit to achieve the maximum power output. In addition, the effective speaker resistance of a radio or television is matched to the last audio stage for maximum power output.

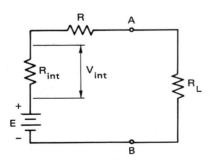

Fig. 5-17

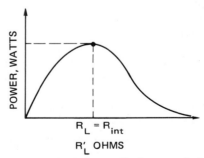

Fig. 5-18 Power versus R_L for a practical power supply

TABLE 5-1 Ratings of Household Appliances

Appliance	Rating, Watts
Air conditioner	2 000
Clock	2
Clothes dryer	6 000
Clothes washer	400
Coffee percolator	1 000
Dishwasher	1 500
Fan	150
Heater	1 500
Stereo sound system	300
Iron	1 200
Record player	75
Radio	30
Range (free-standing)	15 000
Range (wall oven)	8 000
Refrigerator	350
Shaver	10
Tape recorder	60
Toaster	1 200
TV receiver (color)	450
TV receiver (black and white)	200

(R5-12)   A practical power supply delivers 500 W of maximum power to a load. $I = 5$ A. Determine the following: a. R_L; b. R_{int}; c. V_L; d. E (open circuit voltage of the power supply); e. power dissipated internally in the power supply; and f. power from the power supply.

APPLIANCE RATINGS   Typical wattage ratings of some common household appliances are listed in Table 5-1.

LABORATORY EXERCISE 5-1 MAXIMUM POWER TRANSFER

PURPOSE   By completing this exercise, the student will verify experimentally the condition for maximum power transfer from a dc power supply to a resistive load.

EQUIPMENT AND MATERIALS
1 Dc power supply
1 Analog VOM
1 Resistor, 5.1 kΩ, 2 W
1 Rheostat, 10 000 Ω, 5 W
1 Switch, double pole, single throw (DPST)

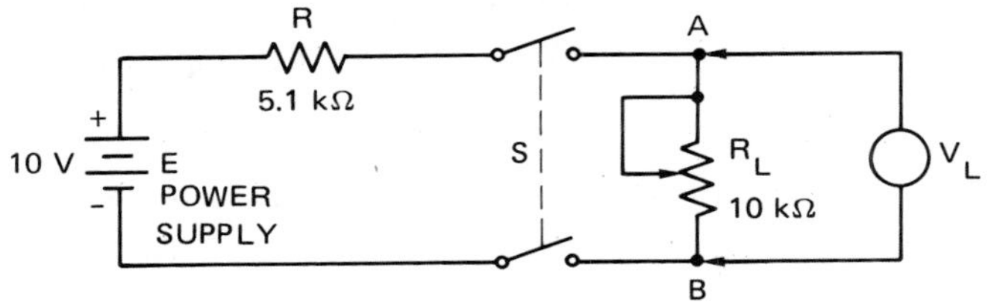

Fig. 5-19 Circuit for Laboratory Exercise 5-1

TABLE 5-2

R_L (ohms)	$(R + R_L)$ (ohms)	V_L (volts)	P (Measured) (watts)	P (Calculated) (watts)
0				
100				
200				
400				
600				
800				
850				
900				
950				
1 000				
1 100				
1 200				
1 500				
1 700				
2 000				
3 000				
4 000				
5 000				
6 000				
8 000				
10 000				

PROCEDURE

1. Connect the components in the circuit shown in figure 5-19.
   a. Open switch S, and set the voltage of the power supply equal to 10 V. Keep this voltage constant throughout the experiment.
   b. Adjust the rheostat so that R_L equals zero ohms, as measured with the VOM.
   c. Close switch S.
   d. Measure voltage V_L across R_L. Record the value in Table 5-2.
   e. Repeat step 1.d. for every value of R_L shown in Table 5-2.

2. a. Calculate the power dissipated in R_L, using the formula $P = V_L^2/R_L$. Record this value in Table 5-2. This is the *measured* value of power.
   b. Calculate the power dissipated in R_L, using the formula
   $$P = \frac{E^2 R_L}{(R + R_L)^2}.$$ Record this value in Table 5-2.
   This is the *calculated* value of power.

c. Draw a graph of P (calculated) versus R_L. Plot values for P on the vertical axis, and values for R_L on the horizontal axis.

ANALYSIS

1. Using the data listed in Table 5-2, compare the measured and calculated values for P. Explain the reasons for any differences in the values.
2. Using the graph plotted in step 2.c., determine the value of R_L for the maximum power transfer. How does this value compare with the value obtained from Table 5-2?

EXTENDED STUDY TOPICS

1. Calculate the work accomplished in moving an object 2 m in 30 s against a force of 17.8 lb.
2. If the object in topic 1 is moved 2 m in 15 s against a force of 17.8 lb, is more or less work done?
3. An electronic device is supplied 960 J of energy over a period of 4 min. Determine the amount of power supplied.
4. The electronic device of topic 3 receives 1 920 J of energy over a period of 8 min. Determine the power supplied.
5. Explain the relationship between the answers to topics 3 and 4.
6. What is the voltage drop across a 1 200-W resistor which draws a current of 10 A?
7. What voltage must be supplied to a 50-Ω resistor so that it will dissipate power at a rate of 200 W?
8. What is the power rating of a 12-Ω heater which draws a current of 10 A?
9. What current flows in a 20-Ω resistor while it dissipates power at the rate of 600 W?
10. What is the resistance of a device which draws 1.5 A from a 300-V source? How much power does the device dissipate?
11. What current does a 150-W lamp draw from a 120-V source? What is the resistance of the lamp
12. What current does a 50-Ω resistor pass if it dissipates power at the rate of 500 W?
13. A 30-V source supplies power to the circuit of figure 5-20. The current through each resistor is 0.4 A. Calculate (a) the power dissipated in each resistor, and (b) the total power supplied by the source.
14. A 1.6-k Ω resistor is connected to a 40-V power source. Determine the power rating of the resistor using a design safety factor of 100%.

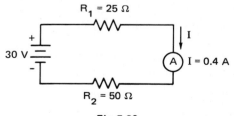

Fig. 5-20

15. What has been learned from the answers to Problem 22 and question (R5-9)?
16. How much will it cost to operate a 4 000-Ω electric clock from a 120-V power line for one month (30 days), if electric energy costs 6¢ per kWh?
17. If electric energy costs 6¢ per kWh, what is the total cost of using:
    a. a 200-W toaster for 30 min;
    b. three 100-W lamps for 4 h;
    c. a 400-W clothes washer for 1 h;
    d. a 6 000-W clothes dryer for 1/2 h?
18. A power supply has an open circuit voltage of 2 V and an internal resistance of 1.2 Ω. Determine the internal voltage drop, when 0.5 A is supplied to the external circuit. What is the load voltage?
19. Which source can deliver more power to a load: a 20-V source with an internal resistance of 10 Ω, or a 20-V source with an internal resistance of 40 Ω?
20. Prove that $P = \dfrac{E^2 R_L}{(R_{int} + R_L)^2}$

<table>
<tr><td>

Chapter

6

</td><td>

Series and Parallel Circuits

</td></tr>
</table>

OBJECTIVES After studying this chapter, the student will be able to

- discuss the principles of series and parallel circuits.
- discuss Kirchhoff' s Voltage Law and Kirchhoff's Current Law.
- explain the voltage divider rule in a series circuit and the current divider rule in a parallel circuit.
- demonstrate an understanding of circuit terminology.

THE SERIES CIRCUIT If the same current flows through two or more electric components, these components are said to be connected in series, figure 6-1.

(R6-1) If the current through resistor R_1 is 2 A, how much current flows through resistor R_2 in figure 6-1?

(R6-2) If resistor R_3 in figure 6-1 develops an open condition, how much current will flow through resistor R_1?

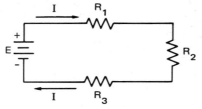

Fig. 6-1 A series circuit

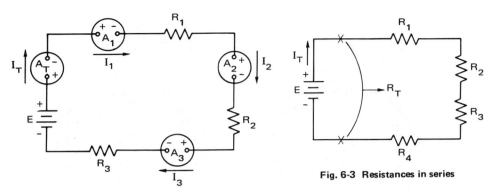

Fig. 6-3 Resistances in series

Fig. 6-2 Ammeter connections in a series circuit

(R6-3) Explain the reason for the answer to question (R6-2).

To measure the current through each component of a series circuit, ammeters are connected in series with each compoennt, figure 6-2. The currents measured by the ammeters are as follows: ammeter A_T measures the total current I_T supplied by the power supply, ammeter A_1 measures the current I_1 through resistor R_1, and ammeter A_2 measures the current I_2 through resistor R_2.

(R6-4) What current is measured by ammeter A_3 in figure 6-2?

(R6-5) What is the direction of current through the power supply in figure 6-2?

Current Flows in One Path in a Series Circuit. Components E, R_1, R_2, and R_3 are connected in series in figure 6-2. There is only one path that current can take through all four components. The following equation expresses this basic characteristic of a series circuit:

$$I_T = I_1 = I_2 = I_3 \qquad \text{Eq. 6.1}$$

Figure 6-2 shows the required polarity connections (+ or -) for ammeters A_T, A_1, A_2, and A_3. If the ammeter leads are reversed, the ammeter pointer will deflect offscale to the left of the zero scale marking.

Resistances in Series. Another characteristic of a series circuit is that the total resistance is the sum of the individual resistances. For the circuit shown in figure 6-3, the total resistance is expressed as follows:

$$R_T = R_1 + R_2 + R_3 + R_4 \qquad \text{Eq. 6.2}$$

The power supply does not feel the effect of the circuit that is composed of four separate resistances in series. The power supply "sees" only one resistor with a magnitude of R_T. That is, the power supply of figure 6-3 "sees" the equivalent circuit shown in figure 6-4. By reducing the original series circuit to a simple equivalent circuit, it is possible to solve for the current using Ohm's Law:

$$I_T = \frac{E}{R_T} \qquad \text{Eq. 6.3}$$

Circuit simplification, or reduction, is an important step in solving complex electric circuits.

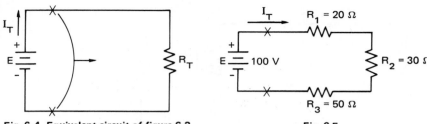

Fig. 6-4 Equivalent circuit of figure 6-3   Fig. 6-5

Problem 1 Calculate the current in a series circuit consisting of 20-Ω, 30-Ω, and 50-Ω resistors connected to a 100-V power supply. Draw the circuit and the simplified equivalent circuit.

Solution The circuit is shown in figure 6-5.

$$R_T = R_1 + R_2 + R_3$$
$$= 20\ \Omega + 30\ \Omega + 50\ \Omega$$
$$= 100\ \Omega$$

The equivalent circuit is shown in figure 6-6.

$$I_T = \frac{E}{R_T}$$
$$= \frac{100\ V}{100\ \Omega}$$
$$= 1\ A$$

Polarity of Voltage Drops. To measure currents and voltages correctly, it is necessary to determine the polarity of the emfs and voltage drops in a circuit. A source of emf is called an *active circuit element.* That is, it generates electric energy at the expense of some other form of energy. Resistors are called *passive circuit elements* because they use electric energy in a closed circuit. Therefore, a complete electric circuit, such as the one in figure 6-7, contains both active and passive circuit elements.

(R6-6) For each element in figure 6-7, indicate whether it is active or passive.

(R6-7) Is current flowing through resistor R_2 in figure 6-7? Explain the answer.

The polarity markings of the power supply shown in figure 6-7 result from the chemical makeup of the battery. The student should remember that the plus (+) terminal has a deficiency of electrons and the minus (–) terminal has an ex-

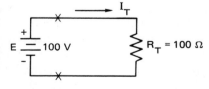

Fig. 6-6 Equivalent circuit of Problem 1

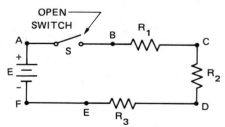

Fig. 6-7 Complete electric circuit

cess of electrons. Recall from previous discussions that the conventional direction of current flow is from the positive terminal of the power supply when there is a closed circuit. There is no flow of current in an open circuit, as shown in figure 6-7. Therefore, for an open circuit, the passive circuit elements do not draw any current, and they do not have any polarity markings. Active circuit elements, however, do have polarity markings for either an open or a closed circuit. In fact, active circuit elements have polarity markings even when they are on a shelf in a storeroom.

Figure 6-8 represents the closed series circuit of figure 6-7. The voltage E_{AF} represents a potential rise. That is, point A is at a higher potential than point F. There is a voltage drop across resistor R_1 due to the flow of current. This voltage drop is shown by the polarity markings in figure 6-8. Ohm's Law is used to find the voltage drop:

$$V_1 = I_T R_1$$

There is a voltage drop, V_2, across resistor R_2, and a voltage drop V_3 across resistor R_3.

Problem 2   The circuit of figure 6-8 has the following values:
$$E = 180 \text{ V}$$
$$R_1 = 15 \text{ }\Omega$$
$$R_2 = 30 \text{ }\Omega$$
$$R_3 = 45 \text{ }\Omega$$

Calculate the voltage drop across each resistor.

Solution
$$R_T = R_1 + R_2 + R_3$$
$$= 15 \text{ }\Omega + 30 \text{ }\Omega + 45 \text{ }\Omega$$
$$= 90 \text{ }\Omega$$

$$I_T = \frac{E}{R_T}$$
$$= \frac{180 \text{ V}}{90 \text{ }\Omega}$$
$$= 2 \text{ A}$$

$$V_1 = I_T R_1$$
$$= (2 \text{ A})(15 \text{ }\Omega)$$
$$= 30 \text{ V}$$

$$V_2 = I_T R_2$$
$$= (2 \text{ A})(30 \text{ }\Omega)$$
$$= 60 \text{ V}$$

$$V_3 = I_T R_3$$
$$= (2 \text{ A})(45 \text{ }\Omega)$$
$$= 90 \text{ V}$$

To measure the current and voltages in a series circuit, meters must be connected, as shown in figure 6-9. As stated earlier, the ammeter is connected in series with the circuit. Thus, the current being measured flows through the am-

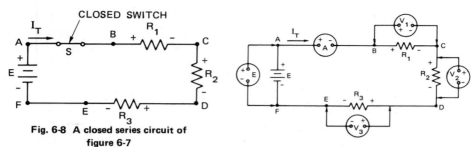

Fig. 6-8 A closed series circuit of
figure 6-7

Fig. 6-9 Meter connections in series circuit

meter. Since a voltmeter reads the potential difference between two points in a
circuit, it is connected across, or in parallel, with a circuit element. The meter
polarity connections shown in figure 6-9 must be followed to obtain an upscale
movement of the pointer.

Kirchhoff's Voltage Law. Kirchhoff's Voltage Law is applied to series circuits
and is stated as follows:

> In any closed electric circuit, the algebraic sum of the emfs must
> equal the algebraic sum of the voltage drops.

For the series circuit shown in figure 6-9, the equation for Kirchhoff's Voltage
Law is:

$$E = V_1 + V_2 + V_3 \qquad\qquad \text{Eq. 6.4}$$

Kirchhoff's Voltage Law offers an alternate method for solving series circuit
problems.

Problem 3   Solve Problem 2 by using Kirchhoff's Voltage Law:

$$E = V_1 + V_2 + V_3$$
$$= I_T R_1 + I_T R_2 + I_T R_3$$

Solution

$$180\ V = 15\ I_T + 30\ I_T + 45\ I_T$$
$$180\ V = 90\ I_T$$
$$90\ I_T = 180\ V$$
$$I_T = \frac{180\ V}{90\ \Omega}$$
$$= 2\ A$$

$$V_1 = I_T R_1$$
$$= (2\ A)\,(15\ \Omega)$$
$$= 30\ V$$

$$V_2 = I_T R_2$$
$$= (2\ A)\,(30\ \Omega)$$
$$= 60\ V$$

$$V_3 = (2\ A)\,(45\ \Omega)$$
$$= 90\ V$$

Check: $E = V_1 + V_2 + V_3$

$$180\ V = 30\ V + 60\ V + 90\ V$$
$$180\ V = 180\ V$$

Voltage Divider Rule. Another characteristic of a series circuit is expressed by the *voltage divider rule.* According to this rule:

> The ratio between any two voltage drops in a series circuit is the same as the ratio of the two resistances across which these voltage drops occur.

To derive this rule mathematically, first solve for the current in the circuit of figure 6-9:

$$I = \frac{V_1}{R_1} = \frac{V_2}{R_2} = \frac{V_3}{R_3} = \frac{E}{R_T}$$

The student should note that Ohm's Law can be applied to *part* of a series circuit, as well as to the *entire* circuit.

$$\frac{V_1}{R_1} = \frac{V_2}{R_2}$$

$$\frac{V_1}{V_2} = \frac{R_1}{R_2} \qquad \text{Eq. 6.5}$$

Also,

$$\frac{V_1}{V_3} = \frac{R_1}{R_3} \qquad \text{Eq. 6.6}$$

and

$$\frac{V_1}{E} = \frac{R_1}{R_T} \qquad \text{Eq. 6.7}$$

$$V_1 = E\left(\frac{R_1}{R_T}\right) \qquad \text{Eq. 6.8}$$

$$V_2 = E\left(\frac{R_2}{R_T}\right) \qquad \text{Eq. 6.9}$$

$$V_3 = E\left(\frac{R_3}{R_T}\right) \qquad \text{Eq. 6.10}$$

For a series circuit, Equations 6.8, 6.9, and 6.10 show that the voltage across any resistor is equal to a fraction of the applied emf. That is, the voltage is equal to the applied emf times a factor which is the ratio of the specific resistance and the total resistance of the circuit.

Problem 4 Use the voltage divider rule to find the voltage drops across the resistors in the circuit shown in figure 6-10.

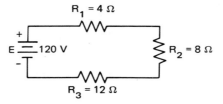

Fig. 6-10

Solution R_T $= R_1 + R_2 + R_3$
$= 4\,\Omega + 8\,\Omega + 12\,\Omega$
$= 24\,\Omega$

$V_1 = E\left(\dfrac{R_1}{R_T}\right)$

$= (120\text{ V})\left(\dfrac{4\,\Omega}{24\,\Omega}\right)$

$= 120\text{ V}\left(\dfrac{1}{6}\right)$

$= 20\text{ V}$

$V_2 = E\left(\dfrac{R_2}{R_T}\right)$

$= (120\text{ V})\left(\dfrac{8\,\Omega}{24\,\Omega}\right)$

$= 120\text{ V}\left(\dfrac{1}{3}\right)$

$= 40\text{ V}$

$V_3 = E\left(\dfrac{R_3}{R_T}\right)$

$= (120\text{ V})\left(\dfrac{12\,\Omega}{24\,\Omega}\right)$

$= 120\text{ V}\left(\dfrac{1}{2}\right)$

$= 60\text{ V}$

Check: $E = V_1 + V_2 + V_3$
120 V $= 20\text{ V} + 40\text{ V} + 60\text{ V}$
120 V $= 120\text{ V}$

Change in Resistance. For any component in a series circuit, a change in the resistance of the component changes the current through all of the components. For example, the current through the motor shown in the circuit of figure 6-11 is controlled by varying the setting of the rheostat (variable resistor).

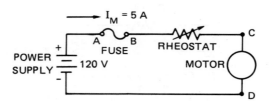

Fig. 6-11 Motor current is controlled by varying the resistance of a rheostat

Problem 5 A certain condition causes the fuse to blow in the circuit shown in figure 6-11.
  a. What is the value of the current I_M?
  b. What is the voltage across terminals A and B?
  c. What is the voltage V_{CD} across the terminals of the motor?

Solution a. When the fuse blows, the circuit is opened and there is no current. Therefore, I_M is zero.

$$I_M = 0 \text{ A}$$

  b. The total source voltage appears across the open circuit at terminals A and B. Terminal A has a positive polarity and Terminal B has a negative polarity.

$$V_{AB} = +120 \text{ V}$$

  c. Since there is no current to the motor windings, there is no voltage across the terminals of the motor.

$$V_{CD} = 0 \text{ V}$$

Circuit Terminology. Figure 6-12 shows four components connected to a voltage source E to form a circuit. The components may be resistors, coils, capacitors, transistors, or some other type of electric or electronic device. Each component is connected by conducting copper wire. The resistance of this copper wire is very small. As a result, this resistance is usually neglected in all calculations. Note that components 2 and 4 have a common terminal (B). Electrically, this terminal is at the negative potential of the power supply.

The current in this circuit flows through paths which are called *branches*. As shown in the figure,

Branch 1 consists of the power supply E and component 1. The same current flows through these two components because they are connected in series.

Branch 2 consists of component 2.

Branch 3 consists of components 3 and 4, which are connected in series.

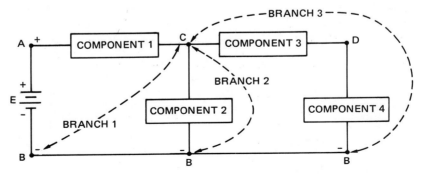

Fig. 6-12 Arrangement of components for circuit terminology analysis

There are four nodes in the circuit of figure 6-12 (A, B, C, D). A node is defined as a junction where two or more current paths come together.

THE PARALLEL CIRCUIT Two or more electric components are connected in *parallel* when one end of each device is tied to a common point, and the other end of *each* device is connected to a second common point. Each component has the *same* voltage impressed across its terminals. Figure 6-13 shows three examples of parallel connections.

The following equation expresses a fundamental characteristic of a parallel circuit:

$$E = V_1 = V_2 = V_3 \qquad \text{Eq. 6-11}$$

where V_1, V_2, and V_3 are the voltages across R_1, R_2, and R_3 respectively.

(R6-8) What is the voltage across R_3 in the circuit of figure 6-13C?

Resistances in Parallel. Another characteristic of a parallel circuit is that the total current is the algebraic sum of all of the branch currents. As applied to figure 6-14, this characteristic is expressed by the following equation:

$$I_T = I_1 + I_2 + I_3 \qquad \text{Eq. 6.12}$$

Problem 6 a. Calculate the total current supplied by the power supply of the circuit shown in figure 6-14.
b. Determine the total resistance seen by the power supply.
c. Draw the equivalent (simplified) circuit for figure 6-14.

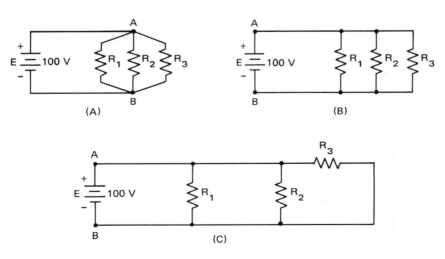

Fig. 6-13 Examples of parallel connections

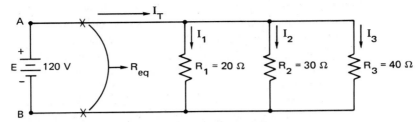

Fig. 6-14 Currents in a parallel circuit

Solution

a. $I_1 = \dfrac{V_1}{R_1}$

$ = \dfrac{120 \text{ V}}{20 \text{ }\Omega}$

$ = 6 \text{ A}$

$I_2 = \dfrac{V_2}{R_2}$

$ = \dfrac{120 \text{ V}}{30 \text{ }\Omega}$

$ = 4 \text{ A}$

$I_3 = \dfrac{V_3}{R_3}$

$ = \dfrac{120 \text{ V}}{40 \text{ }\Omega}$

$ = 3 \text{ A}$

$I_T = I_1 + I_2 + I_3$

$ = 6 \text{ A} + 4 \text{ A} + 3 \text{ A}$

$ = 13 \text{ A}$

b. The total, or equivalent, resistance seen by the power supply may be found by using Ohm's Law:

$$R_{eq} = \frac{E}{I_T} \qquad\qquad \text{Eq. 6.13}$$

$R_{eq} = \dfrac{120 \text{ V}}{13 \text{ A}}$

$\phantom{R_{eq}} = 9.23 \text{ }\Omega$

c. The equivalent circuit for figure 6-14 is given in figure 6-15.

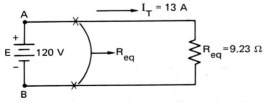

Fig. 6-15 Equivalent resistance circuit for figure 6-14

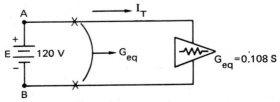

Fig. 6-16 Equivalent conductance for figure 6-14

By analyzing the results of problem 6, the following conclusions can be made for this type of circuit (a dc resistance passive circuit):

1. The total current must be greater than the current passing through any one branch, because it is the sum of all of the branch currents.
2. The total, or equivalent, resistance is always less than the smallest branch resistance.
3. There is a decrease in the total, or equivalent, resistance seen by the power supply as more resistances are connected in parallel.

(R6-9) R_1 in figure 6-14 is replaced with a 40-Ω resistor. Calculate a. the total current, and b. the equivalent resistance of the entire parallel circuit.

Conductance. For a circuit containing resistors in parallel, the total or equivalent conductance of any number of parallel resistors is equal to the sum of the conductances of the individual resistors.

$$G_{eq} = G_1 + G_2 + G_3 + \text{etc.}$$ Eq. 6.14

The total, or equivalent, resistance is found by the following equation:

$$R_{eq} = \frac{1}{G_{eq}}$$ Eq. 6.15

Problem 7   Refer to the circuit shown in figure 6-14.

a. Calculate the total resistance using the conductance method of analysis.
b. Draw the equivalent conductance circuit.
c. Find the total current.

Solution   a. $G_{eq} = G_1 + G_2 + G_3$

$$= \frac{1}{20\ \Omega} + \frac{1}{30\ \Omega} + \frac{1}{40\ \Omega}$$

$$= 0.05\ S + 0.033\ S + 0.025\ S$$

$$= 0.108\ S$$

$$R_{eq} = \frac{1}{G_{eq}}$$

$$= \frac{1}{0.108\ S}$$

$$= 9.23\ \Omega$$

(See figure 6-15 for the equivalent resistance circuit.)
b. See figure 6-16 for the equivalent conductance circuit.

c. $I_T = \dfrac{E}{R_{eq}}$

$$= \frac{120\ V}{9.23\ \Omega}$$

$$= 13\ A$$

(R6-10) R_1 in the circuit of figure 6-14 is replaced with a 40-Ω resistor. a. Calculate the total resistance using the conductance method of analysis. b. Find the total current.

KIRCHHOFF'S CURRENT LAW

Kirchhoff's Current Law states that the algebraic sum of the currents entering and leaving a node is zero. Expressed in other words, the law says that the sum of the currents entering a node is equal to the sum of the currents leaving the node. Kirchhoff's Current Law is expressed mathematically as follows:

$$\Sigma I_{entering} = \Sigma I_{leaving} \qquad Eq. 6.16$$

In figure 6-17, currents I_1, I_3, and I_4 enter the junction. At the same time, currents I_2 and I_5 leave the junction. Applying Equation 6.16 to figure 6-17 yields:

$$I_1 + I_3 + I_4 = I_2 + I_5$$
$$4\,A + 5\,A + 3\,A = 2\,A + 10\,A$$
$$12\,A = 12\,A$$

Figure 6-18 shows the current distribution in the circuit shown in figure 6-14 for the values determined in Problem 6.

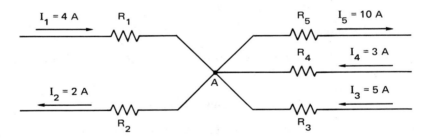

Fig. 6-17 Kirchhoff's Current Law

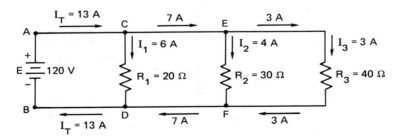

Fig. 6-18 Current distribution in the parallel circuit of figure 6-14

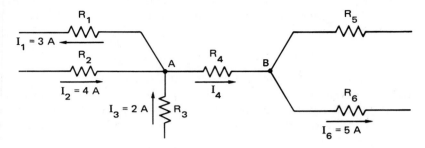

Fig. 6-19

Problem 8   a. Determine the value of I_4 in the circuit of figure 6-19.
b. Does I_5 flow into or out of node B?
c. Calculate the value of I_5.

Solution   a. At node A:

$$I_2 + I_3 = I_1 + I_4$$
$$4 \text{ A} + 2 \text{ A} = 3 \text{ A} + I_4$$
$$I_4 = 3 \text{ A}$$

b. At node B:

$$I_4 = 3 \text{ A (enters the node)}$$
$$I_6 = 5 \text{ A (leaves the node)}$$

Therefore,

I_5 must enter the node to satisfy Kirchhoff's Current Law.
c. $I_4 + I_5 = I_6$
$3 \text{ A} + I_5 = 5 \text{ A}$
$I_5 = 2 \text{ A (entering the node)}$

(R6-11)   Using Kirchhoff's Current Law, determine the value of currents I_3 and I_5 for the circuit of figure 6-20.

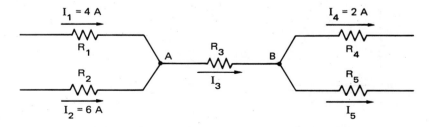

Fig. 6-20

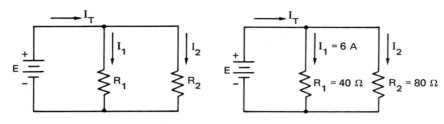

Fig. 6-21 Ratio of branch currents   Fig. 6-22

Ratio of Branch Currents. In a parallel circuit, the ratio between any two branch currents is the inverse of their resistance ratio, and the same as the ratio of their conductances. These relationships are expressed for the currents shown in figure 6-21 by the following equations:

$$\frac{I_1}{I_2} = \frac{R_2}{R_1}$$

Eq. 6.17

$$\frac{I_1}{I_2} = \frac{G_1}{G_2}$$

Eq. 6.18

Problem 9   Using the ratio of the branch currents, calculate the value of current I_2 for the circuit of figure 6-22.

Solution   Method 1 (Using Equation 6.17)

$$\frac{I_1}{I_2} = \frac{R_2}{R_1}$$

$$I_2 = \frac{I_1 R_1}{R_2}$$

$$= \frac{(6 \text{ A})(40 \text{ }\Omega)}{80 \text{ }\Omega}$$

$$= 3 \text{ A}$$

Method 2 (Using Equation 6.18)

$$G_1 = \frac{1}{R_1}$$

$$= \frac{1}{40 \text{ }\Omega}$$

$$= 0.025 \text{ S}$$

$$G_2 = \frac{1}{R_2}$$

$$= \frac{1}{80 \text{ }\Omega}$$

$$= 0.012 \text{ 5 S}$$

$$\frac{I_1}{I_2} = \frac{G_1}{G_2}$$

$$I_2 = \frac{I_1 G_2}{G_1}$$

$$= \frac{(6 \text{ A})(0.012 \text{ 5 S})}{0.025 \text{ S}}$$

$$= 3 \text{ A}$$

(R6-12)   What is the value of voltage E in the circuit of figure 6-22?

(R6-13)   Determine the value of the total, or equivalent, resistance in figure 6-22.

Effect of Change on Parallel Branches. An additional characteristic of a parallel circuit is that each branch is independent of any changes in the other branches, when the voltage across the circuit remains constant.

(R6-14)   If R_3 in the circuit of figure 6-18 develops an open circuit, determine the values of I_1 and I_2.

(R6-15)   Calculate the total, or equivalent, resistance if R_3 in the circuit of figure 6-18 develops an open circuit.

(R6-16)   State the conclusions that can be reached after studying the results of questions (R6-14) and (R6-15).

(R6-17)   In figure 6-23, the refrigerator and the television set are on when the reading lamp is turned on. What is the effect on the voltage and current of the television set and the refrigerator?

When one of the elements of a parallel circuit becomes short-circuited, all of the branch currents are affected. A short circuit across *any* element of a parallel circuit, such as R_3 in figure 6-24, causes a short circuit across *all* of the elements. The voltage across R_1, R_2, and R_3 drops to zero. The currents I_1, I_2, and I_3 drop to zero also. The entire current (I_T) flows through the path of least resistance, which is the short circuit. As I_T approaches infinity, either a fuse blows or the power supply is destroyed.

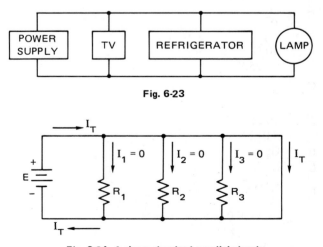

Fig. 6-23

Fig. 6-24 A short-circuited parallel circuit

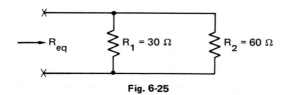

Fig. 6-25

SPECIAL CASES OF PARALLEL CIRCUITS

For any number of resistors in parallel, Equation 6.14 can be expressed as follows:

$$\frac{1}{R_{eq}} = \frac{1}{R_1} + \frac{1}{R_2} + \frac{1}{R_3} + \text{etc.}$$ Eq. 6.19

where

$$G_{eq} = \frac{1}{R_{eq}}$$

When two resistors are connected in parallel, Equation 6.19 becomes:

$$\frac{1}{R_{eq}} = \frac{1}{R_1} + \frac{1}{R_2}$$ Eq. 6.20

Equation 6.20 can be rearranged to solve for R_{eq} directly:

$$R_{eq} = \frac{R_1 R_2}{R_1 + R_2}$$ Eq. 6.21

Equation 6.21 is called the product-over-sum formula because R_{eq} equals the product of the two resistances, divided by the sum of the two resistances.

Problem 10

Find the equivalent resistance of the circuit shown in figure 6-25.

Solution

$$\begin{aligned}
R_{eq} &= \frac{R_1 R_2}{R_1 + R_2} \\
&= \frac{(30\ \Omega)(60\ \Omega)}{30\ \Omega + 60\ \Omega} \\
&= \frac{1\ 800\ (\Omega)^2}{90\ \Omega} \\
&= 20\ \Omega
\end{aligned}$$

Problem 11

Find the equivalent resistance of the circuit shown in figure 6-26, page 112.

Solution

Method 1 (Using conductances)

$$\begin{aligned}
G_{eq} &= G_1 + G_2 + G_3 + G_4 \\
&= \frac{1}{600\ \Omega} + \frac{1}{300\ \Omega} + \frac{1}{400\ \Omega} + \frac{1}{200\ \Omega} \\
&= 0.001\ 7\ \text{S} + 0.003\ 3\ \text{S} + 0.002\ 5\ \text{S} + 0.005\ \text{S} \\
&= 0.012\ 5\ \text{S}
\end{aligned}$$

$$\begin{aligned}
R_{eq} &= \frac{1}{G_{eq}} \\
&= \frac{1}{0.012\ 5\ \text{S}} \\
&= 80\ \Omega
\end{aligned}$$

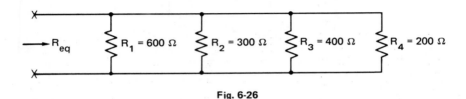

Fig. 6-26

Method 2 (Using the product-over-sum formula)

R'_{eq}, the parallel combination of R_3 and R_4, is determined:

$$R'_{eq} = \frac{R_3 \, R_4}{R_3 + R_4}$$

$$= \frac{(400 \ \Omega) \, (200 \ \Omega)}{400 \ \Omega + 200 \ \Omega}$$

$$= \frac{80 \ 000 \ (\Omega)^2}{600 \ \Omega}$$

$$= 133.3 \ \Omega$$

R''_{eq}, the parallel combination of R_1 and R_2, is determined:

$$R''_{eq} = \frac{R_1 \, R_2}{R_1 + R_2}$$

$$= \frac{(600 \ \Omega) \, (300 \ \Omega)}{600 \ \Omega + 300 \ \Omega}$$

$$= \frac{180 \ 000 \ (\Omega)^2}{900}$$

$$= 200 \ \Omega$$

The circuit reduces to the simplified circuit shown in figure 6-27. This circuit consists of R'_{eq} in parallel with R''_{eq}.

$$R_{eq} = \frac{R'_{eq} \, R''_{eq}}{R'_{eq} + R''_{eq}}$$

$$= \frac{(133.3 \ \Omega) \, (200 \ \Omega))}{133.3 \ \Omega + 200 \ \Omega}$$

$$= \frac{26 \ 660 \ (\Omega)^2}{333.3 \ \Omega}$$

$$= 80 \ \Omega$$

When a number of resistors of equal value are connected in parallel, the equivalent resistance of the parallel combination is found using the following equation:

$$R_{eq} = \frac{R}{N} \qquad\qquad\text{Eq. 6.22}$$

where

R = common resistance value

N = number of resistors

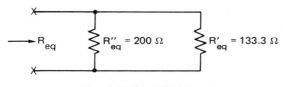

Fig. 6-27 Simplified circuit

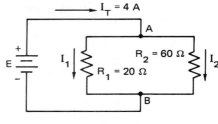

Fig. 6-28

Problem 12 Fifteen 150-Ω resistors are connected in parallel. Find the equivalent resistance of the combination.

Solution
$$R_{eq} = \frac{R}{N}$$

$$R_{eq} = \frac{150\ \Omega}{15}$$

$$= 10\ \Omega$$

(R6-18) Two 3-kΩ resistors are connected in parallel. Find the equivalent resistance.

CURRENT DIVIDER RULE Current divider equations may be used to find the branch currents of the circuit shown in figure 6-28.

$$I_1 = I_T \left(\frac{R_2}{R_1 + R_2} \right) \qquad \text{Eq. 6.23}$$

$$I_2 = I_T \left(\frac{R_1}{R_1 + R_2} \right) \qquad \text{Eq. 6.24}$$

Problem 13 Calculate the branch currents in figure 6-28.

Solution
$$I_1 = I_T \left(\frac{R_2}{R_1 + R_2} \right)$$

$$= (4\ \text{A}) \left(\frac{60\ \Omega}{20\ \Omega + 60\ \Omega} \right)$$

$$= (4\ \text{A}) \left(\frac{60\ \Omega}{80\ \Omega} \right)$$

$$= 3\ \text{A}$$

$$I_2 = I_T \left(\frac{R_1}{R_1 + R_2} \right)$$

$$= (4\ \text{A}) \left(\frac{20\ \Omega}{20\ \Omega + 60\ \Omega} \right)$$

$$= (4\ \text{A}) \left(\frac{20\ \Omega}{80\ \Omega} \right)$$

$$= 1\ \text{A}$$

Check at node A:
$$I_T = I_1 + I_2$$
$$4\ \text{A} = 3\ \text{A} + 1\ \text{A}$$
$$4\ \text{A} = 4\ \text{A}$$

The student should note that the *larger resistor* has the *smaller current*; also, the *smaller resistor* has the *larger current*. Voltage V_1 should equal V_2, since R_1 is in parallel with R_2.

$$V_1 = I_1 R_1$$
$$= (3\text{ A}) (20\ \Omega)$$
$$= 60\text{ V}$$
$$V_2 = I_2 R_2$$
$$= (1\text{ A}) (60\ \Omega)$$
$$= 60\text{ V}$$

(R6-19)   What is voltage E for the circuit of figure 6-28?

POWER IN SERIES AND PARALLEL CIRCUITS   The equations given in Chapter 5 for power can be applied to series and parallel circuits as well.

Problem 14   Calculate the power dissipated by each resistor in the circuit shown in figure 6-29. Find the total power supplied by the power supply.

Solution   Method 1:

$$R_T = R_1 + R_2 + R_3$$
$$= 15\ \Omega + 30\ \Omega + 60\ \Omega$$
$$= 105\ \Omega$$

$$I_T = \frac{E}{R_T}$$
$$= \frac{210\text{ V}}{105\ \Omega}$$
$$= 2\text{ A}$$

$$P_1 = I_T^2 R_1$$
$$= (2\text{ A})^2 (15\ \Omega)$$
$$= 60\text{ W}$$

$$P_2 = I_T^2 R_2$$
$$= (2\text{ A})^2 (30\ \Omega)$$
$$= 120\text{ W}$$

$$P_3 = I_T^2 R_3$$
$$= (2\text{ A})^2 (60\ \Omega)$$
$$= 240\text{ W}$$

$$P_T = P_1 + P_2 + P_3$$
$$= 60\text{ W} + 120\text{ W} + 240\text{ W}$$
$$= 420\text{ W}$$

Check:
$$P_T = I_T^2 R_T = (2\text{ A})^2 (105\ \Omega) = 420\text{ W}$$

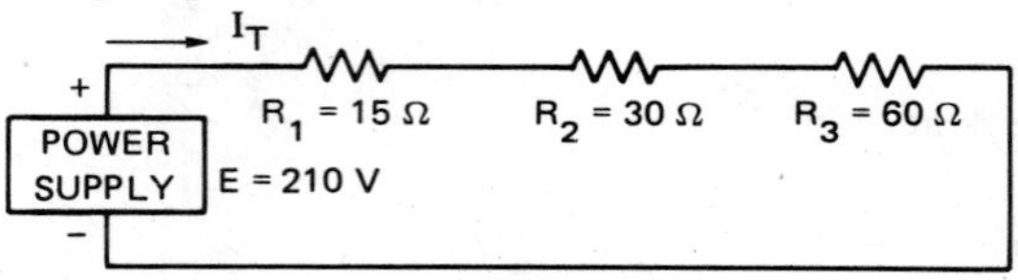

Fig. 6-29

Method 2:

$$V_1 = E\left(\frac{R_1}{R_T}\right)$$

$$= (210 \text{ V})\left(\frac{15 \text{ } \Omega}{105 \text{ } \Omega}\right)$$

$$= 30 \text{ V}$$

$$V_2 = E\left(\frac{R_2}{R_T}\right)$$

$$= (210 \text{ V})\left(\frac{30 \text{ } \Omega}{105 \text{ } \Omega}\right)$$

$$= 60 \text{ V}$$

$$V_3 = E\left(\frac{R_3}{R_T}\right)$$

$$= (210 \text{ V})\left(\frac{60 \text{ } \Omega}{105 \text{ } \Omega}\right)$$

$$= 120 \text{ V}$$

$$P_1 = \frac{V_1{}^2}{R_1}$$

$$= \frac{(30 \text{ V})^2}{15 \text{ } \Omega}$$

$$= 60 \text{ W}$$

$$P_2 = \frac{V_2{}^2}{R_2}$$

$$= \frac{(60 \text{ V})^2}{30 \text{ } \Omega}$$

$$= 120 \text{ W}$$

$$P_3 = \frac{V_3{}^2}{R_3}$$

$$= \frac{(120 \text{ V})^2}{60 \text{ } \Omega}$$

$$= 240 \text{ W}$$

$$P_T = P_1 + P_2 + P_3$$

$$= 60 \text{ W} + 120 \text{ W} + 240 \text{ W}$$

$$= 420 \text{ W}$$

Check:

$$P_T = \frac{E^2}{R_T}$$

$$= \frac{(210 \text{ V})^2}{105 \text{ } \Omega}$$

$$= 420 \text{ W}$$

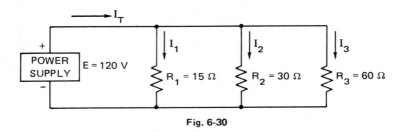

Fig. 6-30

Problem 15   Calculate the power dissipated by each resistor in the circuit shown in figure 6-30. Find the total power supplied by the power supply.

Solution   Method 1:

$$I_1 = \frac{E}{R_1}$$

$$= \frac{120 \text{ V}}{15 \text{ }\Omega}$$

$$= 8 \text{ A}$$

$$I_2 = \frac{E}{R_2}$$

$$= \frac{120 \text{ V}}{30 \text{ }\Omega}$$

$$= 4 \text{ A}$$

$$I_3 = \frac{E}{R_3}$$

$$= \frac{120 \text{ V}}{60 \text{ }\Omega}$$

$$= 2 \text{ A}$$

$$P_1 = I_1{}^2 R_1$$

$$= (8 \text{ A})^2 \,(15 \text{ }\Omega)$$

$$= 960 \text{ W}$$

$$P_2 = I_2{}^2 R_2$$

$$= (4 \text{ A})^2 \,(30 \text{ }\Omega)$$

$$= 480 \text{ W}$$

$$P_3 = I_3{}^2 R_3$$

$$= (2 \text{ A})^2 \,(60 \text{ }\Omega)$$

$$= 240 \text{ W}$$

$$P_T = P_1 + P_2 + P_3$$

$$= 960 \text{ W} + 480 \text{ W} + 240 \text{ W}$$

$$= 1\ 680 \text{ W}$$

Method 2:

$$P_1 = \frac{V_1{}^2}{R_1}$$

$$= \frac{(120 \text{ V})^2}{15 \text{ }\Omega}$$

$$= 960 \text{ W}$$

$$P_2 = \frac{V_2{}^2}{R_2}$$

$$= \frac{(120 \text{ V})^2}{30 \text{ }\Omega}$$

$$= 480 \text{ W}$$

$$P_3 = \frac{V_3{}^2}{R_3}$$

$$= \frac{(120 \text{ V})^2}{60 \text{ }\Omega}$$

$$= 240 \text{ W}$$

$$P_T = P_1 + P_2 + P_3$$

$$= 960 \text{ W} + 480 \text{ W} + 240 \text{ W}$$

$$= 1\ 680 \text{ W}$$

Method 3 (to find P_T):

$$G_T = G_1 + G_2 + G_3$$

$$= \frac{1}{15 \text{ }\Omega} + \frac{1}{30 \text{ }\Omega} + \frac{1}{60 \text{ }\Omega}$$

$$= 0.066\ 7 \text{ S} + 0.033\ 3 \text{ S} + 0.016\ 7 \text{ S}$$

$$= 0.116\ 7 \text{ S}$$

$$R_{eq} = \frac{1}{G_T}$$

$$= \frac{1}{0.116\ 7 \text{ S}}$$

$$= 8.57 \text{ }\Omega$$

$$P_T = \frac{E^2}{R_{eq}}$$

$$= \frac{(120 \text{ V})^2}{8.57 \text{ }\Omega}$$

$$= 1\ 680 \text{ W}$$

LABORATORY EXERCISE 6-1 SERIES DC CIRCUITS

PURPOSE   By completing this exercise, the student will verify experimentally the characteristics of a series dc circuit.

EQUIPMENT AND
MATERIALS

1 Dc power supply
1 Analog VOM
1 Digital multimeter
1 Resistor, 82 Ω, 2 W
1 Resistor, 240 Ω, 2 W
1 Resistor, 360 Ω, 2 W
1 Resistor, 510 Ω, 2 W
1 Resistor, 10 k Ω, 2 W

PROCEDURE

1. Connect the proper components in the circuit shown in figure 6-31. Use the VOM to read the current.
   a. Adjust the power supply voltage to read 30 V.
   b. Measure the voltages E, V_1, V_2, and V_3 with the multimeter. Note the polarity of the voltage across each resistor.
   c. Redraw figure 6-31 and indicate the polarity of the voltages across each resistor.
   d. Is Kirchhoff's Voltage Law satisfied around path E A B C D E?
   e. Measure the current flowing between points A and B.
   f. Insert the VOM at points C, D, and E and measure the current at each point.
   g. How do the currents of step 1.e. and step 1.f. compare?
   h. Calculate the total resistance of the circuit using the readings taken at steps 1.a. and 1.e.
   i. Calculate the total resistance of the circuit from the nominal values of the resistors, as determined from the color code.
   j. Compare the results of steps 1.h. and 1.i.
   k. Disconnect the power supply and the VOM.
   l. Measure the total resistance of the circuit across points B and E using the digital multimeter. Compare this value of R_T with the value calculated in step 1.h.
   m. Calculate the value of each resistor from the readings taken at steps 1.b. and 1.e. Compare the calculated resistance values with the nominal resistor values.
   n. Using an input voltage of 30 V and the resistance values determined in step 1.m., compute the voltage across each resistor by applying the voltage divider rule.

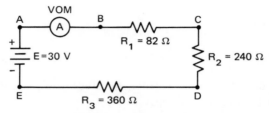

Fig. 6-31 Circuit for Laboratory Exercise 6-1, step 1

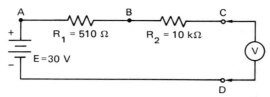

Fig. 6-32 Circuit for Laboratory Exercise 6-1, step 2

      o. Compare the calculated voltage values of step 1.n. with the measured values of step 1.b.
2. Connect the circuit shown in figure 6-32.
      a. Adjust the power supply voltage to read 30 V.
      b. Measure the voltage V_{CD} with the digital multimeter. Note its polarity.
      c. Is Kirchhoff's Voltage Law satisfied by the results of steps 2.a. and 2.b.?

LABORATORY EXERCISE 6-2 PARALLEL DC CIRCUITS

PURPOSE

By completing this exercise, the student will verify experimentally the characteristics of a parallel dc circuit.

EQUIPMENT AND MATERIALS

1 Dc power supply
1 Analog VOM
1 Digital multimeter
1 Resistor, 1.5 k Ω, 2 W
1 Resistor, 2.4 k Ω, 2 W
1 Resistor, 3.6 k Ω, 2 W
1 Resistor, 5.1 k Ω, 2 W

PROCEDURE

1. Connect the circuit shown in figure 6-33.
      a. Measure the resistance of this circuit, using the digital multimeter across terminals A and B. This resistance is called the *input resistance* (or R_{eq}, or R_T) of a parallel circuit.
      b. Using the nominal values of the resistors, calculate the equivalent resistance, R_{eq}, using the conductance method.

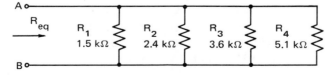

Fig. 6-33 Circuit for Laboratory Exercise 6-2, step 1

TABLE 6-1

Resistor	Measured Value	Nominal Value	% Difference
R_1			
R_2			
R_3			
R_4			
R_T			

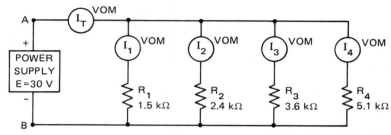

Fig. 6-34 Using a VOM to measure currents in a parallel circuit, Laboratory Exercise 6-2, step 2

    c. Calculate the percent difference between the measured value obtained in step 1.a. and the calculated value obtained in step 1.b. List these results in Table 6-1.

$$\% \text{ difference} = \frac{\text{Measured Value - Calculated Value}}{\text{Calculated Value}} \times 100$$

    d. Remove the resistors from the circuit of figure 6-33. Measure each resistor and complete Table 6-1.

2. Connect the circuit shown in figure 6-34.

    a. Adjust the input voltage to equal 30 V.

    b. Turn off the power supply. Connect the VOM, as shown in figure 6-34.

    c. Turn on the power supply; record the currents I_T, I_1, I_2, I_3, and I_4.

    d. Does the sum of $I_1 + I_2 + I_3 + I_4 = I_T$?

    e. What law does the answer to step 2.d. prove?

EXTENDED STUDY TOPICS

1. A circuit is connected as shown in figure 6-35.

    a. Find R_T.

    b. Find I_T.

    c. Calculate V_1, V_2, V_3, and V_4.

    d. Find the power dissipated by R_1, R_2, R_3, and R_4.

    e. Find the power output of the power supply.

    f. Verify Kirchhoff's Voltage Law.

2. For the circuit of figure 6-35, calculate V_1, V_2, V_3, and V_4 using the voltage divider rule.

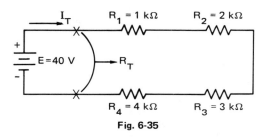

Fig. 6-35

3. Three resistors are connected in parallel across a 12-V source. The resistance values are : $R_1 = 1 \text{ k}\Omega$, $R_2 = 2 \text{ k}\Omega$, and $R_3 = 3 \text{ k}\Omega$.
   a. Find I_1, I_2, I_3, and I_T.
   b. Determine the value or R_{eq} using voltage and current values only.
   c. Calculate R_{eq} using the conductance method.
   d. Find the power dissipated by each resistor.
   e. Find the total power output of the source.
4. Assume that R_1 develops an open condition and repeat topic 3.
5. Four resistors are connected in parallel. The resistance values are: $R_1 = 100 \ \Omega$, $R_2 = 200 \ \Omega$, $R_3 = 300 \ \Omega$, and $R_4 = 620 \ \Omega$. Find R_{eq} using circuit simplification techniques and the product-over-sum formula.
6. Twenty 5.6-kΩ resistors are connected in parallel across a 100-V source.
   a. Calculate the current through each resistor.
   b. Calculate I_T.
   c. Find R_{eq}.
7. A 2-kΩ resistor (R_1) and a 3-kΩ resistor (R_2) are connected in parallel. The total current (I_T) = 20 mA.
   a. Find R_{eq}.
   b. Find I_1 and I_2 using the current divider rule.
   c. Calculate V_1 and V_2.
8. A 40-Ω resistor (R_1) is connected in parallel with resistor R_2 whose value is unknown. $R_{eq} = 20 \ \Omega$. What is the value of R_2?
9. Four resistors are connected in parallel. $R_1 = 1.2 \text{ k}\Omega$, $R_2 = 2.4 \text{ k}\Omega$, $R_3 = 3.6 \text{ k}\Omega$, and $R_4 = 5.6 \text{ k}\Omega$. Calculate R_{eq} using the conductance method.
10. What is the value of R_{eq} if R_3 in topic 9 develops a short circuit?

<table>
<tr><td>Chapter
7</td><td>Series-Parallel Networks</td></tr>
</table>

OBJECTIVES After studying this chapter, the student will be able to

- draw simple equivalent circuits as reductions of complex series-parallel networks.

- solve series-parallel networks by applying Kirchhoff's Laws and Ohm's Laws.

- calculate power in a series-parallel network.

- discuss bridge circuits.

- describe how dc three-wire systems operate.

EQUIVALENT CIRCUITS Circuits containing both series and parallel elements are called *series-parallel networks*. The series-connected elements are governed by the rules for series circuits. The parallel-connected elements are governed by the rules for parallel circuits. These rules were stated and explained in Chapter 6 for series and parallel circuits. Figure 7-1 shows a simple series-parallel circuit.

(R7-1) Which resistors of figure 7-1 are connected in parallel?

(R7-2) Which resistors of figure 7-1 are connected in series?

(R7-3) Explain the reason for the answer to question (R7-2).

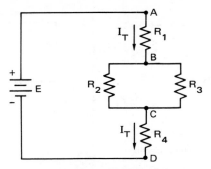

Fig. 7-1 A series-parallel circuit

To solve a circuit such as the one shown in figure 7-1, a step-by-step procedure is used to obtain a final equivalent circuit. The steps in the procedure are:

1. Combine all series resistors to find the total series resistance, R_S.
2. Find the equivalent resistance of all of the parallel branches, R_{eq}.
3. The circuit now consists of R_S in series with R_{eq}. Redraw the circuit.
4. Find R_T:
$$R_T = R_S + R_{eq}$$
5. Redraw the resulting equivalent circuit.
6. Solve for I_T using Ohm's Law.
7. Complete the solution using the rules for series and parallel circuits. The voltage across the parallel branch is equal to $I_T R_{eq}$.

Problem 1 The following values are assigned to the circuit components shown in figure 7-1: $R_1 = 10\ \Omega$, $R_2 = 20\ \Omega$, $R_3 = 30\ \Omega$, $R_4 = 100\ \Omega$, and $E = 100\ V$. Find the total current and V_{BC}.

Solution
1. $R_S = R_1 + R_4$
$= 10\ \Omega + 100\ \Omega$
$= 110\ \Omega$

2. $R_{eq} = \dfrac{R_2 R_3}{R_2 + R_3}$
$= \dfrac{(20\ \Omega)(30\ \Omega)}{20\ \Omega + 30\ \Omega}$
$= \dfrac{600(\Omega)^2}{50\ \Omega}$
$= 12\ \Omega$

3. The series equivalent circuit is given in figure 7-2A.

4. $R_T = R_S + R_{eq}$
$= 110\ \Omega + 12\ \Omega$
$= 122\ \Omega$

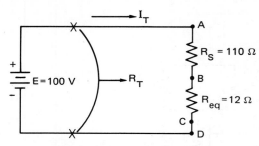

Fig. 7-2A Series equivalent circuit

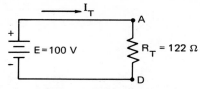

Fig. 7-2B Final equivalent circuit

5. The final equivalent circuit is shown in figure 7-2B.

6. $I_T = \dfrac{E}{R_T}$

$\quad = \dfrac{100 \text{ V}}{122 \text{ }\Omega}$

$\quad = 0.82 \text{ A, or } 820 \text{ mA}$

7. $V_{BC} = I_T R_{eq}$

$\quad = (0.82 \text{ A}) (12 \text{ }\Omega)$

$\quad = 9.84 \text{ V}$

(R7-4) For the circuit of figure 7-1, find V_{AB}.

(R7-5) For the circuit of figure 7-1, find V_{CD}.

(R7-6) By applying Kirchhoff's Voltage Law, prove that $E = V_{AB} + V_{BC} + V_{CD}$ (Refer to figure 7-1.)

Problem 2 For the circuit shown in figure 7-3A, find the current through each resistor. Determine the voltage drop across each resistor.

Solution The circuit is redrawn to the form shown in figure 7-3B. Each current is indicated. All terminal points and nodes are labeled.

$R_S = R_1 + R_4 + R_9$

$\quad = 10 \text{ }\Omega + 51 \text{ }\Omega + 30 \text{ }\Omega$

$\quad = 91 \text{ }\Omega$

$R'_{eq} = R_{BC} = \dfrac{R_2 R_3}{R_2 + R_3}$

$\quad = \dfrac{(20 \text{ }\Omega) (40 \text{ }\Omega)}{20 \text{ }\Omega + 40 \text{ }\Omega}$

$\quad = \dfrac{800(\Omega)^2}{60 \text{ }\Omega}$

$\quad = 13.33 \text{ }\Omega$

$R''_{eq} = R_{DE} = \dfrac{100 \text{ }\Omega}{4}$

$\quad = 25 \text{ }\Omega$

$R_T = R_S + R'_{eq} + R''_{eq}$

$\quad = 91 \text{ }\Omega + 13.33 \text{ }\Omega + 25 \text{ }\Omega$

$\quad = 129.3 \text{ }\Omega$

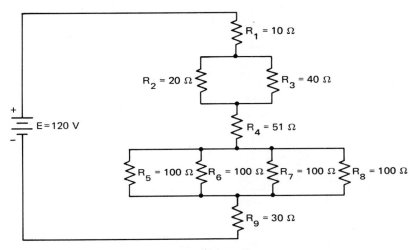

Fig. 7-3A Original circuit

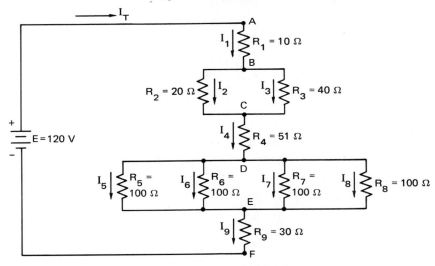

Fig. 7-3B Circuit of figure 7-3A redrawn

The equivalent circuit is shown in figure 7-3C.

$$I_T = \frac{E}{R_T}$$

$$= \frac{120 \text{ V}}{129.3 \ \Omega}$$

$$= 0.928 \text{ A}$$

$$I_T = I_1 = I_4 = I_9 = 0.928 \text{ A}$$

$$I_2 = I_1 \left(\frac{R_3}{R_2 + R_3} \right)$$

$$= (0.928 \text{ A}) \left(\frac{(40 \ \Omega)}{(20 \ \Omega + 40 \ \Omega)} \right)$$

$$= (0.928 \text{ A}) \left(\frac{40 \ \Omega}{60 \ \Omega} \right)$$

$$= 0.619 \text{ A}$$

$$I_3 = I_1 \left(\frac{R_2}{R_2 + R_3} \right)$$

$$= (0.928 \text{ A}) \left(\frac{20 \ \Omega}{60 \ \Omega} \right)$$

$$= 0.309 \text{ A}$$

Check:

$$I_1 = I_2 + I_3$$

$$0.928 \text{ A} = 0.619 \text{ A} + 0.309 \text{ A}$$

$$0.928 \text{ A} = 0.928 \text{ A}$$

$$I_5 = I_6 = I_7 = I_8 = \frac{I_4}{4}$$

$$= \frac{0.928 \text{ A}}{4}$$

$$= 0.232 \text{ A}$$

$$V_1 = V_{AB} = I_1 R_1$$

$$= (0.928 \text{ A})(10 \ \Omega)$$

$$= 9.28 \text{ V}$$

Method 1:

$$V_2 = V_3 = I_2 R_2 = V_{BC}$$

$$= (0.618 \text{ A}) (20 \ \Omega)$$

$$= 12.36 \text{ V}$$

Method 2:

$$V_2 = V_3 = I_1 R'_{eq} = V_{BC}$$

$$= (0.928 \text{ A}) (13.33 \ \Omega)$$

$$= 12.37 \text{ V}$$

$$V_4 = I_4 R_4 = V_{CD}$$

$$= (0.928 \text{ A}) (51 \ \Omega)$$

$$= 47.3 \text{ V}$$

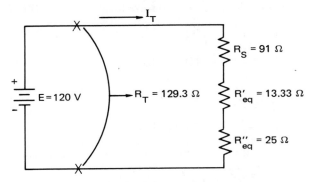

Fig. 7-3C Equivalent circuit

Method 1:
$$V_5 = V_6 = V_7 = V_8 = I_5 R_5 = V_{DE}$$
$$= (0.232\ A)\,(100\ \Omega)$$
$$= 23.20\ V$$

Method 2:
$$V_5 = V_6 = V_7 = V_8 = V_{DE} = I_4 R''_{eq}$$
$$= (0.928\ A)\,(25\ \Omega)$$
$$= 23.20\ V$$

$$V_9 = V_{EF} = I_9 R_9$$
$$= (0.928\ A)\,(30\ \Omega)$$
$$= 27.84\ V$$

Check:
$$E = V_1 + V_2 + V_4 + V_5 + V_9$$
$$120\ V = 9.28\ V + 12.37\ V + 47.3\ V + 23.2\ V + 27.84\ V$$
$$120\ V = 120\ V$$

(R7-7)   Assume that a short circuit occurs across R_5 in figure 7-3B. Solve Problem 2 for this condition.

THE APPLICA-TION OF KIRCHHOFF'S LAWS TO NETWORKS   Kirchhoff's Laws can be applied as an alternate method of solving series-parallel networks. In this method, the circuit is not reduced to a simple circuit, but is left in its original form.

Problem 3   Apply Kirchhoff's Laws and solve for the currents through each resistor in figure 7-4.

Solution   1. $I_1 = I_2 + I_3$ (Kirchhoff's Current Law)

$I_1 = \dfrac{V_1}{R_1}$ (Ohm's Law)

$I_2 = \dfrac{V_2}{R_2}$ (Ohm's Law)

$I_3 = \dfrac{V_3}{R_3}$ (Ohm's Law)

Substitute these values in equation (1) to obtain:

$\dfrac{V_1}{R_1} = \dfrac{V_2}{R_2} + \dfrac{V_3}{R_3}$

2. $\dfrac{V_1}{12\ \Omega} = \dfrac{V_2}{10\ \Omega} + \dfrac{V_3}{40\ \Omega}$

$E = V_1 + V_2$ (Kirchhoff's Voltage Law)

Note that $V_2 = V_3$.

$V_2 = V_3 = E - V_1 = 100\ V - V_1$

Substitute this expression in equation (2):

$\dfrac{V_1}{12\ \Omega} = \dfrac{(100\ V - V_1)}{10\ \Omega} + \dfrac{(100\ V - V_1)}{40\ \Omega}$

Multiply both sides of the equation by 120:

$10\ V_1 = 1\ 200\ V - 12\ V_1 + 300\ V - 3\ V_1$

$25\ V_1 = 1\ 500\ V$

$V_1 = 60\ V$

$V_2 = V_3 = E - V_1 = 100\ V - 60\ V = 40\ V$

$I_1 = \dfrac{V_1}{R_1}$

$= \dfrac{60\ V}{12\ \Omega}$

$= 5\ A$

$I_2 = \dfrac{V_2}{R_2}$

$= \dfrac{40\ V}{10\ \Omega}$

$= 4\ A$

$I_3 = \dfrac{V_3}{R_3}$

$= \dfrac{40\ V}{40\ \Omega}$

$= 1\ A$

Check:

$I_1 = I_2 + I_3$

$5\ A = 4\ A + 1\ A$

$5\ A = 5\ A$

(R7-8)   Solve Problem 3 using the equivalent circuit method.

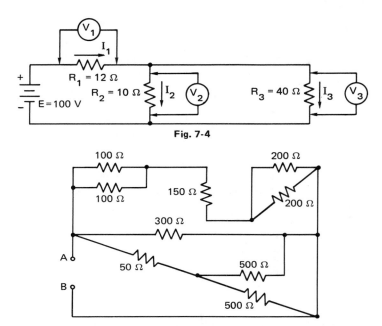

Fig. 7-4

Fig. 7-5

COMPLEX NETWORKS   In some cases, the type of circuit cannot be determined readily. Such a circuit is shown in figure 7-5. Problem 4 illustrates the method used to solve this type of circuit.

Problem 4   Find the total resistance across terminals A and B (R_{AB}) for the circuit of figure 7-5.

Solution   All terminal points and nodes are labeled, as shown in figure 7-6A.

When the circuit is redrawn to the form shown in figure 7-6B, it is apparent that this is a parallel circuit. Two branches of the circuit (No. 1 and No. 3) consist of resistors connected in a series-parallel combination.

Branch 1: $R_{AC} = \dfrac{100\ \Omega}{2} = 50\ \Omega$

$R_{DB} = \dfrac{200\ \Omega}{2} = 100\ \Omega$

$R'_{AB} = 50\ \Omega + 150\ \Omega + 100\ \Omega = 300\ \Omega$

Branch 2: $R''_{AB} = 300\ \Omega$ (by inspection)

Branch 3: $R_{EB} = \dfrac{500\ \Omega}{2} = 250\ \Omega$

$R'''_{AB} = 50\ \Omega + 250\ \Omega = 300\ \Omega$

The original circuit reduces to the parallel circuit of figure 7-6C.

$R_{AB} = \dfrac{300\ \Omega}{3} = 100\ \Omega$

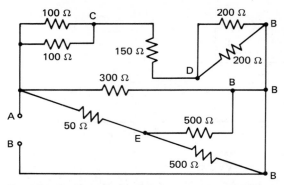

Fig. 7-6A Circuit with terminal points and nodes labeled

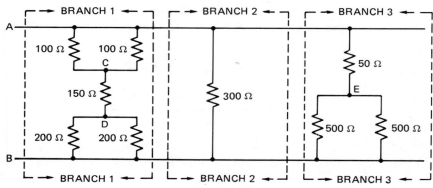

Fig. 7-6B Circuit of figure 7-6A redrawn

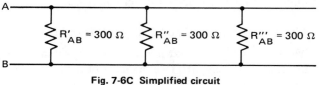

Fig. 7-6C Simplified circuit

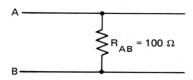

Fig. 7-6D Final simplified circuit

The final simplified circuit is shown in figure 7-6D.

(R7-9)   A voltage of 120 V is connected to terminals A and B of the circuit in figure 7-5. Determine the value of the voltage drop across the 50-Ω resistor.

(R7-10)   One of the 500-Ω resistors in the circuit of figure 7-5 is shorted by accident. Calculate the value of the resistance across terminals A and B.

POWER IN A SERIES-PARALLEL NETWORK The total power is always the arithmetic sum of the individual powers. The total power of an equivalent circuit must equal the total power of the original circuit.

Problem 5

a. Find the power dissipated by each resistor in the circuit of figure 7-4.
b. Find the total power supplied by the power supply.
c. Using the equivalent circuit method, calculate the value of R_T.
d. Determine the total power dissipated by R_T.

Solution From the solution of Problem 3:

$$I_1 = 5A$$
$$I_2 = 4 A$$
$$I_3 = 1 A$$
$$V_1 = 60 V$$
$$V_2 = V_3 = 40 V$$

a. $P_1 = V_1 I_1 = (60 V)(5 A) = 300 W$
   $P_2 = V_2 I_2 = (40 V)(4 A) = 160 W$
   $P_3 = V_3 I_3 = (40 V)(1 A) = 40 W$

b. $P_T = P_1 + P_2 + P_3$
   $= 300 W + 160 W + 40 W$
   $= 500 W$

   Alternate Method:
   $P_T = EI_1$
   $= (100 V)(5 A)$
   $= 500 W$

c. $R_S = R_1 = 12 \ \Omega$

   $R_{eq} = \dfrac{R_2 R_3}{R_2 + R_3}$

   $= \dfrac{(10 \ \Omega)(40 \ \Omega)}{50 \ \Omega}$

   $= \dfrac{400(\Omega)^2}{50 \ \Omega}$

   $= 8 \ \Omega$

   $R_T = R_S + R_{eq}$
   $= 12 \ \Omega + 8 \ \Omega$
   $= 20 \ \Omega$

   $I_T = I_1 = \dfrac{E}{R_T}$

   $= \dfrac{100 V}{20 \ \Omega}$

   $= 5 A$

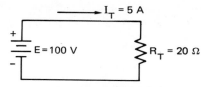

Fig. 7-7 Equivalent circuit

The equivalent circuit is shown in figure 7-7.

d. $P_T = I_T^2 R_T$

$\quad = (5 \text{ A})^2 (20 \text{ }\Omega)$

$\quad = 500 \text{ W}$

Alternate method:

$$P_T = \frac{V_T^2}{R_T}$$

$$= \frac{(100 \text{ V})^2}{20 \text{ }\Omega}$$

$$= 500 \text{ W}$$

(R7-11)   If a circuit is to be the equivalent of another circuit, what requirements must it satisfy?

THREE-WIRE SYSTEMS   The consumer is supplied electric power by means of a three-wire distribution system. For the three-wire system shown in figure 7-8, the two outside conductors are the line wires and the center conductor is the neutral wire. The National Electric Code requires the neutral wire to be grounded. This wire must not be fused.

Three voltages are available from this three-wire system: $V_1 = 120$ V, $V_2 = 120$ V, and $V_3 = 240$ V. Terminal B has a double polarity. That is, it is negative with respect to terminal A, and positive with respect to terminal B.

(R7-12)   What is the voltage V_{AB} in figure 7-8?

(R7-13)   What is the voltage V_{CB} in figure 7-8?

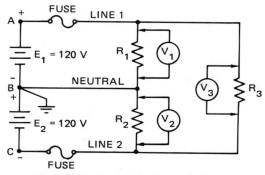

Fig. 7-8 Three-wire distribution system

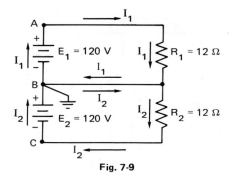

Fig. 7-9

Problem 6   Figure 7-9 shows a three-wire system. The loads on this system are balanced because $R_1 = R_2$. Find the currents in the line conductors and in the neutral wire.

Solution   Load R_1 is connected to E_1. The current I_1 flows as shown in figure 7-9.
Load R_2 is connected to E_2. The current I_2 flows as indicated.

$$I_1 = \frac{E_1}{R_1}$$
$$= \frac{120 \text{ V}}{12 \text{ }\Omega}$$
$$= 10 \text{ A}$$
$$I_2 = \frac{E_2}{R_2}$$
$$= \frac{120 \text{ V}}{12 \text{ }\Omega}$$
$$= 10 \text{ A}$$

Each line conductor has a current of 10 A.
Since I_1 is equal and opposite to I_2, the total neutral current is zero.
$$I_N = I_1 - I_2$$
$$= 10 \text{ A} - 10 \text{ A}$$
$$= 0 \text{ A}$$

(R7-14)   For the circuit of figure 7-10, calculate the currents in the line conductors and in the neutral wire.

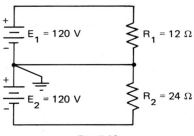

Fig. 7-10

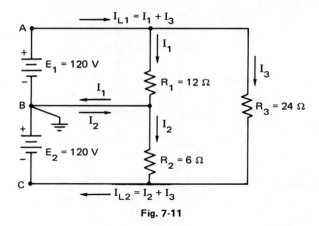

Fig. 7-11

Problem 7   For the circuit shown in figure 7-11, calculate the currents in the line conductors and in the neutral wire.

Solution $I_1 = \dfrac{E_1}{R_1}$

$= \dfrac{120\ V}{12\ \Omega}$

$= 10\ A$

$I_2 = \dfrac{E_2}{R_2}$

$= \dfrac{120\ V}{6\ \Omega}$

$= 20\ A$

$I_N = I_1 - I_2$

$= 10\ A - 20\ A$

$= -10\ A$

(Note: the negative sign indicates that the true direction of I_N is the direction of I_2 and not I_1.

$I_3 = \dfrac{240\ V}{24\ \Omega}$

$= 10\ A$

$I_{L1} = I_1 + I_3$

$= 10\ A + 10\ A$

$= 20\ A$

$I_{L2} = I_2 + I_3$

$= 20\ A + 10\ A$

$= 30\ A$

Check: At node B,

$I_{L2} = 30\ A$ (enters)

$I_{L1} + I_2 = 20\ A + 10\ A = 30\ A$ (leaves).

(R7-15) If loads R_1 and R_2 are interchanged in the circuit of figure 7-11, what is the effect on the current in the neutral wire?

(R7-16) Load R_2 is disconnected from the circuit of figure 7-11. Determine the current in the neutral wire.

(R7-17) Does load R_3 have any effect on the neutral current of the circuit of figure 7-11? Explain the answer.

BRIDGE CIRCUITS The bridge circuit shown in figure 7-12 is often used in measurement circuits and in control systems. This circuit consists of two parallel branches. Each branch has two series resistors: $(R_1 + R_2)$ forms one branch and $(R_3 + R_4)$ forms the other branch.

The source voltage is impressed across each parallel branch. The voltage divides between the two resistors of the branch. The output of the circuit is taken between points C and D. The voltage (V_{CD}) can be positive, negative, or zero, depending upon the resistance values. When $V_{CD} = 0$, the bridge is said to be balanced.

The voltage across R_2 is V_{CB}. It has a positive value and may be written as V_C. When there is only one subscript, the reference point in the circuit is ground. Single subscript notation is used in electronic circuits when the circuit voltages are all measured relative to ground.

(R7-18) What does the voltage V_D represent?

(R7-19) What is the polarity of V_D?

The bridge is balanced when the resistor values are such that $V_C = V_D$. This condition is met for the following relationship:

$$R_1 R_4 = R_2 R_3 \qquad\qquad \text{Eq. 7.1}$$

Equation 7.1 shows that bridge balancing is independent of the supply voltage E.

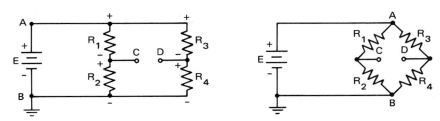

(A) Bridge circuit as two parallel branches (B) Circuit redrawn in conventional bridge arrangement

Fig. 7-12 Bridge circuit

Problem 8 The circuit of figure 7-12 has the following values: $E = 12$ V, $R_1 = 10$ kΩ, $R_2 = 20$ kΩ, $R_3 = 2$ kΩ, and $R_4 = 4$ kΩ. Is the bridge balanced? Determine the voltage V_{CD}.

Solution $R_1 R_4 = R_2 R_3$ (for a balanced bridge)
$(10 \times 10^3 \ \Omega)(4 \times 10^3 \ \Omega) = (20 \times 10^3 \ \Omega)(2 \times 10^3 \ \Omega)$
$$40 \times 10^6 (\Omega)^2 = 40 \times 10^6 (\Omega)^2$$
Therefore, the bridge is balanced and $V_{CD} = 0$

Problem 9 The circuit of figure 7-12 has the following values: $E = 12$ V, $R_1 = 10$ kΩ, $R_2 = 30$ kΩ, $R_3 = 2$ kΩ, and $R_4 = 4$ kΩ. Is the bridge balanced? Determine the voltage V_{CD}.

Solution $R_1 R_4 = R_2 R_3$ (for a balanced bridge)
$(10 \times 10^3 \ \Omega)(4 \times 10^3 \ \Omega) \neq (30 \times 10^3 \ \Omega)(2 \times 10^3 \ \Omega)$
$$40 \times 10^6 \ (\Omega)^2 \neq 60 \times 10^6 \ (\Omega)^2$$
Therefore, the bridge is *not* balanced, and $V_{CD} \neq 0$. (Note: the symbol $\neq$ means "is not equal to".)

$$V_C = E\left(\frac{R_2}{R_1 + R_2}\right)$$
$$= 12 \ V\left(\frac{30 \ k\Omega}{10 \ k\Omega + 30 \ k\Omega}\right)$$
$$= (12 \ V)\left(\frac{30 \ k\Omega}{40 \ k\Omega}\right)$$
$$= +9 \ V \ (V_C \text{ is } +9 \ V \text{ above ground})$$

$$V_D = E\left(\frac{R_4}{R_3 + R_4}\right)$$
$$= (12 \ V)\left(\frac{4 \ k\Omega}{2 \ k\Omega + 4 \ k\Omega}\right)$$
$$= 12 \ V\left(\frac{4 \ k\Omega}{6 \ k\Omega}\right)$$
$$= +8 \ V \ (V_C \text{ is } +8 \ V \text{ above ground})$$

Since V_C is +9 V above ground, and V_D is only +8 V above ground, $V_{CD} = +1$ V. This relationship is shown by the graph in figure 7-13.

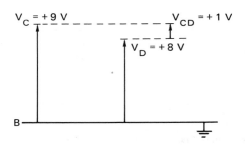

Fig. 7-13 Graphical representation of voltages across a bridge

TABLE 7-1

	R₁	R₂	R₃	R₄	R₅	R₆	R₇	R₈
Nominal Value, ohms	330	470	560	1 200	2 000	3 000	4 300	10 000
Measured Value, ohms								

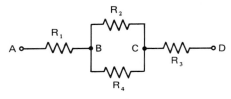

Fig. 7-14

**LABORATORY
EXERCISE 7-1
SERIES-
PARALLEL
NETWORKS**

PURPOSE By completing this exercise, the student will verify experimentally the principles of series-parallel networks.

**EQUIPMENT
AND MATERIALS**
1 Dc power supply
1 Analog VOM
1 Resistor, 330 Ω, 2 W (R_1)
1 Resistor, 470 Ω, 2 W (R_2)
1 Resistor, 560 Ω, 2 W (R_3)
1 Resistor, 1 200 Ω, 2 W (R_4)
1 Resistor, 2 000 Ω, 2 W (R_5)
1 Resistor, 3 000 Ω, 2 W (R_6)
1 Resistor, 4 300 Ω, 2 W (R_7)
1 Resistor, 10 000 Ω, 2 W (R_8)
1 Switch, single pole, single throw (SPST)

PROCEDURE
1. Measure the resistance of each resistor and record the values in Table 7-1.
2. Connect the circuit of figure 7-14.
   a. Measure R_T between terminals A and D. Record this value in Table 7-2.
   b. Measure R_{eq} between terminals B and C with terminals A and D open circuited. Record this value in Table 7-2.

TABLE 7-2

	Measured Value, ohms	Calculated Value, ohms
R_T		
R_{eq}		

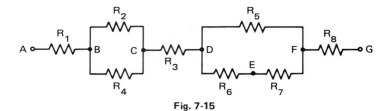

Fig. 7-15

c. Calculate R_{eq}. Record the value in Table 7-2.

d. Calculate R_T. Record the value in Table 7-2.

e. Compare the measured and calculated values of R_T. Discuss any differences in the values.

3. Connect the circuit of figure 7-15.

 a. Measure R_T between terminals A and G. Record this value in Table 7-3.

 b. Measure R'_{eq} between terminals B and C with terminals A and G open circuited. Record the value in Table 7-3.

 c. Measure R''_{eq} between terminals D and F with terminals A and G open circuited. Record the value in Table 7-3.

 d. Calculate R_T, R'_{eq} and R''_{eq}. Record the results in Table 7-3.

 e. Compare the measured and calculated values of R_T. Discuss any differences in the values.

4. Connect the circuit of figure 7-16.

 a. With the SPST switch open, adjust the output of the power supply to 20 V.

TABLE 7-3

	Measured Value, ohms	Calculated Value, ohms
R_T		
R'_{eq}		
R''_{eq}		

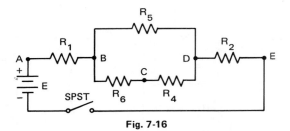

Fig. 7-16

TABLE 7-4

	Measured Value, volts	Calculated Value, volts
V_{AB}		
V_{BD}		
V_{DE}		

    b. Close the SPST switch. Maintain the power supply output at 20 V.

    c. Read V_{AB}, V_{BD}, and V_{DE}. Record the results in Table 7-4.

    d. Calculate V_{AB}, V_{BD}, and V_{DE}. Record these values in Table 7-4.

    e. Compare the measured and calculated values from steps 4.c. and 4.d. Discuss any differences in these values.

    f. Prove how Kirchhoff's Voltage Law works from the measured values of this part of the experiment.

    g. Open the SPST switch. Remove resistor R_2 so that an open circuit remains between terminals D and E. Close the SPST switch. Measure V_{ED}. Does this measured value agree with the expected voltage reading?

EXTENDED STUDY TOPICS

1. Using the equivalent circuit method, find V_1, V_2, V_3, I_1, I_2, and I_3 in figure 7-17.
2. Use Kirchhoff's Laws to solve topic 1.
3. Determine the value of V_5 in figure 7-18.

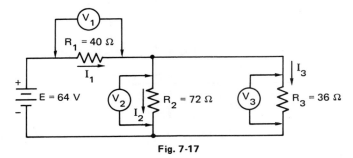

Fig. 7-17

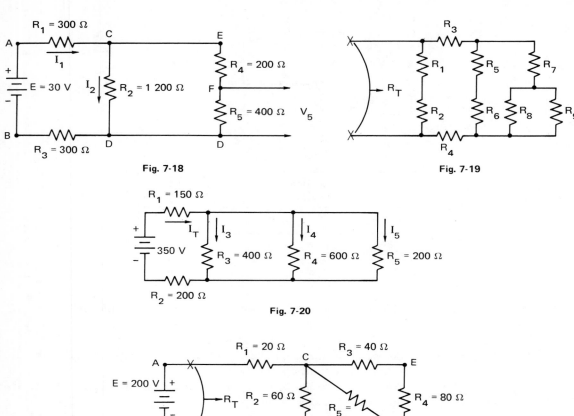

Fig. 7-18

Fig. 7-19

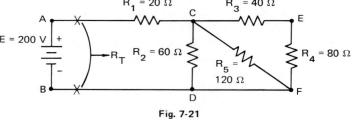

Fig. 7-20

Fig. 7-21

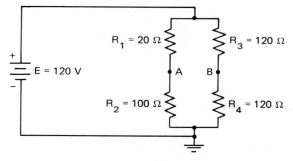

Fig. 7-22

4. Determine the value of R_T in figure 7-19, if all of the resistors have a value of 1 000 Ω.
5. Calculate I_T, I_3, I_4, and I_5 in figure 7-20.
6. Find the power dissipated by each resistor of topic 5. Find the total power output of the power supply.
7. In the circuit of figure 7-21, find: a. R_T; b. the current through each resistor; and c. R_T, if R_5 is shorted by accident.
8. A bridge circuit is shown in figure 7-22.
   a. Is the bridge balanced across terminals A and B?
   b. Calculate V_{AB}.
9. Find the line currents and the current in the neutral wire of the three-wire system shown in figure 7-23. Does the neutral current flow toward node N or away from node N?

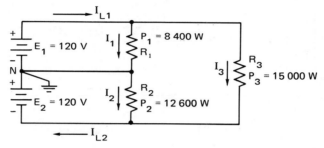

Fig. 7-23

<table>
<tr><td>

Chapter

8

</td><td>

Capacitors and RC Circuits

</td></tr>
</table>

OBJECTIVES   After studying this chapter, the student will be able to

- describe the construction of capacitors and their physical parameters.
- analyze the charging and discharging characteristics of capacitors.
- solve problems with capacitors connected in series or in parallel.
- discuss the charging and discharging characteristics of an RC circuit.

THE CAPACITOR   A capacitor is an electrical device consisting of two metallic conductors (plates) separated from each other by a nonconducting, or insulating, material. This insulating material is called the *dielectric*. Some typical dielectric materials are listed in Table 8-1. Commercial capacitors are shown in figures 8-1 and 8-2.

TABLE 8-1 Dielectric Materials

Air
Bakelite
Ceramic
Glass
Mica
Paraffin-coated paper
Porcelain
Rubber
Teflon
Transformer oil
Water

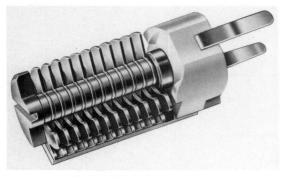

Fig. 8-1 Variable air capacitor

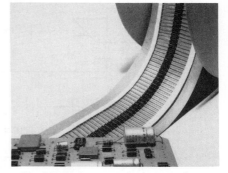

Fig. 8-2 Assortment of mica capacitors

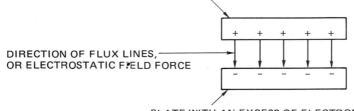

PLATE WITH A DEFICIENCY OF ELECTRONS;
THAT IS, THE PLATE IS POSITIVELY CHARGED

DIRECTION OF FLUX LINES,
OR ELECTROSTATIC FIELD FORCE

PLATE WITH AN EXCESS OF ELECTRONS;
THAT IS, THE PLATE IS NEGATIVELY CHARGED

Fig. 8-3 Electrostatic field in the dielectric of a capacitor

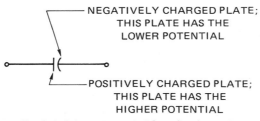

NEGATIVELY CHARGED PLATE;
THIS PLATE HAS THE
LOWER POTENTIAL

POSITIVELY CHARGED PLATE;
THIS PLATE HAS THE
HIGHER POTENTIAL

Fig. 8-4 Schematic symbol for a fixed capacitor

Fig. 8-5 Schematic symbol
for a variable capacitor

A capacitor stores electrical energy in an electric field. The capacitor stores energy when one capacitor plate has an excess of electrons, and the other capacitor plate has a deficiency of electrons. In other words, the two plates become charged bodies and an electric field is set up in the dielectric, figure 8-3. The schematic symbol for a fixed capacitor is shown in figure 8-4. Figure 8-5 gives the schematic symbol for a variable capacitor.

CHARGING THE CAPACITOR The principle of the charging and discharging of a capacitor is shown in figure 8-6. When switch S is closed to a position 1, the battery voltage E is applied across the capacitor plates A and B. The battery acts as an electron pump. Elec-

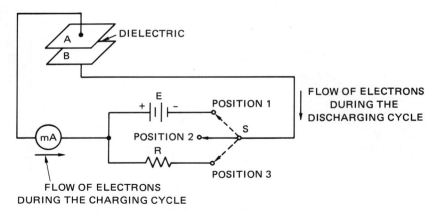

Fig. 8-6 Charging and discharging of a capacitor

trons are removed from plate A, and are deposited on plate B. This movement of electrons from plate A to plate B causes a brief deflection of the pointer of the milliammeter. The movement of electrons continues until the voltage across plates A and B equals the voltage E. There is no further movement of electrons and the current becomes zero. The capacitor is said to be charged. Plate B has an excess of electrons and plate A lacks electrons. The magnitude of the electric charge, Q, is the same for both plate A and plate B.

(R8-1)   What is the polarity of plate A of the charged capacitor shown in figure 8-6?

(R8-2)   What is the polarity of plate B of the charged capacitor shown in figure 8-6?

(R8-3)   E in figure 8-6 is 50 V. What is the voltage V_{AB} after the capacitor is fully charged?

(R8-4)   What prevents the electrons of the charged capacitor from traveling from plate B To plate A?

DISCHARGING THE CAPACITOR   When the switch in figure 8-6 is moved to position 3, the battery is replaced by the resistor R. The excess electrons on plate B now return to plate A through the resistor. The force that causes this return electron flow is the voltage between the charged plates. As the excess electrons from plate B return to plate A, there is a brief movement of the meter pointer in a direction opposite to that of the movement during the charging cycle.

After the surge of current stops, there is no unbalanced charge on either plate. The voltage across the plates is zero and the capacitor is discharged.

(R8-5)   What is responsible for the flow of current during the discharge cycle?

CAPACITANCE   *Capacitance* is the ability of a capacitor to store energy in the form of an electrostatic field. This quantity is a measure of the storage capacity of the capacitor. The SI letter symbol for capacitance is C. The unit of measurement is the farad and its symbol is F. Few practical capacitors have a capacitance greater than a small fraction of a farad. Therefore, two units representing a small part of a farad are commonly used:

1. the microfarad (μF) = 10^{-6} F
2. the picofarad (pF) = 10^{-12} F

Capacitance is determined by the following equation:

$$C = \frac{Q}{V}$$

where   Eq. 8.1

    C = capacitance, in farads
    V = voltage across the capacitor, in volts
    Q = charge on either plate, in coulombs

According to Equation 8.1, a capacitor has a capacitance of one farad if one coulomb of charge is deposited on the plates by a potential difference of one volt across the plates.

(R8-6) What is the capacitance of a capacitor having a charge of 6 mC, when a voltage of 120 V dc is applied to it?

PHYSICAL PARAMETERS The size of a capacitor depends on the following physical parameters:

1. the type of dielectric used.
2. the area of the plates.
3. the distance between the plates.

The following equation for a parallel plate capacitor relates these parameters:

$$C = 8.85 \times 10^{-12} \, K\frac{A}{d}$$

where   Eq. 8.2

    C = capacitance, in farads
    K = dielectric constant, in farads/meter
    A = area of plates, in m^2
    d = distance between plates, in meters

This equation shows that capacitance is directly proportional to the area of the plates, and inversely proportional to the distance between the plates. A change in the dielectric used also has a direct effect on the value of the capacitance. Common dielectrics are listed in Table 8-2 with their dielectric constants.

TABLE 8-2 Dielectric Constants

Dielectric	K = Dielectric Constant (farads/meter)
Vacuum	1.0
Air	1.000 6
Teflon	2.0
Paraffin-coated paper	2.5
Rubber	3.0
Transformer oil	4.0
Mica	5.0
Porcelain	6.0
Bakelite	7.0
Glass	7.5
Water	80.0
Ceramic	7 500.0

Problem 1 A capacitor has the following parameters:
Area of plates = 0.1 m²
Distance between plates = 3 × 10⁻³ m
Dielectric = mica
Find: a. the capacitance, and b. the charge on each plate, if 600 V dc is applied across the plates.

Solution
a. $C = 8.85 \times 10^{-12} \, K \dfrac{A}{d}$

$\quad = (8.85 \times 10^{-12})\left(5\dfrac{F}{m}\right)\left(\dfrac{0.1 \text{ m}^2}{3 \times 10^{-3}\text{m}}\right)$

$\quad = 1\,475 \text{ pF}$

b. $Q = CV$

$\quad = (1\,475 \times 10^{-12} \text{ F})(600 \text{ V})$

$\quad = 0.885 \text{ } \mu C$

Problem 2 If the dielectric of Problem 1 is changed to ceramic and all of the other capacitor parameters remain the same, find: a. the capacitance, and b. the charge.

Solution
a. K for ceramic = 7 500 (K for mica = 5)

$$C = (1\,475 \times 10^{-12} \text{ F})\left(\frac{7\,500}{5}\right)$$

$$= 2.212\,5 \text{ } \mu F$$

b. $Q = CV$

$\quad = (2.212\,5 \times 10^{-6} \text{ F})(600 \text{ V})$

$\quad = 1.327\,5 \text{ mC}$

(R8-7) If the dielectric in Problem 1 is changed to paraffin coated paper, find: a. the capacitance, and b. the charge for the capacitor.

DIELECTRIC STRENGTH When the voltage across a dielectric reaches a large enough value, the bonds within the dielectric are broken and current flows through the dielectric. This voltage is known as the *breakdown voltage*. The value, in volts per mil, required to cause conduction in a dielectric is called its *dielectric strength*. Table 8-3

TABLE 8-3 Dielectric Strength of Dielectric Materials

Dielectric	Dielectric Strength (Volts per Mil)
Air	75
Ceramic	75
Porcelain	200
Bakelite	400
Rubber	700
Paraffin-coated paper	1 300
Teflon	1 500
Glass	3 000
Mica	5 000

lists the dielectric strengths of selected materials. When dielectric breakdown occurs, the capacitor can no longer store energy. In this condition, the capacitor behaves like a conductor.

Problem 3 A capacitor with a ceramic dielectric is designed to oeprate at 300 V. What is the minimum design thickness of the dielectric?

Solution The dielectric strength of ceramic is 75 volts per mil.

$$\frac{300 \text{ V}}{75 \text{ V/mil}} = 4 \text{ mils}$$

Therefore, the ceramic must have a minimum thickness of 4 mils (0.004 in).

(R8-8) If the dielectric of Problem 3 is changed to teflon, what is the minimum required thickness of the dielectric?

ENERGY STORED BY A CAPACITOR The energy stored by a capacitor can be found using the following equation:

$$W = 1/2 \ CV^2 \qquad \text{Eq. 8.3}$$

where

W = energy stored, in joules
C = capacitance, in farads
V = voltage, in volts

Energy can also be determined, in terms of Q and C, from the following equation:

$$W = \frac{Q^2}{2 \ C} \qquad \text{Eq. 8.4}$$

where

Q = charge, in coulombs

Problem 4 Calculate the energy stored by a 120-pF capacitor when 20 volts is applied across the plates of the capacitor.

Solution $W = 1/2 \ CV^2$
$= 1/2(120 \times 10^{-12} \text{ F})(20 \text{ V})^2$
$= 24 \times 10^{-9} \text{ J}$

(R8-9) Find the charge Q on each plate of a 12-μF capacitor if the energy stored is 1 200 J.

CAPACITORS IN SERIES Each capacitor has the same charge when capacitors are connected in series, figure 8.7.

$$Q_T = Q_1 = Q_2 = Q_3 \qquad \text{Eq. 8.5}$$

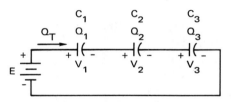

Fig. 8-7 Capacitors in series

Note that Equation 8.5 is similar to the current rule for resistors connected in series.

The following equation is obtained by applying Kirchhoff's Voltage Law:

$$E = V_1 + V_2 + V_3 \qquad \text{Eq. 8.6}$$

The total capacitance is:

$$\frac{1}{C_T} = \frac{1}{C_1} + \frac{1}{C_2} + \frac{1}{C_3} \qquad \text{Eq. 8.7(a)}$$

or,

$$C_T = \frac{1}{\dfrac{1}{C_1} + \dfrac{1}{C_2} + \dfrac{1}{C_3}} \qquad \text{Eq. 8.7(b)}$$

When two capacitors are connected in series, the total capacitance is:

$$C_T = \frac{C_1 C_2}{C_1 + C_2} \qquad \text{Eq. 8.8}$$

When any number of capacitors having the same value are connected in series, the total capacitance is:

$$C_T = \frac{C}{N} \qquad \text{Eq. 8.9}$$

where

C = capacitance of one capacitor, in farads
N = number of capacitors in series

Problem 5   Five 50-μF capacitors are connected in series. Find the total, or equivalent, capacitance.

Solution $C_T = \dfrac{C}{N}$

$\qquad = \dfrac{50 \times 10^{-6}\ \text{F}}{5}$

$\qquad = 10\ \mu\text{F}$

Problem 6   Three capacitors are connected in series, figure 8-7. The data for this circuit is as follows:

$C_1 = 5\ \mu\text{F} \qquad V_1 = 100\ \text{V}$
$C_2 = 10\ \mu\text{F}$
$C_3 = 20\ \mu\text{F}$

Find: a. Q_1, Q_2, Q_3, and Q_T; b. V_2; c. V_3; d. E; and e. C_T.

Solution   a. $Q_1 = C_1 V_1$

$\qquad\qquad = (5 \times 10^{-6}\ \text{F})(100\ \text{V})$

$\qquad\qquad = 500 \times 10^{-6}\ \text{C}$

$\qquad Q_T = Q_1 = Q_2 = Q_3 = 500 \times 10^{-6}\ \text{C}$

b. $V_2 = \dfrac{Q_2}{C_2}$

$\qquad = \dfrac{500 \times 10^{-6}\ \text{C}}{10 \times 10^{-6}\ \text{F}}$

$\qquad = 50\ \text{V}$

c. $V_3 = \dfrac{Q_3}{C_3}$

$= \dfrac{500 \times 10^{-6}\ C}{20 \times 10^{-6}\ V}$

$= 25\ V$

d. $E = V_1 + V_2 + V_3$

$= 100\ V + 50\ V + 25\ V$

$= 175\ V$

e. Method 1:

$C_T = \dfrac{Q_T}{E}$

$= \dfrac{500 \times 10^{-6}\ C}{175\ V}$

$= 2.857\ \mu F$

Method 2:

$\dfrac{1}{C_T} = \dfrac{1}{C_1} + \dfrac{1}{C_2} + \dfrac{1}{C_3}$

$= \dfrac{1}{5 \times 10^{-6}\ F} + \dfrac{1}{10 \times 10^{-6}\ F} + \dfrac{1}{20 \times 10^{-6}\ F}$

$= (200 \times 10^3)\dfrac{1}{F} + (100 \times 10^3)\dfrac{1}{F} + (50 \times 10^3)\dfrac{1}{F}$

$= 350 \times 10^3 \times \dfrac{1}{F}$

$C_T = \dfrac{1}{350 \times 10^3 \times \dfrac{1}{F}}$

$= 2.857\ \mu F$ (This answer is the same as the one obtained by using Method 1.)

(R8-10)   Three capacitors are connected in series. These capacitors have values of 10 μF, 30 μF, and 5 μF. Find the total capacitance of the combination.

By analyzing the previous problems on capacitors in series, the following conclusions can be made:

1. The total capacitance is less than the value of any one capacitor.
2. The total voltage rating is greater than the voltage of any one capacitor.
3. When the voltage rating of a capacitor is less than the circuit voltage, several capacitors may be used in series to obtain a greater overall voltage rating.
4. When capacitors are connected in series, there is an increase in the total thickness of the dielectric across which the circuit voltage is applied. This, in effect, increases the overall voltage rating.

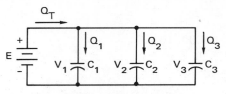

Fig. 8-8 Capacitors in parallel

CAPACITORS IN PARALLEL   Figure 8-8 shows several capacitors connected in parallel. This type of connection has the same effect as increasing the area of the plates.

$$C_T = C_1 + C_2 + C_3 \qquad \text{Eq. 8.10}$$

The voltage is the same across each capacitor.

$$E = V_1 = V_2 = V_3 \qquad \text{Eq. 8.11}$$

The total charge is the sum of the charges on each capacitor.

$$Q_T = Q_1 + Q_2 + Q_3 \qquad \text{Eq. 8.12}$$

Therefore,

$$Q_T = C_T E \qquad \text{Eq. 8.13}$$

Problem 7   Three capacitors are connected in parallel, figure 8-8. The following values are given for this circuit:

$C_1 = 800 \ \mu F \qquad E = 100 \ V$
$C_2 = 1\ 200 \ \mu F$
$C_3 = 500 \ \mu F$

Find: a. C_T; b. Q_1, Q_2 and Q_3; and c. Q_T.

Solution   a. $C_T = C_1 + C_2 + C_3$
$\qquad\qquad = 800 \ \mu F + 1\ 200 \ \mu F + 500 \ \mu F$
$\qquad\qquad = 2\ 500 \ \mu F$

b. $Q_1 = C_1 E = (800 \times 10^{-6} \ F)(100 \ V) = 80 \ mC$
$\quad Q_2 = C_2 E = (1\ 200 \times 10^{-6} \ F)(100 \ V) = 120 \ mC$
$\quad Q_3 = C_3 E = (500 \times 10^{-6} \ F)(100 \ V) = 50 \ mC$

c. $Q_T = Q_1 + Q_2 + Q_3$
$\qquad\quad = 80 \ mC + 120 \ mC + 50 \ mC$
$\qquad\quad = 250 \ mC$

(R8-11)   Calculate the total charge in Problem 7 using Equation 8.13.

Problem 8   Once the capacitor in figure 8-9 is charged to its final value, find: a. V_c, and b. Q.

Solution   a. $V_c = V_2 = \dfrac{(48 \ V)(16 \ \Omega)}{8 \ \Omega + 16 \ \Omega}$
$\qquad\qquad\quad = 32 \ V$

b. $Q = CV_c$
$\qquad = (20 \times 10^{-6} \ F)(32 \ V)$
$\qquad = 640 \ \mu C$

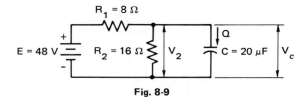

Fig. 8-9

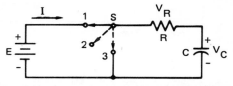

Fig. 8-10 RC circuit for charging and discharging a capacitor

RC CIRCUITS   The circuit in figure 8-10 is designed to charge and discharge the capacitor C. Position 2 of switch S is the *off* position; position 1 is the charging position; and position 3 is the discharging position. The charging and discharging cycles of the capacitor are called the *transient phases*. The resistor opposes the charging of the capacitor. Therefore, the charging cycle follows the exponential curve shown in figure 8-11.

The *time constant* of an RC circuit is the time, in seconds, required to charge the discharged capacitor to 63.2% of the final voltage E. Table 8-4 lists capacitor voltage values at various time constant periods. The time constant can be calculated from the following equation:

$$\text{Time constant} = RC \qquad\qquad \text{Eq. 8.14}$$

where

Time constant is expressed in seconds

R is in ohms

C is in farads

Experiments have shown that for most practical problems, it can be assumed that a capacitor charges to its final value in five time constants.

(R8-12)   A 20-kΩ resistor is connected in series with a 50-μF capacitor. The voltage of the power supply is 100 V dc. Calculate the time constant.

(R8-13)   Determine the voltage across the capacitor described in question (R8-12) after one time constant.

(R8-14)   How long will it take for the capacitor described in question (R8-12) to be fully charged (for practical purposes)?

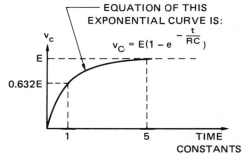

EQUATION OF THIS EXPONENTIAL CURVE IS:

$$v_C = E(1 - e^{-\frac{t}{RC}})$$

Fig. 8-11 A capacitor charging in an RC circuit

TABLE 8-4 Capacitor Voltages at Various Time Constant Periods During the Charging Cycle

Time Constant Periods	Capacitor Voltage
0	0
1	0.632E
2	0.865E
3	0.951E
4	0.981E
5	0.993E

TABLE 8-5 Charging Characteristics

Time	V_R	V_C	I	Remarks
O	E	O	$\dfrac{E}{R}$	Capacitor acts like a short circuit
1 Time constant	0.368E	0.632E	$0.368 \dfrac{E}{R}$	
1 to 5 Time constants	Decreases	Increases	Decreases	
5 Time Constants	0	E	0	Capacitor acts like an open circuit

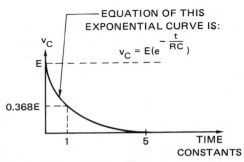

EQUATION OF THIS EXPONENTIAL CURVE IS:

$$v_C = E(e^{-\frac{t}{RC}})$$

Fig. 8-12 A capacitor discharging in an RC circuit

Table 8-5 shows the characteristics of the circuit in figure 8-10 when the capacitor is charging.

When the switch of this circuit is moved to position 2, the capacitor keeps its charge. The discharge cycle begins when the switch is moved to position 3. The graph of figure 8-12 shows that the capacitor discharge follows an exponential curve. Table 8-6 lists the values of the capacitor voltage during discharge for various time constant periods. For all practical purposes, it can be assumed that a capacitor is fully discharged in five time constants. Table 8-7 illustrates the characteristics of the circuit of figure 8-10 when the capacitor is discharging.

TABLE 8-6 Capacitor Voltages at Various Time
Constant Periods During the Discharging Cycle.

Time Constant Periods	Capacitor Voltage
0	E
1	0.368E
2	0.135E
3	0.049E
4	0.019E
5	0.007E

TABLE 8-7 Discharging Characteristics

Time	V_R	V_C	I	Remarks
0	–E	E	$\frac{E}{R}$	Capacitor acts like a short circuit
1 to 5 Time constants	Decreases	Decreases	Decreases	
5 Time constants	0	0	0	Capacitor acts like an open circuit

The capacitor voltage can be determined at any instant of time from the following equations for the exponential curves of figures 8-11 and 8-12:

$$v_C = E(1 - e^{-\frac{t}{RC}})$$

Eq. 8.15

This equation is the *charge* equation shown in figure 8-11 where

e = 2.718 (a constant)

t = any instant of time, in seconds

RC = time constant, in seconds

E = maximum capacitor voltage, in volts

v_C = capacitor voltage at any instant of time, in volts.

The exponential *discharge* curve shown in figure 8-12 has the following equation:

$$v_C = E(e^{-\frac{t}{RC}})$$

Eq. 8.16

Problem 9 A 0.01-μF capacitor is charged by a 30-V dc source through a 5-kΩ series resistor. Find v_C after the capacitor charges for a. 50 μs and b. 100 μs.

Solution a. Time constant = RC

= $(5 \times 10^3 \Omega)(0.01 \times 10^{-6}$ s)

= 50 μs, or 1 time constant

From Table 8-4, it can be determined that the capacitor voltage is 63.2% of the final voltage after one time constant.

Therefore, v_C = 0.632E

= (0.632)(30 V)

= 18.96 V

b. 100 μs = 2 time constant periods

From Table 8-4, it can be determined that the capacitor voltage is 86.5% of the final voltage after two time constant periods.

Therefore, v_C = (0.865)(30 V)

= 25.95 V

(R8-15) Calculate v_C after the capacitor described in Problem 9 charges for 150 μs.

**LABORATORY
EXERCISE 8-1
MEASUREMENT
AND ANALYSIS
OF AN RC
CIRCUIT**

PURPOSE   In completing this experiment, the student will

- observe the effect of the circuit time constant.
- take capacitor charge and discharge measurements.

**EQUIPMENT AND
MATERIALS**

1 Dc power supply
1 Storage oscilloscope
1 10:1 probe
1 Resistor, 0.51 MΩ, 2 W
1 Resistor, 10 kΩ, 2 W
1 Resistor, 1 MΩ, 2 W
2 Capacitors, 1 μF
1 Switch, single-pole, double-throw (SPDT)
1 Digital multimeter

PROCEDURE

1. Refer to the circuit shown in figure 8-13.
   a. Calculate the time constant of this circuit when the switch is in position 1. Calculate the time constant when the switch is in position 2.
   b. Set the power supply for an output of 10 V. Use the digital multimeter to record this voltage. Turn the power supply off.
   c. Construct the circuit shown in figure 8-13.
   d. With the switch in position 1 and the storage oscilloscope connected across the capacitor, turn the power supply on. Use the 10:1 probe for all oscilloscope measurements.
   e. Observe the capacitor charging pattern on the oscilloscope screen. Sketch the charging pattern on the data sheet.
   f. Measure the final voltage and the charging time from the oscilloscope display. How do these values compare with expected values?
   g. Turn the power supply off and observe the discharge pattern. Sketch this pattern on the data sheet. How does the pattern compare with expected results?

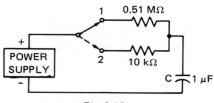

Fig. 8-13

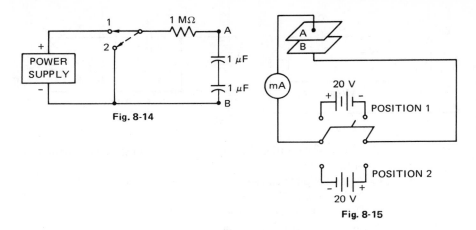

Fig. 8-14

Fig. 8-15

h. Repeat steps 1.d. through 1.g. with the switch in position 2.
i. Explain the reason for the difference in the results for steps 1.d. through 1.g. and step 1.h.
2. Construct the circuit shown in figure 8-14.
  a. With the switch in position 2, turn the power supply on and adjust its output for 10 V. Turn the power supply off.
  b. Connect the oscilloscope across terminals A and B.
  c. Calculate the time constant of the circuit.
  d. With the switch in position 1, turn the power supply on.
  e. Observe the capacitor charging pattern on the oscilloscope screen. Sketch the pattern on the data sheet.
  f. Measure the final voltage, the voltage after one time constant, and the voltage after two time constants. How do these values compare with expected values?
  g. Throw the switch to position 2. Sketch the discharge pattern on the data sheet.
  h. Measure the initial voltage, the final voltage, the voltage after one time constant, and the voltage after two time constants. How do these values compare with expected values?

EXTENDED STUDY TOPICS

1. When the capacitor in the circuit of figure 8-6 is charging, electrons move from plate A to plate B. During this same time period, do the positive charges move from plate B to plate A?
2. In figure 8-15, the DPDT switch is closed to position 1. What happens? The switch is then closed to position 2. Explain the results.
3. A 20-μF capacitor is fully charged. The voltage across its plates measures 120 V. Find the charge.
4. Determine the number of excess electrons on the negative plate of the capacitor in topic 3. (1 coulomb = 6.24 $\times$ 10^{18} electrons)

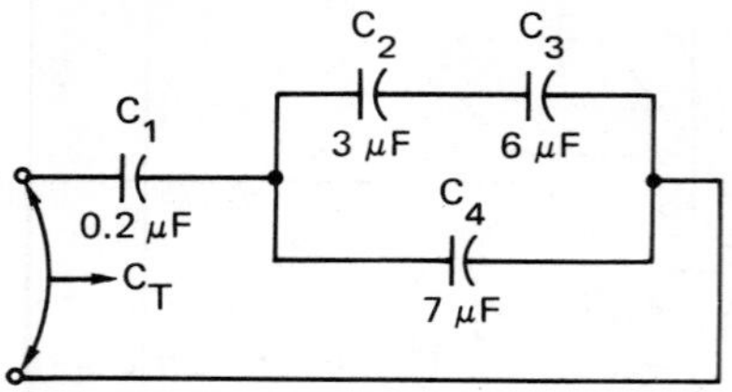

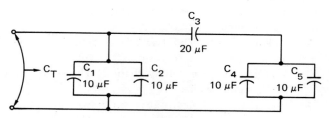

Fig. 8-16

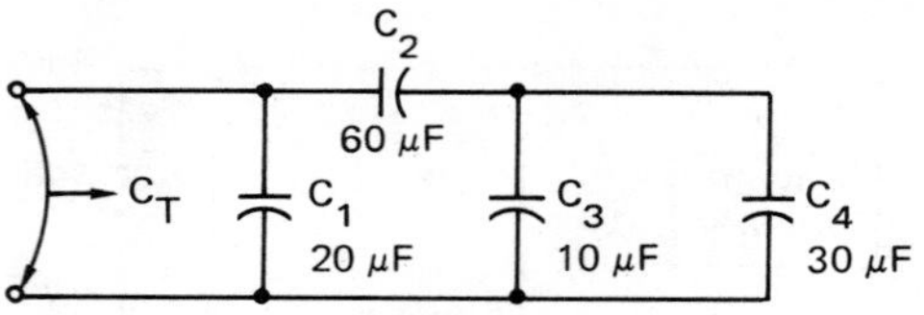

Fig. 8-17

5. How many positive charges are present on the positive plate of the capacitor in topic 3?
6. A parallel plate capacitor has the following parameters:
   Area of plates = 0.8 m^2
   Distance between plates = 10×10^{-3} m
   Dielectric = paraffin coated paper
   Find the capacitance.
7. A potential of 100-V dc is applied across the capacitor described in topic 6. Calculate its charge.
8. A 1 000-μF capacitor is charged to 100 V. Calculate: a. the energy stored by the capacitor and b. the charge on each plate of the capacitor.
9. Four capacitors are to be used. Their values are: 10 μF, 20 μF, 30 μF, and 40 μF. Determine the total capacitance if the four capacitors are connected a. in series and b. in parallel.
10. Find the total capacitance of the circuit shown in figure 8-16.
11. Find the total capacitance of the circuit shown in figure 8-17.
12. A circuit design requires 80 μF at 500 volts. Two 20-μF, 500-V capacitors are available. Calculate the rating of one more capacitor added to fulfill the requirements of the circuit. How are the three capacitors to be connected?
13. Refer to the circuit shown in figure 8-18.
    a. Calculate the total capacitance, C_T.
    b. C_3 becomes shorted. Calculate the new value of C_T.
14. In figure 8-19, each capacitor is charged to its final value. Calculate: a. the voltage across each capacitor, and b. the charge on each capacitor.
15. A 1 000-μF capacitor is charged by a 10-V dc source through a 100-Ω series resistor. Find a. the time constnat and b. v_C after the capacitor charges for 0.3 s.

Fig. 8-18

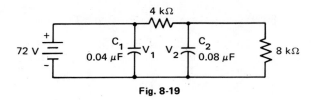

Fig. 8-19

16. Calculate v_C at the instant that the fully charged capacitor of topic 15 discharges for 0.3 s.
17. Find v_C for topic 15.b. using Equation 8.15.
18. Find v_C for topic 16 using Equation 8.16.

Chapter 9

Inductors and RL Circuits

OBJECTIVES After studying this chapter, the student will be able to

- discuss the principles of magnetism and the magnetic field.
- analyze the magnetic field arround a solenoid.
- discuss Lenz's Law and the concept of inductance.
- solve problems with inductors connected in series or in parallel.
- describe the charging and discharging characteristics of an RL circuit.

MAGNETISM The laws of magnetism are applied in many devices, including compasses, generators, motors, electronic equipment, solenoids, meters, speakers, headphones, and relays.

Compass. A compass is a magnetized needle which is free to rotate in a horizontal plane. One end of the compass needle, the north-seeking pole, or north pole, always points to the north magnetic pole of the earth. The north magnetic pole is close to the north geographic pole. The other end of the needle, the south-seeking pole, or south pole, always points to the south magnetic pole of the earth.

The forces between the magnetic poles are controlled by laws similar to those relating positive and negative charges. (Refer to Chapter 2.) These laws are stated as follows:

1. Unlike magnetic poles attract each other.
2. Like magnetic poles repel each other.

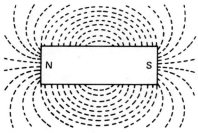

Fig. 9-1 Arrangement of iron filings when sprinkled over a paper placed on top of a bar magnet

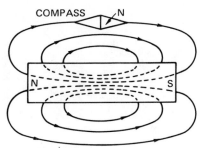

Fig. 9-2 Direction of magnetic lines of force around a bar magnet

(R9-1) What is the magnetic polarity of the north magnetic pole of the earth?

(R9-2) When the south pole of one magnet is brought near the south pole of another magnet, describe the resulting action.

THE MAGNETIC FIELD
A magnet will attract an iron paper clip, or an iron nail. It will not attract a brass screw, an aluminum screw, a piece of copper wire, or a piece of rubber. Any material that can be attracted by a magnet is called a *magnetic material*. Any material that cannot be attracted by a magnet is called a *nonmagnetic material*.

A magnet also affects an area surrounding it. This area is called the *magnetic field*. If iron filings are sprinkled on a sheet of paper placed over a bar magnet, the filings arrange themselves in an orderly manner, figure 9-1. The pattern of the filings indicates the magnetic lines of force around the magnet. The magnetic lines of force have two characteristics which are stated as follows:

1. Magnetic lines of force possess direction, as shown in figure 9-2. Note that the north pole of the compass points toward the south pole of the bar magnet.
2. Magnetic lines of force always form complete loops. These lines of force never intersect each other.

(R9-3) What is the direction of the magnetic lines of force a. within the bar magnet of figure 9-2, and b. outside the bar magnet of figure 9-2?

MAGNETS
A magnetic material can be made into a magnet by rubbing it against a magnet. This process is called *magnetizing*. Another method of magnetizing a material is to place it in a strong magnetic field. After the material is removed from the field, it retains a quantity of magnetic strength. This quantity is called the *residual magnetism*. Figure 9-3A shows a commercial magnet charger. Figures 9-3B, 9-3C, and 9-3D show instruments used to study magnetic fields.

Magnets are either temporary or permanent. They are classified by how long they hold their magnetic strength.

Fig. 9-3B Gaussmeter

Fig. 9-3A Magnet charger

Fig. 9-3C Fluxmeter

Fig. 9-3D Magnetometer

Temporary Magnets. Temporary magnets are made of soft iron. They lose their strength rapidly when the magnetizing force is removed. Electromagnets are temporary magnets which are used in devices such as relays, electric generators, and transformers.

Permanent Magnets. Permanent magnets may occur naturally, or they may be manufactured. A natural magnetic material is the iron ore known as magnetite, or lodestone. Hard metallic alloys such as steel or Alnico (an alloy of aluminum, nickel, iron, and cobalt) are used to make artificial permanent magnets. Once a permanent magnetic material is magnetized, it retains its magnetic properties for a long time. The magnetic properties of artificial magnets are much better than those of natural magnets. Permanent magnets are used in many devices, including microphones, permanent magnet motors, loudspeakers, and meter movements.

CURRENT-CARRYING CONDUCTOR

A magnetic field exists around a current-carrying conductor. The direction of the magnetic lines of force can be determined by applying the right-hand rule, figure 9-4. The following steps indicate how the rule is applied:

1. Grasp the conductor with the right hand, so that the thumb points in the direction of the conventional current flow.

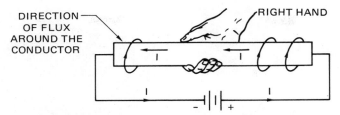

Fig. 9-4 Right-hand rule for determining the direction of the magnetic lines of force around a current-carrying conductor

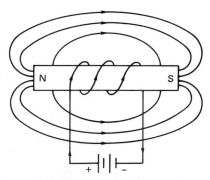

Fig. 9-5 Magnetic field around a solenoid carrying current

2. The fingers encircling the conductor are pointing in the direction of the magnetic lines of force around the conductor.

The magnetic field around a current-carrying conductor has the following characteristics:

1. As the current in the conductor increases, the lines of force increase.
2. When the current is steady, the magnetic lines of force become constant.
3. When the current in the conductor decreases, the lines of force decrease.
4. When the current becomes zero, the lines of force collapse back toward the center of the conductor and decrease to zero.

THE SOLENOID The magnetic field around a conductor can be concentrated by winding the conductor around a magnet, as shown in figure 9-5. The resulting device is called a *solenoid*. The pattern of the magnetic field around the solenoid is similar to the pattern of the magnetic field around the bar magnet in figure 9-2. The solenoid is known as an *electromagnet*. The direction of the magnetic lines of force can be determined by applying the right-hand rule for solenoids. The steps in applying this rule are as follows:

1. Grasp the solenoid with the right hand.
2. The fingers should encircle the solenoid in the direction of the *conventional current*.

(A) Stationary frame of a laminated solenoid

(B) Frame with electrical coil mounted in it

(C) Armature or plunger which moves inside the coil

(D) Assembled laminated linear solenoid

Fig. 9-6 A linear solenoid

A. A magnetic field surrounds single conductor when electric current flows through it

B. An enlarged magnetic field surrounds coil of wire with electric current flowing through it

C. Addition of metal frame further increases magnetic force

D. Seated plunger provides all metallic path for magnetic field of maximum strength

E. Position of armature under load before being energized. Current flowing through coil sets up magnetic field that pulls plunger into seated position

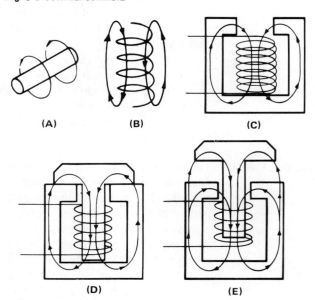

Fig. 9-7 Operation of a linear solenoid

Fig. 9-8 Hand-assembled toroids

3. The thumb then points in the direction of the magnetic lines of force through the center of the solenoid. That is, the thumb points in the direction of the north pole of the solenoid.

PRACTICAL SOLENOIDS AND RELAYS

An example of a practical solenoid is a device that converts electrical energy to linear motion. Figure 9-6 shows the main parts of a linear solenoid. Figure 9-7 illustrates how a linear solenoid operates.

A toroid type of winding is formed when a coil is wound around a ring-shaped core. The toroid winding keeps the magnetic field completely within the coil. Hand-assembled toroids are shown in figure 9-8. These toroids are not machine wound because they use too heavy a wire gauge or the winding configuration is too complicated.

Battery-powered lift magnets are shown in figures 9-9A and 9-9B. Figure 9-10 shows two types of electromagnetic relays: figure 9-10A is a four-pole,

Fig. 9-9A Battery-powered lift magnet

Fig. 9-9B Battery-powered lift magnet

Fig. 9-10A Four-pole, double-throw relay   Fig. 9-10B Two-pole, double-throw relay

double-throw relay and figure 9-10B is a two-pole, double-throw relay. An assortment of relays is shown in figure 9-11: A. general purpose relay, B. latching relay, C. time-delay relay, D. sequence relay, E. sensitive relay for power supplies, F. plug-in reed relay, G. solid-state relay, H. reversing contactor, I. hermetically sealed military relay, and J. general purpose plug-in relay with recycling motor timer.

EDDY CURRENT   Any core material that is an electric conductor will experience eddy currents induced by the changing magnetic field. A high-frequency alternating current induced in a solid metallic core can cause the core to become very hot. The heat results from the power loss ($I^2 R$ loss) that occurs as the eddy current flows through the resistance of the core material.

This power loss due to the eddy current is greatly reduced if the solid core is replaced with one consisting of thin layers of metal. These layers (laminations) are separated from each other by a thin coat of varnish. The laminations are arranged so that the direction in which the eddy current must flow is from lamination to lamination, through the coating of varnish on each lamination. Because

Fig. 9-11 Assortment of relays

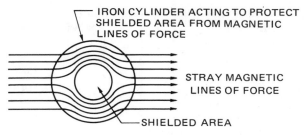

Fig. 9-12 Magnetic shielding

of the high resistance of the varnish, the eddy current in a laminated core, and thus the power loss, are very small.

MAGNETIC SHIELDING
There is no insulator for magnetic lines of force. However, a sensitive meter movement, or other device, can be protected from the stray magnetic fields by enclosing (shielding) it in a container made of high-permeability iron. The presence of iron causes the magnetic lines of force to change their direction to include the low-reluctance path through the iron. Thus, the device is free of stray magnetic lines of force, figure 9-12. Figure 9-13 shows a device that is used to protect data recorded on a cassette tape. This device provides both physical and magnetic protection. Figure 9-14 shows an assortment of shields designed for specific components and systems.

MUTUAL INDUCTION
Figure 9-15 shows the principle of mutual induction. Current flows into the left-hand, or primary, winding on the iron core. The current value is controlled by the magnitude of the source voltage, E. The current produces magnetic lines of force. These lines of force flow through the iron core and link the right-hand, or secondary winding. In other words, a changing voltage E produces a changing current I which, in turn, produces a changing magnetic field. Finally, this changing magnetic field induces an emf in the secondary winding. The term *mutual*

Fig. 9-13 Magnetic shield for cassette tape

Fig. 9-14 Assortment of shields designed for specific components and systems

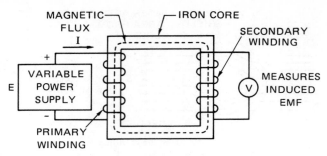

Fig. 9-15 Principle of mutual induction

induction is applied to the process of producing an emf in a secondary winding by a changing current in a primary winding. Mutual induction is the basis of transformer operation. Transformers will be studied in Chapter 18.

The magnitude of the emf induced in the secondary winding is directly proportional to both the number of turns and the rate of change of current. Faraday's Law of Electromagnetic Induction governs the magnitude of the induced emf and is expressed by the following equation:

$$e = \frac{N\Delta\phi}{\Delta t}$$

Eq. 9.1

where

e = induced emf, in volts

N = number of turns of the coil

ϕ = the flux, in webers (Wb). The Greek letter phi represents the flux. (One weber = 10^8 magnetic lines of force.)

t = time, in seconds, during which the conductor is cut by the magnetic field

$\Delta\phi$ = change in flux

Δt = change in time

Faraday's Law will be studied in more detail in Chapter 10.

Problem 1 The secondary winding of the transformer shown in figure 9-15 has 30 turns of wire. The core flux linking the secondary winding is increased from zero to 2 × 10^6 magnetic lines in 0.2 second. Find the magnitude of the voltage induced in the secondary.

Solution $\Delta\phi = \dfrac{2 \times 10^6 \text{ (magnetic lines)}}{10^8 \text{ (magnetic lines per Wb)}}$

$= 2 \times 10^{-2}$ Wb

$e = \dfrac{N\Delta\phi}{\Delta t}$

$= \dfrac{(30 \text{ turns})(2 \times 10^{-2} \text{ Wb})}{0.2 \text{ s}}$

$= 3$ V

Fig. 9-16 Audio transformers

(R9-4) For the transformer shown in figure 9-15, calculate the magnitude of the voltage induced in the 30-turn secondary, if 2×10^6 magnetic lines of force constantly link the secondary.

(R9-5) For the transformer shown in figure 9-15, the core flux linking the secondary winding is increased from zero to 2×10^6 magnetic lines of force in 0.1 second. Calculate the magnitude of the voltage induced across the 30-turn secondary.

The principle of mutual induction is applied to the practical devices shown in figures 9-16 and 9-17. The miniature transformers, figure 9-17, are used in printed circuit applications for power supplies, controls, and instrumentation.

The principle of mutual induction is also put to practical use in the ignition system of the automobile, figure 9-18. The voltage from the automobile battery supplies current to the primary winding of the ignition coil. The distributor

Fig. 9-17 Miniature transformers for printed circuit applications

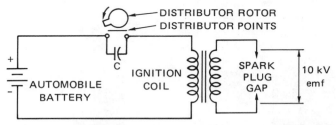

Fig. 9-18 Application of mutual induction principle to an automobile ignition coil

points switch this current on and off. The core flux linking the secondary winding of the ignition coil collapses rapidly when the distributor points open. This collapsing magnetic field induces a voltage of 10 kV. At the same time, a spark jumps across the spark plug gap. A capacitor, also known as a condenser, is connected across the distributor points. The capacitor absorbs energy when the primary circuit of the coil is opened. In this way, there is no arcing at the distributor points.

(R9-6)   If capacitor C in the circuit of figure 9-18 develops a short circuit, what happens to the operation of the automobile?

LENZ'S LAW   Lenz's Law is based on the principle that every action has an equal and opposite reaction. Thus, Lenz's Law is stated as follows:

> The direction of the induced emf is such that any current resulting
> from it develops a flux that opposes any change in the original flux.

In other words, the direction of the induced emf must always be such as to oppose any change in current. Figure 9-19 shows how Lenz's Law is applied:

1. Switch S is closed and the primary flux increases. It flows in a clockwise direction as required by the right-hand rule.
2. The secondary current, I_L, flows in the direction shown. The flux due to this current must oppose the primary flux, as required by Lenz's Law.
3. When switch S is opened, the primary flux collapses.

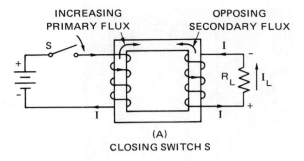

INCREASING
PRIMARY FLUX

OPPOSING
SECONDARY FLUX

(A)
CLOSING SWITCH S

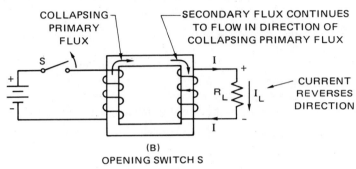

COLLAPSING PRIMARY FLUX

SECONDARY FLUX CONTINUES TO FLOW IN DIRECTION OF COLLAPSING PRIMARY FLUX

CURRENT REVERSES DIRECTION

(B)
OPENING SWITCH S

Fig. 9-19 Application of Lenz's Law

4. As a result, the secondary current, I_L, reverses its direction so that it can oppose the collapse of the primary flux. The secondary flux tries to keep the flow of primary flux in a clockwise direction.

SELF-INDUCTANCE

Self-inductance is the name applied to the generation of an emf in an electric circuit by a changing current. Lenz's Law states that the induced emf must oppose any change in current. Therefore, this induced voltage is called a counter emf (cemf), or back emf.

An electric circuit, or a component in the circuit, having self-inductance opposes any change in the current in the circuit or component. The letter symbol for inductance is L. The basic unit of inductance is the henry (H). A circuit component that is designed to use the property of inductance is called an *inductor*, a *choke coil*, or a *coil*. The following equation relates the factors that affect inductance:

$$E_L = L\frac{\Delta i}{\Delta t}$$

Eq. 9.2

where

L = inductance, in henries

E_L = counter emf induced in the coil, in volts

$\dfrac{\Delta i}{\Delta t}$ = the rate of change of current, in amperes per second

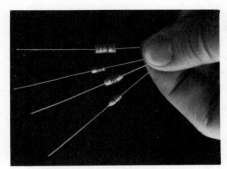

Fig. 9-20 Precision chokes

Four other factors also influence the inductance of a coil:
1. L is directly proportional to the square of the number of turns of the coil.
2. The magnetic field formed around a coil with an iron core is stronger than the field formed around a coil with an air core.
3. L is directly proportional to the cross-sectional area of the magnetic circuit.
4. L is inversely proportional to the length of the magnetic circuit.

(R9-7) The inductance of a coil is 4 H. Determine the inductance if the turns of the coil are doubled.

Problem 2 The current in a coil changes at the rate of 400 mA per second. A counter emf of 0.10 V is induced in the coil. Find the inductance of the coil.

Solution

$$E_L = L\frac{\Delta i}{\Delta t}$$

$$L = \frac{E_L}{\frac{\Delta i}{\Delta t}}$$

$$= \frac{0.10\ V}{400 \times 10^{-3}\ A/s}$$

$$= 0.25\ H$$

(R9-8) The current in a 20-H inductor changes from 10 A to 5 A in 100 ms. What value of cemf is induced in the inductor?

A number of common inductors are shown in figures 9-20 and 9-21. Figure 9-20 shows molded and epoxy precision chokes with tolerances as low as 1%. A radio frequency coil and molded shielded radio frequency chokes are shown in figure 9-21.

INDUCTORS IN SERIES AND PARALLEL

Figure 9-22A shows three iron core inductors in series. The same current flows through each inductor. Therefore, each coil experiences the same rate of change of current. The total inductance equals the sum of the individual inductances, as shown in the following equation:

$$L_T = L_1 + L_2 + L_3 \qquad \text{Eq. 9.3}$$

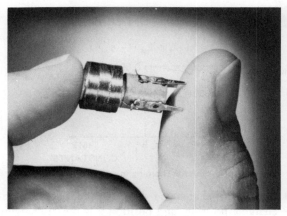

Fig. 9-21A RF coil

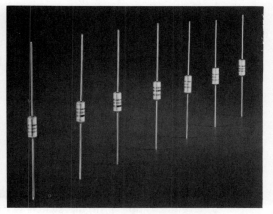

Fig. 9-21B Molded shielded RF chokes

Three air core inductors are connected in parallel in figure 9-22B. Each inductor must develop the same counter emf. The equivalent inductance of inductors in parallel is.

$$L_{eq} = \cfrac{1}{\cfrac{1}{L_1} + \cfrac{1}{L_2} + \cfrac{1}{L_3}}$$ Eq. 9.4

When only two inductors are connected in parallel, Equation 9.4 reduces to the following form:

$$L_{eq} = \frac{L_1 \times L_2}{L_1 + L_2}$$ Eq. 9.5

When any number of inductors with equal values are connected in parallel, the total inductance is:

$$L_{eq} = \frac{L}{N}$$ Eq. 9.6

where

    L = inductance of one inductor, in henries

    N = number of inductors in parallel

(R9-9)   Find the equivalent inductance of four inductors connected in parallel. Each inductor has a value of 8 H.

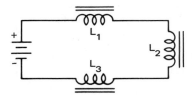

(A) Iron core inductors in series

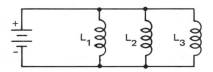

(B) Air core inductors in parallel

Fig. 9-22 Inductor configurations

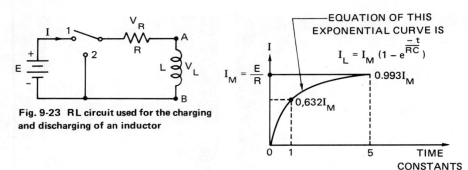

Fig. 9-23 RL circuit used for the charging and discharging of an inductor

Fig. 9-24 An inductor charging in an RL circuit

RL CIRCUITS   Figure 9-23 shows a circuit designed to charge and discharge an inductor. Position 1 of the switch is the charging position, and position 2 is the discharging position. Current through an inductor cannot change instantaneously. Rather, the rise of current follows the exponential curve shown in figure 9-24. The student should note the similarity to the exponential curve for the charging of a capacitor, as shown in Chapter 8.

The *time constant* is a period of time, in seconds, required for the current to reach 63.2% of its final value I_M. Table 9-1 lists the voltages and currents obtained at various time constants for the charging of an inductor. The time constant can be calculated from the following equation:

$$\text{Time constant} = \frac{L}{R} \qquad\qquad \text{Eq. 9.7}$$

where

the time constant is in seconds
L is in henries
R is in ohms

It is assumed that the current in a coil reaches its final value (I_M) in five time constants.

(R9-10)   A 4-Ω resistor is connected in series with an 8-H coil. The voltage of the power supply is 100-V dc. Calculate the time constant.

Questions (R9-11), (R9-12), and (R9-13) refer to question (R9-10).

TABLE 9-1 Energizing Characteristics for an RL Circuit

Time	V_R	V_L	I	Remarks
0	V_R	E	0	L acts as an open circuit
1 Time constant	0.632E	0.368E	$0.632I_M$	
1 to 5 Time constants	Increases	Decreases	Increases	
5 Time constants	E	0	$I_M = \dfrac{E}{R}$	L acts as a short circuit

(R9-11)   What is the voltage across the coil after one time constant?

(R9-12)   What is the voltage across R after one time constant?

(R9-13)   How long will it take for the current through the coil to reach its maximum value (for practical purposes)?

When the switch in the circuit of figure 9-23 is moved to position 2, the coil discharges through resistor R. The coil acts as a source of emf. As required by Lenz's Law, the polarity of V_L causes current to flow in the same direction as it did when the switch was in position 1. The current continues to flow until $V_L = 0$ and $V_R = 0$.

(R9-14)   For the coil shown in figure 9-23, what are the polarities of terminals A and B during the discharge cycle?

ENERGY STORED IN AN INDUCTOR

Energy is stored in an inductor in the magnetic field. Energy is stored only as long as the current continues to flow and the magnetic field is maintained. Energy stored by an inductor differs from energy stored by a capacitor. When a capacitor is charged, the charging current becomes zero. Energy is stored in a static form by a capacitor in the electric field. In an inductor, energy is stored in the magnetic field in a dynamic form.

The energy stored by an inductor can be calculated from the following equation:

$$W = 1/2\ LI^2 \qquad \text{Eq. 9.8}$$

where

W = stored energy, in joules
L = inductance, in henries
I = steady-state value of current, in amperes

Problem 3   An inductor has an inductance of 8 H and a resistance of 2 Ω. Calculate the energy stored in the magnetic field of the inductor if the inductor is connected to a 30-V dc power supply.

Solution $I = \dfrac{E}{R}$

$= \dfrac{30\ V}{2\ \Omega}$

$= 15\ A$

$W = 1/2 LI^2$

$= 1/2 \times 8\ H \times (15\ A)^2$

$= 900\ J$

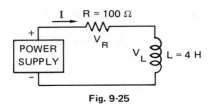

Fig. 9-25

LABORATORY EXERCISE 9-1 MEASUREMENT AND ANALYSIS OF AN RL CIRCUIT

PURPOSE   By completing this exercise, the student will study the characteristics of an RL circuit.

EQUIPMENT AND MATERIALS
1 Dc power supply
1 Resistor, 100 Ω, 2 W
1 Digital multimeter
1 Inductor, 4 H

PROCEDURE
1. Connect the circuit shown in figure 9-25.
   a. Turn the power supply to *ON*. Adjust the output voltage to measure 20 V.
   b. Measure I, V_R, and V_L.
2. Perform the following calculations, and answer the following questions:
   a. What is the time constant of the RL circuit?
   b. Calculate the value of the current I, the voltage V_R, and the voltage V_L. Compare the calculated values with the values measured in step 1.b.
   c. What is the value of V_L at the instant the circuit is energized?
   d. What is the value of V_L after one time constant?

EXTENDED STUDY TOPICS
1. The core flux linking the secondary winding of a transformer is increased from zero to 0.05 Wb in four seconds. Calculate the number of turns required in the secondary winding to induce a voltage of 6 V.
2. How long will it take the current in a 4-H inductor to rise from zero to 10 A if the cemf induced in the inductor is 5 V?
3. Calculate the effective inductance of three inductors when connected a. in series, and b. in parallel. The values of the inductors are: $L_1 = 20$ mH, $L_2 = 40$ mH, and $L_3 = 50$ mH.
4. Two inductors are connected in parallel. If L_1 equals 4 H and L_2 equals 8 H, what is the equivalent inductance?
5. What inductance must be placed in parallel with an inductance of 60 mH to reduce the equivalent inductance to 20 mH?

6. A series circuit has a resistance of 10 Ω and an inductance of 4 H. Once the switch in this circuit is closed, how long will it take the current to reach its final value?
7. A solenoid has an inductance of 8 H. What is the resistance of this solenoid if it takes 0.2 second for the current to reach its final value? (The student should note that coils and solenoids also have inherent internal series resistance.)
8. Determine the amount of energy stored in a 100-mH coil if the current is 5 A.
9. A choke coil has an inductance of 40 H and a resistance of 20 Ω. The coil is connected to a 40-V dc power supply.
   a. Calculate the time constant.
   b. How long does it take the current to reach its final value?
   c. What is the initial current?
   d. What is the final steady-state current?
10. What is the value of the cemf in topic 9, at the instant the choke coil is energized?

Chapter 10

Alternating Current

OBJECTIVES   After studying this chapter, the student will be able to

- discuss the sinusoidal waveform.
- use alternating current and voltage terminology correctly.
- describe the meaning of the frequency and period of a sinusoidal wave.
- explain the concept of the effective value of a sinusoidal wave.
- diagram and explain the phase relationships of sinusoidal waves of the same frequency.
- diagram nonsinusoidal waveforms and explain the meaning of their average value.

DC VOLTAGES   A dc current always flows in the same direction. A dc voltage has a constant polarity. Figure 10-1 shows the graph of the voltage from a +50-V dc source.

(R10-1)   What is the value of the dc voltage when a. t = 5 seconds, b. t = 10 seconds, and c. t = 15 seconds?

If the leads to the power supply shown in figure 10-1 are reversed, the voltage measured by the voltmeter is negative. The graph for this reversed polarity condition is shown in figure 10-2.

(R10-2)   For the dc voltage shown in figure 10-2, what is the value at a. t = 5 seconds, b. t = 10 seconds, and c. t = 15 seconds?

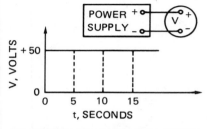

Fig. 10-1 Positive dc voltage versus time

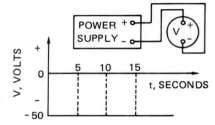

Fig. 10-2 Negative dc voltage versus time

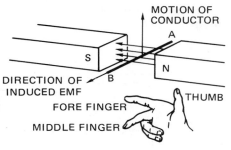

MOTION OF
CONDUCTOR

A

S

N

B

DIRECTION OF
INDUCED EMF

THUMB

FORE FINGER

MIDDLE FINGER

Fig. 10-3 Fleming's right-hand rule

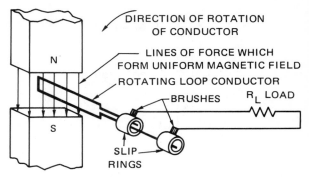

DIRECTION OF ROTATION
OF CONDUCTOR

N

LINES OF FORCE WHICH
FORM UNIFORM MAGNETIC FIELD

ROTATING LOOP CONDUCTOR

BRUSHES R_L LOAD

S

SLIP
RINGS

Fig. 10-4 Basic ac generator

ELECTRO-MAGNETIC INDUCTION
When a conductor is moved in a magnetic field, a voltage is induced in the conductor. The direction of the induced voltage follows Fleming's Right-hand Rule. According to this rule:

1. The forefinger points in the direction of the lines of flux.
2. The thumb points in the direction of relative motion of the conductor. (Note: a voltage is also induced if the conductor is stationary and the magnetic field moves.)
3. The middle finger points in the direction of the induced voltage.

Figure 10-3 shows how this rule is applied. When the conductor is moved upward, a voltage is induced in the conductor in the direction A to B.

(R10-3) If the conductor in figure 10-3 is moved in a downward direction, what is the direction of the induced voltage?

The magnitude of the induced voltage depends upon several factors which are related as shown by the following equation:

$$e = KN\frac{\Delta\phi}{\Delta t}$$

Eq. 10.1

where

e   = induced voltage, in volts
K   = a constant which converts magnetic units to volts
N   = number of turns of the conductor
$\frac{\Delta\phi}{\Delta t}$ = the rate of change of flux with respect to time.

Equation 10.1 is based on Faraday's discovery of the principle of electromagnetic induction.

ROTATING GENERATOR
Figure 10-4 shows a basic ac generator. The rotating conductor is formed into a loop. This loop is rotated at a constant speed in a uniform magnetic field. As the conductor rotates, it delivers its induced voltage to the load terminals through the slip rings and brushes.

The two sides of the loop cut across magnetic lines of force in opposite directions as the loop rotates. The voltages induced in each side of the loop are additive. Therefore, the total emf is twice the emf induced in each side of the loop.

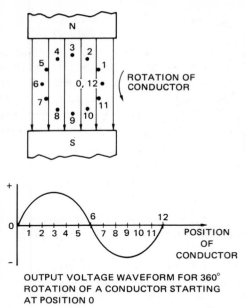

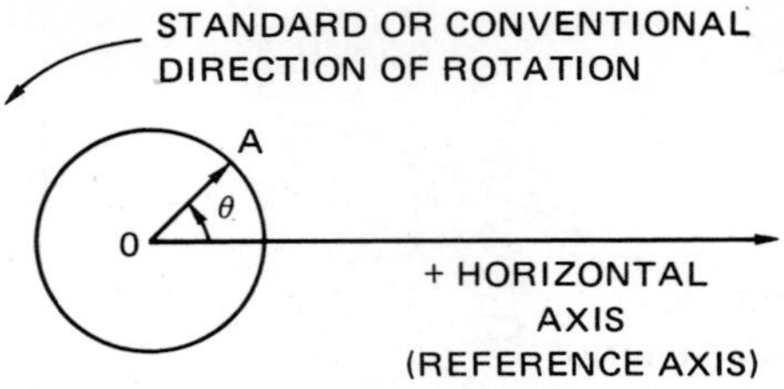

Fig. 10-6 Rotating vector, or phasor, representation of a rotating conductor

OUTPUT VOLTAGE WAVEFORM FOR 360°
ROTATION OF A CONDUCTOR STARTING
AT POSITION 0

Fig. 10-5 Induced emf for one cycle (360°)

ANALYSIS OF THE INDUCED VOLTAGE

Figure 10-5 shows one side of the loop of figure 10-4 as it rotates through twelve positions at 30° intervals to complete a 360° revolution. At position 0, the conductor is moving parallel to the magnetic lines of force. No lines of force are being cut by the conductor at this position and the voltage induced is zero. The conductor cuts some lines of force as it moves to position 1. The resulting induced voltage is determined by Equation 10.1. The rate at which the conductor cuts magnetic lines increases as it moves to position 2 (60°). Thus, the induced voltage is greater. At position 3 (90°), the conductor is cutting across the magnetic lines of force at right angles. The rate of cutting magnetic lines of force is greatest at this point. This means that the maximum emf is induced when the conductor is at 90°.

From positions 3 to 6 (90° to 180°), the rate of cutting decreases and the induced voltage decreases. When the conductor is at position 6, the induced voltage is zero.

From positions 6 to 12 (180° to 360°), the conditions of positions 1 to 6 are repeated. However, the conductor is now cutting across the magnetic lines of force in the opposite direction. Thus, the polarity of the induced voltage is reversed. The output voltage waveform shown in figure 10-5 is the result of a complete 360° rotation of the conductor.

(R10-4) If the prime mover rotates the conductor of figure 10-5 at twice the speed, what is the effect on the magnitude of the induced voltage?

(R10-5) Explain the reason for the answer given to question (R10-4).

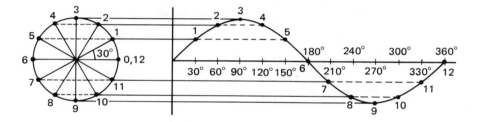

Point	Angle	Sine of Angle	Instantaneous Value of Voltage
0	0°	0	0
1	30°	0.5	50
2	60°	0.866	86.6
3	90°	1.0	100
4	120°	0.866	86.6
5	150°	0.5	50
6	180°	0	0
7	210°	−0.5	− 50
8	240°	−0.866	− 86.6
9	270°	−1.0	−100
10	300°	−0.866	− 86.6
11	330°	−0.5	− 50
12	360°	0	0

Fig. 10-7 Development of a sine wave by a rotating vector (phasor) having a value of 100 V

ROTATING VECTOR OR PHASOR It is possible to analyze the nature of the induced voltage. To do this, the rotating conductor shown in figures 10-4 and 10-5 is represented by a rotating vector, or phasor, indicated as OA in figure 10-6. Point A is free to rotate around the pivot, point O, at the origin. The standard direction of rotation is counterclockwise, starting from the reference axis which is the positive horizontal axis.

The emf induced by the rotating vector or phasor is directly proportional to the sine of the angle through which the vector rotates from the reference axis. In figure 10-7, a sine wave is developed as a vector of 100 volts rotates through 360°. The voltage at any instant is determined from the following equation:

$$e = E_m \sin\theta \qquad \text{Eq. 10.2}$$

where

e = instantaneous value of voltage, in volts

E_m = maximum value of voltage, in volts

θ = angle of rotation of the vector from the reference axis

The student should note the similarity between the sinusoidal wave shown in figure 10-7, and the output waveform shown in figure 10-5 for the rotating generator.

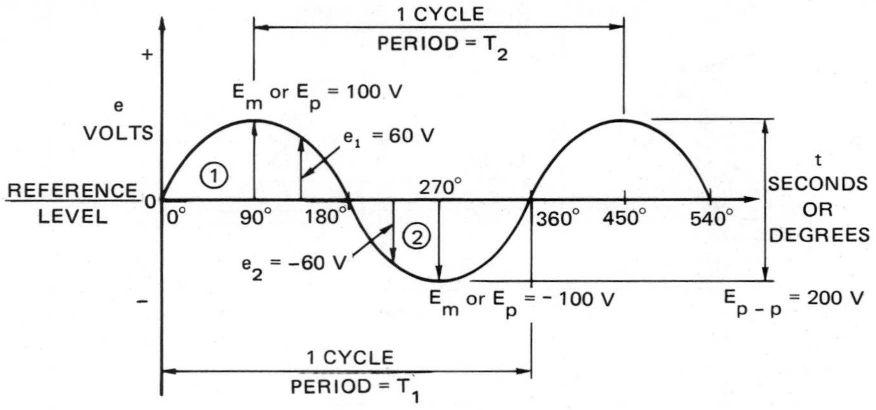

Fig. 10-8 The sine wave

Problem 1   Determine the instantaneous value of voltage when the angle θ in figure 10-8 equals $300°$.

Solution $e = E_m \sin \theta$
$= 100 \text{ V} \sin 300°$
$= -(100 \text{ V} \sin 60°)$
$= -(100 \text{ V})(0.866)$
$= -86.6 \text{ V}$

(R10-6)   For the wave shown in figure 10-8, what are the angular values in degrees at which magnitude of the wave reaches its maximum?

Problem 2   At what degree values does the magnitude of the voltage wave of figure 10-8 reach one-half its maximum value?

Solution $e = E_m \sin \theta$

$\dfrac{e}{E_m} = \sin \theta$

$\sin \theta = \dfrac{e}{E_m} = 0.5$

$\theta = \sin^{-1} 0.5 = \text{arc } \sin (0.5)$
$\quad = 30°$

By inspecting the waveform shown in figure 10-8, it can be seen that the instantaneous value of the wave is one-half the maximum value at four points in time. These points are listed as follows:

$30°$
$180° - 30°$ (or $150°$)
$180° + 30°$ (or $210°$)
$360° - 30°$ (or $330°$)

DEFINITIONS   An ac voltage is any voltage that varies in amplitude and reverses its polarity with respect to time. A periodic ac voltage repeats itself in fixed intervals of time.

The voltage generated by the rotating vector in figure 10-7 is both a sinusoidal voltage and a periodic ac voltage. That is, it varies in amplitude and polarity with respect to time.

Instantaneous voltage values are written as lowercase letters, such as e_1 and e_2, figure 10-8.

(R10-7)   What is the difference between e_1 and e_2?

The maximum value of voltage is labeled E_m. This value is also called the peak value and is labeled E_p. The maximum value is defined as the value from the reference level to the peak. For a sine wave, the peak-to-peak value is equal to twice the maximum value:

$$E_{p-p} = 2\, E_m$$

(R10-8)   For the curve shown in figure 10-8, what is the value of E_{p-p}?

R10-9)   What is the difference between the two values of E_m shown in figure 10-8?

The time interval (T) in seconds between any successive repetitions of a periodic waveform is called the *period*. For the wave in figure 10-8, the period is the time interval T_1 or T_2.

The portion of the waveform contained in one period of time is called a *cycle*.

(R10-10)   How many degrees does the period of a sinusoidal wave contain?

(R10-11)   How many cycles are represented in figure 10-8?

An *alternation* is that part of the waveform contained in one-half cycle. Each sinusoidal wave contains two alternations, a positive alternation and a negative alternation.

The number of cycles that occur in one second is called the frequency (f). The SI frequency unit is the hertz (Hz). The standard frequency of the electric power in the United States is 60 Hz.

This value may be checked using the frequency meters shown in figure 10-9.

Fig. 10-9A Frahm® frequency meter

Fig. 10-9B Frequency meter

Frequency and period have an inverse relationship as shown by the following equations:

$$f = \frac{1}{T} \qquad \text{Eq. 10.3}$$

$$T = \frac{1}{f} \qquad \text{Eq. 10.4}$$

where

T = time interval, in seconds
f = frequency, in Hz

Problem 3   a. What is the period of the standard electric power voltage wave in the United States?
b. How long does it take this wave to reach its maximum value from its zero reference level?

Solution   a. f = 60 Hz

$$T = \frac{1}{f}$$

$$= \frac{1}{60} \text{ s} = 16.7 \text{ ms}$$

b. The wave reaches its maximum value from the zero level in one-quarter cycle.

$$1/4 \times \frac{1}{60} \text{ s} = \frac{1}{240} \text{ s} = 4.17 \text{ ms}$$

ALTERNATING CURRENT   An alternating voltage can be supplied from a number of sources, including an ac generator, a function generator, an audio oscillator, a signal generator, and an ac power supply. As shown in figure 10-10, a sinusoidal ac voltage applied to a resistive load results in a sinusoidal ac current. If the equation of the ac voltage is $e = E_m \sin \theta$, the equation for the resulting ac current is:

$$i = I_m \sin \theta \qquad \text{Eq. 10.5}$$

Both waves are at the same frequency.

During the first half of the ac voltage cycle, or the first alternation, the polarity of the ac power supply is as shown in figure 10-10A. The current flows through the load R from A to B. During the second half of the ac voltage cycle, or the second alternation, the polarity of the ac power supply reverses, figure 10-10B. The current reverses through the load R and flows from B to A.

DIRECTION OF CURRENT DURING
FIRST ALTERNATION OF VOLTAGE

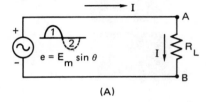

(A)

CURRENT DIRECTION DURING
SECOND ALTERNATION OF VOLTAGE

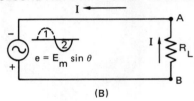

(B)

Fig. 10-10 Ac current in a resistive load

(R10-12) Is the current through load R_L of figure 10-10 constant during each alternation?

(R10-13) Explain the answer to question (R10-12).

EFFECTIVE VALUES

During the positive and negative alternations of the current cycle in figure 10-10, power is delivered at each instant of time to the resistive load. The magnitude of the ac current varies according to Equation 10.5. Therefore, the power delivered at each instant varies.

The equivalent dc value of a sinusoidal current, or voltage, is a constant value which produces the same power dissipation as the varying sinusoidal value. The equivalent dc value is called the *effective value*, or rms value. Rms is the abbreviation for root-mean-square, which describes how the effective value is derived mathematically. The following equations show the relationship between the effective and maximum values of a sinusoidal waveform:

$$I \text{ (equivalent dc)} = I \text{ (effective)} = 0.707 \, I_m \qquad \text{Eq. 10.6}$$
$$I_m = \sqrt{2} \, I = 1.414 \, I \qquad \text{Eq. 10.7}$$
$$E \text{ (equivalent dc)} = E \text{ (effective)} = 0.707 \, E_m \qquad \text{Eq. 10.8}$$
$$E_m = \sqrt{2} \, E = 1.414 \, E \qquad \text{Eq. 10.9}$$

Generally, the subscripts are not used for the effective values of voltage and current. Most ac instruments are calibrated to indicate the effective values of voltage and current. The maximum and peak-to-peak values may be determined with the use of an oscilloscope.

Problem 4

An ac voltmeter reads 120 V. a. What voltage does this value represent? b. What is the peak value of the voltage wave? c. What is the peak-to-peak value of the voltage wave?

Solution

a. 120 V is the equivalent dc voltage.

That is, it is the effective, or rms, voltage.

b. $E_m = \sqrt{2} \, E$
$= (1.414)(120 \text{ V})$
$\cong 170 \text{ V}$

Alternate solution:

$E_m = \dfrac{E}{0.707}$

$= \dfrac{120 \text{ V}}{0.707}$

$\cong 170 \text{ V}$

c. $E_{p-p} = 2 \, E_m$
$= 2(170 \text{ V})$
$\cong 340 \text{ V}$

The meter shown in figure 10-11 can be used to measure rms amperes, rms volts, period, and frequency. Figure 10-12 shows an alternating current having a peak value of 10 amperes and its equivalent rms value of 7.07 A. The student should note in figure 10-12 that the sinusoidal current, with a peak value of 10 amperes, *changes* in magnitude and direction. On the other hand, the effective, or rms, current has a *constant* value which is the equivalent value.

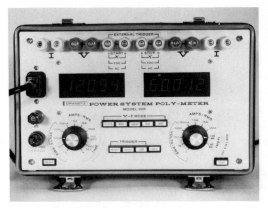

Fig. 10-11 Power system poly-meter

(R10-14) A sinusoidal current has a peak value of 15 mA. What is the multimeter indication for this current?

(R10-15) The voltage output of an ac power supply is shown by an oscilloscope to be 400 V peak-to-peak. What is the multimeter indication for this voltage?

ANGULAR Sinusoidal voltages and currents may be expressed as functions of time, or degrees.
VELOCITY Another unit of measurement is the radian. As shown in figure 10-13, a *radian* is the angle cut by an arc equal in length to the radius of the circle. The rotating vec-

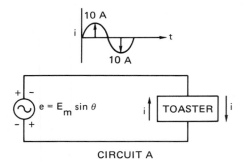

CIRCUIT A

The sinusoidal current has a peak valve of 10 A. The current direction changes with each alternation. (Note the *double set* of polarity marks on the ac generator.)

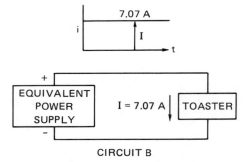

CIRCUIT B

The rms or effective valve of current equals 7.07 A. This current flows in *one* direction only. The effect of the current in this circuit is the same as the effect of the sinusoidal current in circuit A. (Note the *single set* of polarity marks on the equivalent power supply.) The alternating current with a 10-A peak valve produces the same degree of toasting as the 7.07-A direct current from the equivalent power supply.

Fig. 10-12 A sinusoidal ac current and its equivalent effective value (showing current flow through a toaster)

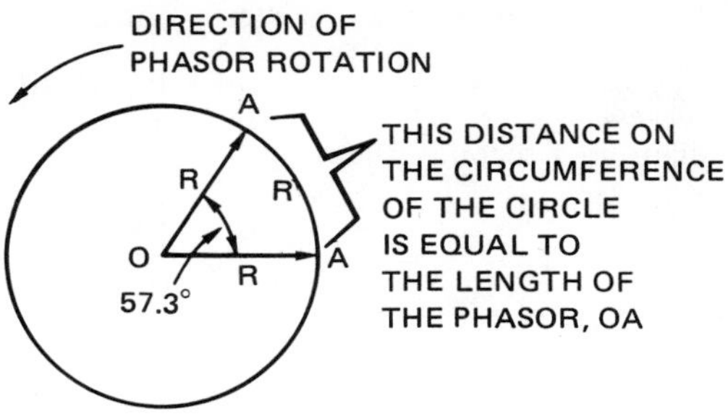

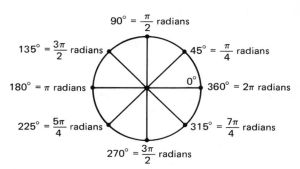

DIRECTION OF PHASOR ROTATION

THIS DISTANCE ON THE CIRCUMFERENCE OF THE CIRCLE IS EQUAL TO THE LENGTH OF THE PHASOR, OA

Fig. 10-13 The radian

Fig. 10-14 Comparison of radians and degrees for 360° cycle

tor, or phasor, in figure 10-13 is the radius of a circle. The circumference of a circle is equal to $2\pi R$ and there are 2π radians in one complete cycle (360°).

$$2\pi \text{ radians} = 360° \qquad\qquad \text{Eq. 10.10}$$

$$1 \text{ radian } = \frac{360°}{2\pi} = 57.3° \qquad\qquad \text{Eq. 10.11}$$

By setting up the simple proportion, $\dfrac{\text{radians}}{\pi} = \dfrac{\text{degrees}}{180°}$, the following equations relating degrees and radians are obtained:

$$\text{Radians} = \frac{\pi}{180°} \times \text{degrees} \qquad\qquad \text{Eq. 10.12}$$

$$\text{Degrees} = \frac{180°}{\pi} \times \text{radians} \qquad\qquad \text{Eq. 10.13}$$

Problem 5 a. Convert $\dfrac{3\pi}{2}$ radians to degrees.

b. Convert 60° to radians.

Solution a. Degrees $= \dfrac{180°}{\pi} \times$ radians

$$= \frac{180°}{\pi} \times \frac{3\pi}{2}$$

$$= 270°$$

b. Radians $= \dfrac{\pi}{180°} \times$ degrees

$$= \frac{\pi}{180°} \times 60°$$

$$= \frac{\pi}{3} = 1.047 \text{ radians}$$

(R10-16) Express: a. 30° in radians, and b. $\dfrac{\pi}{2}$ radians in degrees.

Figure 10-14 is a comparison of radians and degrees for a full 360° cycle.
The rotating vector of figure 10-6 rotates at a velocity that is called the *angular*

velocity. The symbol for angular velocity is the lower case Greek letter omega, ω.

$$\omega = \frac{\text{angular distance, in radians}}{\text{time, in seconds}} \qquad \text{Eq. 10.14}$$

If a phasor rotates $360°$, or 2π radians, in T seconds, then:

$$\omega = \frac{2\pi}{T} \qquad \text{Eq. 10.15}$$

Since $T = \frac{1}{f}$,

then

$$\omega = \frac{2\pi}{\frac{1}{f}}$$

$$\omega = 2\pi f \qquad \text{Eq. 10.16}$$

where

ω = angular velocity, in radians per second

f = frequency, in Hz.

However, $\theta = \omega t$ so that $e = E_m \sin \theta = E_m \sin \omega t$.

Problem 6   Express a voltage of 120 V at a frequency of 60 Hz using angular velocity time notation.

Solution   E   = 120 V

E_m = (120 V)(1.414)

$\cong$ 170 V

e   = 170 V sin ωt

ω = $2\pi f$

= (2)(3.1416)(60 Hz)

= 377 radians per second

e   = 170 V sin 377 t

(R10-17)   Solve Problem 6 for a frequency of 25 Hz.

PHASE RELATIONSHIPS   Figure 10-15 shows the relationship between the voltage and current waveforms of a resistive circuit. These waveforms are in phase, or in step, because they pass through zero at the same time and reach maximum values at the same time. The equations for the two waves in angular velocity time notation are:

$$e = E_m \sin \omega t$$
$$i = I_m \sin \omega t$$

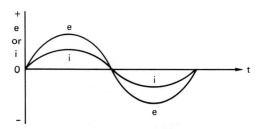

Fig. 10-15   Current and voltage in phase

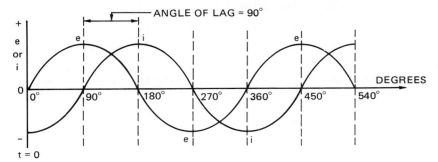

Fig. 10-16 I lags E by 90°

In figure 10-16, there is a lagging relationship between the current and voltage waveforms. Voltage is the standard used for reference. Therefore, this configuration of waveforms is said to be *lagging* because the current passes through zero after the voltage passes through zero as they both increase in the positive direction. The amount by which one wave leads, or lags, another wave is called the *phase difference*, or phase angle, between the two waves. The phase angle meter shown in figure 10-17 can be used to measure the phase angle, in degrees, between the waves.

(R10-18)   What is the phase angle between the waveforms shown in figure 10-16?

For the two waves in figure 10-16, the equations in time notation are:

$$e = E_m \sin \omega t$$
$$i = I_m \sin(\omega t - 90°)$$

A leading relationship between current and voltage is shown in figure 10-18. The equations for these waves in time notation are:

$$e = E_m \sin \omega t$$
$$i = I_m \sin (\omega t + 90°)$$

(R10-19)   What is the phase angle between the waveforms shown in figure 10-18?

The student must realize and remember that phase is meaningful only when comparing two waves having the same frequency.

Fig. 10-17 Phase angle meter

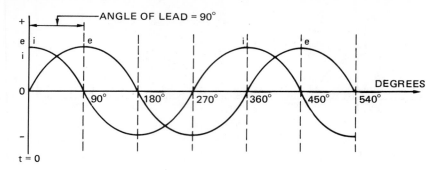

Fig. 10-18 I leads E by 90°

Problem 7   What is the phase relationship between the following sinusoidal waveforms?

$$e = 100 \sin (\omega t + 30°)$$
$$i = 50 \sin (\omega t + 70°)$$

Solution   When t = 0, the voltage wave is +30° from the reference axis, and the current wave is +70° from the reference axis. Since the standard direction of rotation is counterclockwise, the current wave is +40° ahead of the voltage wave. In other words, the current leads the voltage by 40°.

(R10-20)   Determine the phase relationship between the following sinusoidal waveforms.

$$e = 10 \sin (\omega t - 10°)$$
$$i = 5 \sin (\omega t - 40°)$$

NONSINUSOIDAL WAVEFORMS

Any waveform that is not a pure sinusoidal waveform is termed a nonsinusoidal wave. A dc voltage, or a dc current, are two examples of nonsinusoidal waveforms.

A function generator is shown in figure 10-19. This device electronically generates the three waveforms shown in figure 10-20. Wave A is a pure sinusoidal ac waveform. Wave B is an alternating square wave, but it is a nonsinusoidal, or complex wave. Mathematically, it can be shown that this wave consists of many sinusoidal waves of harmonically related frequencies. Since this wave is symmetrical around the reference axis, it has no dc component. Wave C

Fig. 10-19 Function generator which produces the three waveforms shown in figure 10-20

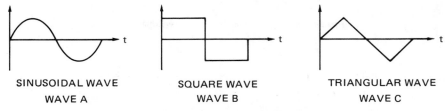

SINUSOIDAL WAVE   SQUARE WAVE   TRIANGULAR WAVE
WAVE A   WAVE B   WAVE C

Fig. 10-20 Waveforms generated by the function generator of figure 10-19

is a triangular alternating wave, but it too is a nonsinusoidal, or complex, wave. Mathematically it can be shown that this wave is the composite of many sinusoidal waves of different frequencies.

(R10-21)   Is there a dc component in wave C of figure 10-20?

The dc component of a waveform is called the *average value*, or *dc value*. The function generator shown in figure 10-19 has a *DC OFFSET* control. A dc component can be added to any of the three output waveforms generated by this device. The dc component of a wave may be calculated from the following equation:

$$E_{dc}, \text{ or } E_{avg} = \frac{\text{Net Area For One Period}}{T}$$

Problem 8   Calculate the dc component of the voltage waveform shown in figure 10-21.
Solution   The period of this offset wave is 4 seconds.
Positive area = 100 V × 2 s = +200 volt seconds.
Negative area = –(50 V × 2 s) = –100 volt seconds.
Net Area for One period = +200 volt seconds + (–100 volt seconds)
= +100 volt seconds

$$E_{dc}, \text{ or } E_{avg} = \frac{\text{Net Area}}{T}$$
$$= \frac{+100 \text{ volt seconds}}{4 \text{ seconds}}$$
$$= +25 \text{ V}$$

The proper operation of electronic circuits requires a unidirectional wave, figure 10-22. This wave is a composite, or complex, wave consisting of a dc component and an ac sinusoidal component. The value of the dc component is 50

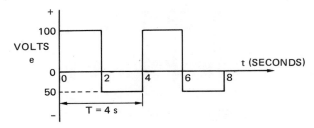

Fig. 10-21

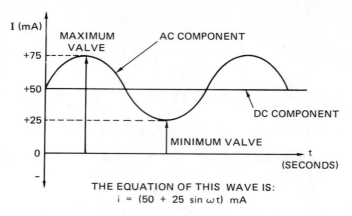

THE EQUATION OF THIS WAVE IS:
$$i = (50 + 25 \sin \omega t) \text{ mA}$$

Fig. 10-22 Unidirectional wave

mA. The maximum value of the unidirectional wave is +75 mA. The minimum value of the unidirectional wave is +25 mA. The wave is positive at all times, and the current flows in one direction only.

(R10-22)   What is the maximum value of the ac component of the unidirectional wave shown in figure 10-22?

The wave analyzer shown in figure 10-23 is used to determine the components of complex waves.

LABORATORY EXERCISE 10-1 MEASUREMENT AND ANALYSIS OF SINUSOIDAL AND NON SINUSOIDAL WAVEFORMS

PURPOSE   After completing this exercise, the student will be able to apply measurement and analysis techniques to sinusoidal and nonsinusoidal waveforms to learn their characteristics.

Fig. 10-23 Wave analyzer

EQUIPMENT
AND MATERIALS

1 Oscilloscope
1 Dc power supply
1 Analog VOM
1 Sine-wave generator
1 Frequency counter

PROCEDURE

1. Set the sine-wave generator to deliver an output signal at a frequency of 1 000 Hz.
   a. Display at least one complete cycle of the waveform on the oscilloscope with the input coupling control set to AC.
   b. Adjust the amplitude control of the sine-wave generator unitl the wave displayed has a peak-to-peak value of 5 V.
   c. What is the maximum value of the wave?
   d. Calculate the effective or rms value of the wave.
   e. Calculate the period of the wave. How does this value compare with the value displayed on the oscilloscope?
   f. Connect the frequency counter to the sine-wave generator. Measure the frequency of the wave displayed. Compare this value with 1 000 Hz.
   g. Measure the voltage output of the sine-wave generator using the analog VOM. Compare this reading with the value calculated in step 1.d.

2. Connect the circuit shown in figure 10-24.
   a. Adjust the dc power supply to give an output voltage of 5 V.
   b. Set the sine-wave generator for an output of 5 V peak-to-peak.
   c. Set the input coupling control of the oscilloscope to GROUND. Observe the reference level trace.
   d. Move the input coupling control of the oscilloscope to DC. Describe the result.
   e. Calculate the dc component of the wave obtained in step 2.d.
   f. Why is the waveform displayed in step 2.d. called a nonsinusoidal waveform?
   g. Move the input coupling control of the oscilloscope to AC. Measure the peak value of the wave.

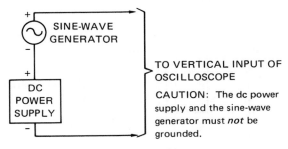

Fig. 10-24

h. Explain the difference between the dc and ac input coupling controls of the oscilloscope.

EXTENDED
STUDY TOPICS

1. Convert $90°$ to radians.
2. Convert 1.25π radians to degrees.
3. A sinusoidal voltage has a peak value of 4 mV and a period of 2×10^{-6} seconds. Express this voltage in time notation.
4. A sinusoidal function follows the equation:
$$i = 140 \sin (157t + 60°) \text{ mA}$$
   a. Calculate the rms value of the current
   b. What is the maximum value of the current?
   c. What is the peak-to-peak current value?
5. A sinusoidal current has an rms value of 10 mA, a frequency of 40 Hz, and a $+50°$ phase shift when $t = 0$. Express this current in time notation.
6. Calculate the instantaneous value of current in topic 5 when $\omega t = 40°$.
7. At what degree values does the following sinusoidal wave reach a magnitude equal to 0.707 times its maximum value?
$$e = 100 \sin 377t$$
8. Calculate the period of a periodic waveform which has a frequency of 1 000 Hz.
9. A sinusoidal voltage is expressed in time notation as:
$$e = 141.4 \sin 628.32t$$
Calculate: a. the voltage indicated by a multimeter; b. f; and c. the period.
10. The voltage and current waveforms of a circuit have the following equations:
$$e = 100 \sin (\omega t - 70°)$$
$$i = 10 \sin (\omega t - 120°)$$
What is the phase angle of the circuit?
11. Calculate the dc component or the average value of the waveform shown in figure 10-25.
12. Find the dc component or average value of the following waveform:
$$i = 50 \sin (\omega t - 50°)$$
13. A current wave in a transformer amplifier has the following equation:
$$i = 100 + 20 \sin (2\pi\, 10^{6}t) \text{ mA}$$
Determine: a. the dc component; b. the peak value of the ac component; c. the rms value of the ac component; d. f; and e. T.

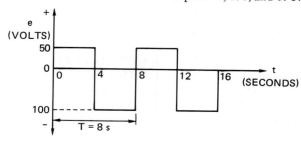

Fig. 10-25

Chapter 11

Phasor Algebra

OBJECTIVES

After studying this chapter, the student will be able to

- apply the mathematical concepts of real, imaginary, and complex numbers.
- use the operator j.
- perform mathematical operations using complex numbers.
- solve ac circuit problems using the principles of vector and phasor algebra.

PHASOR REPRE-SENTATION

It was shown in Chapter 10 that a sinusoidal voltage wave can be generated by a rotating vector called a phasor. A phasor is defined as a radius vector having a constant magnitude (its length) at a fixed angle from a reference axis. One end of the phasor is fixed at the origin. Phasors represent the position of the sine wave when time equals zero (t = 0). Sinusoidal voltage or current waves may be represented by phasors. A sinusoidal voltage wave and its phasor representation are shown in figure 11-1.

The standard direction of rotation for phasors is counterclockwise. The position of a phasor indicates its phase relationship with reference to another phasor, or to the reference axis. In a problem, all of the phasors must represent sine waves of the same frequency. Therefore, these phasors must rotate at the same speed and keep the same phase angle relationship. The magnitude of a phasor in vector algebra is the effective value of the voltage or current. Figure 11-2 shows the phase relationships between voltage and current for three separate circuits.

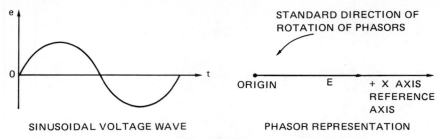

SINUSOIDAL VOLTAGE WAVE   PHASOR REPRESENTATION

Fig. 11-1 Phasor representation of a sinusoidal voltage wave

(R11-1) Three circuits are represented by phasors as shown in figure 11-2. Find the values of E and I. Determine the phase angle between E and I.

(R11-2) Which circuit shown in figure 11-2 has a leading current? Which circuit has a lagging current? What can be said about the relationship between the current and voltage in circuit 1?

Problem 1 Express the phasors of figure 11-2 mathematically.
Solution Circuit 1: E $= 100$ V $\underline{/0°}$
      I $=$ 5 A $\underline{/0°}$
    Circuit 2: E $= 150$ V $\underline{/0°}$
      I $=$ 10 A $\underline{/45°}$
    Circuit 3: E $= 200$ V $\underline{/0°}$
      I $=$ 20 A $\underline{/-45°}$

   To express a phasor mathematically, two values are required: its magnitude and its phase angle with respect to the +X-axis as a reference.

(R11-3) What is the phase angle difference between the current and voltage of figure 11-3?

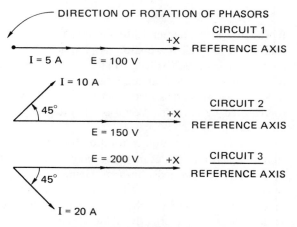

Fig. 11-2 Phasor representation of three separate circuits

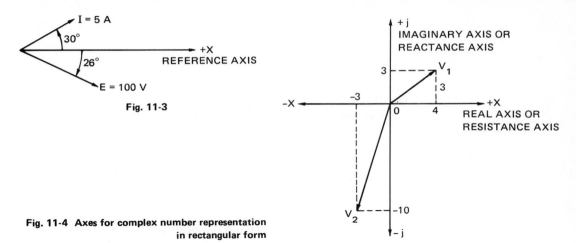

Fig. 11-3

Fig. 11-4 Axes for complex number representation in rectangular form

RECTANGULAR NOTATION	A complex number represents a point which is positioned with reference to two axes, figure 11-4. This figure also shows that a vector can be drawn from the origin of the axes to the point. Mathematically, the horizontal, or X, axis is known as the *real axis*. The vertical, or Y, axis is called the *imaginary*, or *j, axis*. The value j is equal to $\sqrt{-1}$. The electronics engineer calls the horizontal axis the *real* , or *resistance, axis* and the vertical axis the *reactance axis*. These designations are explained in detail in Chapter 12.

As shown in Problem 2, phasors can be expressed as complex numbers in the rectangular form. In this form, they can be added, subtracted, multiplied, or divided using the principles of vector or phasor algebra.

Problem 2 Express voltage V_1, figure 11-4, in rectangular notation.
Solution The horizontal component = +4
The vertical component   = +j3
Therefore,
$V_1 = 4 + j3$

(R11-4) Express voltage V_2, figure 11-4, in rectangular notation.

POLAR NOTATION A complex number can also be represented by a radius vector which is drawn from the origin at a fixed angle from a reference axis, figure 11-5. A phasor may be expressed as a complex number in the polar form. For example, voltage V_1 in figure 11-5 is represented in polar form as:
$$V_1 = 100 \text{ V } \underline{/30°}$$
The magnitude of the voltage, or its effective value, is 100 V. The angle 30° is measured from the reference axis, which is the positive real axis.

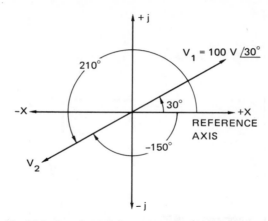

Fig. 11-5 Complex numbers represented in polar form

Problem 3   Express voltage V_2 of figure 11-5 in polar notation. The magnitude of V_2 is 200 volts.

Solution $V_2 = 200 \text{ V } \underline{/210^\circ}$
or
$V_2 = 200 \text{ V } \underline{/-150^\circ}$

The angle is measured from the positive real axis. If the angle is measured in the counterclockwise direction, it is positive. The angle is negative when it is measured in the clockwise direction.

THE OPERATOR j   A phasor can be rotated by plus or minus 90° by applying the appropriate j-operator. For example, +j is a 90° operator which rotates a phasor 90° in the counterclockwise direction (+90°). The –j operator rotates the phasor 90° in the clockwise direction (–90°). Figure 11-6 shows the positions of voltage V after it is multiplied, or operated upon, by the factors j, j^2, j^3, and j^4.

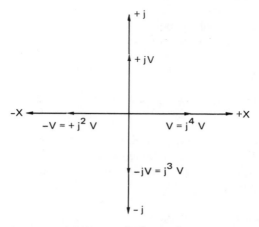

Fig. 11-6 Effect of operators j, j^2, j^3, and j^4 on a voltage phasor V

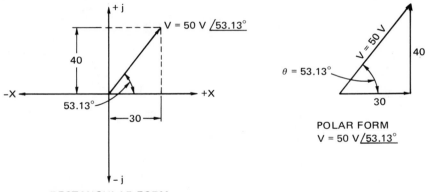

RECTANGULAR FORM
V = (30 + j 40) V

POLAR FORM
V = 50 V/53.13°

Fig. 11-7 Rectangular to polar conversion

Problem 4 Evaluate the factors j, j², j³, and j⁴.

Solution $j = \sqrt{-1}$

$j^2 = (\sqrt{-1})(\sqrt{-1}) = -1$

(Note: in effect, the operator j² rotates a phasor 180°.)

$j^3 = (j^2)(j) = (-1)(j) = -j$

$j^4 = (j^2)(j^2) = (-1)(-1) = +1$

CONVERSION FROM RECTANGULAR TO POLAR NOTATION A voltage phasor is expressed in both the rectangular form and the polar form, figure 11-7. In the rectangular form, V = 30 + j40. These values form a right triangle. The base, or horizontal, component equals 30. The altitude, or vertical, component, equals 40. V is obtained by applying the Pythagorean Theorem:

$$V = \sqrt{(30)^2 + (40)^2}$$
$$= \sqrt{2\ 500}$$
$$= 50\ V$$

In polar notation, V is expressed as follows:

$$\theta = \arctan \frac{40}{30}, \text{ or } \tan^{-1} \frac{40}{30}$$
$$= \arctan 1.333$$
$$\theta = 53.13°$$
$$V = 50\ V\underline{/53.13°}$$

Problem 5 Convert I = (−40 − j100)A to polar notation. Refer to figure 11-8.

Solution $I = \sqrt{(40)^2 + (100)^2}$

$= \sqrt{11\ 600}$

= 107.7A (magnitude only)

$\phi = \arctan \dfrac{100}{40}$

$= \arctan 2.5$

$= 68.2°$

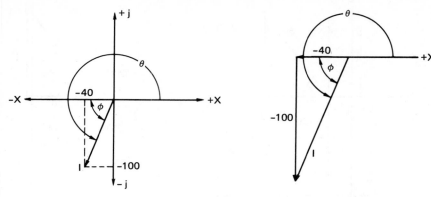

Fig. 11-8

But, θ must be measured from the positive real axis.
Therefore,

$\theta = 180° + \phi$
$= 180° + 68.2°$
$= 248.2°$
$I = 107.7 \text{ A } \underline{/248.2°}$

(R11-5)   Express the result of Problem 5 in polar form, but with an angle less than 180°.

CONVERSION FROM POLAR TO RECTANGULAR NOTATION   Figure 11-9 shows the conversion from polar to rectangular notation. The horizontal component equals $I \cos \theta$. The vertical component equals $I \sin \theta$. The general conversion formula is:

$$I \underline{/\theta°} = I \cos \theta + j\, I \sin \theta$$

Therefore,

$I = 100 \text{ A } \underline{/45°}$
$= 100 \text{ A } \cos 45° + j100 \text{ A } \sin 45°$
$= 100 \text{ A}(0.707) + j(100 \text{ A})(0.707)$
$= 70.7 \text{ A} + j70.7 \text{ A}$

Problem 6   Convert $V = 67 \text{ V } \underline{/153.5°}$ to rectangular notation. Refer to figure 11-10.
Solution $\phi = 180° - \theta$
$= 180° - 153.5°$
$= 26.5°$

The horizontal component is negative: $-67 \text{ V} \cos 26.5° = -(67 \text{ V})(0.895) \cong -60 \text{ V}$.
The vertical component is positive: $67 \text{ V} \sin 26.5° = (67\text{V})(0.446\ 2) \cong 30 \text{ V}$.
Therefore,

$V = (-60 + j30)V$

(R11-6)   Convert $I = 72.2 \text{ A } \underline{/123.6°}$ to rectangular notation.

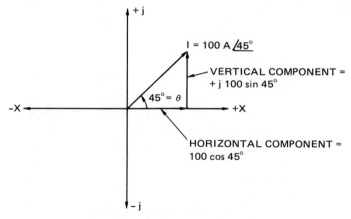

Fig. 11-9 Polar to rectangular conversion

ADDITION OF PHASORS   The sum of two or more sine waves having the same frequency is a sine wave of that same frequency. Simple algebraic addition cannot be used when sine waves are out of phase. Instead, vector algebra is applied. Thus, the horizontal component of the resultant is the algebraic sum of the horizontal components of *each* wave. The vertical component of the resultant is the algebraic sum of the vertical components of *each* wave.

Problem 7 $V_1 = (20 + j40)$ V
$V_2 = (30 + j10)$ V
Find the sum of V_1 and V_2.

Solution $V_1 = (20 + j40)$ V
$\underline{V_2 = (30 + j10)\ V}$
$V_T = (50 + j50)$ V

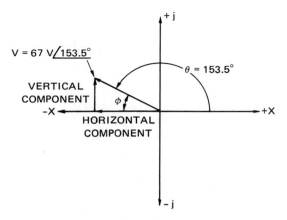

Fig. 11-10

Problem 8 $I_1 = (0.3 + j0.6)$ A
$I_2 = (-0.6 + j0.3)$ A
Find the sum of I_1 and I_2.

Solution $I_1 = (0.3 + j0.6)$ A
$\underline{I_2 = (-0.6 + j0.3)}$ A
$I_T = (-0.3 + j0.9)$ A

When phasors are expressed in the polar form, they are converted to rectangular form and then are added by vector algebra.

Problem 9 $V_1 = 80$ V $\underline{/60°}$
$V_2 = 80$ V $\underline{/-135°}$
Refer to figure 11-11 and find V_T, the sum of V_1 and V_2.

Solution $V_1 = 80$ V $\underline{/60°}$
$= 80 \cos 60° + j \sin 60°$
$= (40 + j69.3)$ V
$V_2 = 80$ V $\underline{/-135°}$
$= -80 \cos 45° - j80 \sin 45°$
$= (-56.6 - j56.6)$ V
$V_1 = (40 + j69.3)$ V
$\underline{V_2 = (-56.6 - j56.6)}$ V
$V_T = (-16.6 + j12.7)$ V

(R11-7)   Convert the answer to Problem 9 to the polar form.

SUBTRACTION OF PHASORS   The difference between two sine waves having the same frequency is a sine wave of that same frequency. Two sinusoidal waves are subtracted by first changing the sign of the subtrahend. Then, the procedure for vector algebra addition is used.

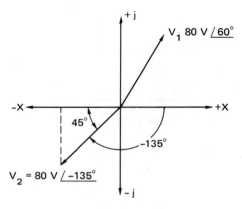

Fig. 11-11

Problem 10   Subtract I_2 from I_1:
$$I_2 = (4 - j5) \text{ A}$$
$$I_1 = (3 + j6) \text{ A}$$

Solution
$$I_1 \quad = 3 + j6 \quad = (3 \quad + j6) \text{ A}$$
$$-I_2 = -(4 - j5) = \underline{(-4 + j5) \text{ A}}$$
$$(-1 + j11) \text{ A}$$

(R11-8)   Convert the answer to Problem 10 to the polar form.

Problem 11   Subtract V_2 from V_1:
$$V_2 = 80 \text{ V} \underline{/-135°}$$
$$V_1 = 80 \text{ V} \underline{/60°}$$

Solution   The phasors are expressed in polar form. Thus, they must be converted to the rectangular form. Then they are subtracted by vector algebra. (See Problem 9 for the procedure used in making the conversions from polar to rectangular notation.)
$$V_1 \quad = 40 + j69.3 \quad = (40 \quad + j69.3) \text{ V}$$
$$-V_2 = -(-56.6 - j56.6) = \underline{(56.6 + j56.6) \text{ V}}$$
$$(96.6 + j125.9) \text{ V}$$

(R11-9)   Subtract $V_2 = -2 + j5$ from $V_1 = 3 + j3$

(R11-10)   Subtract $I_2 = 4 - j5$ from $I_1 = 3 + j6$

(R11-11)   Express the answer to question (R11-10) in polar form.

MULTIPLICATION OF PHASORS   Complex quantities expressed in polar form are multiplied as follows:
1. multiply the magnitudes of the quantities.
2. add the phase angles algebraically.

Problem 12
$$I = 8 \text{ A} \underline{/40°}$$
$$V = 10 \text{ V} \underline{/-20°}$$
Find the product of V and I.

Solution
$$V \times I = (8 \text{ A} \underline{/40°})(10 \text{ V} \underline{/-20°})$$
$$= (8 \times 10) \underline{/40° - 20°}$$
$$= 80 \text{ VA} \underline{/20°}$$

(R11-12) $V = 6 \text{ V} \underline{/-30°}$ and $I = 5 \text{ A} \underline{/-40°}$. Find the product of V and I.

Complex numbers   expressed in rectangular form can also be multiplied. This type of multiplication follows the rules for the product of two binomial expressions.

Problem 13
$$V = (3 + j5) \text{ V}$$
$$I = (5 + j6) \text{ A}$$
Find the product of V and I.

Solution
$$(3 + j5)(5 + j6) = 15 + j25 + j18 + j^2 \, 30$$
$$= 15 + j43 - 30$$
$$= (-15 + j43) \text{ VA}$$

(R11-13) $V = (5 + j6)$ V and $I = (-4 - j7)$ A
Find the product of V and I.

As shown by comparing the solutions to Problems 12 and 13, it is easier to multiply quantities expressed in polar form.

DIVISION
OF PHASORS

Two complex numbers expressed in polar form are divided as follows:
1. the magnitudes of the two numbers are divided.
2. the phase angles are subtracted algebraically.

Problem 14

$E_1 = 10$ V $\underline{/50°}$
$I_1 = 5$ A $\underline{/30°}$
Divide E_1 by I_1.

Solution

$\dfrac{E_1}{I_1} = \dfrac{10 \text{ V} \underline{/50°}}{5 \text{ A} \underline{/30°}}$

$= \dfrac{10}{5} \underline{/50° - 30°}$

$= 2 \underline{/20°}$

Problem 15

$V_1 = 20$ V $\underline{/60°}$
$I_1 = 5$ A $\underline{/-30°}$
Divide V_1 by I_1.

Solution

$\dfrac{V_1}{I_1} = \dfrac{20 \text{ V} \underline{/60°}}{5 \text{ A} \underline{/-30°}}$

$= \dfrac{20}{5} \underline{/60° - (-30°)}$

$= 4 \underline{/60° + 30°}$

$= 4 \underline{/90°}$

(R11-14)   If $V = 10 \underline{/-10°}$ and $I = 5 \underline{/-30°}$, what is the value of $\dfrac{V}{I}$?

Complex numbers expressed in rectangular form can also be divided by following the rules for finding the quotient of two binomial expressions.

Problem 16

$V_1 = (6 + j12)$ V
$I_1 = (-4 - j10)$ A
Divide V_1 by I_1.

Solution

This problem can be solved by rationalizing, or eliminating, the j-factor from the denominator. To do this, both the numerator and the denominator are multiplied by the conjugate of the denominator. The *conjugate* is obtained by changing the sign of the imaginary term. The solution to this problem shows that when a denominator is multiplied by its conjugate, a real number is obtained.

$$\frac{V_1}{I_1} = \frac{6 + j12}{-4 - j10}$$

$$= \frac{(6 + j12)(-4 + j10)}{(-4 - j10)(-4 + j10)}$$

$$= \frac{-24 - j48 + j60 + j^2 120}{16 - j^2 100}$$

$$= \frac{-24 + j12 - 120}{16 + 100}$$

$$= \frac{-144 + j12}{116}$$

$$= -1.24 + j0.103$$

(R11-15) If $V = (8 + j6)$ V and $I = (2 - j4)$ A, what is the value of $\frac{V}{I}$?

An analysis of the solutions to Problems 14, 15, and 16 shows that division is easier when the complex numbers are in polar form.

CONVERSION BETWEEN TIME NOTATION AND PHASOR NOTATION

The angular velocity, or frequency, does not affect the conversion from time notation to phasor notation. Phasor notation uses the effective value of the voltage or current. Time notation generally is expressed in terms of the maximum value of the voltage or current. The angle θ in time notation also appears in phasor notation.

Problem 17 A sinusoidal voltage wave is expressed in time notation as e = 100 sin (157t – 60°). Express this wave in phasor notation.

Solution The maximum value of voltage = 100 V
The effective value of voltage = (0.707)(100) = 70.7 V
The angle θ = –60°
E = 70.7 V $\underline{/-60°}$

(R11-16) Express I = 5 A $\underline{/100°}$ in time notation. Assume the frequency is 60 Hz.

EXTENDED STUDY TOPICS
1. Express the phasors of figure 11-12 in polar notation.
2. What is the phase angle between the current and voltage of circuit 3 in figure 11-12? Is this a leading or lagging circuit?
3. Convert V = 33.9 – j33.9 to polar notation.
4. Convert V = 10 V $\underline{/215°}$ to rectangular notation.
5. Find the sum of I_1 and I_2, if $I_1 = (10 + j20)$ A and $I_2 = (20 - j30)$ A.
6. Find I_2, the difference between I_T and I_1 if $I_T = (40 + j60)$ A and $I_1 = (10 - j10)$A.
7. Express 18 $\underline{/33.7°}$ in rectangular notation.
8. Convert 40 $\underline{/240°}$ to rectangular notation.
9. Convert 33.9 – j33.9 to polar notation.
10. Convert –60 + j30 to polar notation.

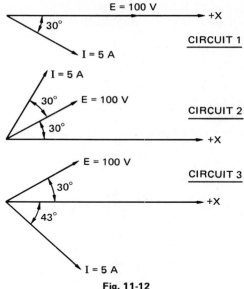

Fig. 11-12

11. Subtract I_1 from I_T, if $I_T = 100$ A $\underline{/120°}$ and $I_1 = 100$ A $\underline{/60°}$.
12. $I = 10$ A $\underline{/-30°}$. Find I^2.
13. $V = (2 + j3)$ V and $I = (6 + j8)$ A. Find $V \times I$.
14. $V = 42$ V $\underline{/10°}$ and $I = 7$ A $\underline{/60°}$. Find $\dfrac{V}{I}$.

15. Evaluate $\dfrac{80 + j80}{20 + j20}$

16. Convert $e = 50 \sin(377\,T - 90°)$ to phasor notation.
17. Express $I = 40$ A $\underline{/-20°}$ in time notation. Assume $f = 60$ Hz.

<table>
<tr><td>

Chapter

12

</td><td>

Series AC Circuits

</td></tr>
</table>

OBJECTIVES   After studying this chapter, the student will be able to

- explain phase relationships in series ac circuits.

- describe inductive reactance and capacitive reactance.

- discuss the concept of impedance and the use of impedance triangles.

- solve series ac circuit problems.

CIRCUIT WITH   Figure 12-1 shows a simple ac circuit with resistance (R) only. The ac voltage
R ONLY   of the power supply is expressed by the equation:

$$e = E_m \sin \omega t$$

This voltage causes an ac current of the same frequency to flow through the resistive load.   The equation for the current is:

$$i = I_m \sin \omega t \qquad \qquad \text{Eq. 12.1}$$

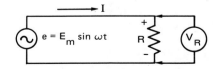

Fig. 12-1 Simple ac circuit with resistance only

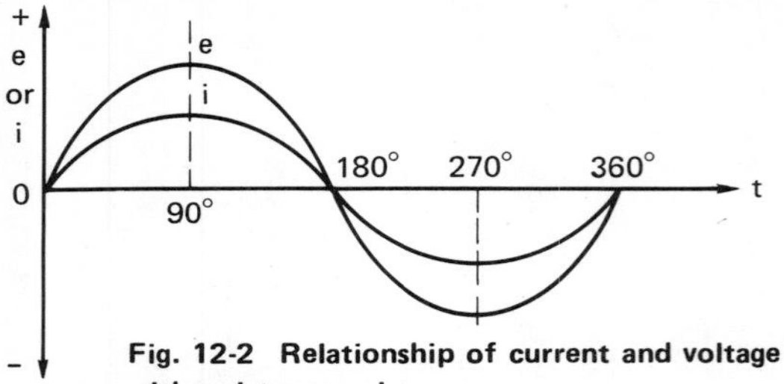

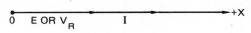

Fig. 12-2   Relationship of current and voltage in a circuit with resistance only

Fig. 12-3 Phasor representation of figure 12-22

The current is in step, or in phase, with the applied voltage, figure 12-2. Thus, the maximum value of current occurs at the same instant of time as the maximum value of voltage. The maximum current is determined as follows:

$$I_m = \frac{E_m}{R}$$ Eq. 12.2

Equation 12.2 can also be expressed in terms of rms or effective values:

$$I = \frac{E}{R}$$ Eq. 12.3

Boldface type is used from this point on to indicate phasors, vectors, and complex numbers having both magnitude and angular direction. The magnitude of phasors, vectors, or complex numbers is indicated in normal type. Normal type is also used to represent scalars, or quantities that have magnitude but no angular direction.

Figure 12-3 shows the phasor representation of the waveforms given in figure 12-2. In polar notation:

$$\mathbf{E} = E\,\underline{/0^\circ}$$
or   Eq. 12.4
$$\mathbf{V_R} = V_R\,\underline{/0^\circ}$$
$$\mathbf{I} = I\,\underline{/0^\circ}$$ Eq. 12.5
$$\mathbf{R} = R\,\underline{/0^\circ}$$ Eq. 12.6

Problem 1   A circuit containing pure resistance has a voltage e = 100 V sin (377t + 60°). a. Calculate the current if R = 50 ohms. b. Express the current in time notation. c. Express E and I in polar form.

Solution   a. $I_m = \dfrac{E_m}{R}$

$= \dfrac{100\ V}{50\ \Omega}$

$= 2\ A$

$I = (0.707)(2\ A)$

$= 1.414\ A$

b. i $= 2\ A \sin (377t + 60^\circ)$

c. E $= (100\ V)(0.707)$

$= 70.7\ V$

$\mathbf{E} = 70.7\ V\,\underline{/60^\circ}$

$\mathbf{I} = 1.414\ A\,\underline{/60^\circ}$

(R12-1)   Solve Problem 1 if the voltage is e = 200 sin (157t − 30°) and R = 20 ohms.

CIRCUIT WITH INDUCTANCE ONLY The circuit in figure 12-4 contains an ideal coil having inductance (L) only (no resistance). To satisfy Kirchhoff's Voltage Law at every instant of time, the inductive voltage across this coil must equal the applied voltage. Thus, the instantaneous voltage across the coil is:

$$v_L = e = E_m \sin \omega t$$

As a result of this voltage, an ac current of the same frequency flows through the coil. The equation for the current is:

$$i = I_m \sin (\omega t - \frac{\pi}{2}) \qquad \text{Eq. 12.7}$$

The current through a pure inductive coil lags the voltage by $\frac{\pi}{2}$ radians (90°), figure 12-5. Inductive reactance (X_L) is the opposition to the flow of alternating current due to an inductance. The unit of inductive reactance is ohms when E is expressed in volts and I is expressed in amperes.

$$\left. \begin{array}{c} X_L = \dfrac{E}{I} \\ \text{or} \\ X_L = \dfrac{V_L}{I} \end{array} \right\} \qquad \text{Eq. 12.8}$$

The waveforms of figure 12-5 are represented by phasors in figure 12-6. In polar notation:

$$\left. \begin{array}{c} \mathbf{E} = E \underline{/0°} \\ \text{or} \\ \mathbf{V_L} = V_L \underline{/0°} \end{array} \right\} \qquad \text{Eq. 12.9}$$

$$\mathbf{I} = I \underline{/-90°} \qquad \text{Eq. 12.10}$$

$$X_L = \frac{E}{I}$$

$$= \frac{E \underline{/0°}}{I \underline{/-90°}}$$

$$= X_L \underline{/90°} \qquad \text{Eq. 12.11}$$

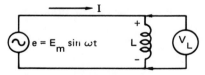

Fig. 12-4 Circuit with an ideal coil (inductance only)

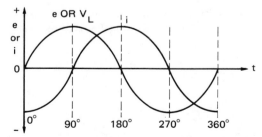

Fig. 12-5 Relationship of current and voltage in a circuit with inductance only

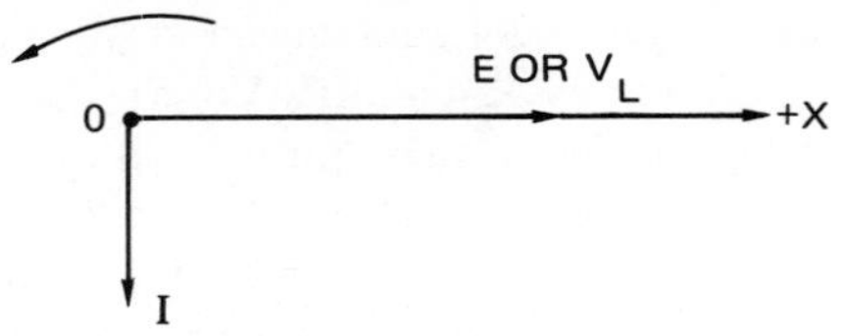

E OR V$_L$

0   +X

I

Fig. 12-6 Phasor representation of the wave forms in figure 12-5

(R12-2) A circuit with an ideal coil has a voltage e = 100 V sin $(\omega t + 50°)$ and a current I = 2 A. Find: a. X_L, b. i expressed in time notation, and c. the polar expressions for E and I.

ANALYSIS OF INDUCTIVE REACTANCE

A mathematical analysis of a circuit with pure inductance shows that:
$$X_L = 2\pi fL \qquad\qquad \text{Eq. 12.12}$$
where

X_L = inductive reactance, in ohms
f   = frequency, in Hz
L   = inductance, in henries

Since $\omega = 2\pi f$,
$$X_L = \omega L \qquad\qquad \text{Eq. 12.13}$$

Equations 12.12 and 12.13 show that inductive reactance is directly proportional both to frequency and the inductance L. Figure 12-7 shows the linear relationship of X_L versus frequency (f) for a constant value of inductance (L).

Equation 12.8 can be expressed as follows:
$$I = \frac{E}{X_L} = \frac{E}{2\pi fL} \qquad\qquad \text{Eq. 12.14}$$

According to Equation 12.14, the current through a coil is inversely proportional to both the frequency and the inductance L. Coils are often called *choke coils* because they "choke," or block, the flow of high-frequency ac currents.

Problem 2 What is the reactance of a 250-mH radio frequency choke coil at a frequency of 5 mHz?

Solution
$X_L = 2\pi fL$
$= (2)(3.141\ 6)(5 \times 10^6\ \text{Hz})(250 \times 10^{-3}\ \text{H})$
$= 7.854\ \text{M}\Omega$

(R12-3) a. At what frequency will a 0.5-H coil have a reactance of 1 000 ohms? b. What is X_L if f = 0 (that is, direct current)?

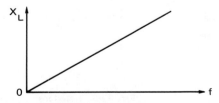

Fig. 12-7 X_L versus f for a constant value of L

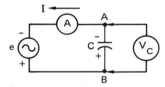

Fig. 12-8 Circuit with capacitance only

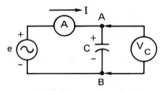

(A) POSITIVE ALTERNATION
OF INPUT VOLTAGE

(B) NEGATIVE ALTERNATION·
OF INPUT VOLTAGE

Fig. 12-9 Capacitor charging on ac

CIRCUIT WITH CAPACITANCE ONLY

Figure 12-8 shows a circuit with a capacitor only. During the positive alternation of the input voltage, the capacitor is charged, figure 12-9A. The charge flows from plate B to plate A of the capacitor. During the negative alternation of the input voltage, the charge across the capacitor is reversed, figure 12-9B. The charge flows from plate A to plate B.

Note: the student should recall that conventional current is opposite to electron flow.

(R12-4) What is the direction of electron flow in figure 12-9A? In figure 12-9B?

The ac ammeter of figure 12-9 will indicate a steady value because it is reading the effective value. This does not mean that charge is flowing *through* the dielectric of the capacitor. A dc ammeter will indicate zero. In an ac circuit, the current is in one direction for one alternation, or half-cycle, and in the opposite direction for the other alternation, or half-cycle. The capacitor alternately charges, discharges, and recharges with opposite polarities, figure 12-9.

To satisfy Kirchhoff's Voltage Law, the potential drop across the capacitor must equal the applied voltage at every instant of time.

$$v_C = e = E_m \sin \omega t$$

This voltage causes an ac current of the same frequency in the capacitor circuit. The equation for the current is:

$$i = I_m \sin (\omega t + \frac{\pi}{2})$$ Eq. 12.15

The current in a circuit with pure capacitance leads the voltage by $\frac{\pi}{2}$ radians, or 90°, figure 12-10. *Capacitive reactance* (X_C) is the term applied to the opposition created by a capacitance to the flow of alternating current. The unit of capacitive reactance is ohms when E is in volts and I is in amperes.

$$\left.\begin{array}{c} X_C = \frac{E}{I} \\ \text{or} \\ X_C = \frac{V_C}{I} \end{array}\right\}$$ Eq. 12.16

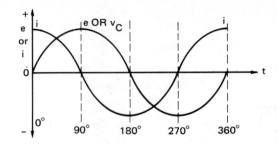

Fig. 12-10 Relationship of current and voltage in a circuit with capacitance only

The waveforms shown in figure 12-10 are represented by phasors in figure 12-11. In polar notation, these phasors are written as follows:

$$\left.\begin{aligned} \mathbf{E} &= E\,\underline{/0^\circ} \\ \text{or} \\ \mathbf{V_C} &= V_C\,\underline{/0^\circ} \end{aligned}\right\} \quad \text{Eq. 12.17}$$

$$\mathbf{I} = I\,\underline{/90^\circ} \quad \text{Eq. 12.18}$$

$$X_C = \frac{E}{I}$$

$$= \frac{E\,\underline{/0^\circ}}{I\,\underline{/90^\circ}}$$

$$\mathbf{X_C} = X_C\,\underline{/-90^\circ} \quad \text{Eq. 12.19}$$

(R12-5) A circuit contains only pure capacitance. The circuit voltage is e = 100 V sin (ωt –60°) and the current is 4 A. Find: a. capacitive reactance, b. i expressed in time notation, and c. E and I expressed in polar notation.

ANALYSIS OF CAPACITIVE REACTANCE A mathematical analysis of a capacitive circuit shows that:

$$X_C = \frac{1}{2\pi fC} \quad \text{Eq. 12.20}$$

where

X_C = capacitive reactance, in ohms
f   = frequency, in Hz
C   = capacitance, in farads

Since $\omega = 2\pi f$,

$$X_C = \frac{1}{\omega C} \quad \text{Eq. 12.21}$$

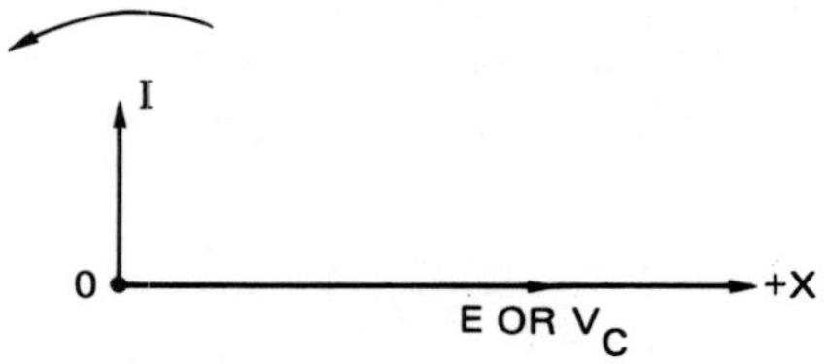

Fig. 12-11 Phasor representation of wave forms in figure 12-10

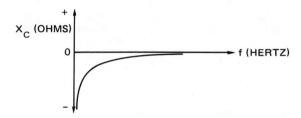

Fig. 12-12 X_C versus f for constant C

Equations 12.20 and 12.21 show that capacitive reactance is inversely proportional to both frequency and capacitance. Figure 12-12 shows the relationship of X_C versus f for a constant value of C. Capacitive reactance has a negative value, as indicated by the negative angle in Equation 12.19.

Equation 12.16 can be rewritten in the form shown:

$$I = \frac{E}{X_C}$$

$$= \frac{E}{\dfrac{1}{2\pi fC}}$$

$$= 2\pi fCE \qquad\qquad \text{Eq. 12.22}$$

According to Equation 12.22, the current in a capacitive circuit is directly proportional to both the frequency and the capacitance.

Problem 3   An 8-μF capacitor is connected to a 120-V ac power supply. a. At what frequency will the capacitor have a reactance of 160 ohms? b. Calculate the current at that frequency.

Solution   a. $X_C = \dfrac{1}{2\pi fC}$

$$f = \frac{1}{2\pi CX_C}$$

$$= \frac{1}{(2)(3.141\,6)(8 \times 10^{-6}\ F)(160\ \Omega)}$$

$$= 124.4\ Hz$$

b. $\mathbf{I} = \dfrac{E\ \underline{/0^\circ}}{X_C\ \underline{/-90^\circ}}$

$$= \frac{120\ V\ \underline{/0^\circ}}{160\ \Omega\ \underline{/-90^\circ}}$$

$$= 0.75\ A\ \underline{/90^\circ}$$

(R12-6)   a. If the frequency determined in part a. of Problem 3 is doubled, determine the value of I. b. If f = 0 (dc), find I.

SERIES RL CIRCUIT   Figure 12-13 shows a series RL ac circuit. The same current flows through all of the components shown in this circuit.

$$I_T = I_R = I_L \qquad\qquad \text{Eq. 12.23}$$

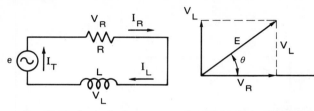

Fig. 12-13 Series RL circuit

Fig. 12-14 Phasors representing the current and voltages of figure 12-13

The phasor sum of the voltage drops across the components is equal to the applied voltage. The voltage drop, V_R, across resistor R is in phase with the current. The voltage drop, V_L, across the pure inductance leads the current by 90°. In figure 12-14, phasors represent the current and voltages shown in figure 12-13. The current is common to all components. Therefore, it is used as the reference phasor. The relationship of the voltages can be expressed in rectangular coordinates as:

$$E = V_R + j\,V_L \qquad \text{Eq. 12.24}$$

In polar form, the relationship is expressed as:

$$E = E\,\underline{/\theta}$$

or

$$E = \sqrt{V_R{}^2 + V_L{}^2}\,\underline{\bigg/\arctan\left(\frac{V_L}{V_R}\right)} \qquad \text{Eq. 12.25}$$

According to figure 12-14, the phase angle between the applied voltage and the current is θ. In addition, figure 12-14 shows how Equation 12.24 is constructed geometrically.

(R12-7)   Is the circuit shown in figure 12-13 a leading or lagging circuit?

IMPEDANCE   The total opposition to the flow of current in an ac circuit is called *impedance* (Z). The unit of impedance is the ohm. Ohm's Law for ac circuits is written as follows:

$$Z = \frac{E}{I} \qquad \text{Eq. 12.26}$$

where

   Z = the impedance of the ac circuit, in ohms

   E = the applied emf, in volts

   I = the current taken from the source, in amperes.

The relationship between Z, R, and X_L can be determined by the following procedure.

Express Equation 12.24 in rectangular form,

$$E = V_R + j\,V_L$$

Divide both sides of the equation by I,

$$\frac{E}{I} = \frac{V_R}{I} + j\frac{V_L}{I}$$

and

$$Z = R + j\,X_L \qquad \text{Eq. 12.27}$$

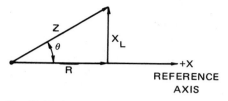

Fig. 12-15 Impedance triangle of impedance diagram of the circuit shown in figure 12-13

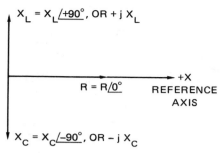

Fig. 12-16 Relationship of R, X_L, and X_C

Equation 12.27 shows that the impedance of a series circuit is the vector sum of the resistance and the reactance. Impedance does not vary sinusoidally. That is, it does not vary with time and is not a phasor. Impedance is a vector quantity having magnitude and direction. In polar notation, impedance is expressed as:

$$\mathbf{Z} = Z\,\underline{/\theta}$$

or

$$\mathbf{Z} = \sqrt{R^2 + X_L{}^2}\,\Big/\!\arctan\!\left(\frac{X_L}{R}\right) \qquad \text{Eq. 12.28}$$

Figure 12-15 shows the impedance diagram for the circuit given in 12-13. In making an impedance diagram, the student must realize that resistance is always drawn in the direction of the reference axis (at an angle of $0°$), inductive reactance is drawn in the direction of $+ j$ (at an angle of $+ 90°$), and capacitive reactance is drawn in the direction of $-j$ (at an angle of $-90°$). Figure 12-16 shows the relationship of R, X_L, and X_C.

Problem 4   The circuit of figure 12-13 has the following values:

E = 120 V
f   = 60 Hz
R = 100 ohms
L = 0.5 H

Find the current **I**, and express it in the polar form using **E** as the reference.

Solution X_L = ω L
            = (377)(0.5 H)
            = 188.5 Ω

   Z   = R + j X_L
        = 100 Ω + j 188.5
        = 213 $\Omega\,\underline{/62°}$

   I   = $\dfrac{E}{Z}$

        = $\dfrac{120\text{ V}\,\underline{/0°}}{213\ \Omega\,\underline{/62°}}$

        = 0.563 A $\underline{/\!-62°}$

(R12-8)   Draw the phasor diagram of E and I in Problem 4, using **E** as the reference.

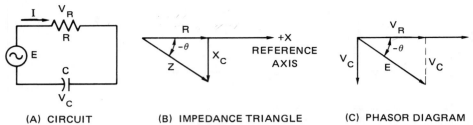

(A) CIRCUIT   (B) IMPEDANCE TRIANGLE   (C) PHASOR DIAGRAM

Fig. 12-17 Series RC ac circuit

(R12-9) What conclusion can be reached by comparing the impedance angle and the angle between the applied voltage and the resulting current in a series ac circuit? This angle is called the circuit phase angle.

(R12-10) For Problem 4 and question (R12-8), why is the circuit phase angle negative and the impedance angle positive?

SERIES RC CIRCUIT Figure 12-17 shows a series RC ac circuit, its voltage-current phasor diagram, and its impedance triangle. The following relationships can be obtained from this figure.

$$\mathbf{E} = V_R - j \, V_C \text{ (rectangular coordinates)} \qquad \text{Eq. 12.29}$$

$$\mathbf{E} = E \, \underline{/-\theta^\circ} \qquad \text{Eq. 12.30}$$

$$\mathbf{E} = \sqrt{V_R{}^2 + V_C{}^2} \, \Big/ {-}\arctan\left(\frac{V_C}{V_R}\right) \qquad \text{Eq. 12.31}$$

$$\mathbf{Z} = R - j \, X_C \qquad \text{Eq. 12.32}$$

$$\mathbf{Z} = Z \, \underline{/-\theta^\circ} \qquad \text{Eq. 12.33}$$

$$\mathbf{Z} = \sqrt{R^2 + X_C{}^2} \, \Big/ {-}\arctan\left(\frac{X_C}{R}\right) \qquad \text{Eq. 12.34}$$

Problem 5 The circuit shown in figure 12-17 has the following values:

E = 120 V
f = 60 Hz
R = 50 ohms
I = 0.5 A

Find C.

Solution

$$Z = \frac{E}{I}$$

$$= \frac{120 \text{ V}}{0.5 \text{ A}}$$

$$= 240 \ \Omega$$

$$Z = \sqrt{R^2 + X_C{}^2}$$

$$Z^2 = R^2 + X_C{}^2$$

$$X_C = \sqrt{Z^2 - R^2}$$

$$= \sqrt{(240 \ \Omega)^2 - (50 \ \Omega)^2}$$

$$= 235 \ \Omega$$

$$X_C = \frac{1}{\omega C}$$

$$C = \frac{1}{\omega X_C}$$

$$= \frac{1}{(377)(235 \ \Omega)}$$

$$= 11.4 \ \mu F$$

(R12-11)   What is the difference between the circuit phase angles for an RL series circuit and an RC series circuit?

(R12-12)   Calculate the phase angle of the circuit given in Problem 5, using **I** as a reference.

SERIES RLC   The circuit in figure 12-18 is a series RLC ac circuit. Also shown is the voltage-
CIRCUIT   current phasor diagram for this circuit and its impedance triangle where V_L is greater than V_C. The following relationship can be obtained from figure 12-18.

$$\mathbf{E} = \mathbf{V_R} + \mathbf{V_L} + \mathbf{V_C} \qquad \text{Eq. 12.35}$$

It can be seen from figure 12-18B that voltage V_L leads the current by 90°, and voltage V_C lags the current by 90°. Therefore, these voltages are 180° out of phase. V_L and V_C may be subtracted arithmetically to obtain the following equation from Equation 12.35:

$$\mathbf{E} = \mathbf{V_R} + j(V_L - V_C) \text{ (rectangular notation)}$$
$$= \mathbf{V_R} + j \, V_X \qquad \text{Eq. 12.36}$$

where

$$V_X = V_L - V_C = \text{net reactive voltage}$$

The net reactive voltage may be 0 (when $V_L = V_C$), inductive (when $V_L >$ V_C), or capacitive (when $V_L < V_C$). The current will be in phase with, will lag, or will lead the applied emf.

$$\mathbf{Z} = R + j \, X_L - j \, X_C$$
$$= R + j \, (X_L - X_C)$$
$$= R + j \, X_{eq} \qquad \text{Eq. 12.37}$$

where

$$X_{eq} = X_L - X_C = \text{net equivalent reactance}$$

In polar form, the impedance for an RLC circuit is expressed as:

$$\mathbf{Z} = \sqrt{R^2 + X_{eq}^2} \Big/ \arctan\left(\frac{X_{eq}}{R}\right) \qquad \text{Eq. 12.38}$$

$$\mathbf{Z} = Z \underline{/\theta^\circ} \qquad \text{Eq. 12.39}$$

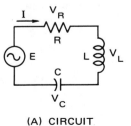
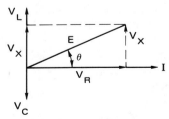
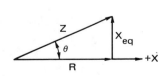

(A) CIRCUIT   (B) PHASOR DIAGRAM   (C) IMPEDANCE TRIANGLE

Fig. 12-18 Series RLC ac circuit

Problem 6   The circuit of figure 12-18 has the following values:

$E = 120$ V
$f = 60$ Hz
$R = 100$ Ω
$L = 0.5$ H
$C = 26.5$ μF

Find Z in polar form.

Solution $\omega = 377$

$X_L = \omega L$
$\quad = (377)(0.5$ H$)$
$\quad = 188.5$ Ω

$X_C = \dfrac{1}{\omega C}$

$\quad = \dfrac{1}{(377)(26.5 \times 10^{-6}$ F$)}$

$\quad = 100$ Ω

$Z = R + j(X_L - X_C)$
$\quad = 100$ $\Omega + j(188.5$ $\Omega - j$ 100 $\Omega)$
$\quad = 100$ $\Omega + j$ 88.5 Ω
$\quad = 133.5$ Ω $\underline{/41.5°}$

The following questions refer to Problem 6.

(R12-13)   Calculate the current in this circuit.

(R12-14)   Draw an equivalent circuit to represent this circuit.

(R12-15)   Calculate the equivalent inductance of the circuit.

LABORATORY EXERCISE 12-1 MEASUREMENT AND ANALYSIS OF A SERIES RLC AC CIRCUIT

PURPOSE   By completing this exercise, the student will determine experimentally the characteristics of a series RLC ac circuit.

EQUIPMENT AND MATERIALS
1 Dual trace oscilloscope
1 Sinusoidal function generator
1 Digital multimeter
1 Resistor, 12 kΩ, 2 W
1 Inductor, 4 H
1 Capacitor, 0.01 μF

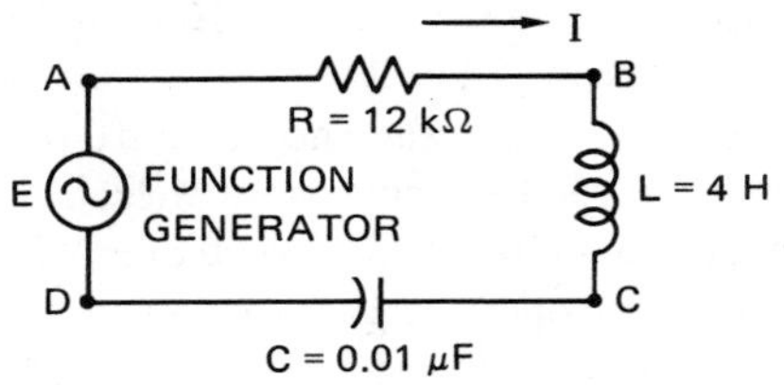

Fig. 12-19

TABLE 12-1

E(V_{AD})	I	V_{AB}	V_{BC}	V_{CD}	V_{AD}	θ
10 V_{p-p} f = 1 kHz						

PROCEDURE

1. Connect the circuit shown in figure 12-19.
   a. Turn the function generator ON. Adjust the output voltage to 10 V peak-to-peak at a frequency of 1 kHz, as measured on the oscilloscope.
   b. Use the digital multimeter and measure the current I. Record this value in Table 12-1.
   c. Measure the voltages V_{AB}, V_{BC}, V_{CD}, and V_{AD} using the digital multimeter. Record these values in Table 12-1.
   d. Connect one input of the dual trace oscilloscope to terminals A and D. Connect the other input of the dual trace oscilloscope to terminals A and B. Set both voltage waves to the same reference axis on the oscilloscope. Measure the circuit phase angle, θ, and record the value in Table 12-1.
2. Perform the following calculations and answer the following questions.
   a. Explain why the procedure of step 1.d. measures the circuit phase angle.
   b. Calculate the inductive reactance, capacitive reactance, and circuit impedance at a frequency of 1 kHz.
   c. Using the data from Table 12-1, determine the inductive reactance, capacitive reactance, and circuit impedance.
   d. Compare the results of steps 2.b. and 2.c.
   e. Show that $V_{AD} = V_{AB} + V_{BC} + V_{CD}$. (Use phasor algebra.)
   f. Construct an impedance triangle. Label all three sides. Calculate the impedance angle of the triangle.
   g. Compare the impedance angle with the circuit phase angle recorded in Table 12-1.

EXTENDED STUDY TOPICS

1. Calculate the capacitive reactance of a 15-pF capacitor at a frequency of 30 MHz.
2. A capacitor has a reactance of 10 ohms at a frequency of 1 kHz. Determine its reactance at a frequency of 2 kHz.

3. A capacitor has a reactance of 100 ohms at a frequency of 1 kHz. Calculate its capacitance.

4. Calculate the total reactance of a 10-pF capacitor in series with a 15-pF capacitor, if the frequency is 30 MHz.

5. Calculate the inductive reactance of a 2.5-mH coil at a frequency of 1 MHz.

6. A choke coil operates at a frequency of 30 MHz. If the desired inductive reactance is 100 k Ω, calculate the inductance of the coil.

7. The voltage drop across a coil measures 40 V, when its current is 120 mA at a frequency of 400 Hz. Find its inductance.

8. The voltage across a 0.08-μF capacitor is 220 V at a frequency of 20 kHz. Calculate the current.

9. A coil with a reactance equal to 650 ohms is connected in series with a 900-ohm resistor. The source voltage is 120 V $\underline{/0°}$. Find:
   a. Z in polar and rectangular notation.
   b. I in polar notation.
   c. V_R in polar notation.
   d. V_L in polar notation.
   e. Draw the impedance triangle.
   f. What is the circuit phase angle?

10. A capacitor with a reactance equal to 200 ohms is connected in series with a 150-ohm resistor. The source voltage is 100 V $\underline{/0°}$. Find:
    a. Z in polar and rectangular notation.
    b. I in polar notation.
    c. V_R in polar notation.
    d. V_C in polar notation.
    e. Draw the impedance triangle.
    f. What is the circuit phase angle?

11. A series RLC circuit is connected to a source voltage equal to 50 V $\underline{/0°}$. If $X_C = 150$ Ω, $X_L = 400$ Ω, and R = 400 Ω, find:
    a. Z in polar and rectangular notation.
    b. I in polar notation.
    c. V_R in polar notation.
    d. V_L in polar notation.
    e. V_C in polar notation.
    f. Draw the impedance triangle.
    g. What is the circuit phase angle?

<table>
<tr><td>Chapter
13</td><td>Complex AC Circuits</td></tr>
</table>

OBJECTIVES After studying this chapter, the student will be able to

- discuss impedances in series and in parallel ac circuits.

- apply the current divider rule and the voltage divider rule in ac circuits.

- solve for the series equivalent of a complex ac circuit.

- summarize the characteristic conditions of parallel and series-parallel ac circuits.

PRACTICAL INDUCTORS A practical inductor is *not* a pure reactance having a $+90°$ impedance angle. Rather, there is some resistance due to the wire used to wind the coil. An ac voltmeter connected across the two terminals of a practical inductor indicates the total voltage drop across the coil. The voltmeter cannot show just the inductive voltage drop across the inductance and it cannot indicate just the IR drop across its resistance.

(R13-1) What is the normal impedance angle of a practical inductor?

Problem 1 The choke coil shown in the circuit of figure 13-1 has an impedance of $(15 + j100)$ Ω. (Note that R is usually small with respect to X_L.) What is the total impedance of the circuit when the choke coil is connected in series with an 85-Ω resistor?

Solution The impedance is determined using the following formula:

$$Z_T = R_T + j\, X_T \qquad\qquad \text{Eq. 13.1}$$

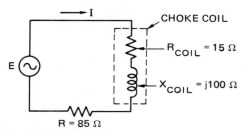

Fig. 13-1 Practical inductor in series with a resistor

where

R_T = the sum of all of the series resistances, in ohms
X_T = the algebraic sum of all of the series reactances, in ohms
Z_T = $(15 + 85)\ \Omega + j100\ \Omega$
   = $(100 + j100)\ \Omega$
   = $141.4\ \Omega\underline{/45°}$

(R13-2) If the voltage **E** in figure 13-1 equals 100 V $\underline{/0°}$, find the current drawn by the circuit.

IMPEDANCES IN SERIES The total impedance of a number of impedances connected in series can be found as shown in Problem 2

Problem 2 Two impedances are connected in series. Their values are Z_1 = 60 $\Omega\underline{/60°}$ and Z_2 = 80 $\Omega\underline{/-45°}$. Find the total impedance of the series combination.

Solution Z_1 = 60 $\Omega\underline{/60°}$ = $(30 + j52)\ \Omega$
Z_2 = 80 $\Omega\underline{/-45°}$ = $(56.5 - j56.5)\ \Omega$
Z_T = $R_T + j\ X_T$ (Eq. 13.1)
   = $(30 + 56.5)\ \Omega + j(52 - 56.5)\ \Omega$
   = $(86.5 - j4.5)\ \Omega$
   = $86.6\ \Omega\underline{/-3°}$

(R13-3) Draw the equivalent circuit of Problem 2 and label each component.

VOLTAGE DIVIDER RULE An earlier chapter showed how the voltage divider rule is applied to dc circuits. It can also be applied in a similar manner to ac circuits.

Problem 3 Find V_C for the circuit in figure 13-2 using the voltage divider rule.

Solution $V_C = \dfrac{E\ X_C}{R + X_C}$

$= \dfrac{(100\ V\underline{/0°})(4\ \Omega\underline{/-90°})}{3\ \Omega\underline{/0°} + 4\ \Omega\underline{/-90°}}$

$= \dfrac{400\ V\Omega\underline{/-90°}}{(3 - j4)\ \Omega}$

$= \dfrac{400\ V\Omega\underline{/-90°}}{5\ \Omega\underline{/-53°}}$

$= 80\ V\underline{/-37°}$

(R13-4) Find V_R for the circuit of figure 13-2 using the voltage divider rule.

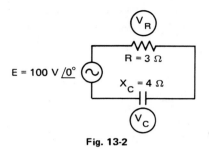

Fig. 13-2

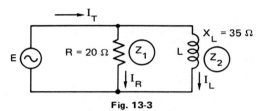

Fig. 13-3

PARALLEL RL CIRCUIT Parallel RL circuits can be solved using either one of the two methods shown in Problem 4.

Problem 4 Find the total impedance of the RL parallel circuit shown in figure 13-3.

Solution

Method 1: This method is known as the *total current method*. Since the voltage E is not specified, any value may be assumed. It is convenient to assume that **E** = 140 V $\underline{/0°}$.

$$I_R = \frac{E}{R}$$
$$= \frac{140 \text{ V} \underline{/0°}}{20 \text{ Ω} \underline{/0°}}$$
$$= 7 \text{ A} \underline{/0°}$$

$$I_L = \frac{E}{X_L}$$
$$= \frac{140 \text{ V} \underline{/0°}}{35 \text{ Ω} \underline{/90°}}$$
$$= 4 \text{ A} \underline{/-90°}$$

$$I_T = I_R + I_L$$
$$= 7 \text{ A} \underline{/0°} + 4 \text{ A} \underline{/-90°}$$
$$= (7 + j0) \text{ A} + (0 - j4) \text{ A}$$
$$= (7 - j4) \text{ A}$$
$$= 8.06 \text{ A} \underline{/-29.75°}$$

$$Z_T = \frac{E}{I_T}$$
$$= \frac{140 \text{ V} \underline{/0°}}{8.06 \text{ A} \underline{/-29.75°}}$$
$$= 17.4 \text{ Ω} \underline{/29.75°}$$

Method 2:   Two impedances are connected in parallel. Therefore, the total impedance can be found using the product over the sum rule.

$$Z_T = \frac{Z_1 \, Z_2}{Z_1 + Z_2} \qquad\qquad \text{Eq. 13.2}$$

$Z_1 = 20 \ \Omega \underline{/0^\circ}$
$Z_2 = 35 \ \Omega \underline{/90^\circ}$

$$Z_T = \frac{(20 \ \Omega \underline{/0^\circ})(35 \ \Omega \underline{/90^\circ})}{(20 \ \Omega \underline{/0^\circ}) + (35 \ \Omega \underline{/90^\circ})}$$

$$= \frac{700 \ \Omega^2 \ \underline{/90^\circ}}{(20 + j0)\Omega + (0 + j35)\Omega}$$

$$= \frac{700 \ \Omega^2 \ \underline{/90^\circ}}{(20 + j35)\Omega}$$

$$= \frac{700 \ \Omega^2 \ \underline{/90^\circ}}{40.3 \ \Omega \underline{/60.25^\circ}}$$

$$= 17.4 \ \Omega \underline{/29.75^\circ}$$

(R13-5)   Two impedances are connected in parallel. Find Z_T using the product over the sum rule. $Z_1 = j10 \ \Omega$ and $Z_2 = -j20 \ \Omega$

(R13-6)   What is the series equivalent of the circuit described in question (R13-5)?

CURRENT DIVIDER RULE   The current divider rule may be applied to two ac impedances connected in parallel.

$$I_1 = I_T \left(\frac{Z_2}{Z_1 + Z_2} \right) \qquad\qquad \text{Eq. 13.3}$$

$$I_2 = I_T \left(\frac{Z_1}{Z_1 + Z_2} \right) \qquad\qquad \text{Eq. 13.4}$$

Problem 5   Figure 13-4 shows two impedances connected in parallel. The following values are given for this circuit: $I_T = 20 \ A \underline{/0^\circ}$, $Z_1 = 3 \ \Omega \underline{/0^\circ}$, $Z_2 = 4 \ \Omega \underline{/90^\circ}$. Find I_1 using the current divider rule.

Solution $I_1 = I_T \left(\dfrac{Z_2}{Z_1 + Z_2} \right)$

$$= \frac{(20 \ A \underline{/0^\circ})(4 \ \Omega \underline{/90^\circ})}{(3 \ \Omega \underline{/0^\circ}) + (4 \ \Omega \underline{/90^\circ})}$$

$$= \frac{80 \ A \ \Omega \underline{/90^\circ}}{(3 + j4)\Omega}$$

$$= \frac{80 \ A \ \Omega \underline{/90^\circ}}{5 \ \Omega \underline{/53^\circ}}$$

$$= 16 \ A \underline{/37^\circ}$$

(R13-7)   Find I_2 in Problem 5 using the current divider rule.

(R13-8)   Draw a current phasor diagram for the circuit of Problem 5 showing I_T, I_1, and I_2.

(R13-9)   What is the phase angle between I_1 and I_2 of Problem 5?

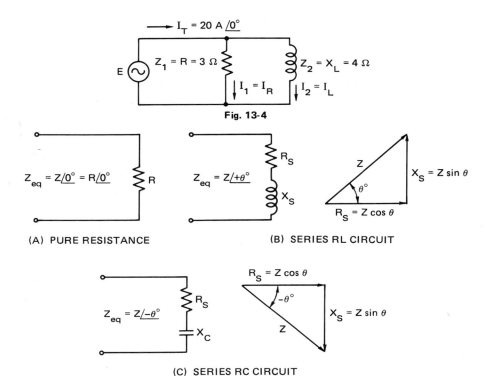

Fig. 13-4

(A) PURE RESISTANCE $\quad$ (B) SERIES RL CIRCUIT

(C) SERIES RC CIRCUIT

Fig. 13-5 Series equivalents of parallel circuits

SERIES EQUIVALENT OF A PARALLEL CIRCUIT

A parallel circuit may have a total equivalent impedance that is similar to one of the following:

1. Where $Z_{eq} = Z\underline{/0^\circ}$, the series equivalent of the circuit is a pure resistance, figure 13-5A.

2. Where $Z_{eq} = Z\underline{/+\theta^\circ}$, the series equivalent of the circuit is a series RL circuit, figure 13-5B, with

$$R_S = Z \cos \theta \qquad\qquad \text{Eq. 13.5}$$
$$X_S = Z \sin \theta \qquad\qquad \text{Eq. 13.6}$$

3. Where $Z_{eq} = Z\underline{/-\theta^\circ}$, the series equivalent of the circuit is a series RC circuit, figure 13-5C, with

$$R_S = Z \cos \theta$$
$$X_S = Z \sin \theta$$

(R13-10) What is the series equivalent of five impedances connected in parallel if $Z_{eq} = 10 \ \Omega\underline{/0^\circ}$?

(R13-11) What is the series equivalent of a parallel circuit if Z_{eq}, or $Z_T = 100 \ \Omega\underline{/45^\circ}$?

(R13-12) What is the series equivalent of a parallel circuit if $Z_{eq} = 100 \ \Omega\underline{/-45^\circ}$?

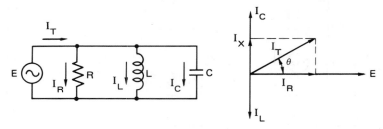

Fig. 13-6 Parallel RLC circuit

**PARALLEL
RLC CIRCUIT**   Figure 13-6 shows a parallel RLC circuit, where I_C is greater than I_L. The voltage-current phasor diagram of this circuit is also shown. The same voltage appears across each branch of a parallel circuit. Therefore, this voltage is used as the reference phasor. The currents are found as follows:

$$I_R = \frac{E \underline{/0^\circ}}{R \underline{/0^\circ}}$$

$$I_L = \frac{E \underline{/0^\circ}}{X_L \underline{/+90^\circ}} = I_L \underline{/-90^\circ}$$

$$I_C = \frac{E \underline{/0^\circ}}{X_C \underline{/-90^\circ}} = I_C \underline{/90^\circ} \qquad \text{Eq. 13.7}$$

$$\begin{aligned} I_T &= I_R - j\,I_L + j\,I_C \\ &= I_R - j(I_L - I_C) \\ &= I_R - j\,I_X \qquad \text{Eq. 13.8} \end{aligned}$$

where

$$I_X = I_L - I_C = \text{net reactive current}$$

The equivalent impedance of the parallel circuit is:

$$Z_{eq} = \frac{E}{I_T} \qquad \text{Eq. 13.9}$$

Problem 6   The circuit shown in figure 13-6 has the following values:

E = 200 V $\underline{/0^\circ}$
f = 60 Hz
R = 100 Ω
L = 0.5 H
C = 26.5 μF
Find Z_{eq}.

Solution $X_L = (377)(0.5 \text{ H}) = 188.5 \; \Omega$

$$X_C = \frac{1}{(377)(26.5 \times 10^{-6} \text{ F})} = 100 \; \Omega$$

$$\begin{aligned} I_R &= \frac{E}{R} \\ &= \frac{200 \text{ V} \underline{/0^\circ}}{100 \; \Omega \underline{/0^\circ}} \\ &= 2 \text{ A} \underline{/0^\circ} \end{aligned}$$

$$I_L = \frac{E}{X_L}$$
$$= \frac{200 \text{ V} \underline{/0^\circ}}{188.5 \ \Omega \underline{/90^\circ}}$$
$$= 1.06 \text{ A} \underline{/-90^\circ} = (-j1.06) \text{ A}$$

$$I_C = \frac{E}{X_C}$$
$$= \frac{200 \text{ V} \underline{/0^\circ}}{100 \ \Omega \underline{/-90^\circ}}$$
$$= 2 \text{ A} \underline{/90^\circ} = (+j2) \text{ A}$$

$$I_T = I_R - j I_L + j I_C$$
$$= (2 - j1.06 + j2) \text{ A}$$
$$= (2 + j0.94) \text{ A}$$
$$= 2.21 \text{ A} \underline{/25.2^\circ}$$

$$Z_{eq} = \frac{E}{I_T}$$
$$= \frac{200 \text{ V} \underline{/0^\circ}}{2.21 \text{ A} \underline{/25.2^\circ}}$$
$$= 90.5 \ \Omega \underline{/-25.2^\circ}$$

(R13-13) Draw the series equivalent circuit of the parallel circuit of Problem 6.

(R13-14) Calculate the value of each component of the equivalent circuit drawn for question (R13-13). The components should be expressed in terms of their R, L, or C parameters.

(R13-15) Draw the equivalent impedance triangle for Problem 6. Completely label all parts of the triangle.

(R13-16) What is the phase angle of the circuit in Problem 6? Does this angle represent a leading or lagging circuit?

The solutions to questions (R13-13), R13-14), and (R13-15) determined the equivalent resistance for Problem 6. This resistance is not the resistance R of this circuit. The equivalent resistance is the resistive (real) component of the complex impedance Z. The equivalent reactance is the reactive (imaginary) component of the complex impedance Z.

IMPEDANCES IN PARALLEL

There are three methods for solving for impedances connected in parallel.

1. The total current method may be used to find Z_{eq} for any number of impedances connected in parallel.
2. The second method is used for the special case of two impedances connected in parallel. The total impedance, or the equivalent impedance, is found by the product over the sum rule.
3. This method is known as the reciprocal method and applies to any number of impedances connected in parallel:

$$\frac{1}{Z_{eq}} = \frac{1}{Z_1} + \frac{1}{Z_2} + \frac{1}{Z_3} + \ldots \text{etc.} \qquad \text{Eq. 13.10}$$

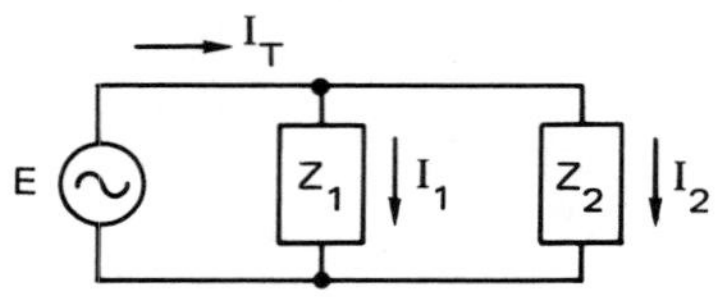

Fig. 13-7

Problem 7 Two impedances are connected in parallel, figure 13-7. $Z_1 = 60\ \Omega\ \underline{/60°}$ and $Z_2 = 80\ \Omega\ \underline{/-45°}$. The voltage of the power supply is 240 V $\underline{/0°}$.

Find Z_{eq} using the total current method.

Solution

$$I_1 = \frac{E}{Z_1}$$

$$= \frac{240\ V\ \underline{/0°}}{60\ \Omega\ \underline{/60°}}$$

$$= 4\ A\ \underline{/-60°} = (2 - j3.46)\ A$$

$$I_2 = \frac{E}{Z_2}$$

$$= \frac{240\ V\ \underline{/0°}}{80\ \Omega\ \underline{/-45°}}$$

$$= 3\ A\ \underline{/45°} = (2.12 + j2.12)\ A$$

$$I_T = I_1 + I_2$$

$$= (2 - j3.46)\ A + (2.12 + j2.12)\ A$$

$$= (4.12 - j1.34)\ A$$

$$= 4.33\ A\ \underline{/-18°}$$

$$Z_{eq} = \frac{E}{I_T}$$

$$= \frac{240\ V\ \underline{/0°}}{4.33\ A\ \underline{/-18°}}$$

$$= 55.5\ \Omega\ \underline{/18°}$$

Note: Method 3 (reciprocal method) is similar to method 1 (total current method). However, method 3, in effect, assumes that the supply voltage is one volt.

(R13-17) Solve Problem 7 using the product over the sum rule.

SERIES-PARALLEL CIRCUITS The exact methods used to solve series-parallel ac circuits depend upon a number of factors, including the actual circuit diagram, the information given, and the solution required. A general procedure can be used to solve many problems of this type. The steps in this procedure are as follows:

1. All series impedances are added by vector algebra.
2. All parallel impedances are reduced to their equivalent series impedance. As a result, the circuit is composed of a group of series-connected impedances.
3. The single equivalent impedance for the entire circuit is calculated.
4. The total current taken from the power supply is determined.
5. The individual branch currents are found by applying the current divider rule.

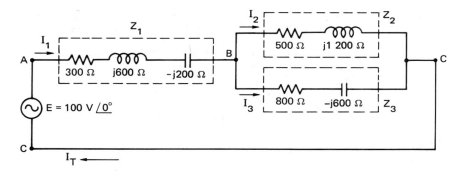

Fig. 13-8

Problem 8   A series-parallel circuit is shown in figure 13-8. Find: a. Z_{eq}, and b. I_1, I_2, I_3, and I_T.

Solution   a. Z_1 $= (300 + j600 - j200)\ \Omega$
$= (300 + j400)\ \Omega$
$= 500\ \Omega\ \underline{/53.1^\circ}$

Z_2 $= (500 + j1\ 200)\ \Omega$
$= 1\ 300\ \Omega\ \underline{/67.4^\circ}$

Z_3 $= (800 - j600)\ \Omega$
$= 1\ 000\ \Omega\ \underline{/-36.9^\circ}$

Z_{BC} $= \dfrac{Z_2\,Z_3}{Z_2 + Z_3}$

$= \dfrac{(1\ 300\ \Omega\ \underline{/67.4^\circ})(1\ 000\ \Omega\ \underline{/-36.9^\circ})}{(500 + j1\ 200)\ \Omega + (800 - j600)\ \Omega}$

$= 908\ \Omega\ \underline{/5.7^\circ} = (901 + j90.2)\ \Omega$

Z_{eq} $= Z_1 + Z_{BC}$
$= (300 + j400)\ \Omega + (901 + j90.2)\ \Omega$
$= (1\ 201 + j490)\ \Omega = 1\ 290\ \Omega\ \underline{/22.4^\circ}$

b. $I_1 = I_T = \dfrac{E}{Z_{eq}}$

$= \dfrac{100\ V\ \underline{/0^\circ}}{1\ 290\ \Omega\ \underline{/22.4^\circ}}$

$= 0.77\ 5\ A\ \underline{/-22.4^\circ}$

$I_2 = \dfrac{I_T\,Z_3}{Z_2 + Z_3}$

$= \dfrac{(0.077\ 5\ A\ \underline{/-22.4^\circ})(1\ 000\ \Omega\ \underline{/-36.9^\circ})}{(500 + j1\ 200)\ \Omega + (800 - j600)\ \Omega}$

$= 0.054\ 1\ A\ \underline{/-84.1^\circ}$

$I_3 = \dfrac{I_T\,Z_2}{Z_2 + Z_3}$

$= \dfrac{(0.077\ 5\ A\ \underline{/-22.4^\circ})(1\ 300\ \Omega\ \underline{/67.4^\circ})}{(500 + j1\ 200)\ \Omega + (800 - j600)\ \Omega}$

$= 0.070\ 9\ A\ \underline{/20.2^\circ}$

(R13-18)   What is the phase angle of the circuit in Problem 8?

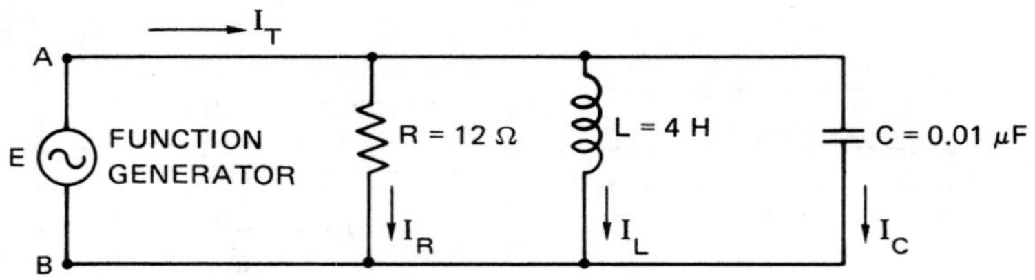

Fig. 13-9

**LABORATORY
EXERCISE 13-1
MEASUREMENT
AND ANALYSIS
OF A PARALLEL
RLC AC CIRCUIT**

PURPOSE By completing this exercise, the student will determine experimentally the characteristics of a parallel RLC ac circuit.

**EQUIPMENT
AND MATERIALS**
1 Dual trace oscilloscope
1 Sinusoidal function generator
1 Digital multimeter
1 Resistor, 12 kΩ, 2 W
1 Inductor, 4 H
1 Capacitor, 0.01 μF

PROCEDURE 1. Connect the circuit shown in figure 13-9.
   a. Turn the function generator to the ON position. Adjust the output voltage to 10 V peak-to-peak at a frequency of 1 kHz as measured on the oscilloscope.
   b. Use the digital multimeter to measure the voltage V_{AB}. Record this value in Table 13-1.
   c. Measure I_T, I_R, I_L, and I_C using the digital multimeter. Record these values in Table 13-1.
2. Perform the following calculations, and answer the following questions:
   a. Calculate the inductive reactance, capacitive reactance, and circuit impedance at a frequency of 1 kHz.
   b. Using the data of Table 13-1, determine the circuit impedance.
   c. Compare the results of steps 2.a. and 2.b. for circuit impedance.
   d. Calculate I_T, I_R, I_L, and I_C.
   e. Compare the calculated values of step 2.d. with the values measured in step 1.c.
   f. Show that $I_T = I_R + I_L + I_C$. (Use phasor algebra)

TABLE 13-1

E(V_{AB})	V_{AB} (rms)	I_T	I_R	I_L	I_C
10 V_{p-p} at f = 1 kHz					

EXTENDED STUDY TOPICS

1. A practical inductor is connected in series with a resistor. The voltage from the power supply is 120 V $\underline{/45°}$. The voltage across the resistor measures 80 V. The voltage across the terminals of the inductor also measures 80 V. Calculate the resistance of the coil winding if the circuit current is 0.5 A.

2. Three impedances are connected in series across a voltage of 100 V $\underline{/0°}$.
   Z_1 = 70.7 $\Omega \underline{/45°}$
   Z_2 = 92.4 $\Omega \underline{/-30°}$
   Z_3 = 67 $\Omega \underline{/60°}$
   Find: a. the total impedance, and b. the current taken from the power supply.

3. Find V_1, V_2, and V_3 for topic 2 using the voltage divider rule.

4. Two impedances are connected in parallel. Find Z_T and express the quantity in polar notation, using a. the total current method, and b. the product over the sum rule.
   Z_1 = 60 $\Omega \underline{/-30°}$
   Z_2 = 120 $\Omega \underline{/+60°}$

5. Two impedances are connected in parallel. The total current taken from the power supply is 20 mA $\underline{/0°}$.
   Z_1 = (900 + j3 000) Ω
   Z_2 = (750 – j1 500) Ω
   Calculate the two branch currents using the current divider rule.

6. An RLC circuit has the following components in parallel: R = 100 Ω, L = 20 mH, C = 10 μF
   The supply voltage is 35 V $\underline{/0°}$ and the frequency is 500 Hz. Find:
   a. I_R, I_L, I_C, and I_T
   b. Z_{eq} in polar notation
   c. the circuit phase angle
   d. Draw the series equivalent circuit.
   e. Draw the equivalent impedance triangle and label all parts.

7. Three impedances are connected in parallel across a power supply having a voltage equal to 33 V $\underline{/0°}$.
   Z_1 = 1 606 $\Omega \underline{/51°}$
   Z_2 = 977 $\Omega \underline{/-33°}$
   Z_3 = 953 $\Omega \underline{/-19°}$
   Find: a. I_1, I_2, I_3, and I_T, and b. Z_{eq} in polar notation.

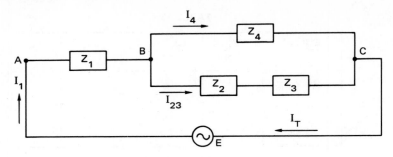

Fig. 13-10

8. The series-parallel circuit shown in figure 13-10 has the following values:

f = 250 kHz
Z_1 = (40 + j280) Ω
Z_2 = (30 + j150) Ω
Z_3 = (50 – j240) Ω
Z_4 = (90 + j400) Ω
I_{23} = 0.4 A $\underline{/0°}$

Find: a. **E** in polar notation
b. Z_{eq}
c. Draw the series equivalent circuit and label the components with their R, L, or C values.

<table>
<tr><td>

Chapter

14

</td><td>

Circuit Analysis and Network Theorems

</td></tr>
</table>

OBJECTIVES   After studying this chapter, the student will be able to

- solve circuits by the branch current method of circuit analysis.

- apply the network superposition theorem.

- apply Thévenin's theorem to reduce circuits to their Thévenin equivalent.

INTRODUCTION   The techniques and theorems used in dc circuit analysis are covered in this chapter. The techniques may be applied to ac circuits, but vector algebra must be used to determine the currents, voltages, and impedances. The maximum power transfer theorem was studied in chapter 5.

BRANCH CURRENT METHOD   The branch current method may be used to solve a network with two or more sources of emf. Such a network is shown in figure 14-1. The branch current is applied in the following steps to solve this network:

1. Label all terminal points with letters to identify the various paths. The terminal points of figure 14-1 are marked A, B, C, D, and E as shown.

2. Indicate the fixed polarity of all emf sources. Note the polarities of E_1 and E_2 as indicated in the figure.

3. Assign a current to each branch of the network. An arbitrary direction may be given to each current. In figure 14-1, the currents I_1, I_2, and I_3 are assumed to flow as indicated.

4. Show the polarity of the voltage developed across each resistor as determined by the direction of the assumed currents.

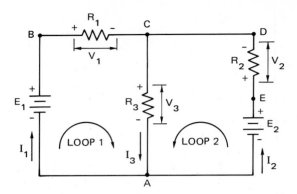

Fig. 14-1 Network solution using the branch current method

5. Apply Kirchhoff's Voltage Law around each closed loop. This network has two closed loops. Loop 1 is AEDCA. Loop 2 is ADCA. This step yields two simultaneous equations.
6. Apply Kirchhoff's Current Law at a node that includes all of the branch currents of the network. The three branch currents are included at node C. This step gives a third simultaneous equation.
7. Solve the resulting simultaneous equations for the assumed branch currents. Either of the following algebraic procedures may be used to solve the equations: a. the elimination method, or b. using determinants.
8. If the solution of the equations yields positive current values, then the directions assumed in step 3 are the actual directions. If any current value is negative, then the actual direction of the current is opposite to the assumed direction. It is not necessary to resolve the network using the true direction of current.

Problem 1 Use the branch current method to solve for the branch currents of the circuit shown in figure 14-2.

Solution
1. Loop 1 (ABCA): $+6 \text{ V} - 120 I_1 - 200 I_3 = 0$
2. Loop 2 (AEDCA): $+12 \text{ V} - 240 I_2 - 200 I_3 = 0$
3. At Node C: $I_1 + I_2 = I_3$

The three simultaneous equations are solved to obtain the following current values:

$I_1 = 2.4 \text{ mA}$
$I_2 = 26.2 \text{ mA}$
$I_3 = 28.6 \text{ mA}$

(R14-1) Write the three equations required to solve the network shown in figure 14-1.

(R14-2) Write Kirchhoff's Current Law at node A of figure 14-1.

(R14-3) Use the branch current method to solve for the branch currents of the circuit shown in figure 14-3.

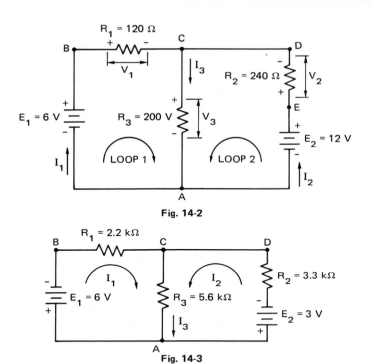

Fig. 14-2

Fig. 14-3

THE SUPERPOSITION THEOREM

The superposition theorem simplifies the analysis of a network having more than one emf source. The theorem is stated as follows:

> In any network containing more than one source of emf, the current through any branch is the algebraic sum of the currents produced by each source acting independently.

This theorem is applied in the following steps:

1. Select one source of emf. Replace all of the other sources with their internal resistances. If the internal resistance is zero, replace the source with a short circuit.
2. Calculate the magnitude and direction of the current in each branch due to the source of emf acting alone.
3. Repeat steps 1 and 2 for each source of emf until the branch current components are calculated for all sources.
4. Find the algebraic sum of the component currents to obtain the true magnitude and direction of each branch current.

Problem 2   Use the superposition theorem to solve for the current through R_1 in the circuit shown in figure 14-4A.

Solution   Consider the effect of E_1 alone. See figure 14-4B for the resulting circuit.

$$R_T = 600 \ \Omega + \frac{(200 \ \Omega)(500 \ \Omega)}{200 \ \Omega + 500 \ \Omega}$$
$$= 600 \ \Omega + 142.9 \ \Omega$$
$$= 742.9 \ \Omega$$

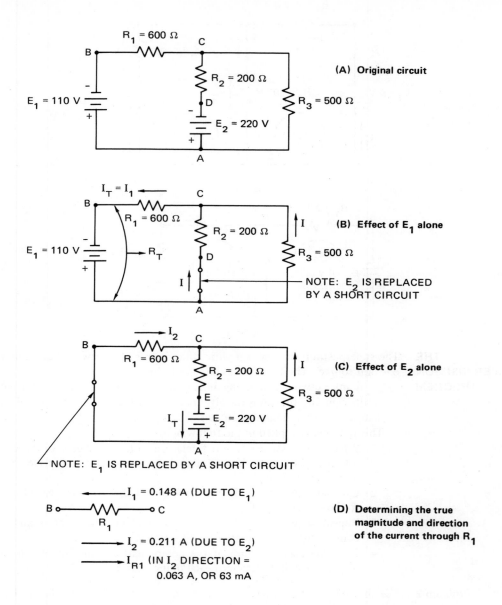

Fig. 14-4 Application of the superposition theorem

$$I_T = I_1 = \frac{E_1}{R_T}$$

$$= \frac{110 \text{ V}}{742.9 \text{ }\Omega}$$

$$= 0.148 \text{ A}$$

Now consider the effect of E_2 alone. See figure 14-4C for the resulting circuit. Note that R_1 is connected in parallel with R_3. This parallel combination is in series with R_2.

$$R_T = 200 \text{ }\Omega + \frac{(500 \text{ }\Omega)(600 \text{ }\Omega)}{500 \text{ }\Omega + 600 \text{ }\Omega}$$

$$= 200 \text{ }\Omega + 272.7 \text{ }\Omega$$

$$= 472.7 \text{ }\Omega$$

$$I_T = \frac{E_2}{R_T}$$

$$= \frac{220 \text{ V}}{472.7 \text{ }\Omega}$$

$$= 0.465 \text{ A}$$

Use the current divider rule to obtain I_2:

$$I_2 = I_T \left(\frac{R_3}{R_1 + R_3} \right)$$

$$= 0.465 \text{ A} \left(\frac{500 \text{ }\Omega}{600 \text{ }\Omega + 500 \text{ }\Omega} \right)$$

$$= 0.211 \text{ A}$$

Figure 14-4D shows how the true magnitude and direction of the current through R_1 are determined. R_1 has two components of current, one from each source of emf. I_1 (from E_1) flows from C to B. I_2 (from E_2) flows from B to C. The true direction of current through R_1 is in the direction of the I_2 component. Its magnitude is:

$$I_{R_1} = I_2 \text{ component} - I_1 \text{ component}$$

$$= 0.211 \text{ A} - 0.148 \text{ A}$$

$$= 0.063 \text{ A, or } 63 \text{ mA}$$

(R14-4) Use the superposition theorem to solve for the current through R_2 of figure 14-4A.

(R14-5) Solve for the current through R_3 of figure 14-4A using the superposition theorem.

THÉVENIN'S THEOREM   Thévenin's theorem states that:

> Any two-terminal network containing resistances and sources of emf may be replaced by a single source of emf in series with a single resistance. The emf of the single source of emf, called E_{TH}, is the open circuit emf at the network terminal. The single series resistance, called R_{TH}, is the resistance between the network terminals when all of the sources are replaced by their internal resistances.

When the Thévenin equivalent circuit is determined for a network, the process is known as "Thévenizing" the circuit. Thévenin's theorem is applied according to the following procedure:

1. Remove the load resistor and calculate the open circuit terminal voltage of the network. This value is E_{TH}.
2. Redraw the network with each source of emf replaced by a short circuit in series with its internal resistance.
3. Calculate the resistance of the redrawn network as seen by looking back into the network from the output terminals. This value is R_{TH}.
4. Draw the Thévenin equivalent circuit. This circuit consists of the series combination of E_{TH} and R_{TH}.
5. Connect the load resistor across the output terminals of the series circuit. This Thévenin circuit is equivalent to the original network. The advantage of Thévenizing a circuit is that the load resistor can be varied without changing the Thévenin equivalent circuit. Changes in R_L affect only the simple series Thévenin circuit. Therefore, the load current can be calculated easily for any number of values of R_L.

Problem 3 Use Thévenin's theorem to find the value of R_L for the circuit in figure 14-5A.
Solution Remove resistor R_L. Calculate the open circuit voltage, or E_{TH}, at terminals A-B, figure 14-5B. There is no current through R_3 in figure 14-5B. Thus, the voltage E_{TH} (E_{AB}) is the same as the voltage across resistor R_2. Use the voltage divider rule to find E_{TH}:

$$E_{TH} = 100 \text{ V} \times \frac{100 \ \Omega}{100 \ \Omega + 100 \ \Omega} = 50 \text{ V}$$

Redraw the network as shown in figure 14-5C. The source of emf (E) is replaced by a short circuit. R_{TH} is the resistance looking into the network from terminals A-B. R_3 is in series with the parallel combination of R_1 and R_2.

$$R_{TH} = 50 \ \Omega + \frac{(100 \ \Omega)(100 \ \Omega)}{100 \ \Omega + 100 \ \Omega}$$
$$= 100 \ \Omega$$

The Thévenin equivalent circuit is shown in figure 14-5D with the load resistor connected across terminals A-B.

The total resistance seen by E_{TH} is R_T.
But,

$$R_T = \frac{E_{TH}}{I_L}$$
$$= \frac{50 \text{ V}}{\frac{1}{3} \text{ A}}$$
$$= 150 \ \Omega$$
$$R_L = R_T - R_{TH}$$
$$= 150 \ \Omega - 100 \ \Omega$$
$$= 50 \ \Omega$$

(R14-6) For the circuit shown in figure 14-6, solve for the current through R_L using Thévenin's theorem.

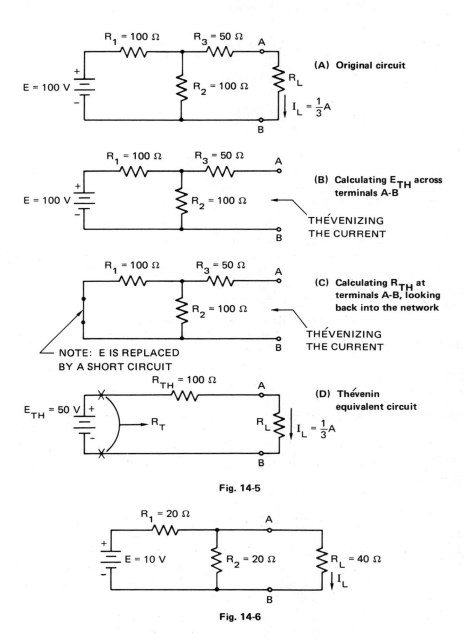

(A) Original circuit

(B) Calculating E_{TH} across terminals A-B

THÉVENIZING THE CURRENT

(C) Calculating R_{TH} at terminals A-B, looking back into the network

THÉVENIZING THE CURRENT

NOTE: E IS REPLACED BY A SHORT CIRCUIT

(D) Thévenin equivalent circuit

Fig. 14-5

Fig. 14-6

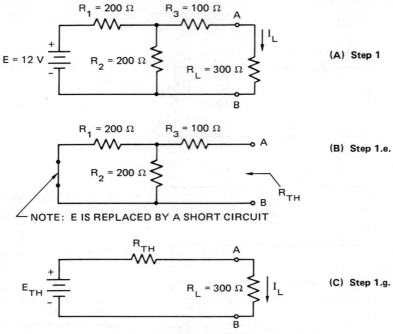

Fig. 14-7 Verifying Thévenin's Theorem

LABORATORY EXERCISE 14-1 THÉVENIN'S THEOREM

PURPOSE   By completing this exercise, the student will prove Thévenin's theorem experimentally.

EQUIPMENT AND MATERIALS

1 Variable dc power supply
1 VOM
2 Resistors, 200 Ω, 2 W
1 Resistor, 100 Ω, 2 W
1 Resistor, 300 Ω, 2 W

PROCEDURE

1. Connect the circuit shown in figure 14-7A.
   a. Turn the dc power supply ON. Adjust the output voltage to 12 V.
   b. Measure I_L and V_{AB}.
   c. Remove R_L.
   d. Measure V_{AB}. This value is E_{TH}, the Thévenin voltage.
   e. Connect the circuit shown in figure 14-7B.
   f. Measure the resistance at terminals A and B. This value is R_{TH}, the Thévenin resistance.

g. Connect the circuit shown in figure 14-7C. This circuit is the Thévenin equivalent circuit of the original network. Adjust the power supply to the value of E_{TH} measured in step 1.d. R_{TH} should be the value measured in step 1.f.

h. Measure I_L and V_{AB}.

2. Perform the following calculations and answer the following questions:

a. Calculate E_{TH} and R_{TH} of the network of figure 14-7A.

b. Compare the calculated values of E_{TH} and R_{TH} with the values measured in steps 1.d. and 1.f.

c. Compare the values of I_L and V_{AB} for the original circuit as measured in step 1.b. with the values of I_L and V_{AB} for the Thévenin equivalent circuit as measured in step 1.h.

d. What conclusions can be made from the results of this experiment?

e. What assumption is made in figure 14-7B?

EXTENDED STUDY TOPICS

1. Use the branch current method to solve for the branch currents for the circuit in figure 14-8.

2. For the circuit of figure 14-9, solve for the ac branch currents using the branch current method. Express the branch currents in rectangular notation.

3. For the circuit of topic 2, do the real components of the currents at node C add algebraically to zero? Do the reactive (imaginary) components of the currents at node C add algebraically to zero?

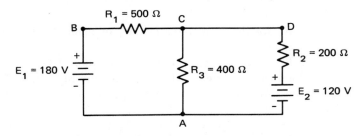

Fig. 14-8

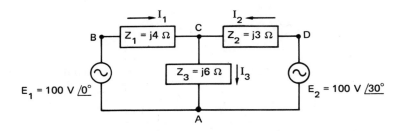

Fig. 14-9

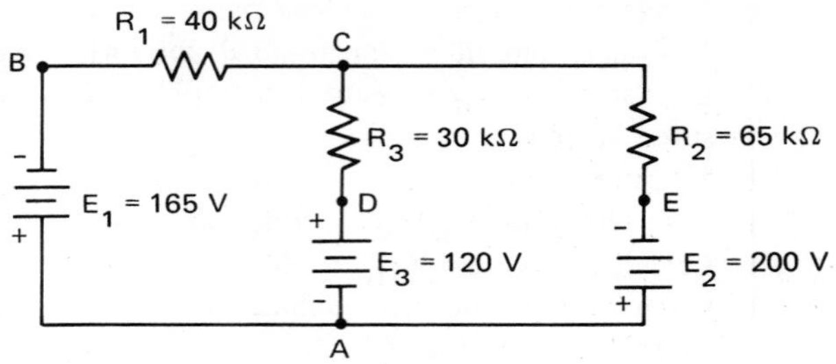

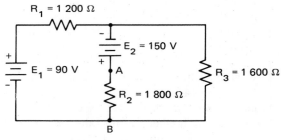

Fig. 14-10

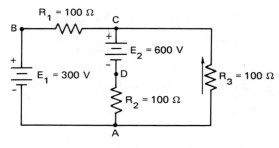

Fig. 14-11

4. Solve for the current through R_1 of figure 14-10 using the superposition theorem.
5. Solve for the current through R_2 of figure 14-10 using the superposition theorem.
6. Solve for the current through R_3 of figure 14-10 using the superposition theorem.
7. Solve for the current through R_2 of figure 14-11 using Thévenin's theorem.
8. Solve for the current through R_3 of figure 14-12 using Thévenin's theorem.
9. Solve for the current through R_3 of figure 14-12 using the branch current method.
10. Solve for the current through R_3 of figure 14-12 using the superposition theorem.

Fig. 14-12

<table>
<tr>
<td>

Chapter
15

</td>
<td>

Power in AC Circuits

</td>
</tr>
</table>

OBJECTIVES   After studying this chapter, the student will be able to

- discuss power in ac circuits with various combinations of R, L, and C.

- apply the power triangle to problems.

- solve for the power factor and reactive factor of a circuit.

- apply power factor correction techniques.

POWER IN A CIR-CUIT WITH PURE RESISTANCE   The instantaneous power dissipated in a resistor connected to an ac power supply may be calculated as follows:

$$p = ei = \frac{e^2}{R} = i^2 R \qquad \text{Eq. 15.1}$$

where

p = instantaneous power dissipated in resistor R, in watts
e = instantaneous value of the ac emf, in volts
i = instantaneous value of the current flowing through resistor R, in amperes
R = resistance of resistor R, in ohms

The waveforms of the voltage and current through R are sine waves. These waves are in phase, as explained in Chapter 10. Since R is a constant, the instantaneous power is a sine-squared wave. Figure 15-1 shows the relationship of e, i, and p in a circuit with pure resistance.

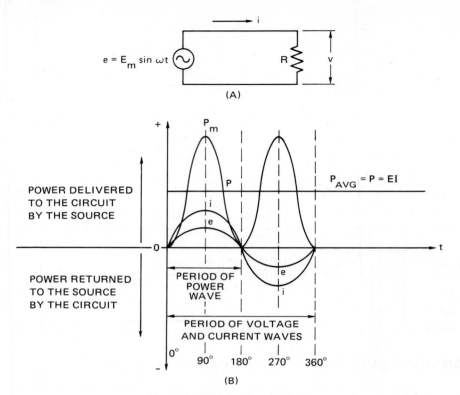

Fig. 15-1 Power in a pure R circuit

A number of conclusions can be made by analyzing figure 15-1.

1. The power wave is a sine wave. It is symmetrical around an average value of power dissipation, P. The same letter symbol, P, is used to express both the true power, or real power, in an ac circuit and the power in a dc circuit. The power dissipated by the resistor, as measured by a wattmeter, is P.
2. The power wave has twice the frequency of the voltage wave.
3. The instantaneous power is always positive. Positive power represents the conversion of electric energy from one form into another. In figure 15-1, the energy conversion is in the form of heat.
4. P_m is the instantaneous peak power and is expressed in the following forms:

$$P_m = E_m I_m \qquad \text{Eq. 15.2}$$
$$P_m = I_m{}^2 R \qquad \text{Eq. 15.3}$$
$$P_m = \frac{E_m{}^2}{R} \qquad \text{Eq. 15.4}$$

5.
$$P = \frac{P_m}{2} \qquad \text{Eq. 15.5}$$
$$P = VI \qquad \text{Eq. 15.6}$$
$$P = I^2 R \qquad \text{Eq. 15.7}$$
$$P = \frac{V^2}{R} \qquad \text{Eq. 15.8}$$

where

P = average power dissipation, in watts

V = rms voltage across the resistor, in volts

I = rms current through the resistor, in amperes

6. The area under the power curve represents energy since:

$$W = Pt$$

(see Chapter 5)

Problem 1 For the circuit shown in figure 15-1A, the wattmeter reading is 200 W and the current reading is 2 A. Determine the value of the resistor R.

Solution P = 200 W

I = 2 A

$P = I^2 R$

$R = \dfrac{P}{I^2}$

$= \dfrac{200 \text{ W}}{(2 \text{ A})(2 \text{ A})}$

$= 50 \ \Omega$

Refer to Problem 1 to answer questions (R15-1) through (R15-4).

(R15-1) Use Ohm's Law to calculate the voltage across resistor R.

(R15-2) Calculate the voltage across R using Equation 15.6.

(R15-3) Calculate the voltage across R using Equation 15.8.

(R15-4) What is the instantaneous peak power?

(R15-5) Why is the power wave of figure 15-1B always positive?

POWER IN AN IDEAL INDUCTOR WITH L ONLY Figure 15-2 shows the relationship of e, i, and p in an ideal inductor with L only. The conclusions that can be made from this figure are as follows:

1. The power wave is a sine wave. It is symmetrical around the horizontal, or reference, axis.

2. The power wave has twice the frequency of the current wave.

3. In the intervals from 0° to 90°, and from 180° to 270°, the following are true:

   a. i is increasing.

   b. the magnetic field builds up around the inductor.

   c. the inductor takes or absorbs energy from the power supply.

   d. the energy taken from the power supply is stored in the magnetic field of the inductor.

   e. the instantaneous power is positive; thus, energy is delivered to the inductor by the power supply.

4. In the intervals from 90° to 180°, and from 270° to 360°, the following are true:

   a. i is decreasing.

   b. the magnetic field around the inductor collapses.

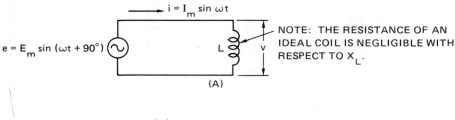

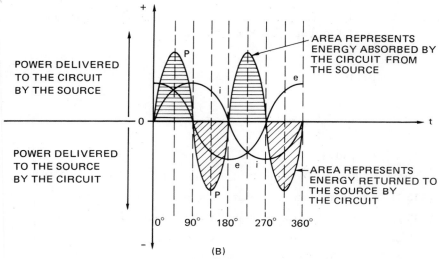

Fig. 15-2 Power in an ideal inductor

      c. the instantaneous power is negative; thus, energy is returned by the inductor to the source.

   5. The average, or true, power in a pure inductor is zero and a wattmeter will indicate zero watts. An ideal inductor in an ac circuit alternately takes and then returns equal amounts of energy to the power supply.

The product of the rms voltage and the rms current in an ideal inductor is called the *reactive power* of the inductor. The letter symbol for inductive reactive power is Q_L. The unit of reactive power is the voltampere reactive which is abbreviated var.

$$Q_L = V_L I_L \qquad\qquad \text{Eq. 15.9}$$

where

     Q_L = inductive reactive power, in vars
     V_L = rms voltage across the inductor, in volts
     I_L = rms current through the inductor, in amperes
In Chapter 10 it was shown that:

$$X_L = \frac{V_L}{I_L}$$

Therefore

$$Q_L = I^2_L X_L \qquad \text{Eq. 15.10}$$

and

$$Q_L = \frac{V_L{}^2}{X_L} \qquad \text{Eq. 15.11}$$

Problem 2   A 1.0-H inductor draws a current of 2.0 A from a 60-Hz source, figure 15-2. Find: a. the inductive reactance, and b. the reactive power.

Solution   a. $X_L = \omega L$
$$= (377)(1.0 \text{ H})$$
$$= 377 \ \Omega$$
b. $Q_L = I_L{}^2 X_L$
$$= (2.0 \text{ A})^2 (377 \ \Omega)$$
$$= 1 \ 508 \text{ vars}$$

Questions (R15-6) through (R15-8) refer to Problem 2.

(R15-6)   Use Ohm's Law and calculate the voltage across the inductor.

(R15-7)   Calculate the voltage across the inductor using Equation 15.9.

(R15-8)   Calculate the voltage across the inductor using Equation 15.11.

POWER IN AN IDEAL CAPACITOR WITH C ONLY   Figure 15-3 shows the relationship of e, i, and p in an ideal capacitor with C only. A study of this figure yields the following conclusions:

1. The power wave is a sine wave. It is symmetrical around the horizontal, or reference, axis.
2. The power wave is 180° out of phase with the power wave shown in figure 15-2.
3. The power wave has twice the frequency of the applied emf.
4. In the intervals from 90° to 180°, and from 270° to 360°, the following conditions are observed:
   a. e is increasing in magnitude and the potential drop across C is increasing.
   b. the capacitor takes energy from the source as it stores a charge on its plates.
   c. the energy taken from the power supply is stored in the electrostatic field in the dielectric of the capacitor.
   d. the instantaneous power is positive, indicating that energy is delivered to the capacitor by the power supply.
5. In the intervals from 0° to 90°, and from 180° to 270°, the following are true:
   a. e is decreasing and the potential drop across the capacitor is decreasing.
   b. the instantaneous power is negative; that is, energy is returned by the capacitor to the source.
6. The average, or true, power in a pure capacitor is zero and a wattmeter will indicate zero watts. An ideal capacitor in an ac circuit alternately takes energy from the power supply and returns an equal amount of energy to the supply.

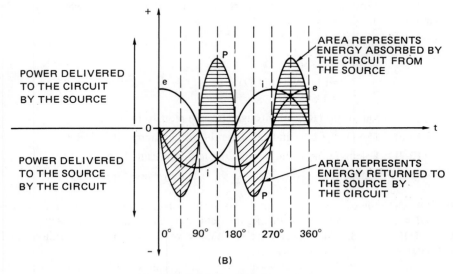

Fig. 15-3 Power in an ideal capacitor

The product of the rms voltage and the rms current in an ideal capacitor is called the *reactive power* of the capacitor (Q_C). The unit of reactive power is the voltampere reactive and is abbreviated var.

$$Q_C = V_C I_C \qquad \text{Eq. 15.12}$$

where

Q_C = capacitive reactive power, in vars
V_C = rms voltage across the capacitor, in volts
I_C = rms current drawn by the capacitor, in amperes

In Chapter 12, it was shown that

$$X_C = \frac{V_C}{I_C}$$

Therefore,

$$Q_C = I_C{}^2 X_C \qquad \text{Eq. 15.13}$$

$$Q_C = \frac{V_C{}^2}{X_C} \qquad \text{Eq. 15.14}$$

Problem 3 In the circuit of figure 15-3, a 10-μF capacitor draws a current of 2.0 A from a 60-Hz source. Find: a. the capacitive reactance, and b. the value of reactive power.

Solution a. $X_C = \dfrac{1}{\omega C}$

$$= \frac{1}{(377)(10 \times 10^{-6}\ F)}$$

$$= 265\ \Omega$$

  b. $Q_C = I_C{}^2 X_C$

$$= (2.0\ A)^2 (265\ \Omega)$$

$$= 1\ 060\ \text{vars}$$

Questions (R15-9) through (R15-11) refer to Problem 3.

(R15-9) Calculate the voltage across the capacitor using Ohm's Law.

(R15-10) Calculate the voltage across the capacitor using Equation 15.12

(R15-11) Calculate the voltage across the capacitor using Equation 15.14.

POWER IN AN RL CIRCUIT Figure 15-4 shows the relationship of e, i, and p in an RL circuit where X_L equals R.

(R15-12) What is the phase relationship between E and I in figure 15-4?

(R15-13) What is the relationship between V_R and V_L in figure 15-4?

A study of figure 15-4 leads to the following conclusions:

1. The power wave is a sine wave. It is not symmetrical around the horizontal, or reference, axis. However, it is symmetrical around a value of P that is determined as follows:

$$P = VI \cos \theta \qquad \text{Eq. 15.15}$$

  where

  P = true power, in watts
  V = rms voltage across the series combination of R and L, in volts
  I   = rms current in the series combination of R and L, in amperes
  θ = phase angle between V and I

2. The frequency of the power wave is twice that of the applied emf or the circuit current.

3. In the intervals from 0° to 135°, and from 180° to 315°, the power is positive. This means that the power supply provides power to the circuit during these intervals.

4. In the intervals from 135° to 180°, and from 315° to 360°, the power is negative. During these intervals, therefore, the circuit returns power to the power supply.

5. The instantaneous power is more positive than negative. The average power is the true power of the circuit. Equation 15.15 can be used to determine the true power input to the circuit. This value is the power dissipated by resistor R.

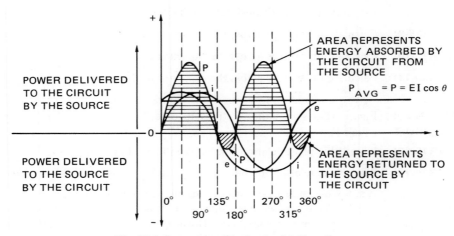

Fig. 15-4 Power in an RL circuit with $X_L = R$

The apparent power (S) of the circuit is the product of the total voltage and the total current. The voltampere (VA) is the unit of apparent power.

$$S = E_T I_T \qquad \text{Eq. 15.16}$$

where

S = apparent power, in VA
E_T = total rms voltage, in volts
I_T = total rms current, in amperes

Problem 4   A series RL circuit is similar to the circuit shown in figure 15-4. The following values are given for this circuit:

E_T = 100 V
I_T = 100 mA
θ = 60°

a. Calculate the apparent power.
b. Find the true power supplied to the circuit.

Fig. 15-5 Phase meter

Solution   a. $S = E_T I_T$
$= (100 \text{ V})(100 \times 10^{-3} \text{ A})$
$= 10 \text{ VA}$
b. $P = E I \cos \theta$
$P = (100 \text{ V})(100 \times 10^{-3} \text{ A})(\cos 60°)$
$= (100 \text{ V})(100 \times 10^{-3} \text{ A})(0.5)$
$= 5 \text{ W}$

(R15-14)   What is the value of R in Problem 4?

(R15-15)   A series RL circuit is similar to the one shown in figure 15-4. The following values are given for this circuit: $S = 5\ 000 \text{ VA}$, $E_T = 100 \text{ V}$, and $\theta = 53°$. Find the value of I_T.

The circuit phase angle, θ, can be measured using an instrument called a *phase meter*, figure 15-5.

THE POWER TRIANGLE   Figure 15-6 shows power triangles for ac circuits. Power triangles are made following the standard of drawing inductive reactive power in the +j direction and capacitive reactive power in the –j direction.

Two expressions are obtained by applying the Pythagorean theorem to these power triangles:

$$S = \sqrt{P^2 + Q_L^2}$$ Eq. 15.17

$$S = \sqrt{P^2 + Q_C^2}$$ Eq. 15.18

Equations 15.17 and 15.18 can be applied to series circuits, parallel circuits, and series-parallel circuits.

The net reactive power supplied by the source to an RLC circuit is the difference between the positive inductive reactive power and the negative capacitive reactive power.

$$Q_X = Q_L - Q_C$$ Eq. 15.19

where Q_X = net reactive power, in vars

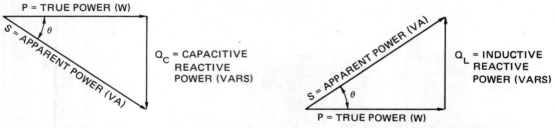

(A) CAPACITIVE POWER TRIANGLE   (B) INDUCTIVE POWER TRIANGLE

Fig. 15-6 Power triangles

Problem 5   For the circuit shown in figure 15-7, find: a. total watts, b. total vars, and c. total voltamperes.

Solution   a. $P_T = P_1 + P_2$
$ = 200\ W + 500\ W$
$ = 700\ W$

Arithmetic addition can be used to find the total power because power is always a real quantity on the horizontal axis.

b. $Q_X = Q_L - Q_C$
$ = 1\ 200\ vars - 500\ vars$
$ = 700\ vars\ (inductive)$

c. $S = \sqrt{P_T{}^2 + Q_X{}^2}$
$ = \sqrt{(700\ W)^2 + (700\ vars)^2}$
$ = 989.9\ VA$

(R15-16)   Calculate the total current in the circuit for Problem 5.

(R15-17)   What is the phase angle between E_T and I_T in Problem 5?

POWER FACTOR AND REACTIVE FACTOR   The *power factor* of an ac load is the numerical ratio between the true power (P) and the apparent power (S) of the load. It can be seen by referring to the power triangles of figure 15-6 that this ratio is equal to the cosine of the power factor angle, θ. The power factor angle is the same as the phase angle between the voltage across the load and the current through the load.

$$\text{Power factor} = \cos \theta = \frac{P}{S} \qquad \text{Eq. 15.20}$$

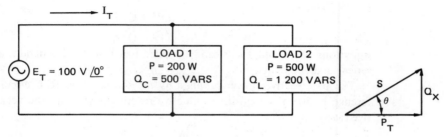

Fig. 15-7

Inductive loads have a lagging power factor. Capacitive loads have a leading power factor. The value of the power factor is expressed either as a decimal or as a percentage. This value is always less than 1.0 or less than 100%. The majority of industrial loads, such as motors and air conditioners, are inductive loads.

The numerical ratio between the reactive power and the apparent power of an ac load is called the *reactive factor*. By referring to the power triangles of figure 15-6, it can be seen that this ratio is equal to the sine of the power factor angle.

$$\text{Reactive factor} = \sin \theta = \frac{Q_L}{S} \text{ or } \frac{Q_C}{S} \qquad \text{Eq. 15.21}$$

Problem 6 Calculate the power factor and the reactive factor for the circuit shown in figure 15-8.

Solution $Z_1 = R + jX_L$
$= 100\ \Omega + j100\ \Omega$
$= 141.4\ \Omega\ \underline{/+45°}$

$I_1 = \dfrac{E}{Z_1}$
$= \dfrac{120\ V\ \underline{/0°}}{141.4\ \Omega\ \underline{/+45°}}$
$= 0.849\ A\ \underline{/-45°}$
$= (0.6 - j0.6)\ A$

$I_2 = \dfrac{E}{X_C}$
$= \dfrac{120\ V\ \underline{/0°}}{60\ \Omega\ \underline{/-90°}}$
$= 2\ A\ \underline{/+90°}$
$= (0 + j2)\ A$

$I_T = I_1 + I_2$
$= (0.6 - j0.6)\ A + (0 + j2)\ A$
$= (0.6 + j1.4)\ A$
$= 1.523\ A\ \underline{/66.8°}$

$S = E_T\ I_T$
$= (120\ V)(1.523\ A)$
$= 182.8\ VA$

Power factor $= \cos \theta$
$= \cos 66.8°$
$= 0.394,\ \text{or } 39.4\%$

Reactive factor $= \sin \theta$
$= \sin 66.8°$
$= 0.92,\ \text{or } 92\%$

Questions (R15-18) through (R15-21) refer to Problem 6.

(R15-18) Calculate the true power of the circuit.

(R15-19) Calculate the power factor of the circuit using the calculated values of P and S.

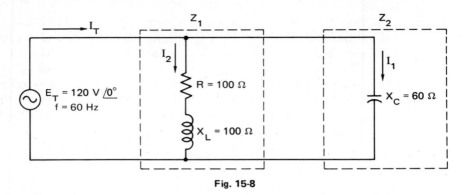

Fig. 15-8

(R15-20)   Calculate the power dissipated in Z_1.

(R15-21)   Compare and explain the results of questions (R15-18) and (R15-20).

POWER FACTOR CORRECTION   The effects of a power factor with a value less than unity are shown in figure 15-9. For figure 15-9A, the motor current is calculated as follows:

$$S = \frac{P}{\cos \theta}$$
$$= \frac{1\ 200\ \text{W}}{0.7}$$
$$= 1\ 714\ \text{VA}$$
$$I = \frac{S}{E}$$
$$= \frac{1\ 714\ \text{VA}}{120\ \text{V}}$$
$$= 14.29\ \text{A}$$

The active component of this current is the component in phase with the voltage. This component, which results in true power consumption, is:

$$I \cos \theta = (14.29\ \text{A})(0.7) = 10\ \text{A}$$

In other words, the 70% power factor of the motor means that the circuit must supply 14.29 A to obtain a useful current of 10 A.

The amount of current required by the load determines the sizes of the wire used in the windings of the generator and in the conductors connecting the motor to the generator. The copper losses due to the wire in the generator and conductors depend upon the square of the load current. For this reason, a public utility company finds it more economical to supply 10 A at a power factor of 100% than to supply 14.29 A at a power factor of 70%.

Problem 7   For the circuit shown in figure 15-9B, calculate the value of the capacitor needed to obtain a circuit power factor of 100%.

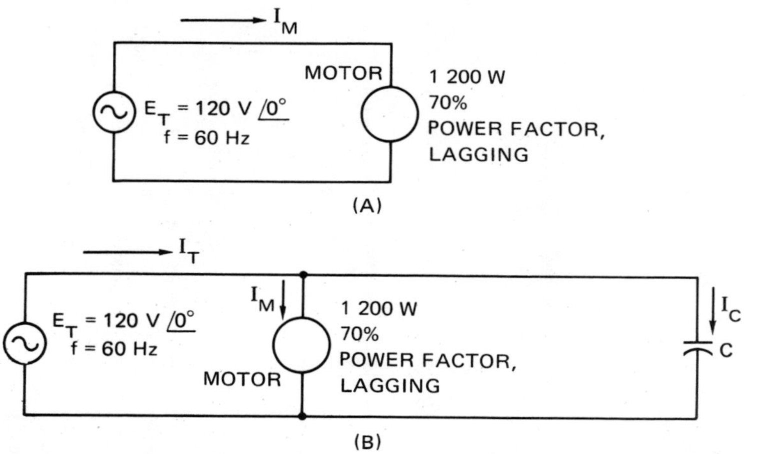

Fig. 15-9 Effect of low power factor and its correction

Solution   In this parallel circuit, the motor draws 14.29 A at a 70% lagging power factor. To obtain a circuit power factor of 100%, the inductive vars of the motor and the capacitive vars of the capacitor must be equal.

$$Q_L = EI\sqrt{1 - \cos^2\theta} \qquad\qquad \text{Eq. 15.22}$$

where

$$\sqrt{1 - \cos^2\theta} = \text{reactive factor}$$

$$\begin{aligned} Q_L &= (120\ \text{V})(14.29\ \text{A})\sqrt{1 - (0.7)^2} \\ &= 1\ 714\sqrt{0.51} \\ &= 1\ 224\ \text{vars (inductive)} \end{aligned}$$

Q_C (capacitive) must equal 1 224 vars (inductive) to obtain a 100% power factor.

$$\begin{aligned} X_C &= \frac{V_C^2}{Q_C} \\ &= \frac{(120\ \text{V})^2}{1\ 224\ \text{vars (capacitive)}} \\ &= 11.76\ \Omega\ \text{(capacitive)} \end{aligned}$$

$$\begin{aligned} C &= \frac{1}{\omega X_C} \\ &= \frac{1}{(377)(11.76\ \Omega)} \\ &= 225.5\ \mu\text{F} \end{aligned}$$

(R15-22)   What is the value of the current I_C in Problem 7?

(R15-23)   What is the value of the reactive component of the current I_M in figure 15-9A?

(R15-24)   Draw the current phasor diagram for the circuit of figure 15-9B. Use E_T as a reference and label all parts of the diagram.

(R15-25)   What is the value of the total current in the circuit shown in figure 15-9B?

(R15-26)   Find the apparent power for the circuit of figure 15-9A.

(R15-27)   Calculate the apparent power for figure 15-9B.

Fig. 15-10 High-voltage power capacitors used for power factor correction

(A) Bank of capacitors in a substation

(B) Pole-mounted bank of capacitors

Fig. 15-11 Capacitors used for power factor correction

CONCLUSIONS A number of conclusions can be drawn from the preceding discussion on power factor correction:

1. To reduce the apparent power drawn from a power supply, a capacitor is added in parallel with an inductive load.
2. The current through the inductive load is not changed when the capacitor is added in parallel with the load.
3. The total current drawn from the source is reduced.
4. When the load is inductive, a circuit power factor of 100% is obtained by connecting a capacitor in parallel with the load. This capacitor must have a reactive power equal to the reactive power of the load.
5. The best power factor correction is obtained by connecting the capacitor as close to the inductive load as possible.
6. There is less voltage drop in the line wires between the source and load.

High-voltage power capacitors used for power factor correction are shown in figure 15-10. Figure 15-11A shows a bank of capacitors installed in a substation. A pole-mounted bank of capacitors is shown in figure 15-11B. Figure 15-12 shows pole-mounted capacitors which have been installed in a residential area by a public utility company.

Fig. 15-12 Pole-mounted capacitors in a residential area

LABORATORY EXERCISE 15-1 MEASUREMENT AND ANALYSIS OF POWER IN AC CIRCUITS

PURPOSE By completing this exercise, the student will be able to understand and apply power factor corrections in ac circuits.

EQUIPMENT AND MATERIALS

1 Dual trace oscilloscope
1 Sinusoidal function generator
1 Digital multimeter
1 Inductor, 4 H (low resistance)
1 Capacitor, 0.01 μF
1 Resistor, 1 kΩ, 2 W
1 Resistor, 100 Ω, 2 W

PROCEDURE

1. Connect the circuit shown in figure 15-13.
   a. Turn the function generator ON. Adjust the output voltage to 10 V (rms), as measured with the digital multimeter across terminals A and B at a frequency of 600 Hz.
   b. Measure I_T using the digital multimeter.
   c. Measure the phase angle between E_{AB} and I_T on the oscilloscope.
   d. Connect the capacitor as shown in the figure. Measure the phase angle between E_{AB} and I_T on the oscilloscope.
2. Perform the following calculations and answer the following questions:
   a. Using the data from steps 1.a., 1.b., and 1.c., calculate the apparent power, the true power, the reactive power, and the power factor of the circuit.

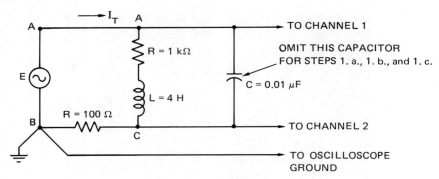

Fig. 15-13

b. Using the data from step 1.d., repeat the calculations of step 2.a.
c. What conclusions can be reached by analyzing the results of steps 2.a. and 2.b.?
d. Calculate the size of the capacitor required to obtain a 100% power factor.

EXTENDED STUDY TOPICS

1. An electric appliance draws 15 A from a 120-V, 60-Hz source.
   a. Calculate the power consumption of the appliance.
   b. What is the peak value of the instantaneous power input to the appliance?
2. A 100-Ω resistor is connected to a voltage having a value of e = 200 sin ωt.
   a. Find the peak value of the instantaneous current through the resistor.
   b. Calculate the true power dissipated by the resistor.
3. An ideal inductor having a value of 5 mH draws a 30-mA current from a 1.0-kHz power supply.
   a. Calculate the true power dissipated by the inductor.
   b. Calculate the inductive reactive power.
4. A capacitor having a value of 25 μF draws 1.5 A from a 120-V, 60-Hz source.
   a. Calculate the true power dissipated by the capacitor.
   b. Calculate the capacitive reactive power.
5. For the circuit shown in figure 15-14, calculate: a. the apparent power, b. the power factor of the circuit, and c. the reactive factor of the circuit.

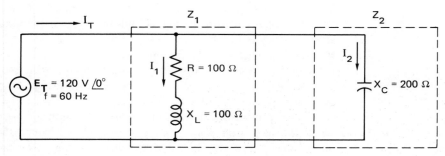

Fig. 15-14

6. For the circuit of topic 5, figure 15-14, find: a. the inductive vars, b. the capacitive vars, and c. the net reactive vars supplied by the source to the circuit.
7. For the circuit of topic 6, explain how the circuit operates in relation to the flow of reactive power.
8. Repeat topic 5 with the capacitor removed from the circuit.
9. Discuss the effects of including the capacitor in the circuit of figure 15-14.

<table>
<tr><td>

Chapter
16

</td><td>

Resonance and Filters

</td></tr>
</table>

OBJECTIVES

After studying this chapter, the student will be able to

- explain the performance of series and parallel circuits at resonance.
- explain the resonant rise in voltage in a series circuit at resonance.
- discuss the bandwidth of a circuit.
- describe the figure of merit (Q).
- solve for the current and impedance magnification in a parallel circuit at resonance.
- discuss filter networks.

THE IDEAL
SERIES LC
CIRCUIT

It was shown in Chapter 12 that the reactances of capacitors and inductors are frequency sensitive. Figure 16-1 shows an ideal series LC circuit connected to a variable frequency source. When $X_L = X_C$ in such a circuit, a condition known as *resonance* occurs.

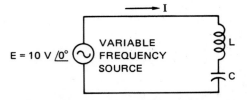

Fig. 16-1 Ideal series LC circuit (with no resistance)

The frequency at which resonance occurs is called the *resonant frequency*, f_r. The following derivation yields an equation for f_r.

$$X_L = X_C$$

$$2\pi f_r L = \frac{1}{2\pi f_r C}$$

$$f_r^2 = \frac{1}{4\pi^2 LC}$$

$$f_r = \frac{1}{2\pi \sqrt{LC}} \qquad \text{Eq. 16.1}$$

where

f_r = resonant frequency, in Hz
L = inductance, in henries
C = capacitance, in farads

(R16-1) Determine the resonant frequency of a 0.2-μH inductor connected in series with a 0.2-μF capactitor to the variable frequency source shown in figure 16-1.

(R16-2) What is the value of I in the ideal series circuit of question (R16-1)?

THE PRACTICAL SERIES RLC CIRCUIT

Figure 16-2 shows a practical series RLC circuit. R represents the entire series resistance, including the resistance of the windings of the inductor.

Problem 1 Assume the following values for the circuit shown in figure 16-2:

E = 100 V $\underline{/0°}$
R = 1 000 Ω
L = 0.2 H
C = 0.05 μF
Find: a. f_r; b. Z_T at f_r; and c. I at f_r.

Solution a. $f_r = \dfrac{1}{2\pi\sqrt{LC}}$

$$= \frac{1}{2\pi\sqrt{(0.2 \text{ H})(0.05 \times 10^{-6} \text{ F})}}$$

$$\cong 1\ 590 \text{ Hz}$$

b. $Z_T = R + j(X_L - X_C)$
But,
$X_L = X_C$ at f_r
Therefore,
$Z_T = R$
$\quad = 1\ 000\ \Omega\underline{/0°}$

c. $I = \dfrac{E}{Z_T} = \dfrac{E}{R}$

$$= \frac{100 \text{ V} \underline{/0°}}{1\ 000\ \Omega\underline{/0°}}$$

$$= 0.1 \text{ A} \underline{/0°}$$

The only opposition to current at series resonance is the resistance R, since $X_L = X_C$ at f_r.

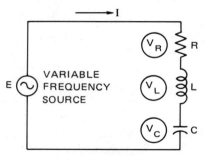

Fig. 16-2 Practical series RLC circuit

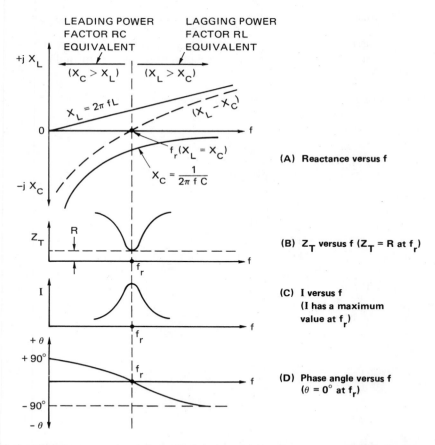

Fig. 16-3 Effect of frequency variation on the reactances, total impedance, current, and phase angle for a series RLC circuit

Figure 16-3 shows the effect of a varying frequency on the reactances, total impedance, current, and phase angle for the RLC series circuit of figure 16-2.

RESONANT RISE IN VOLTAGE

At the resonant frequency of the circuit of figure 16-2, the following expressions are true:

$$V_L = j\,IX_L \qquad\qquad \text{Eq. 16.2}$$
$$V_C = -j\,X_C \qquad\qquad \text{Eq. 16.3}$$
$$V_R = I\,R \qquad\qquad \text{Eq. 16.4}$$

These voltages depend upon the current, reactance, and resistance of the individual components. X_L and X_C may have very high values at the resonant frequency, but the total reactance is zero. The voltages across L and C are 180° out of phase and may have values much higher than the voltage of the power source. For this reason, the electronics technician must be very careful when working with series resonant circuits. This effect is called the *resonant rise in voltage* in a series RLC circuit, figure 16-4.

(R16-3) Calculate the values of the following quantities for the circuit of Problem 1 at the resonant frequency: $X_L, X_C, V_R, V_L,$ and V_C.

(R16-4) Draw the voltage and current phasor diagrams for Problem 1 and question (R16-3). Label all parts of the diagrams.

APPLICATION OF THE SERIES RESONANT CIRCUIT

Figure 16-5 shows a simplified schematic of the first radio frequency (RF) stage of a radio receiver. Voltages at many frequencies are induced into the antenna. These voltages cause currents in the primary of the transformer. These currents, in turn, induce voltages in the secondary winding, as shown in the equivalent circuit of figure 16-5B. The components L_2 and C are in series and there is only one path which the current can follow. Note that this series network only appears to be a parallel circuit because of the location of the output signal voltage, V_C.

The tuning capacitor C can be adjusted until the circuit resonates at the signal of the selected radio station. The frequency at resonance causes a resonant rise in voltage, V_C, across the capacitor C. This voltage is then amplified by succeeding stages of the radio receiver.

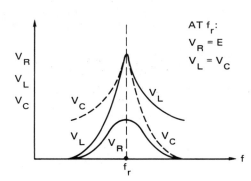

Fig. 16-4 Resonant rise in voltage for the series RLC circuit of figure 16-2: $V_R, V_L,$ and V_C versus f

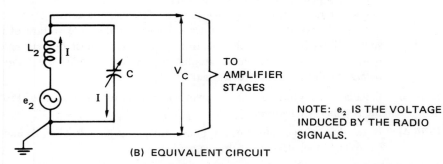

(A) SCHEMATIC

(B) EQUIVALENT CIRCUIT

NOTE: e_2 IS THE VOLTAGE
INDUCED BY THE RADIO
SIGNALS.

Fig. 16-5 Simplified schematic of the first radio frequency (RF) stage of a radio receiver and
its equivalent circuit.

Problem 2   A radio receiver is to be tuned to a radio station broadcasting at a frequency of
1 000 Hz. If $L_2 = 100\ \mu H$, calculate the value of the tuning capacitor. (Refer to
figure 16-5A.)

Solution $f_r = \dfrac{1}{2\pi \sqrt{LC}}$

$C = \dfrac{1}{4\pi^2\, f_r^{\,2}\, L}$

$C = \dfrac{1}{4\pi^2 (1\ 000 \times 10^3\ Hz)^2\ (100 \times 10^{-6}\ H)}$

$= 253\ pF$

(R16-5)   Calculate the value of C in Problem 2 if the radio frequency is 650 kHz.

THE FIGURE OF   The *figure of merit* is the ratio of the voltage V_C, or V_L, at resonance to the voltage
MERIT (Q)   of the power source. This value is an indication of the quality of a series resonant
circuit. The ratio is also known as the *Q factor* of the circuit and as the *voltage
magnification factor*.

At series resonance:

$$V_L = I\, X_L$$

Since

$$I = \frac{E}{R}$$

Thus,

$$V_L = \left(\frac{E}{R}\right) X_L$$

$$\frac{V_L}{E} = \frac{X_L}{R} = Q \qquad\qquad \text{Eq. 16.5}$$

$$V_L = Q\,E \qquad\qquad \text{Eq. 16.6}$$

Equation 16.6 shows the voltage magnification effect of a series resonant circuit. The voltage across the coil is Q times that of the applied signal E, but it is 90° out of phase with this applied signal.

At resonance:

$$V_L = V_C$$

According to Equation 16.6:

$$V_C = Q\,E \qquad\qquad \text{Eq. 16.7}$$

But,

$$X_L = X_C$$

Therefore,

$$Q = \frac{X_L}{R} = \frac{X_C}{R}$$

Substitute $X_C = \dfrac{1}{\omega C}$ in the previous equation to obtain:

$$Q = \frac{1}{\omega CR} \qquad\qquad \text{Eq. 16.8}$$

But,

$$Q = \frac{X_L}{R}$$

$$= \frac{2\pi f_r L}{R}$$

Substituting for f_r yields:

$$Q = \frac{2\pi L \left(\dfrac{1}{2\pi \sqrt{LC}}\right)}{R}$$

This expression reduces to:

$$Q = \frac{1}{R}\sqrt{\frac{L}{C}} \qquad\qquad \text{Eq. 16.9}$$

Q is a measure of the ability of a resonant circuit to select or reject a band of frequencies. As the value of Q increases, the selectivity of a series resonant circuit also increases. Figure 16-6 shows the frequency response of a series RLC circuit with three different values of Q.

In many cases, the resistance of the coil is the only resistance in a series resonant circuit. For these cases, the Q factor is called the Q factor of the coil.

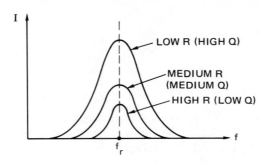

Fig. 16-6 Frequency response of a series RLC circuit for different values of Q

Problem 3   A series circuit operates at the following conditions:

E = 10 V $\underline{/0°}$

R = 100 Ω

X_C = 2 kΩ, at resonance

X_L = unknown

Find: a. X_L at resonance, b. Q of the circuit, c. I at resonance, and d. V_L using Equation 16.6.

Solution   a. X_L = X_C at resonance

X_L = 2 kΩ

b. Q = $\dfrac{X_L}{R}$

= $\dfrac{2\ 000\ \Omega}{100\ \Omega}$

= 20

c. I = $\dfrac{E}{Z_T}$

= $\dfrac{E}{R}$ (at series resonance)

= $\dfrac{10\ V}{100\ \Omega}$

= 0.1 A, or 100 mA

d. V_L = $Q\ E$

= (20)(10 V)

= 200 V

(R16-6)   If the resonant frequency of the circuit in Problem 3 is 3 180 Hz, find the value of L.

(R16-7)   Find the value of C in Problem 3 and question (R16-6).

(R16-8)   Use Equation 16.9 to calculate the value of Q for Problem 3 and for questions (R16-6) and (R16-7).

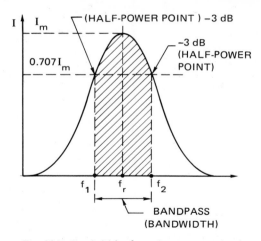

Fig. 16-7 Bandwidth of a series resonant circuit

BANDWIDTH   The bandwidth (BW) of a circuit is the total number of cycles below and above the resonant frequency that produce almost the same voltage gain as the value at f_r. The width of this band of frequencies is also known as the *bandpass* of the circuit. By convention, the bandpass is limited to those points on the resonance curve which give at least 0.707 times the maximum current or voltage. Figure 16-7 shows the bandwidth of a series resonant circuit. The shaded area in this figure represents the band of frequencies for which I is greater than 0.707 I_m. The frequencies f_1 and f_2 are called the *half-power points,* or the upper and lower cutoff points. At these frequencies, the power delivered to the circuit is one-half the power delivered at resonance. The half-power points are also known as the −3 dB points. The concept of decibels (dB) is explained in Chapter 25.

The bandwidth is expressed in terms of f_r and the Q of the circuit as follows:

$$BW = \frac{f_r}{Q} \qquad \text{Eq. 16.10}$$

One-half of the bandpass is above the resonant frequency, from f_r to f_2, and the other half is below the resonant frequency, from f_r to f_1.

Therefore,

$$f_2 = f_r + \frac{BW}{2} \qquad \text{Eq. 16.11}$$

$$f_1 = f_r - \frac{BW}{2} \qquad \text{Eq. 16.12}$$

Problem 4   A series RLC circuit has the following values at resonance:
E = 20 V $\underline{/0°}$
R = 100 Ω
Q = 15
f_r = 3 000 Hz
Find: a. the bandwidth of the circuit, and b. the upper and lower cutoff points.

Solution   a. $\text{BW} = \dfrac{f_r}{Q}$

$\qquad\qquad = \dfrac{3\ 000\ \text{Hz}}{15}$

$\qquad\qquad = 200\ \text{Hz}$

b. $f_2 = f_r + \dfrac{\text{BW}}{2}$

$\qquad\quad = 3\ 000\ \text{Hz} + \dfrac{200\ \text{Hz}}{2}$

$\qquad\quad = 3\ 100\ \text{Hz}$

$\quad f_1 = f_r - \dfrac{\text{BW}}{2}$

$\qquad\quad = 3\ 000\ \text{Hz} - \dfrac{200\ \text{Hz}}{2}$

$\qquad\quad = 2\ 900\ \text{Hz}$

(R16-9)   Calculate the power dissipated at resonance by the circuit specified in Problem 4.

(R16-10)   Find the power dissipated when $f = f_1$ for the circuit specified in Problem 4.

THE IDEAL PARALLEL RESONANT CIRCUIT   An ideal parallel resonant circuit is shown in figure 16-8A. The phasor relationships for this circuit are shown in figure 16-8B. This circuit is known as a *tank circuit*. That is, it can store energy in either one of two ways: 1. as an electrostatic charge in the capacitor, or 2. as an electromagnetic field in the inductor. I_C and I_L are 180° out of phase with each other, and 90° out of phase with the applied emf.

When this circuit is at resonance, the following conditions exist:

1. $X_L = X_C$
2. $I_L = I_C$ (180° out of phase with each other)
3. I_{line} (I_T) equals zero.
4. Z_T is infinite; thus, the tank circuit acts as an open circuit to the power supply.
5. $f_r = \dfrac{1}{2\pi\sqrt{LC}}$

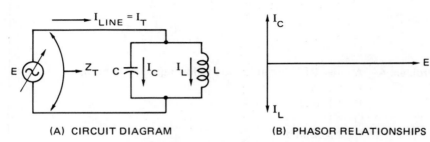

(A) CIRCUIT DIAGRAM   (B) PHASOR RELATIONSHIPS

Fig. 16-8 Ideal parallel resonant circuit

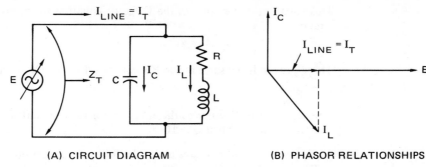

(A) CIRCUIT DIAGRAM (B) PHASOR RELATIONSHIPS

Fig. 16-9 Practical parallel resonant circuit

THE PRACTICAL PARALLEL RESONANT CIRCUIT

A practical parallel resonant circuit and its phasor relationships are shown in figure 16-9. For this circuit, the following conditions exist at resonance:

1. the phasor addition of I_L and I_C gives a line current (I_{line} or I_T) in phase with the applied voltage, E.

2. $Z_T = \dfrac{L}{CR}$ Eq. 16.13

   At resonance, the line current and the applied voltage are in phase, and the impedance is resistive. The impedance is a maximum at f_r.

$$Z_T = Q\,X_L \qquad \text{Eq. 16.14}$$

3. I_{line} (I_T) is a minimum at f_r. Figure 16-10 shows how Z_T and I_{line} (I_T) vary with frequency.

4. The current magnification that occurs is expressed as follows:

$$Q = \frac{I_L}{I_{line}}$$

$$I_L = Q\,I_{line}\,(I_T) \qquad \text{Eq. 16.15}$$

5. $Q = \dfrac{X_L}{R}$ Eq. 16.16

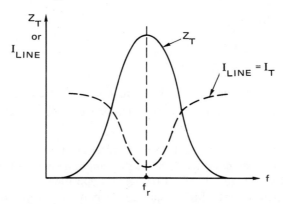

Fig. 16-10 Variation of Z_T and I_{line} (I_T) with frequency in a practical RLC parallel circuit

This equation is the same as the Q factor equation for a series resonant circuit.

6. $BW = \dfrac{f_r}{Q}$ Eq. 16.17

This equation is the same as the BW equation for a series resonant circuit.

7. $f_r = \dfrac{1}{2\pi \sqrt{LC}}$ Eq. 16.18

This equation is the same as the equation for the resonant frequency of a series RLC circuit when $Q > 10$.

Problem 5   A parallel RL and C circuit operates under the following conditions at resonance (figure 16-9):

f_r = 100 kHz
X_L = 10 kΩ
R = 200 Ω
E = 100 V $\underline{/0^\circ}$

Find: a. Q of the circuit, b. Z_T, c. I_{line} (I_T), and d. BW.

Solution   a. $Q = \dfrac{X_L}{R}$

$= \dfrac{10\ 000\ \Omega}{200\ \Omega}$

$= 50$

b. $Z_T = Q\,X_L$

$= (50)(10\ 000\ \Omega)$

$= 500\ k\Omega$

This is an example of how a parallel RL and C circuit magnifies the impedance at resonance.

c. $I_{line}\ (I_T) = \dfrac{E}{Z_T}$

$= \dfrac{100\ V}{(500)(10^3)\ \Omega}$

$= 0.2\ mA$

d. $BW = \dfrac{f_r}{Q}$

$= \dfrac{(100 \times 10^3)\ Hz}{50}$

$= 2\ kHz$

(R16-11)   Calculate the value of the current in the tank circuit of Problem 5.

(R16-12)   Find the power dissipated at resonance in the tank circuit of Problem 5.

LOW-PASS FILTERS   One use of low-pass filters is to filter out the ripple frequency (ac component) in power supplies. A schematic diagram of a low-pass filter is shown in figure 16-11. The inductor L offers negligible opposition to low-frequency current and high opposition to high-frequency current. The capacitor offers a low-impedance path back to the source for the high frequencies attenuated (blocked) by the in-

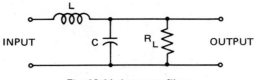

Fig. 16-11 Low-pass filter

ductor. When X_C is low compared to R_L, the high-frequency components in the input are bypassed around the load.

The frequency response curve for a typical low-pass filter is shown in figure 16-12. Note that all of the low frequencies are passed. However, at a point known as the cutoff frequency (f_C), the output begins to roll off. The cutoff frequency occurs when the output is 0.707 times the output at the lower frequencies. The following equations may be used to determine component values for the low-pass filter circuit shown in figure 16-11.

$$L = \frac{R_L}{\pi f_C} \qquad \text{Eq. 16.19}$$

$$C = \frac{1}{\pi f_C R_L} \qquad \text{Eq. 16.20}$$

$$f_C = \frac{1}{\pi \sqrt{LC}} \qquad \text{Eq. 16.21}$$

where

L = inductance, in henries
C = capacitance, in farads
R_L = load or terminating resistance, in ohms
f_C = cutoff frequency, in Hz

HIGH-PASS FILTERS When the reactive components (L and C) of a low-pass filter circuit, figure 16-12, are interchanged, the resulting circuit is a high-pass filter, figure 16-13. The fre-

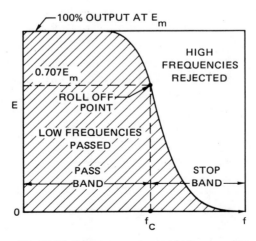

Fig. 16-12 Frequency response of a low-pass filter

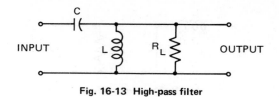

Fig. 16-13 High-pass filter

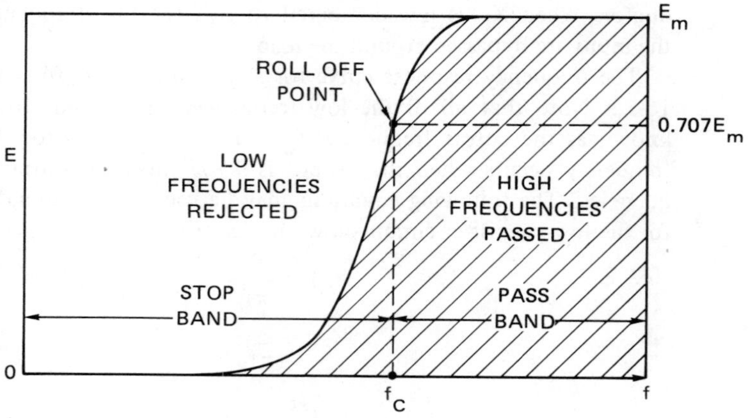

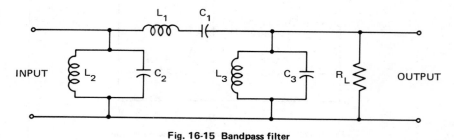

Fig. 16-14 Frequency response of a high-pass filter

quency response curve for a high-pass filter is shown in figure 16-14. Applications of high-pass filters include radio tone-control circuits, and installation between the antenna and the radio receiver to attenuate (reject) low-frequency interference. The following equations may be used to determine the values of L and C for this type of high-pass filter.

$$L = \frac{R_L}{4\pi f_C}$$ Eq. 16.22

$$C = \frac{1}{4\pi f_C R_L}$$ Eq. 16.23

BANDPASS FILTERS   Figure 16-15 shows a bandpass filter using series and parallel resonant circuits. The parallel resonant circuits form high-impedance paths for the resonant band of frequencies. The series resonant circuit (L_1 and C_1) provides a low-impedance

Fig. 16-15 Bandpass filter

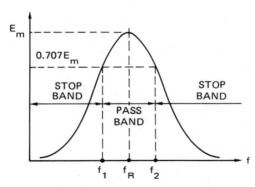

Fig. 16-16 Frequency response of a bandpass filter

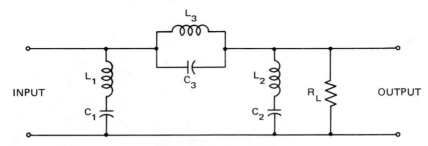

Fig. 16-17 Band stop filter

path to the resonant band of frequencies. Therefore, these frequencies are passed from the input to the output with very little opposition.

(R16-13)   Refer to figure 16-16 and describe the action of the bandpass filter for frequencies outside the pass band.

Bandpass filters are used in electronic and communication circuits. If variable capacitors are used in the filter circuits, the various resonant circuits pass one band of frequencies as required, and reject all other frequencies. When the capacitors are adjusted to another broadcasting station, a different band of frequencies is passed.

BAND STOP FILTERS   The circuit shown in figure 16-17 is a band stop filter, or band rejection filter, using series and parallel resonant circuits. The two series resonant circuits form low-impedance paths to the resonant band of frequencies. The parallel resonant circuit provides a high-impedance path to the resonant band of frequencies. Therefore, the resonant band of frequencies is rejected, or suppressed, as shown in the frequency response curve, figure 16-18.

(R16-14)   Describe the action of the band stop filter for frequencies outside the stop band.

COMMERCIAL FILTERS   Low-pass and high-pass filters for CB (citizens band) interference are shown in figure 16-19. Figure 16-20 is an assortment of power line filters, audio interference filters, low-pass filters, and high-pass filters. A multichannel analog filter

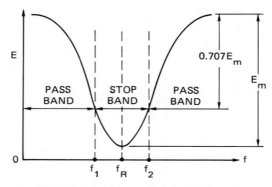

Fig. 16-18 Frequency response of a band stop filter

Fig. 16-20 Assortment of filters

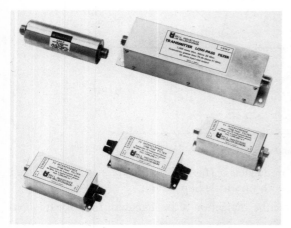

Fig. 16-19 Low-pass and high-pass filters for CB interference

is shown in figure 16-21. This filter is compatible with a computer and is used for automated data acquisition and signal processing in fields such as sonar, vibration, speech, medicine, geophysics, and process control. The plug-in card for the analog filter of figure 16-21 is shown in figure 16-22.

**LABORATORY
EXERCISE 16-1
MEASUREMENT
AND ANALYSIS
OF A SERIES
RESONANT
CIRCUIT**

PURPOSE   In completing this exercise, the student will analyze the characteristics of a series resonant circuit.

Fig. 16-21 Multichannel analog filter designed to be compatible with a computer

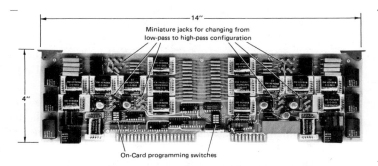

Fig. 16-22 Plug-in card for analog filter of figure 16-21

EQUIPMENT AND MATERIALS
1 Sinusoidal signal generator
1 Analog multimeter
1 Digital multimeter
1 Resistor, 10 kΩ, 2 W
1 Inductor, 4 H
1 Capacitor, 0.01 μF

PROCEDURE
1. Connect the circuit shown in figure 16-23.
   a. Turn the sinusoidal signal generator ON. Adjust the output voltage to 10 V (rms), as measured across terminals A and D using the digital multimeter.
   b. Vary the signal generator frequency to obtain a maximum value for the current I, as measured by the analog multimeter. Use Table 16-1 to record

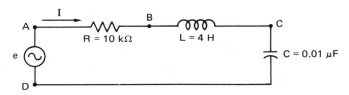

Fig. 16-23

TABLE 16-1

R	V_{AD}	I	f_r	V_{AB}	V_{BC}	V_{CD}	f_2	f_1	Q	BW
10 kΩ	10 V									

the maximum value of the current and the resonant frequency (f_r) at which this current is obtained.

c. Use the digital multimeter to measure the voltages V_{AB}, V_{BC}, V_{CD}, and V_{AD}. Record these values in Table 16-1.

d. Increase the frequency of the signal generator until the current is 0.707 times the value measured in step 1.b. Record this frequency (f_2) in Table 16-1.

e. Decrease the frequency of the signal generator until the current is 0.707 times the value measured in step 1.b. Record this frequency (f_1) in Table 16-1.

f. Vary the frequency in steps of 100 Hz above and below the resonant frequency (f_r) over a range of frequencies from 20 Hz to 2 kHz. Measure and record the current I at each frequency setting.

2. Perform the following calculations and answer the following questions:
   a. Calculate I, f_r, V_{AB}, V_{BC}, V_{CD}, Q, and BW, at the resonant frequency.
   b. Compare the calculated values with the measured values recorded in Table 16-1.
   c. Draw a graph of I versus f using the data of steps 1.d., 1.e., and 1.f.
   d. On this graph, label: f_r, f_1, f_2, and BW.
   e. Compare the value of BW obtained from the graph with the value recorded in Table 16-1.
   f. Discuss how the resistance of the circuit of figure 16-23 affects the resonant frequency (f_r), the circuit Q, the bandwidth, and the impedance at resonance.

EXTENDED STUDY TOPICS

1. Calculate the resonant frequency of a series circuit having the following values: R = 100 Ω, L = 250 μH, and C = 350 pF.
2. What is the effect of resistor R in topic 1 on the resonant frequency?
3. A series LC circuit is resonant at 500 kHz. If the capacitor has a value of 250 pF, find the value of the inductor.
4. A voltage of 100 V is supplied to a series RLC circuit at a frequency of 600 kHz. R = 50 Ω, L = 200 μH, and C = 350 pF.
   a. Is this circuit in series resonance for practical purposes?
   b. Calculate the current at f = 600 kHz.
   c. Calculate the power dissipated in the circuit at resonance.
   d. What is the Q of the circuit?
5. Calculate V_L and V_C for topic 4, using the value of I calculated in topic 4.
6. Explain how a series resonant circuit magnifies the voltage. Calculate V_L and V_C, using this concept and the data from topic 4.

7. A parallel RL and C circuit has the following values: R = 100 Ω, X_L (at f_r) = 3 000 Ω, f_r = 600 kHz. Find the bandwidth.

8. Repeat topic 7 with R = 200 Ω. Analyze the change in the bandwidth.

9. Calculate the Q of a series circuit if: f_r = 50 MHz, f_2 = 52.5 MHz, and f_1 = 47.5 MHz.

10. Calculate the impedance of an RL and C parallel circuit at resonance if R = 100 Ω, L = 400 μH, and C = 200 pF.

11. A low-pass filter has a value of L = 200 mH and C = 0.08 μF. Calculate the cutoff frequency.

12. Explain the meaning of the cutoff frequency for topic 11.

Chapter
17

Polyphase Circuits

OBJECTIVES

After studying this chapter, the student will be able to

- explain how a two-phase, three-wire system operates.
- solve problems in three-phase systems.
- discuss the wye-delta system.
- calculate power in balanced three-phase circuits.
- define phase sequence.

INTRODUCTION

The circuits used to study alternating emf in Chapter 10 are known as single-phase circuits ($1\,\phi$). Single-phase systems are not practical when large amounts of electric energy must be generated and then distributed over large distances.

Polyphase systems are preferred to single-phase systems for the following reasons:

1. The generators are less expensive because the entire surface of the armature can be used.
2. The efficiency of the generators is higher.
3. The generators have better voltage regulation; that is, the output voltage remains constant regardless of changes in the load requirements.
4. Polyphase motors cost more than single-phase motors but have better operating characteristics.
5. Less copper wire is required to supply power to a given load at a given voltage.

6. When each phase of a polyphase source has the same load, the instantaneous power output of the generator is constant. In large alternators, this condition means that there is a steady conversion of mechanical energy into electrical energy.

7. A polyphase system supplies a rotating magnetic field which rotates at the frequency of the emf. The fact that this field is present simplifies the design of ac motors. In a single-phase system, the resulting magnetic field increases and decreases in flux density, and it reverses direction each 180°, but it does not rotate. The advantages of a rotating magnetic field are discussed in more detail in Chapter 20.

THE TWO-PHASE GENERATOR

A generator with two rotating loops, aA and bB, is shown in figure 17-1. The two loops produce two separate output voltages that are 90° out of phase with each other.

(R17-1) Draw the phasor diagram of the output voltages for the two-phase generator shown in figure 17-1.

(R17-2) What is the instantaneous output voltage of loop aA in figure 17-1 when t = 0? What is the instantaneous output voltage of loop bB in figure 17-1 when t = 0?

(R17-3) What is the direction of the instantaneous output voltage of loop aA in figure 17-1A?

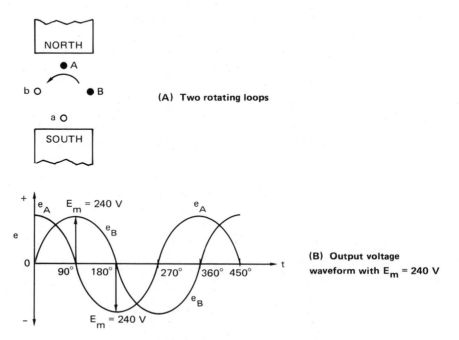

(A) Two rotating loops

(B) Output voltage waveform with E_m = 240 V

Fig. 17-1 The two-phase generator

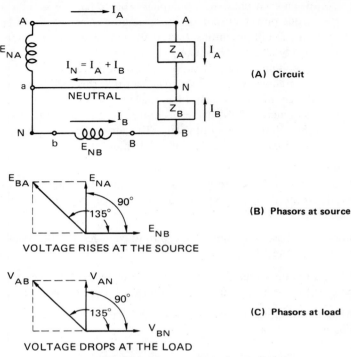

(A) Circuit

(B) Phasors at source

VOLTAGE RISES AT THE SOURCE

(C) Phasors at load

VOLTAGE DROPS AT THE LOAD

Fig. 17-2 The two-phase, three-wire system

Analysis of Two-phase Generator Action. A number of statements can be made about the action of the two-phase generator.

1. Both loops are mounted on the same rotor assembly, and thus rotate at the same angular velocity.
2. Both induced emfs have the same frequency.
3. Both loops have the same number of turns. Thus, both emfs have the same maximum values and the same effective values.
4. Loop aA always cuts a certain magnetic line of force 90° before loop bB cuts the same line of force. As a result, the emf in loop aA leads the emf in loop bB by 90°.
5. By convention, the voltage induced in loop aA is called E_{NA}. This loop is known as phase A. Therefore, a two-phase generator has two outputs, phase A and phase B.

TWO-PHASE, THREE-WIRE SYSTEM   A two-phase, three-wire system is formed when two-phase voltages are interconnected with a common neutral lead, figure 17-2. The rms voltages between the phases of a two-phase system are always equal to $\sqrt{2}$ times the rms voltage of a single phase. The current in the neutral line is the phasor sum of the load currents. When the system has balanced loads, the load of phase A is equal to the load of phase B. Thus, the current in the neutral, I_N, is equal to $\sqrt{2}$ times the phase current.

$$\mathbf{V_{AN}} = V_{AN} \underline{/90^\circ} \qquad \text{Eq. 17.1}$$

$$\mathbf{V_{BN}} = V_{BN} \underline{/0^\circ} \qquad \text{Eq. 17.2}$$

$$\mathbf{V_{AB}} = \sqrt{2}\, V_{AN} \underline{/135^\circ} \qquad \text{Eq. 17.3}$$

where

$\mathbf{V_{AB}}$ = phase-to-phase voltage
$\mathbf{V_{AN}}$ = phase-to-neutral voltage

Therefore,

$$I_N = \sqrt{2}\, I_A \qquad \text{Eq. 17.4}$$

where

I_N = neutral current
I_A = phase current

Note that Equation 17.4 can be used only when the load is balanced, or $I_A = I_B$.

$$I_N = I_A + I_B \qquad \text{Eq. 17.5}$$

where I_N is the phasor sum of I_A and I_B.

Problem 1 A balanced 240-V, two-phase, three-wire system has a load impedance of 12 ohms per phase. The power factor of the load is 0.707 lagging.

Find: a. the line currents, b. the neutral current, and c. the total power supplied to the circuit.

Solution a. $I_A = \dfrac{V_{AN} \underline{/90^\circ}}{Z_{AN} \underline{/45^\circ}}$

$\qquad = \dfrac{240 \text{ V} \underline{/90^\circ}}{12\ \Omega \underline{/45^\circ}}$

$\qquad = 20 \text{ A} \underline{/45^\circ}$

$\quad I_B = \dfrac{V_{BN} \underline{/0^\circ}}{Z_{BN} \underline{/45^\circ}}$

$\qquad = \dfrac{240 \text{ V} \underline{/0^\circ}}{12\ \Omega \underline{/45^\circ}}$

$\qquad = 20 \text{ A} \underline{/-45^\circ}$

Therefore,

$\quad I_A = I_B = 20 \text{ A}$

b. $I_N = \sqrt{2}\, I_A$

$\qquad = (1.414)(20 \text{ A})$

$\qquad = 28.3 \text{ A}$

c. $P_A = V_{AN}\, I_A \cos\theta$

$\qquad = (240 \text{ V})(20 \text{ A})(0.7)$

$\qquad = 3\ 360 \text{ W}$

$\quad P_T = 2\, P_A$

$\qquad = (2)(3\ 360 \text{ W})$

$\qquad = 6\ 720 \text{ W}$

Problem 2 A two-phase, three-wire system has the following values:

$$V_{AN} = 240 \text{ V} \underline{/90°}$$
$$V_{BN} = 240 \text{ V} \underline{/0°}$$
$$Z_{AN} = 10 \text{ }\Omega \underline{/45°}$$
$$Z_{BN} = 10 \text{ }\Omega \underline{/60°}$$

Find the line currents, and express the answers in polar notation.

Solution $I_A = \dfrac{V_{AN}}{Z_{AN}}$

$$= \frac{240 \text{ V} \underline{/90°}}{10 \text{ }\Omega \underline{/45°}}$$
$$= 24 \text{ A} \underline{/45°}$$

$I_B = \dfrac{V_{BN}}{Z_{BN}}$

$$= \frac{240 \text{ V} \underline{/0°}}{10 \text{ }\Omega \underline{/60°}}$$
$$= 24 \text{ A} \underline{/-60°}$$

(R17-4) Calculate the neutral current in the system described in Problem 2.

(R17-5) Find the total power supplied to the circuit of Problem 2.

Three-phase systems are more efficient than two-phase systems. For this reason, electric power is rarely distributed by means of a two-phase system. However, many industries still use two-phase motors which must be operated on two-phase power. This means that these companies must change the three-phase power input to two-phase power. Chapter 18 will explain how transformers are used to change three-phase power to two-phase power.

Another application of a two-phase system is the automatic pilot control of an aircraft. In this case, the two-phase power operates servomotors which automatically keep the aircraft on course by moving the control surfaces.

THE THREE-PHASE GENERATOR Figure 17-3 shows a generator with three coils, aA, bB, and cC. These coils are mounted on the rotor at 120° intervals. This alternator generates three sinusoidal voltages which are 120° out of phase with one another.

(R17-6) Draw the phasor diagram of the output voltages for the three-phase generator shown in figure 17-3.

(R17-7) For the generator of figure 17-3, find the instantaneous output voltage of a. loop aA, b. loop bB, and c. loop cC, when t = 0.

(R17-8) When t = 0, find the algebraic sum of the instantaneous output voltages of loops aA, bB, and cC.

Analysis of Three-phase Generator Action. The following statements can be made about the action of a three-phase generator.

    1. All three loops, or coils, rotate at the same angular velocity, since they are mounted on the same rotor assembly.

    2. All three induced emfs have the same frequency.

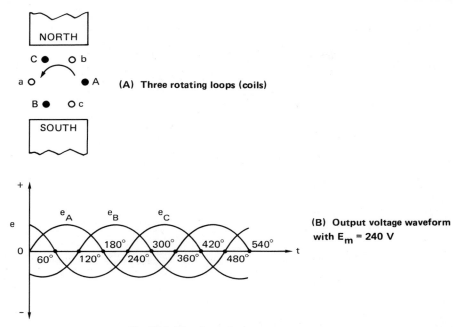

Fig. 17-3 The three-phase generator

3. Each of the three coils has the same number of turns. Therefore, all three emfs have the same maximum values and the same effective values.

4. Loop aA always cuts a certain magnetic line of force 120° before loop bB cuts the same line of force. Loop aA also cuts this line of force 240° before loop cC. The emf in loop aA leads the emf in loop bB by 120° and the emf in loop cC by 240°.

5. The voltage induced in loop aA is called E_{NA} by convention. This winding is known as phase A. Therefore, a three-phase generator produces three outputs: phase A, phase B, and phase C. These outputs are equal in magnitude and 120° out of phase with each other.

THREE-PHASE, FOUR-WIRE, WYE-CONNECTED SYSTEM

A three-phase, four-wire, wye-connected system is shown in figure 17-4. The neutral current is the phasor sum of the three load currents, figure 17-4A. When the load is balanced, the currents in each load are equal in magnitude and 120° out of phase with one another, Figure 17-4D. Therefore, the neutral current is zero when the three loads are balanced. For the wye-connected system, which is also known as the star-connected system, the following equations apply:

$$V_{AN} = V_{AN} \, \underline{/0°} \qquad \text{Eq. 17.6}$$

$$V_{BN} = V_{BN} \, \underline{/-120°} \qquad \text{Eq. 17.7}$$

$$V_{CN} = V_{CN} \, \underline{/+120°} \qquad \text{Eq. 17.8}$$

$$I_N = I_A + I_B + I_C \qquad \text{Eq. 17.9}$$

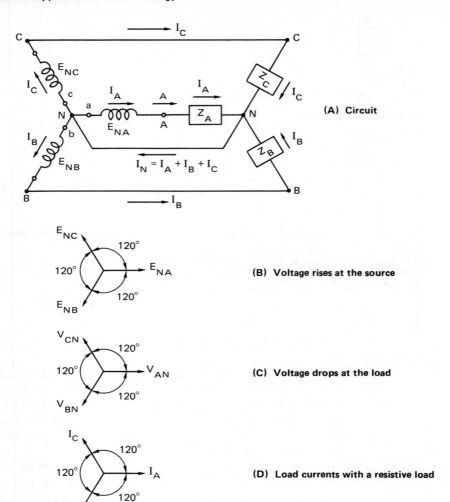

(A) Circuit

(B) Voltage rises at the source

(C) Voltage drops at the load

(D) Load currents with a resistive load

Fig. 17-4 Three-phase, four-wire, wye-connected system

Problem 3   A balanced three-phase, four-wire, wye-connected system has the following values: $\mathbf{V_{AN}} = 240\ V\ \underline{/0°}$, $\mathbf{V_{BN}} = 240\ V\ \underline{/-120°}$, $\mathbf{V_{CN}} = 240\ V\ \underline{/+120°}$, $\mathbf{Z_A} = \mathbf{Z_B} = \mathbf{Z_C} = 24\ \Omega\ \underline{/0°}$.

Find: a. the line currents, and b. the total power output of the power supply.

Solution   a. $\mathbf{I_A} = \dfrac{\mathbf{V_{AN}}}{\mathbf{Z_A}}$

$= \dfrac{240\ V\ \underline{/0°}}{24\ \Omega\ \underline{/0°}}$

$= 10\ A\ \underline{/0°}$

$$I_B = \frac{V_{BN}}{Z_B}$$

$$= \frac{240 \text{ V} \underline{/-120^\circ}}{24 \text{ } \Omega \underline{/0^\circ}}$$

$$= 10 \text{ A} \underline{/-120^\circ}$$

$$I_C = \frac{V_{CN}}{Z_C}$$

$$= \frac{240 \text{ V} \underline{/+120^\circ}}{24 \text{ } \Omega \underline{/0^\circ}}$$

$$= 10 \text{ A} \underline{/+120^\circ}$$

b. $P_A = V_{AN} I_A \cos \theta$

$$= (240 \text{ V})(10 \text{ A})(1)$$

$$= 2\ 400 \text{ W}$$

$$P_T = 3 P_A$$

$$= (3)(2\ 400 \text{ W})$$

$$= 7\ 200 \text{ W}$$

Note: the phase angle between the voltage and current of each phase is zero degrees. Therefore,

$$\cos \theta = \cos 0^\circ = 1$$

(R17-9) What is the value of I_N in Problem 3?

Problem 4 A balanced three-phase, four-wire, wye-connected system operates under the following conditions: $V_{AN} = 220 \text{ V} \underline{/0^\circ}$, $V_{BN} = 220 \text{ V} \underline{/-120^\circ}$, $V_{CN} = 220 \text{ V} \underline{/+120^\circ}$, $Z_A = Z_B = Z_C = 20 \text{ } \Omega \underline{/36.9^\circ}$.

Find: I_A, I_B, and I_C. Express each current value in polar notation.

Solution $I_A = \frac{V_{AN}}{Z_A}$

$$= \frac{220 \text{ V} \underline{/0^\circ}}{20 \text{ } \Omega \underline{/36.9^\circ}}$$

$$= 11 \text{ A} \underline{/-36.9^\circ}$$

$$I_B = \frac{V_{BN}}{Z_B}$$

$$= \frac{220 \text{ V} \underline{/-120^\circ}}{20 \text{ } \Omega \underline{/36.9^\circ}}$$

$$= 11 \text{ A} \underline{/-156.9^\circ}$$

$$I_C = \frac{V_{CN}}{Z_C}$$

$$= \frac{220 \text{ V} \underline{/120^\circ}}{20 \text{ } \Omega \underline{/36.9^\circ}}$$

$$= 11 \text{ A} \underline{/83.1^\circ}$$

(R17-10) Calculate I_N for Problem 4.

(R17-11) Draw the phasor diagram of the load voltages and the load currents for Problem 4.

(R17-12) Calculate the total power supplied to the system described in Problem 4.

THREE-PHASE, THREE-WIRE, DELTA-CONNECTED SYSTEM

A three-phase, three-wire, delta-connected system is shown in figure 17-5. According to the convention adopted in industry, the line currents are labeled I_A, I_B, and I_C, and the load or phase currents are labeled I_{AC}, I_{CB}, and I_{BA}.

In comparing a wye-connected system and a delta-connected system, the following points can be made:

1. When a load is wye connected, each arm of the load is connected from a line to the neutral. The impedance (Z) is shown with a single subscript, such as Z_A.

2. If a load is delta connected, each arm of the load is connected from line to line. In this case, the impedance is written with a double subscript, such as Z_{AC}.

3. When Kirchhoff's Voltage Law is applied, the emf of the source voltage is equal to the voltage drop across the load, or $E_{AB} = V_{BA}$. Note that the letter E is normally used to indicate voltage rise at the source. The letter V is used for the voltage drop at the load.

4. The phase current of the source is the current in a particular coil of the source of emf. The phase current of the load is the current in a particular arm of the load.

5. In a wye-connected system, the phase current of the source, the line current, and the phase current of the load are all equal.

6. For the delta-connected system shown in figure 17-5, each line must carry components of current for two arms of the load. One current component moves toward the source, and the other current moves away from the source. The line current to a delta-connected load is the phasor difference between the two load currents at the entering node.

7. The following equations apply in a delta-connected system:

$$E_{CA} = V_{AC}$$ Eq. 17.10
$$E_{BC} = V_{CB}$$ Eq. 17.11
$$E_{AB} = V_{BA}$$ Eq. 17.12

Note that the voltage drop at the load is equivalent to the voltage rise at the source.

$$I_A = I_{AC} - I_{BA}$$ Eq. 17.13
$$I_B = I_{BA} - I_{CB}$$ Eq. 17.14
$$I_C = I_{CB} - I_{AC}$$ Eq. 17.15
$$I_A + I_B + I_C = 0$$ Eq. 17.16

In a delta-connected system, there is no neutral by which the current can return to the source. Therefore, Equation 17.16 shows that the phasor sum of all three line currents is always zero. This condition is true even if the load is not balanced.

When a delta-connected load is balanced, the impedances of the three arms are equal in magnitude and phase angle. The phase voltages are equal and 120° out of phase with each other, figure 17-5D. The phase currents are also equal in magnitude and 120° out of phase with one another. The line current in the balanced delta load has a magnitude of $\sqrt{3}$ times the phase current in each arm of the load. The line current is 30° out of phase with the phase current. Figure

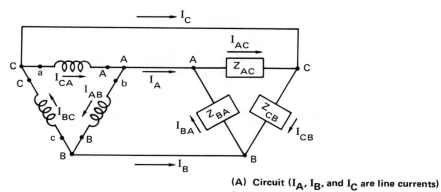

(A) Circuit (I_A, I_B, and I_C are line currents)
(I_{AC}, I_{CB}, and I_{BA} are phase or load currents)

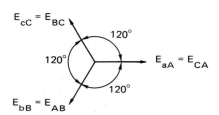

(B) Source and load voltages in a delta arrangement

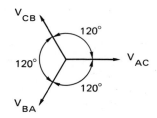

(C) Source voltages drawn from a common origin

(D) Load or phase voltages drawn from a common origin

Fig. 17-5 Three-phase, three-wire, delta-connected system

17-6 shows the relationship between the phase currents in each arm of the delta load and the line currents for a balanced load. The following equation expresses this relationship:

$$I_L = \sqrt{3}\, I_P \qquad\qquad \text{Eq. 17.17}$$

where
I_L = line current
I_P = phase current

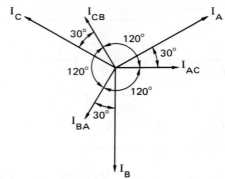

**Fig. 17-6   Relationship between the phase and line
currents in a balanced delta-connected load**

(R17-13) I_{AC} = 10 A $\underline{/0°}$ in a balanced delta-connected system. Express the value of the
line current (I_A) in polar notation.

Problem 5   The delta-connected system shown in figure 17-5 has the following values: V_{AC}
= 240 V $\underline{/0°}$, V_{CB} = 240 V $\underline{/+120°}$, V_{BA} = 240 V $\underline{/-120°}$, Z_{AC} = 120 Ω $\underline{/0°}$,
Z_{CB} = 60 Ω $\underline{/-30°}$, Z_{BA} = 60 Ω $\underline{/+45°}$.
Find: a. the three load currents, and b. the three phase currents.

Solution   a. $I_{AC} = \dfrac{V_{AC}}{Z_{AC}}$

$\qquad = \dfrac{240 \text{ V} \underline{/0°}}{120 \text{ Ω} \underline{/0°}}$

$\qquad = 2 \text{ A} \underline{/0°}$

$\quad I_{CB} = \dfrac{V_{CB}}{Z_{CB}}$

$\qquad = \dfrac{240 \text{ V} \underline{/+120°}}{60 \text{ Ω} \underline{/-30°}}$

$\qquad = 4 \text{ A} \underline{/+150°}$

$\quad I_{BA} = \dfrac{V_{BA}}{Z_{BA}}$

$\qquad = \dfrac{240 \text{ V} \underline{/-120°}}{60 \text{ Ω} \underline{/+45°}}$

$\qquad = 4 \text{ A} \underline{/-165°}$

b. $I_A = I_{AC} - I_{BA}$
$\qquad = 2 \text{ A} \underline{/0°} - 4 \text{ A} \underline{/-165°}$
$\qquad = (2 + j0) \text{ A} - (-3.86 - j1.04) \text{ A}$
$\qquad = (5.86 + j1.04) \text{ A}$
$\qquad = 6 \text{ A} \underline{/+10°}$

$\quad I_B = I_{BA} - I_{CB}$
$\qquad = 4 \text{ A} \underline{/-165°} - 4 \text{ A} \underline{/+150°}$
$\qquad = (-3.86 - j1.04) \text{ A} - (-3.46 + j2) \text{ A}$
$\qquad = (-0.4 - j3.04) \text{ A}$

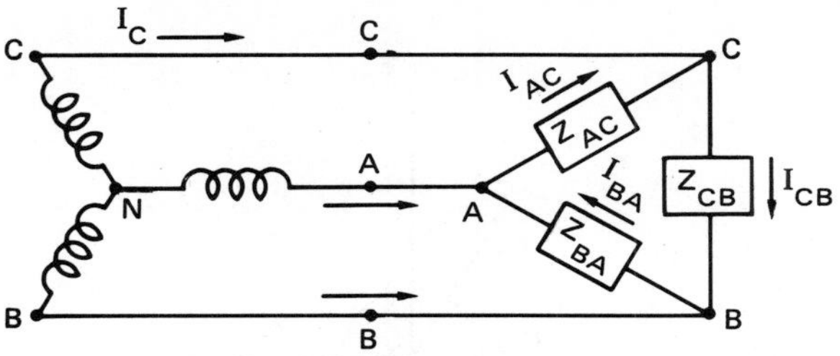

Fig. 17-7 Wye-delta system

$$= 3.07 \text{ A} \underline{/-97.5°}$$
$$\mathbf{I_C} = \mathbf{I_{CB}} - \mathbf{I_{AC}}$$
$$= 4 \text{ A} \underline{/+150°} - 2 \text{ A} \underline{/0°}$$
$$= (-3.46 + j2) \text{ A} - (2 + j0) \text{ A}$$
$$= (-5.46 + j2) \text{ A}$$
$$= 5.8 \text{ A} \underline{/+159.9°}$$

(R17-14)   Prove the answers obtained in Problem 5.

(R17-15)   A delta-connected system operates under the following conditions: $\mathbf{E_{CA}} = 200$ V $\underline{/0°}$, $\mathbf{E_{AB}} = 200$ V $\underline{/+120°}$, $\mathbf{E_{BC}} = 200$ V $\underline{/-120°}$. The load is balanced and consists of three 4-Ω resistors. Find: a. the three load currents, and b. the three line currents.

(R17-16)   Calculate the total power supplied to the load of question (R17-15).

WYE-DELTA SYSTEM

A wye-delta system is shown in figure 17-7. According to a convention adopted by industry, the voltages measured between two lines are known as line voltages, unless specified otherwise. In a delta-connected system, the phase voltage and the line voltage are the same value.

In a wye-connected source, the phase voltages are always equal in magnitude and 120° out of phase with one another. Each line voltage, however, is $\sqrt{3}$ times the phase voltage, as shown in the following equation:

$$\mathbf{E_L} = \sqrt{3} \, \mathbf{E_p} \qquad \text{Eq. 17.18}$$

where

$\mathbf{E_L}$ = line voltage
$\mathbf{E_p}$ = phase voltage

The line voltages are 30° out of phase with the phase voltages. Figure 17-8 shows the relationship of the line voltages and the phase voltages for a wye-connected three-phase source of emf. Also shown is the application of the following equations:

$$E_{AB} = E_{NB} - E_{NA} \qquad \text{Eq. 17.19}$$
$$E_{BC} = E_{NC} - E_{NB} \qquad \text{Eq. 17.20}$$
$$E_{CA} = E_{NA} - E_{NC} \qquad \text{Eq. 17.21}$$

(R17-17)   If $\mathbf{E_{NA}} = 120$ V $\underline{/0°}$ in figure 17-8, express the value of $\mathbf{E_{AB}}$ in polar notation.

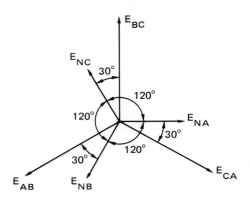

Fig. 17-8 Relationship between the line and phase voltages for a wye-connected, three-phase source of emf

The wye-connected, three-phase source is commonly used as a distribution system to supply two different three-phase voltages:

- 120 V, three phase (ϕ) is supplied to a four-wire, wye-connected load.
- 208 V, three phase (ϕ) is supplied to a three-wire, delta-connected load.

POWER IN BALANCED THREE-PHASE CIRCUITS The total power is equal to three times the power in each phase of a balanced three-phase load.

$$P_T = 3 \, E_p \, I_p \cos \theta \qquad \text{Eq. 17.22}$$

P_T can also be expressed in terms of the line voltage and the line current.

$$P_T = \sqrt{3} \, E_L \, I_L \cos \theta \qquad \text{Eq. 17.23}$$

Equation 17.23 applies to both wye-connected and delta-connected systems.

Problem 6 A balanced delta-connected load ($Z_L = 50 \ \Omega\,\underline{/30°}$) is supplied from a 120-V/208-V, three-phase source. Calculate the total power supplied to the load.

Solution
$$I_p = \frac{E_p}{Z_p}$$
$$= \frac{208 \text{ V}}{50 \ \Omega}$$
$$= 4.16 \text{ A}$$
$$I_L = \sqrt{3} \, I_p$$
$$= (1.732)(4.16 \text{ A})$$
$$= 7.2 \text{ A}$$
$$P_T = \sqrt{3} \, E_L \, I_L \cos \theta$$
$$= (1.732)(208 \text{ V})(7.2 \text{ A})(0.866)$$
$$= 2 \ 246 \text{ W}$$

(R17-18) Calculate the total power supplied to the load in Problem 6 for the case where the load is connected in wye. (Note: $E_p = 120$ V.)

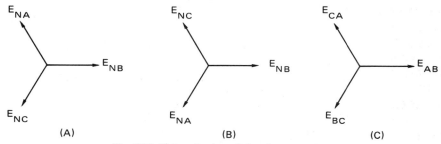

Fig. 17-9 Determination of the phase sequence

PHASE SEQUENCE   The phase sequence, or phase rotation, of a three-phase system determines the lead or lag relationship of the three voltages. The phase sequence is important because it determines the direction of rotation of three-phase motors. The phasor diagram is examined to determine if the phase sequence is ABC or CBA. Recall that phasors rotate counterclockwise. Therefore, the phase sequence is determined by reading the phasor diagram in a clockwise direction. The phase sequence will be shown either by all of the first subscript letters, or by all of the second subscript letters.

(R17-19)   What is the phase sequence in figure 17-9A, figure 17-9B, and figure 17-9C?

Figure 17-10 shows a phase sequence tester that is used to determine the phase rotation of a three-phase system. It can also be used to determine the direction of rotation of a motor.

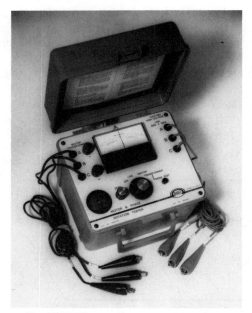

Fig. 17-10 Phase and motor rotation tester

**LABORATORY
EXERCISE 17-1
MEASUREMENT
AND ANALYSIS
OF VOLTAGES
AND CURRENTS
IN BALANCED
THREE-PHASE
SYSTEMS**

PURPOSE This exercise enables the student to study the voltage and current relationships in three-phase systems with balanced wye-connected and delta-connected loads.

EQUIPMENT AND MATERIALS
3 Resistors, 200 Ω, 200 W
1 Ac voltmeter
1 Ac ammeter
1 Power supply, 120 V/208 V, 60 Hz, three phase, four wire
1 Phase sequence tester

PROCEDURE
1. Turn the power supply ON. Record the phase sequence of the line voltages. Turn the power supply OFF. Connect the circuit shown in figure 17-11.
   a. Turn the power supply ON.
   b. Measure the voltages at the load: V_{AC}, V_{CB}, and V_{BA}. Record these values in Table 17-1.
   > CAUTION: Before installing the ammeter in the circuit, the student must turn the power supply OFF since there is a shock hazard.
   c. Measure the load phase currents: I_{AC}, I_{CB}, and I_{BA}. Record these values in Table 17-1.
   d. Measure the line currents: I_A, I_B, and I_C. Record the values in Table 17-1.
   e. Turn the power supply OFF.

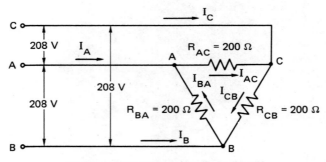

Fig. 17-11 Balanced delta-connected load

TABLE 17-1

Delta-connected Load		Wye-connected Load	
Voltages at load	V_{AC} = V_{CB} = V_{BA} =	Voltages at load	V_{AN} = V_{BN} = V_{CN} =
Load phase currents	I_{AC} = I_{CB} = I_{BA} =	Line voltages	V_{AC} = V_{CB} = V_{BA} =
Line currents	I_A = I_B = I_C =	Load phase currents	I_A = I_B = I_C =
		Neutral current	I_N =

2. Connect the circuit shown in figure 17-12. Record all measured values in Table 17-1.

   a. Turn the power supply ON.

   b. Measure and record the voltages at the load: V_{AN}, V_{BN}, and V_{CN}.

   c. Measure and record the line voltages: V_{AC}, V_{CB}, and V_{BA}.

   d. Measure and record the load phase currents: I_A, I_B, and I_C.

   e. Measure and record the neutral current I_N.

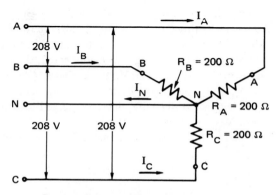

Fig. 17-12 Balanced wye-connected load

3. Perform the calculations required and answer the following questions for the delta-connected load of step 1:

a. Calculate the load phase currents: I_{AC}, I_{CB}, and I_{BA}. Compare these values with the values measured in step 1.c.

b. Calculate the line currents: I_A, I_B, and I_C. Compare these values with the values measured in step 1.d.

c. What is the relationship between the line voltages and the phase voltages?

d. What is the relationship between the line currents and the phase currents?

e. Calculate the power delivered to the load.

4. Perform the calculations required and answer the questions which follow for the wye-connected load of step 2:

a. Calculate the load currents: I_A, I_B, I_C, and I_N. Compare these values with the values measured in steps 2.d. and 2.e.

b. What is the relationship between the line voltages and the phase voltages?

c. What is the relationship between the line currents and the phase currents?

d. Calculate the power delivered to the load.

EXTENDED STUDY TOPICS

1. A two-phase, three-wire system operates at the following conditions:

V_{AN} = 240 V $\underline{/90°}$ Z_A = 12 Ω $\underline{/53.1°}$
V_{BN} = 240 V $\underline{/0°}$ Z_B = 12 Ω $\underline{/36.9°}$

Find: a. I_A, I_B, and I_N, and b. draw the phasor diagram of the load voltages and the load currents.

2. Find the total power supplied to the system of topic 1.

3. A three-phase, four-wire, wye-connected system has the following values:
V_{AN} = 220 V $\underline{/0°}$, V_{BN} = 220 V $\underline{/-120°}$, V_{CN} = 220 V $\underline{/120°}$, Z_A = 20 Ω $\underline{/0°}$, Z_B = 10 Ω $\underline{/36.9°}$, and Z_C = 20 Ω $\underline{/-53.1°}$.
Find: I_A, I_B, I_C, and I_N. Express the answers in polar notation.

4. Draw the phasor diagram of the load voltages and the load currents (including I_N) for topic 3.

5. Find the total power supplied to the system of topic 3.

6. A delta-connected system, figure 17-5, operates at the following conditons:
E_{CA} = 110 V $\underline{/0°}$, E_{AB} = 110 V $\underline{/+120°}$, E_{BC} = 110 V $\underline{/-120°}$, Z_{AC} = 10 Ω $\underline{/45°}$, Z_{BA} = 22 Ω $\underline{/0°}$, and Z_{CB} = 20 Ω $\underline{/-36.9°}$.
Find: a. the three load currents, and b. the three line currents.

7. What is the total power supplied to the load of topic 6?

8. Make a diagram showing a wye-connected, three-phase source supplying the following loads: a 120-V, three-phase, 4-W, wye-connected load, and a 208-V, three-phase, 3-W, delta-connected load.

9. A three-phase, three-wire, delta-connected load has the following load voltages: V_{AC} = 208 V $\underline{/0°}$, V_{BA} = 208 V $\underline{/120°}$, V_{CB} = 208 V $\underline{/-120°}$. a. Draw the load voltage phasor diagram. b. What is the phase sequence of the power supply?

<table>
<tr><td>

┌──────────┐
│ Chapter │
│   **18**   │
└──────────┘

</td><td>

Transformers

</td></tr>
</table>

OBJECTIVES After studying this chapter, the student will be able to

- describe transformer operation for the cases of no load and with a load.

- explain the use of polarity markings.

- discuss special transformers, such as autotransformers, polyphase transformers, Scott-connected transformers, and multiple secondary transformers, in terms of how the connections are made and list applications.

- describe how instrument transformers are used and list applications.

- use transformers for impedance matching.

INTRODUCTION A transformer is an electrical device consisting of coils, or windings, that transfer energy from one circuit to another by means of electromagnetic induction. Transformers are used to

    1. step up a voltage.
    2. step down a voltage.
    3. provide electrical isolation between two circuits.
    4. provide impedance matching.

Figure 18-1 shows a cutaway view of a transformer.

TRANSFORMER OPERATION WITH NO LOAD The basic parts of a transformer, as shown in figure 18-2, consist of a laminated steel core, an input, or primary, winding having N_1 turns, and an output, or secondary winding having N_2 turns.

Fig. 18-1 Cutaway view of a transformer

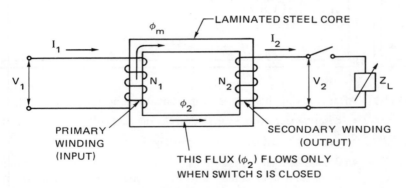

Fig. 18-2 Basic transformer

Operation. When switch S in the transformer circuit is in the OPEN position, the following sequence of operation takes place.

1. Voltage V_1 causes a small sinusoidal current in the primary winding. This current is known as the *no-load current*, the *magnetizing current*, or the *exciting current*. It is less than 10% of the rated maximum primary current of the transformer.

2. The exciting current produces a sinusoidal magnetic flux ϕ_m in the core of the transformer.

3. The changing flux cuts the secondary winding. A voltage V_2 is induced across the terminals of the secondary winding.

4. The same emf is induced in each turn of a transformer winding. This means that the voltage per turn for all of the turns of a transformer winding is a constant value.

(R18-1)   What factors determine the value of the exciting current of a transformer?

Fig. 18-3 Transformer turn ratio test set

Transformer Turn Ratio. The *turn ratio* of a transformer is the ratio of the number of primary turns to the number of secondary turns.

$$\text{TR} = \frac{N_1}{N_2} = \frac{V_1}{V_2} \qquad \text{Eq. 18.1}$$

where TR = turn ratio

Note that the turn ratio is also equal to the ratio of the voltages of the windings because the voltage in each winding is proportional to the number of turns in the winding. A transformer turn ratio test set is shown in figure 18-3.

(R18-2) What is the turn ratio of a transformer if V_1 = 120 V and V_2 = 40 V?

(R18-3) A transformer has the following number of turns: N_1 = 1 000 turns and N_2 = 500 turns. a. If 120 V is impressed on the primary, what is the value of the secondary voltage? b. What is the voltage per turn of the primary? c. What is the voltage per turn of the secondary?

When V_2 is less than V_1, the transformer is called a *step-down transformer*. When V_2 is greater than V_1, the transformer is called a *step-up transformer*. If V_2 is equal to V_1, the turn ratio is unity, and the transformer is called an *isolating transformer* because there are no metallic conductors directly connecting the primary and secondary circuits.

OPERATION WITH A LOAD

When switch S of figure 18-2 is in the CLOSED position, a current, I_2, flows in the load.

(R18-4) What formula may be used to find the value of I_2?

For the transformer in figure 18-2, the sequence of operation is as follows:

1. The current I_2 through the secondary coil produces a flux ϕ_2. This flux opposes the primary flux, ϕ_m.
2. As stated previously, the voltage per turn is a constant value. The primary current I_1 automatically increases, since the secondary flux opposes the

primary flux and thereby reduces the counter emf in the primary. The net flux remains constant. The increase in the primary flux always makes up for the counter flux caused by I_2. As a result, the transformer acts as a constant voltage, or constant potential, device.

The following equations are based on the law of the conservation of energy. This law states that the power supplied to an ideal transformer equals the power delivered to the load. The phase angle of the primary is the same as the phase angle of the secondary.

$$V_1 I_1 = V_2 I_2 \qquad\qquad \text{Eq. 18.2}$$
$$P_1 = P_2 \qquad\qquad \text{Eq. 18.3}$$
$$P_1 = V_1 I_1 \cos\theta \qquad\qquad \text{Eq. 18.4}$$
$$P_2 = V_2 I_2 \cos\theta \qquad\qquad \text{Eq. 18.5}$$
$$\mathbf{V}_1 = V_1 \:\underline{/\theta^\circ} \qquad\qquad \text{Eq. 18.6}$$
$$\mathbf{V}_2 = V_2 \:\underline{/\theta^\circ} \qquad\qquad \text{Eq. 18.7}$$

Equation 18.2 can be rearranged as follows:

$$\frac{I_1}{I_2} = \frac{V_2}{V_1} = \frac{1}{TR} \qquad\qquad \text{Eq. 18.8}$$

Equation 18.8 is called the *current ratio* of the transformer.

Two conclusions can be made after a study of Equation 18.1 and Equation 18.8.

1. When a transformer steps up the voltage, it steps down the current.
2. When a transformer steps down the voltage, it steps up the current.

It can also be seen that the ampere-turns of the primary equals the ampere-turns of the secondary:

$$N_1 I_1 = N_2 I_2 \qquad\qquad \text{Eq. 18.9}$$

Problem 1   An overhead distribution transformer is shown in figure 18-4. This transformer is rated at 2 400 V/240 V, 25 kVA, 60 Hz. It is designed to have an induced emf of 2.4 volts per turn. Calculate: a. the number of turns, and b. the full-load current of the primary and secondary windings.

Fig. 18-4 Overhead distribution transformer

Solution   a. $N_1 = \dfrac{V_1}{\text{volts per turn}}$ Eq. 18.10

$= \dfrac{2\ 400\ \text{V}}{2.4\ \text{V/turn}}$

$= 1\ 000\ \text{turns}$

$N_2 = \dfrac{V_2}{\text{volts per turn}}$ Eq. 18.11

$= \dfrac{240\ \text{V}}{2.4\ \text{V/turn}}$

$= 100\ \text{turns}$

b. $I_1 = \dfrac{V_1 I_1}{V_1}$

$= \dfrac{25\ 000\ \text{VA}}{2\ 400\ \text{A}}$

$= 1.04\ \text{A}$

$I_2 = \dfrac{V_2 I_2}{V_2}$

$= \dfrac{25\ 000\ \text{VA}}{240\ \text{A}}$

$= 104\ \text{A}$

(R18-5)   Calculate the turn ratio for the transformer described in Problem 1.

(R18-6)   The primary current equals 4.0 A when 1-Ω load is connected to the 40-V secondary of a step-down transformer. Find: a. the primary voltage, b. the volt-ampere value of the primary, and c. the volt-ampere value of the secondary.

IMPEDANCE MATCHING   Transformers are commonly used to match impedances between parts of a circuit.

(R18-7)   What is the requirement for maximum power transfer?

Figure 18-5 shows how an output transformer is used to match the output impedance of a transistor amplifier to the 8-Ω impedance of a loudspeaker. One of the design criteria of the output, or matching, transformer is expressed by the following equation:

$$Z_1 = (TR)^2\ Z_L \qquad\qquad \text{Eq. 18.12}$$

where

Z_L = load impedance, in ohms
Z_1 = secondary reflected impedance (or, the impedance "seen" by the primary of the transformer), in ohms
TR = turn ratio of the transformer

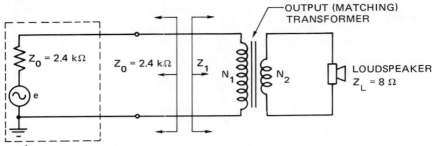

THÉVENIN EQUIVALENT
OF TRANSISTOR AMPLIFIER

Fig. 18-5 Output (matching) transformer used to match transistor amplifier output imped-ance to loudspeaker impedance

Problem 2   For the output transformer shown in figure 18-5, calculate the turn ratio re-quired to obtain the maximum power transfer to the loudspeaker.

Note: The output impedance of the transistor amplifier is 2.4 kΩ. This impe-dance value is the Thévenin impedance looking back into the amplifier. To obtain the maximum power transfer, Z_1 must equal 2.4 kΩ.

Solution Z_1 = $(TR)^2\, Z_L$

$(TR)^2$ = $\dfrac{Z_1}{Z_L}$

TR = $\sqrt{\dfrac{Z_1}{Z_L}}$

= $\sqrt{\dfrac{2\,400\ \Omega}{8\ \Omega}}$

= $\sqrt{300}$

= 17.32 to 1

Figure 18-6 shows wide-band transformers used to achieve an impedance match between a load impedance and the impedance of a driving source.

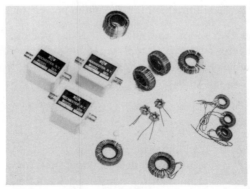

Fig. 18-6 Assortment of wide-band transformers used for impedance matching

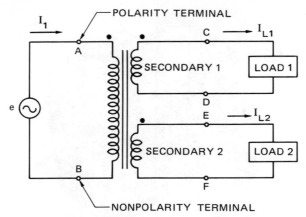

Fig. 18-7 Polarity markings — dot convention

(R18-8)   Assume that the circuit shown in figure 18-5 has two identical loudspeakers connected in parallel. Calculate the turn ratio required to obtain the maximum
power transfer.

POLARITY   Figure 18-7 shows the dot convention used to indicate the polarity of the trans-
MARKINGS   former leads. One end of each winding is marked with a dot. The dots indicate
that these points are at the same polarity.

(R18-9)   What is the polarity of terminal C, when terminal A is positive?

(R18-10)   What is the polarity of terminal E, when terminal A is negative?

As a result of the dot convention, the following statements can be made:

• When current enters the primary terminal with the marked polarity, the current leaves the secondary at the polarity terminal.

• When current leaves the polarity terminal of the primary, the current enters the secondary at the polarity terminal.

Figure 18-8 shows a machine tool control transformer. Note that the manufacturer has supplied a plate which indicates the leads leaving the transformer.
The leads on the high-voltage side are marked H_1, H_2, etc. The leads on the low-
voltage side are marked X_1, X_2, etc. H_1 has the same polarity as X_1, and H_2 has
the same polarity as X_2. The double primary winding can be operated at 480 V
or 240 V. The single secondary winding operates at 120 V.

(R18-11)   Indicate the connections that are to be made on the transformer shown in figure
18-8 if: a. the primary supply voltage is 480 V, and b. the primary supply voltage
is 240 V.

MULTIPLE   Many iron-core transformers have more than one secondary winding. For each
SECONDARIES   winding, the output voltage is determined by its turn ratio with the primary. The
total primary current is obtained by combining the effects of each secondary
current.

Fig. 18-8 H- and X- markings for the transformer leads

Problem 3   The following data is given for the transformer shown in figure 18-9:
V_1 = 2 000 V
V_2 = 1 000 V
V_3 = 500 V
Find: a. the ratio of each secondary winding with respect to the primary, b. the primary current I_1, and c. $(kVA)_1$ supplied by the primary.

Solution   a. $(TR)_2$ $= \dfrac{V_1}{V_2} = \dfrac{N_1}{N_2}$

$= \dfrac{2\ 000\ V}{1\ 000\ V}$

$= 2$

$(TR)_3$ $= \dfrac{V_1}{V_3} = \dfrac{N_1}{N_3}$

$= \dfrac{2\ 000\ V}{500\ V}$

$= 4$

b. I_2 $= \dfrac{V_2}{R_2}$

$= \dfrac{1\ 000\ V}{500\ \Omega}$

$= 2\ A$

I_3 $= \dfrac{V_3}{R_3}$

$= \dfrac{500\ V}{500\ \Omega}$

$= 1\ A$

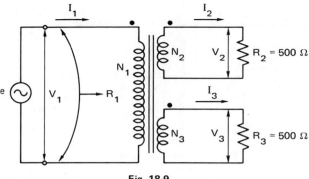

Fig. 18-9

$$N_1 I_1 = N_2 I_2 + N_3 I_3$$ Eq. 18.13

$$I_1 = \frac{N_2}{N_1} I_2 + \frac{N_3}{N_1} I_3$$

$$= \frac{1}{2}(2\ A) + \frac{1}{4}(1\ A)$$

$$= 1.25\ A$$

Alternate solution:

Power in = power out

$$V_1 I_1 = V_2 I_2 + V_3 I_3$$

$$(2\ 000\ V)I_1 = (1\ 000\ V)(2\ A) + (500\ V)(1\ A)$$

$$I_1 = 1.25\ A$$

c. $(VA)_1 = V_1 I_1$

$$= (2\ 000\ V)(1.25\ A)$$

$$= 2\ 500\ VA$$

$$(kVA)_1 = \frac{2\ 500\ VA}{1\ 000}$$

$$= 2.5\ kVA$$

(R18-12)   Refer to Problem 3 and find the imput impedance seen by the source. Use the calculated value of I_1.

**AUTO-
TRANSFORMERS**   The autotransformer consists of a single winding on an iron core, figure 18-10. The coil terminal C is common to both the input and the output. The winding is tapped somewhere along its length to form Terminal B. This means that an auto-transformer is a three-terminal device. The total winding between terminals A and C is the primary winding of the transformer. The section of the winding between terminals B and C is the secondary winding.

(R18-13)   What is the disadvantage of an autotransformer?

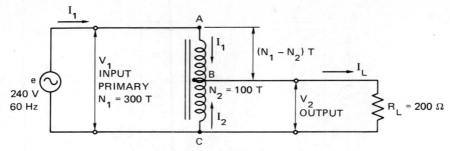

Fig. 18-10 Step-down autotransformer

Problem 4 For the autotransformer shown in figure 18-10, calculate: a. the primary current, and b. the secondary current.

Solution a. $TR = \dfrac{N_1}{N_2}$

$= \dfrac{300}{100}$

$= 3$

$V_2 = \dfrac{V_1}{TR}$

$= \dfrac{240\ V}{3}$

$= 80\ V$

$I_L = \dfrac{V_2}{R_L}$

$= \dfrac{80\ V}{20\ \Omega}$

$= 4\ A$

Power delivered to load $= V_2 I_L$

$= (80\ V)(4\ A)$

$= 320\ W$

The input power (from the source) must equal the power delivered to the load, or 320 W.

$V_1 I_1 = 320\ W$

$I_1 = \dfrac{320\ W}{V_1}$

$= \dfrac{320\ W}{240\ V}$

$= 1.33\ A$

b. I_2 (secondary current) $= I_L - I_1$

$= 4\ A - 1.33\ A$

$= 2.67\ A$

The secondary winding does not conduct the full load current. To satisfy Kirchhoff's Current Law, therefore, the load current ($I_L = 4$ A) is made up of the primary current ($I_1 = 1.33$ A) and the current through the secondary winding ($I_2 = 2.67$ A).

Fig. 18-11   500 000-kVA, 500 000/345 000-V autotransformer

(A)   Assortment of Variac® autotransformers

(B)   Eighteen Variac® units ganged for a special control application

Fig. 18-12   Variable autotransformers

Figure 18-11 shows a 500 000-kVA, single-phase, 500 000/345 000-volt autotransformer. Several continuously adjustable autotransformers are shown in figure 18-12A. These autotransformers provide smooth control of ac voltage to regulate light, heat, power, current, motion, and speed. Eighteen variable transformers are "ganged" for a special control application in figure 18-12B.

POLYPHASE TRANSFORMERS   The voltage of a three-phase system can be transformed by installing a separate single-phase transformer in each of the three phases of the system. However, a single three-phase transformer is commonly used because it is compact, lighter,

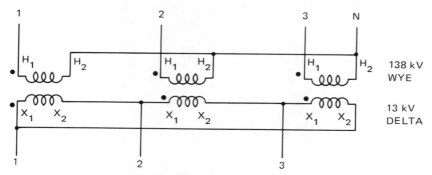

Fig. 18-13 Wye-delta connection of three-phase transformers

Fig. 18-14 A 1 300 000-kVA, three-phase, 24 500/
345 000-V generator step-up transformer

Fig. 18-15 Substation polyphase transformer

Fig. 18-16 Coils and internal wiring of a polyphase transformer

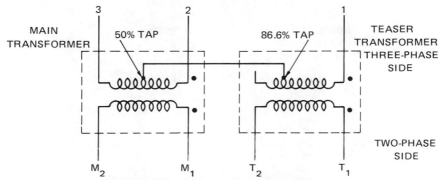

Fig. 18-17 Scott connection used to transform a voltage from three phase to two phase, or from two phase to three phase

and cheaper than three single-phase transformers. In a three-phase transformer, all of the primary and secondary windings are placed in the same housing and they share a common magnetic core. The connection of the transformer windings is the same for both a three-phase transformer and three single-phase transformers.

Figure 18-13 shows how three-phase transformers are connected in wye-delta. The proper polarity must be observed when connecting these transformers. Figure 18-14 shows a 1 300 000-kVA, three-phase, 24 500/345 000-V generator step-up transformer. A substation polyphase transformer is shown in figure 18-15. The coils and internal wiring of a polyphase transformer are shown in figure 18-16.

Two transformers can be used in a method known as the Scott connection, figure 18-17, to transform a voltage from three phase to two phase, or from two phase to three phase. The main transformer in this method has a 50% tap. The second, or teaser, transformer has a tap at 86.6% of the coil.

TYPES OF TRANSFORMERS

Transformers are classified according to the method by which heat due to internal losses is dissipated. The most common types of transformers are:
1. Dry, self-cooled.
2. Dry, forced air-cooled.
3. Liquid-immersed, self-cooled.
4. Oil-immersed, forced air-cooled.
5. Oil-immersed, water-cooled.
6. Oil-immersed, forced oil-cooled.

An auxiliary cooler for a transformer is shown in figure 18-18. The use of such a cooler means that the transformer can carry an additional load without exceeding its design temperature limit.

INSTRUMENT TRANSFORMERS

Instrument transformers are used with ac instruments, meters, and relays when high voltages or large currents are present. These transformers are classified into two groups: current transformers and potential transformers.

Fig. 18-18 Auxiliary cooler installation

Current transformers are used when the current is to be measured at a considerable distance from the main circuit. Current transformers also insulate the instrument from the high-voltage system, and eliminate the electrical shock hazard. The primary winding of the current transformer consists of one or more turns of wire placed directly in the circuit to be measured. The total load current flows in the primary winding. The secondary winding contains a larger number of turns. Normally this winding is designed to operate at 5 A for full-scale deflection.

> CAUTION. Do NOT operate a current transformer if the secondary winding is open-circuited. Serious overheating of the transformer core can occur when there is no secondary current to oppose the core flux generated by the primary current. The open-circuit voltage of the secondary can reach a dangerously high level.

Several types of current transformers are shown in figures 18-19, 18-20, and 18-21:

- a bar-type transformer, figure 18-19

- a lightweight, split-core transformer, figure 18-20

- a molded epoxy transformer rated for operation at 15 kV, figure 18-21.

Potential transformers reduce the voltage to a nominal value of 120 V. A typical potential transformer is shown in figure 18-22. A molded epoxy potential transformer, figure 18-23, is rated for operation at 34.5 kV. Resistance-type potential sensing units, figure 18-24, are used for relaying and nonrevenue metering applications up to 230 kV.

Fig. 18-19 Bar-type current transformer

Fig. 18-20 Split-core-type current transformer

Fig. 18-21 High-voltage current transformer

Fig. 18-22 Potential transformer

Fig. 18-23 High-voltage potential transformer

Fig. 18-24 138-kV resistance-type potential sensing units

LABORATORY EXERCISE 18-1 MEASUREMENT AND ANALYSIS OF A TRANSFORMER

PURPOSE This exercise allows the student to study the characteristics of a transformer.

EQUIPMENT AND MATERIALS
1 Transformer, 120 V/12.6 V
1 Resistor, 100 Ω, 2 W
1 Resistor, 1 kΩ, 2 W
1 Digital VOM
1 Oscilloscope
1 Power supply, 120 V ac

PROCEDURE
1. Connect the circuit shown in figure 18-25.
   a. Measure and record V_1 and V_2.
   b. Measure and record I_1 and I_2.
   c. Connect V_{AB} to the horizontal input of the oscilloscope. Connect V_{CD} to the vertical input of the oscilloscope. Sketch the pattern observed on the oscilloscope screen.
2. Perform the following calculations, and answer the following questions:
   a. Calculate the transformer turn ratio from the values recorded in step 1.a. Compare this value with the nominal value given by the manufacturer.
   b. Calculate the transformer turn ratio from the values recorded in step 1.b. Compare this value with the value calculated in step 2.a.
   c. Calculate $(kVA)_1$ and $(kVA)_2$ from the values recorded in steps 1.a. and 1.b. Are these two calculated values equal? What conclusion may be drawn from this set of readings?
   d. Calculate the input impedance seen at terminals A and B, when $R_L = 100\ \Omega$. Calculate the input impedance, using the measured values of V_1 and I_1. Are these two values equal?

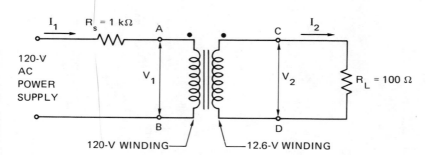

Fig. 18-25

e. What is the phase angle relationship between the primary voltage and the secondary voltage, as calculated from the pattern observed in step 1.c.?

EXTENDED STUDY TOPICS

1. The primary winding of a transformer is designed to operate at 120 V, 60 Hz. What will happen if 120 V dc is applied to this winding?

2. A step-down transformer has the following values:
   $V_1 = 10$ V, $V_2 = 2$ V, and $R_L = 2$ Ω.
   Find: a. TR, b. I_2, c. I_1, d (VA)$_1$, e. (VA)$_2$, and f. Z_1 (transformer input impedance).

3. Repeat topic 2, but reverse the input and output connections. The transformer now operates as a step-up transformer with $V_1 = 2$ V, $V_2 = 10$ V, and $R_L = 2$ Ω.

4. What conclusion can be drawn from the results of topics 2 and 3 regarding the impedance seen by the primary source?

5. A transformer with a center-tapped secondary is shown in figure 18-26. Calculate: a. I_2, b. I_3, c. I_N, d. I_1, and e. Z_1. (NOTE: This problem may be solved using a method similar to that used in Problem 3, but with multiple secondaries.)

6. A step-down autotransformer has the following values: $V_1 = 120$ V, $N_1 = 600$ T, $N_2 = 200$ T, $R_L = 20$ Ω.
   Find: a. I_L, b. I_1, and c. I_2.

7. A step-up autotransformer has the following values:
   $N_1 = 200$ T, $N_2 = 2\,000$ T, $V_1 = 120$ V, and $R_L = 400$ Ω.
   Find: a. V_2 (voltage across R_L), b. I_L, and c. I_1.

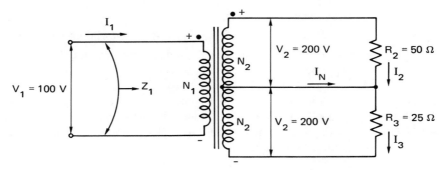

Fig. 18-26

<table>
<tr><td>

```
┌─────────────┐
│  Chapter    │
│     19      │
└─────────────┘
```

</td><td>

Generators—Ac and Dc

</td></tr>
</table>

OBJECTIVES After studying this chapter, the student will be able to

- discuss basic generator concepts.
- discuss alternators, including the frequency of the output voltage and alternator voltage regulation.
- describe a basic dc generator, its construction, and its excitation.
- analyze the characteristics and construction of series, shunt, and compound dc generators.

BASIC GENERATOR CONCEPTS A generator is a machine that converts mechanical energy into electrical energy by means of electromagnetic induction. (The concept of electromagnetic induction was presented in Chapter 10.) The magnitude of the voltage generated is determined by three factors:

1. The speed at which the conductor passes through the magnetic field.
2. The strength of the magnetic field.
3. The number of loops of wire.

Voltage is generated by means of the electromagnetic induction that exists:

1. when a conductor is moved through a magnetic field, or
2. when a magnetic field is moved with relation to a stationary conductor in that field.

The mechanical energy that provides the required motion may be supplied by wind, waterfalls, high-pressure steam, or engines powered by natural gas, diesel oil, or gasoline. Figures 19-1 and 19-2 illustrate a variety of machines which pro-

Fig. 19-1B Emergency power system

Fig. 19-1C Trailer-mounted switchable voltage generator

Fig. 19-1A Wind-powered electric generator

Fig. 19-2 Cross section of a vertical-hydraulic turbine drive

vide power. Figure 19-1A shows an electric generator which produces 200 W to charge a 12-V battery. An emergency power system which is packaged at the factory is shown in figure 19-1B. This system uses natural gas and can supply loads ranging from 3 kW to 65 kW. The trailer-mounted switchable voltage regulator of figure 19-1C supplies ac power at three voltages: 115/230 V single-phase, 115/230 V three-phase, and 277/480 V three-phase. The capacity of the voltage regulator may be 12 kW, 15 kW, 20 kW, 25 kW, or 30 kW. It may be powered by a diesel or gasoline engine running at 1 800 r/min. Figure 19-2 shows a cross section of a vertical-hydraulic turbine drive with an enclosure for outdoor installation.

(R19-1) What device or force supplies the mechanical energy for the generator shown in figure 19-1A?

ALTERNATORS   Ac generators are called alternators. An ac generating system consists of an armature, a magnetic field, and a prime mover.

(R19-2)   Why is an ac generator called an alternator?

An output voltage is induced in the armature of a generator. In a generator (alternator), the armature consists of a group of coils or loops. When the armature rotates, it is called a *rotor*. When the armature is stationary, it is called a *stator*. (The current flowing through the load also flows through the armature.) Figure 19-3 shows the core iron being assembled for the stator of a large generator/motor rated at 135 000 kVA, 90 r/min. Such a machine is used in a pumped storage hydraulic turbine unit.

The magnetic field for an alternator is supplied by a dc current through the field windings. A separate dc source is used for the field. In alternators supplying up to 50 kW, the field winding is the stator. In larger machines, the field winding is the rotor. The revolving-field alternator is preferred for most applications because it does not require slip rings to carry the output current from the generator. For large machines at full load, the output current is many times larger than the field current. Figure 19-4 shows a rotor weighing four million pounds, which is to be installed in the world's largest hydroelectric generator at Grand Coulee Dam.

Alternators with rotating and stationary fields are shown in figure 19-5. The field current is supplied by a dc generator which is known as an *exciter*. This machine operates at either 125 V or 250 V dc. The generator may be connected to the same shaft as the alternator. If the machine speed is low, the generator may be driven from the alternator shaft by a belt connection. Industrial power plants maintain spare exciters that can be used if an operating exciter fails.

(R19-3)   What is missing from figure 19-5?

Fig. 19-3 Assembly of the stator core iron for a generator/ motor

Fig. 19-4 Rotor for 600 000-kW hydroelectric generator at the Grand Coulee Dam

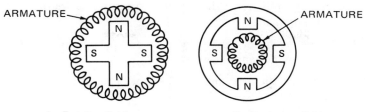

A. Rotating field   B. Stationary field

Fig. 19-5 Alternators and their fields

(R19-4) What happens to the output voltage of a 13-kV generator when the exciter fails?

FREQUENCY OF OUTPUT VOLTAGE The frequency of the output voltage of an alternator is controlled by the speed of the rotor. The relationship between frequency and the speed is expressed as follows:

$$f = \frac{PS}{60} \qquad \text{Eq. 19.1}$$

where

f = frequency of output voltage, in Hz (cycles per second)
P = number of pairs of magnetic poles
S = speed of rotor, in revolutions per minute (r/min)
60 = seconds per minute

Problem 1 An alternator is driven at a constant speed of 3 600 r/min. If the frequency of the output voltage is constant at 60 Hz, how many pairs of poles does the field contain?

Solution
$$f = \frac{PS}{60}$$
$$P = \frac{60\,f}{S}$$
$$= \frac{(60)(60\ \text{Hz})}{3\ 600\ \text{r/min}}$$
$$= 1$$

(R19-5) What is the speed of the rotor for the alternator in Problem 1 if the output frequency is 25 Hz?

Many electrical devices operate properly only when the frequency of the supply is accurate. For example, electric clocks depend upon a frequency of exactly 60 Hz to maintain the correct time. The constant speed of a synchronous motor depends upon a constant frequency input. Once an alternator is designed and built, its frequency depends upon the speed of the prime mover. To insure that the frequency remains constant, the prime mover driving the generator must have reliable speed regulation. In large power generating plants, recording instruments are installed to monitor the frequency. Automatic control devices are also used to compensate for any change in the speed. For example, if the frequency drops for a short time, the control device overspeeds the generator to make up the loss.

(R19-6)   If the frequency output increases to 60.2 Hz for one minute, what action is taken by the control device?

(R19-7)   For what length of time will the control device act under the conditions of question (R19-6)?

ALTERNATOR VOLTAGE REGULATION   The voltage output of an alternator must be varied during operation to compensate for voltage drops in the lines supplying the load. Ideally, it is desired to provide a constant output to the customer despite changes in the load current.

The voltage output of an alternator is controlled by adjusting the dc exciter voltage, which thus controls the current flowing in the alternator magnetic field. Refer to the schematic diagram of the alternator shown in figure 19-6.

(R19-8)   Is it possible to obtain voltage regulation by changing the alternator speed? Explain the answer.

The percentage voltage regulation is the percentage change in voltage from a no-load condition to full load when compared to the full-load voltage. The percentage voltage regulation is expressed as follows:

$$VR = \frac{V_{NL} - V_{FL}}{V_{FL}} \times 100\% \qquad \text{Eq. 19.2}$$

where

VR = voltage regulation
V_{NL} = no-load voltage
V_{FL} = full-load voltage

A small number for VR represents the desirable condition of a low variation in load voltage.

(R19-9)   A customer has a no-load voltage of 2 400 V. At full load, the voltage drops to 2 300 V. Calculate the percentage regulation at the customer's plant.

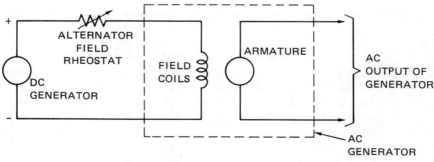

Fig. 19-6   Alternator voltage regulation

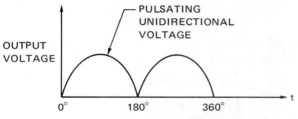

Fig. 19-7 Output of a single coil dc generator

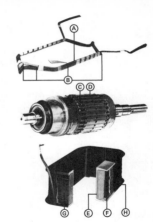

Fig. 19-8 Electrical elements of a dc generator

BASIC DC GENERATOR The ac output voltage of the armature coils of a dc generator can be changed to a unidirectional voltage. A commutator is used in place of the slip rings in a single coil dc generator to obtain the output shown in figure 19-7. The output voltage is pulsating and unidirectional because the brushes reverse the connections to the armature coil every half cycle when the voltage is zero and when it is ready to change its polarity. The electrical elements of a dc generator are shown in figure 19-8.

A commutator is equivalent to a slip ring which is split into two semicircular halves called segments. These segments are insulated from both the generator shaft and each other. The two ends of the armature coil are connected to separate commutator segments, figure 19-9. A conductive brush is positioned on each of the commutator segments. The brushes are placed so that they are on opposite sides of the commutator. As the coil is rotated in the magnetic field, the stationary brushes reverse the connections to the armature coil. In this way, terminal A is always positive with respect to terminal B.

(R19-10) At what point in the output cycle must the commutator reverse the coil connections?

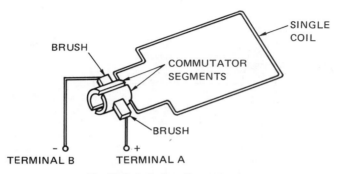

Fig. 19-9 A single coil commutator

Fig. 19-10 Dc generator

**DC GENERATOR
CONSTRUCTION**

The major components of a dc generator, figure 19-10, are: 1. the frame, 2. end bells, 3. pole pieces, 4. the shaft, 5. the armature, 6. the commutator, and 7. the brush assembly.

The frame (or yoke) is made of steel because this material is an excellent conductor of magnetic lines of force. The frame and pole arrangements for a two-pole dc generator and a four-pole dc generator are shown in figure 19-11.

The end bells are bolted to each end of the generator frame. These bells contain the bearings that support the shaft of the rotating armature.

The pole pieces, or pole shoes, are always used in pairs and are bolted to the generator frame. They support the field windings which are used to produce the magnetic field.

When a second coil is added to the armature of a dc generator, it is placed at right angles to the first coil. The output of a two-coil dc generator is shown in figure 19-12. In a practical dc generator, many coils are added to the armature. As a result, the actual output approaches the smooth ideal dc output shown in figure 19-13.

A dc generator with many coils also has many copper commutator segments. Mica is used to insulate the commutator segments from each other.

Graphite or carbon is used to make the brushes. Another material used is a mixture of powdered copper and powdered graphite, which is pressed and baked

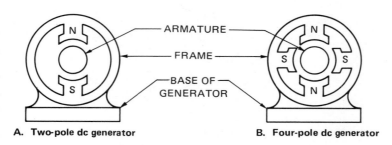

A. Two-pole dc generator   B. Four-pole dc generator

Fig. 19-11 Frame and pole arrangement of a dc generator

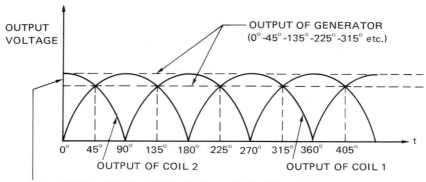

OUTPUT VOLTAGE

OUTPUT OF GENERATOR
(0°-45°-135°-225°-315° etc.)

0°   45°   90°   135°   180°   225°   270°   315°   360°   405°   t

OUTPUT OF COIL 2   OUTPUT OF COIL 1

THE OUTPUT OF THE GENERATOR VARIES BETWEEN THE
TWO DOTTED LINES AS THE COMMUTATOR CONNECTS THE
OUTPUT ALTERNATELY FROM COIL 2 TO COIL 1 AND BACK.
NOTE THAT THE UNIDIRECTIONAL PULSATIONS FOR THE
TWO COILS ARE SMOOTHER THAN THOSE FOR A SINGLE COIL.

Fig. 19-12 Output of a two-coil dc generator

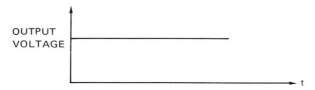

OUTPUT VOLTAGE

t

Fig. 19-13 Ideal dc output voltage. When many coils are added
to the armature, the actual output approaches this ideal output.

at a low temperature. A complete brush assembly consists of the brush, the brush holder, adjustable springs to hold the brush, and pigtailed connections.

Armature coils are wound on drum-type cores in most cases. The coils are preformed, and are wound to their final shape before they are placed in the armature.

EXCITATION OF A DC GENERATOR

A dc generator is said to be *separately excited* when the current for the field is supplied by a separate source of direct current. This source may be a battery, or it may be a separate dc generator, figure 19-14. The magnetic field of the separately excited dc generator is independent of the output circuit of the generator.

(R19-11)   What is the purpose of the field rheostat shown in figure 19-14?

Current is supplied to the field coils of a self-excited dc generator from the commutator by means of the brushes. Three types of circuits are used for self-excited dc generators: 1. series circuit, 2. shunt circuit, and 3. compound circuit.

The voltage in a self-excited dc generator builds up as indicated by the following steps:

1. Residual magnetism is retained in the electromagnetic field poles even when the coil is not energized.

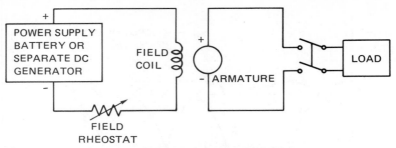

Fig. 19-14 Separately excited dc generator

2. As the generator armature rotates, a small dc voltage is induced in the armature coils.
3. This small dc voltage causes a small amount of current in the field windings.
4. The magnetic field increases in strength.
5. Both the output voltage and the magnetic field strength increase until they reach their maximum values.

SERIES DC GENERATOR

A series dc generator is shown in figure 19-15.

(R19-12)   If the external load is not connected, can a series dc generator supply its output voltage?

The output voltage-load current characteristics of a series dc generator is shown in figure 19-16. When this generator operates in Region 2, as shown in figure 19-16, it acts like a constant-current generator. Note that a slight increase in the load current produces a sharp decrease in the load voltage. Welding generators normally operate in this region. On the other hand, dc railroads operate in Region 1, where the generator acts like a voltage booster.

SHUNT DC GENERATOR

A self-excited shunt dc generator is shown in figure 19-17. Note that the generator circuit is connected in parallel with respect to the load. Part of the armature current is supplied to the electromagnetic field windings, but the main current is delivered to the external load.

(R19-13)   If a self-excited shunt generator is to build up its rated voltage, is it necessary to connect the external load?

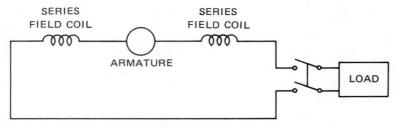

Fig. 19-15 Series dc generator

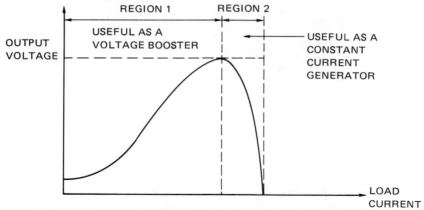

Fig. 19-16 Output voltage-load current characteristic of a series dc generator

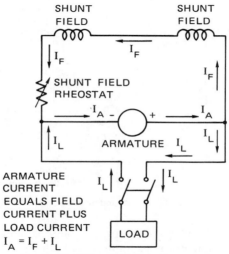

Fig. 19-17 Self-excited shunt dc generator

The output voltage-load current characteristic of a shunt dc generator is shown in figure 19-18. This type of generator acts like a constant-voltage generator up to the rated current. When the load current is increased beyond the rated value, the output voltage falls off rapidly.

COMPOUND DC GENERATOR A compound dc generator contains both series field windings and shunt field windings mounted on the same pole pieces. As the load increases or decreases, the strength of the magnetic field also increases or decreases. The schematic diagrams of a long-shunt compound generator and a short-shunt compound generator are shown in figure 19-19. The operation and output of the long-shunt and short-shunt compound generators are the same. The only difference between the generators is the point at which the series field is connected.

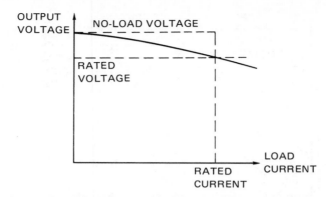

Fig. 19-18 Output voltage-load current characteristic of a shunt dc generator

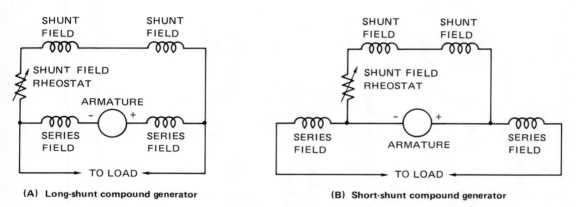

(A) Long-shunt compound generator

(B) Short-shunt compound generator

Fig. 19-19 Compound dc generators

The short-shunt compound generator is more commonly used in industry because it combines the advantages of both the series and shunt dc generators.

(R19-14)   Is it necessary to connect the external load of figure 19-19B when starting the generator?

The series field windings control the output voltage by helping to overcome the decrease in the output voltage of a shunt generator. A compound generator is said to be *flat compounded* if the series winding produces the same terminal voltage at both the rated load and no load. The generator is said to be *under-compounded* if the series winding produces a lower terminal voltage at the rated load than it does at no load. The generator is said to be *overcompounded* if the series winding produces a terminal voltage that is greater at the rated load than at no load.

Compound generators are usually wound so that they are slightly overcompounded. Figure 19-20 is the output voltage-load current characteristic of compound dc generators.

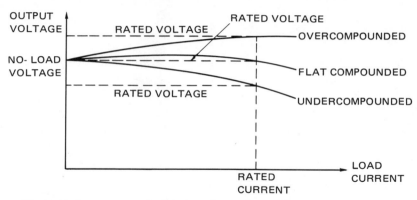

Fig. 19-20 Output voltage-load current characteristic of compound dc generators

EXTENDED STUDY TOPICS

1. List several types of prime movers that may be used to supply mechanical power to drive the rotor of an alternator.
2. The output voltage of an ac generator is induced in which component?
3. An alternator is driven at a speed of 1 800 r/min. If the alternator has an output frequency of 60 Hz, how many pairs of poles does the field have?
4. An alternator has four poles and an output frequency of 25 Hz. What is the speed of this alternator?
5. What adjustment can be made, in addition to that shown in figure 19-6, to regulate the ac output voltage of an alternator?
6. The no-load output voltage of a generator is 13 800 V. At full load, its output voltage is 14 600 V. Calculate the percentage voltage regulation of the generator.
7. What is the purpose of the end bells of a dc generator?
8. What factor(s) govern the building up of voltage in a self-excited dc generator?
9. Name two applications of a series dc generator.
10. What can be said about the voltage regulation of a series dc generator?
11. What is the effect of varying the resistance of the shunt-field rheostat shown in figure 19-17?
12. Does a compound dc generator characteristic resemble that of a series or a shunt dc generator?
13. Which type of compound dc generator has a rated load voltage which is equal to its no-load voltage?

Chapter 20

Motors—Ac and Dc

OBJECTIVES

After studying this chapter, the student will be able to

- discuss basic dc motor concepts.
- describe the characteristics of dc motors.
- discuss the basic principles of ac synchronous motors.
- solve for the slip of ac motors.
- analyze various ac motors.
- describe how SCRs are used to control the speed of motors.

BASIC DC MOTOR CONCEPTS

A motor is a machine that converts electrical energy into mechanical energy. A dc motor operates on the principle that a current-carrying conductor in a magnetic field tends to move at right angles to the direction of the magnetic field. Figure 20-1 illustrates this statement. Note in figure 20-1A that the direction of the conductor field at the bottom of the conductor is the same as that of the magnetic field between the pole pieces. Thus, the conductor field aids, or strengthens, the field at the bottom of the pole pieces. The fields are in opposite directions at the top of the pole pieces and thus oppose each other. As a result, the conductor is forced to move up. In figure 20-1B, the current is reversed in the conductor, causing it to move down. The interaction of the conductor field and the field around the poles creates the twisting force, or *torque*, which causes the shaft of the motor to rotate. Miniature dc motors, such as the one shown in figure 20-2, have many applications, including driving cassettes and reel recorders.

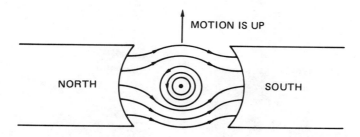

(A) Current in the conductor is coming out of the paper

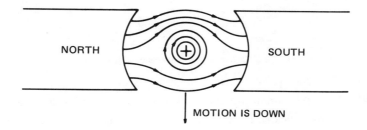

(B) Current in the conductor is heading into the paper

Fig. 20-1 A current-carrying conductor in a magnetic field

Fig. 20-2 Miniature dc motor

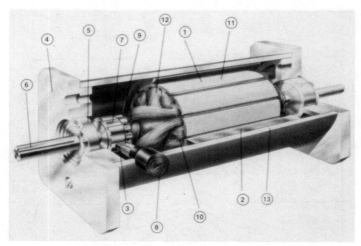

Fig. 20-3 Construction features of a dc motor

(R20-1)   How does a motor differ from a generator?

(R20-2)   What two conditions must exist before motor action can occur?

As indicated in figure 20-3, the components of a dc motor are similar to dc generator parts. Direct current is supplied to both the stator (the field windings on the pole pieces) and the rotor (the commutator and armature). The direct current to the rotor is converted to alternating current so that continuous motor rotation is achieved.

Dc Motor Operation. The operation of a dc motor is described in the following steps.
1. When a direct current passes through the stator field windings, a magnetic field is created.
2. The north pole of the rotor is attracted to the south pole of the magnetic field of the stator. The south pole of the rotor is attracted to the north pole of the magnetic field of the stator. This attraction causes the rotor to rotate.
3. The commutator reverses the direction of the current through the rotor when the rotor reverses its position. This sequence is repeated and continuous motor rotation is obtained.

COUNTER EMF   Recall that a voltage is induced in any conductor rotating in a magnetic field. Thus, a running dc motor is also a generator. The voltage induced in the spinning armature opposes the applied voltage. The induced voltage is also less than the applied voltage. Figure 20-4 illustrates the three factors that govern the current drawn by a dc motor. These factors are:
1. the applied voltage,
2. the resistance of the motor, and
3. the counter emf (cemf).

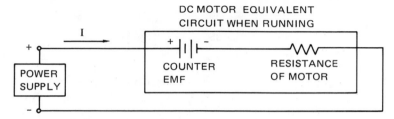

Fig. 20-4 Factors affecting the current drawn by a dc motor

(R20-3)   Why does a dc motor draw a very large starting current?

The amount of current drawn by a running motor depends upon the load of the motor. As the load is increased, the motor speed decreases, the cemf decreases, and the motor current increases.

DC SHUNT MOTORS   A dc shunt motor is shown in figure 20-5. Many turns of small wire make up the shunt field winding which is connected in parallel with the armature winding. The speed and torque characteristics of a dc shunt motor are shown in figure 20-6. The dc shunt motor has a fairly constant speed as the load varies. Although

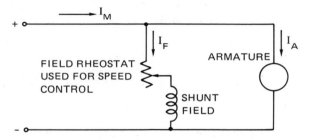

Fig. 20-5 Schematic diagram of a dc shunt motor

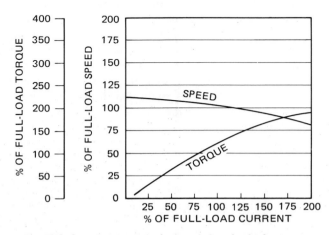

Fig. 20-6 Speed and torque characteristics of a dc shunt motor

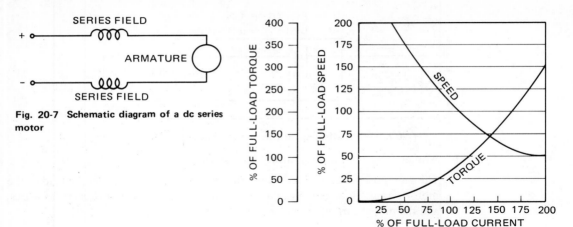

Fig. 20-7 Schematic diagram of a dc series motor

Fig. 20-8 Speed and torque characteristics of a dc series motor

the speed of this type of motor decreases slightly when the load increases, it is called a *constant-speed motor*. This motor is unsuitable for starting under a heavy load because the torque at a low value of current is low.

(R20-4)   What are the two main characteristics of a dc shunt motor?

(R20-5)   If the direction of the current through the field windings is reversed, what will be the effect on the direction of rotation of the motor?

(R20-6)   Does a dc shunt motor have a low or high starting torque?

DC SERIES MOTORS   A schematic diagram of a dc series motor is shown in figure 20-7. With an increasing load, the operation of a series motor is as follows:

1. An increase in the motor load causes the armature to draw more current.
2. The strength of the series magnetic field increases.
3. As a result, the armature does not have to turn as fast to develop a cemf nearly equal to the applied voltage, and the motor slows down.

The speed and torque characteristics are given in figure 20-8.

A large series motor should not be operated without a load. The no-load speed of such a motor is so great that the motor may be damaged. A series motor develops a high torque. This means that the motor is ideal for heavy loads that are to be started or reversed rapidly. An automobile starter is one application of a dc series motor.

(R20-7)   Compare the speed regulation of a dc shunt motor with that of a dc series motor.

DC COMPOUND MOTORS   A dc compound motor has the characteristics of both a dc shunt motor and a dc series motor. The motor is a *cumulative-compound motor* when the shunt field coil and the series field coil help each other. In a *differential-compound motor*, the shunt field coil and the series field coil oppose each other. Compound motors

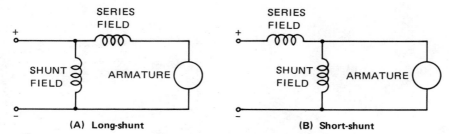

(A) Long-shunt   (B) Short-shunt

Fig. 20-9 Dc compound motors

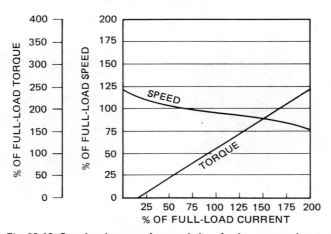

Fig. 20-10 Speed and torque characteristics of a dc compound motor

are divided into two groups according to how the series field coil is connected with respect to the shunt field coil. The two groups are:
1. long-shunt compound motors, and
2. short-shunt compound motors.

Figure 20-9 shows field connections for the two groups of dc compound motors. The speed and torque characteristics of a typical dc compound motor are given in figure 20-10. The characteristics of the dc compound motor can be varied between those of a dc series motor and those of a dc shunt motor by controlling the relative strengths of the series and shunt field coils.

(R20-8)   State how a short-shunt compound motor differs from a long-shunt compound motor.

DC STEPPING MOTORS   A stepping motor translates electrical pulses into mechanical movements. The output shaft moves, or rotates, through a specific angular rotation for each electrical pulse. This type of motor can control velocity, distance, and direction. The stepping motor, figure 20-11, receives power from a dc supply and is controlled by electronic drive-logic circuitry. The preset indexer module, figure 20-12, provides bidirectional control of the SLO-SYN® stepping motor. The

Fig. 20-11 SLO-SYN® stepping motor, type M092

Fig. 20-12 SLO-SYN® preset indexer, type SP153

Fig. 20-13 Internal circuitry of SLO-SYN® preset indexer module, type PIM151

Fig. 20-14 SLO-SYN® synchronous motor, type SS250

module drives the motor in half-steps of 0.9° or in full steps of 1.8°. The internal circuitry of the indexer module is shown in figure 20-13.

AC SYNCHRONOUS MOTORS

An ac synchronous motor runs at one speed only. This type of motor must always be in step with the changing field. Therefore, the speed of the motor is synchronized to the frequency of its ac supply. Synchronous motors are used in timing devices to maintain a high degree of timing accuracy. Permanent magnets provide the fields in small synchronous motors, figure 20-14.

(R20-9) What is the most important characteristic of a synchronous motor?

The synchronous speed of an ac motor is expressed by the following equation:

$$S = \frac{60\,f}{P} \qquad\qquad \text{Eq. 20.1}$$

where
 S = synchronous speed, in r/min
 f = frequency of input voltage, in Hz
 P = number of pairs of poles

Problem 1 Find the synchronous speed of an ac motor connected to a 60-Hz power line if the motor has four poles.

Solution $P = \dfrac{4}{2} = 2$ pairs of poles

$S = \dfrac{60 \ f}{P}$

$= \dfrac{(60)(60 \ Hz)}{2}$

$= 1 \ 800 \ r/min$

SLIP IN AC INDUCTION MOTORS The speed of an ac induction motor approaches the synchronous speed, but never equals this speed. Assume that the rotor of an induction motor does move at the synchronous speed. As a result, the conductors of the rotor will not cut magnetic lines of force and a voltage will not be induced in the rotor.

Slip is the difference between the synchronous speed of the magnetic field and the actual speed of the motor.

$$\text{Slip} = S - R_s \qquad \text{Eq. 20.2}$$

where

S = synchronous speed, in r/min

R_s = actual running speed, in r/min

In most ac induction motors, the slip at full load varies from 4% to 6%. The percentage of slip of an induction motor is expressed as follows:

$$\% \text{ Slip} = \frac{S - R_s}{S} \times 100 \qquad \text{Eq. 20.3}$$

Problem 2 The synchronous speed of the magnetic field of a 60-Hz induction motor is 3 600 r/min. The motor has an actual speed of 3 425 r/min at full load. Calculate: a. the slip, and b. % slip.

Solution a. Slip $= S - R_s$

$3 \ 600 \ r/min - 3 \ 425 \ r/min$

$= 175 \ r/min$

b. % Slip $= \dfrac{S - R_s}{S} \times 100$

$= \dfrac{3 \ 600 \ r/min - 3 \ 425 \ r/min}{3 \ 600 \ r/min} \times 100$

$= \dfrac{175 \ r/min}{3 \ 600 \ r/min} \times 100$

$= 4.86\%$

(R20-10) What is synchronous speed?

INDUCTION MOTORS There is no electrical connection to the rotor of an induction motor, other than for starting in some types. As the name of the motor implies, current is induced in the rotor. Figures 20-15A and 20-15B show components of the assembled induction motor of figure 20-15C.

The stator core is constructed of laminations punched from high grade silicon steel. This steel is selected for its magnetic properties and is pre-coated to minimize eddy current losses in the core. Laminations are stacked, keyed, compressed and secured with a heavy retaining ring and a tack-welded snap ring. This construction provides a strong, rigid assembly which minimizes vibration and noise and assures vital air gap accuracy.

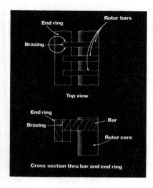

The yoke (motor frame) is constructed of either cast iron or fabricated steel depending on frame size and type of enclosure. Both types of heavy duty construction provide strong, sturdy bearing support to assure accurate alignment of the rotor and shaft.

The entire stator assembly with coils inserted in the slots is dipped in insulating varnish and baked. This process results in a unit which is physically and electrically strong and one which is protected against moisture, heat, and corrosion.

(A) Stator and frame arrangement

Shafts are machined from medium carbon steel bar stock and are generously sized for transmitting torques to the driven load. Ample fillets are provided at each change in shaft diameter to reduce stress concentration and all bearing fits are polish ground to close tolerances.

The rotors are either die cast or fabricated bar construction.

Die cast rotor laminations are stacked on a mandrel, compressed, and cast into a solid rugged rotor. The cast rotor is keyed to the shaft or spider before turning and balancing. With this construction the bars and end rings are cast in one solid piece eliminating the possibility of bond faults or flexing failures at the joints.

Fabricated rotor laminations are shrink fitted and keyed to the shaft or spider. These laminations are rigidly secured by heavy metallic end heads. Tooth supports prevent flaring of the teeth.

Copper bar conductors are tightly fitted in the slots and brazed to the end rings. The end rings are extra thick and slotted so that the end of the bar is embedded in the slot. Brazing joins the bar to the end rings on both sides and the end of the bar, thus assuring a solid electrical and mechanical joint. The extra thick ring minimizes cracking around the joint. See sketch.

On larger sizes, motor performance can be closely matched to the load characteristics by using extruded aluminum rotor bars. The end rings are also aluminum and the two are joined by inert gas welding process to assure the integrity of the weld.

Cooling air passageways and vents are built into the rotor core at the proper locations to assure cool operation. Axial vents parallel to the shaft carry the cooling air to radial vents built into the core. Air passing through these vents picks up heat close to its source and carries it away from the rotor.

(B) Rotor and shaft construction
Fig. 20-15 An ac induction motor

(C) Induction motor

Fig. 20-15 An ac induction motor (cont'd)

Single-phase induction motors are not self-starting. The field of the stator winding is stationary and pulsates in step with the applied frequency. The rotor field is also stationary. As a result, an auxiliary starting system is required to turn the rotor of a single-phase ac motor. Once the rotor is moving, the stationary field of the stator keeps it spinning. Many types of induction motors with several starting methods are available and will be analyzed later in this chapter.

(R20-11)   Does the magnetic field rotate in a 60-Hz ac, single-phase induction motor?

SHADED-POLE MOTORS   Figure 20-16 is the schematic diagram of a shaded-pole motor. A slot is cut into each pole piece and a copper ring is placed over each slotted section. The main flux of the ac field induces a current in the copper rings. The resulting flux opposes the main flux. Most of the main flux passes through the unshaded portions of the pole pieces. The inductance in the copper rings causes the flux in the shaded portion to lag behind the flux in the unshaded portion. As a result, the magnetic field rotates in the direction of the shaded poles. This field supplies the torque required for motor rotation. Shaded-pole motors are simple in construction, low in cost, and very reliable. Shaded-pole stators are used for the fractional horsepower motors, figure 20-17.

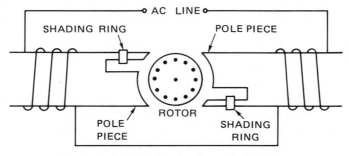

Fig. 20-16 Schematic diagram of a shaded-pole motor

(A) Two-pole shaded-pole motor

(B) Four-pole shaded-pole motor

(C) Six-pole shaded-pole motor

Fig. 20-17 Shaded-pole motors

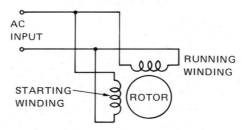

Fig. 20-18 Schematic diagram of a split-phase motor

(R20-12)   What is the relationship between the field of the copper shading ring and the main field?

SPLIT-PHASE MOTORS   The stator of a split-phase, fractional horsepower motor contains two separate coils. There is a main (running) coil made of large wire and a starting coil which uses smaller wire. Because of the smaller wire, the starting coil has a higher resistance than the main coil.

The schematic diagram of a split-phase motor is shown in figure 20-18. The coils are designed so that the current in the main (running) coil lags behind the current in the starting coil. As a result, the physical arrangement of the two windings produces a rotating magnetic field. This field accounts for the starting torque. A centrifugally operated switch disconnects the starting winding when the motor reaches 60% of its full-load speed.

(R20-13)   What is the purpose of the starting coil?

CAPACITOR MOTORS   Split-phase motors have a low starting torque because of the small phase angle between the currents in the starting and running windings. If a capacitor is used in series with one of the windings, as in a capacitor induction motor, there is a phase shift of nearly 90°. Thus, a high starting torque is produced. Figure 20-19 shows two types of capacitor motors. A motor-starting electrolytic capacitor is shown in figure 20-20.

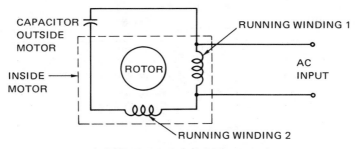

(A) Capacitor-start, capacitor-run motor

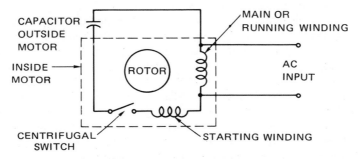

(B) Capacitor-start motor with centrifugal switch

Fig. 20-19 Capacitor motors

Fig. 20-20 Motor-starting electrolytic capacitor

(R20-14) What is the phase angle between the fields of winding 1 and winding 2 in figure 20-19A?

When the motor reaches 75% of its rated speed, the centrifugal switch shown in figure 20-19B cuts out the capacitor and the starting winding. The motor then runs as a single-phase induction motor.

REPULSION MOTORS A single-phase ac repulsion motor has a commutator, brushes, and field windings. As shown in figure 20-21, two brushes are placed 180° apart and are short circuited. The expanding and contracting single-phase field of the stator induces

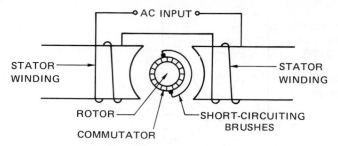

Fig. 20-21 Repulsion-induction motor

a current in the rotor. The short-circuited brushes furnish a path for the induced current and magnetic poles are formed around the rotor. The rotor is repelled by the stator. This action gives the motor its name. The repulsion between the stator field and the rotor causes the magnetic poles of the rotor to develop a torque as the rotor turns. As a result, the rotation continues. The speed of the motor varies inversely with the applied load. This means that the motor will race if the load is removed. This type of motor has a good starting torque. Thus, when there is a heavy starting load, a repulsion motor is used.

The repulsion-induction motor has a wound rotor and a commutator. Shorting brushes contact the commutator and the motor starts as a repulsion motor. When the motor approaches its rated speed, a device short circuits all of the commutator bars. As a result, the motor runs as an induction motor at a constant speed.

(R20-15)   Compare a repulsion motor and a repulsion-induction motor in construction and operation.

SQUIRREL-CAGE MOTORS   The three-phase squirrel-cage induction motor, figure 20-22, is the most widely used polyphase motor. The rotor consists of a laminated iron core mounted on a shaft. Copper bars are placed in slots along the length of the core. These bars are welded to copper end rings. As a result, the rotor becomes a short circuit.

The rotor turns because of the mutual induction between the rotor and the rotating field due to the stator windings. Magnetic fields are set up by the cur-

Fig. 20-22 Squirrel-cage induction motor

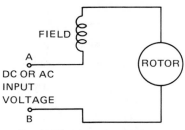

Fig. 20-23 The universal motor

rents produced in the conductors of the rotor. The poles of these fields are opposite to the poles of the rotating magnetic field. The rotor is attracted by the rotating magnetic field of the stator and rotates in the same direction as this field.

The speed of rotation of the field at full load is 3% to 20% greater than the rotor speed so that the motor will develop the required torque. At no load, the speed of the rotor is nearly the same as that of the rotating magnetic field.

(R20-16) What is the slip at no load?

UNIVERSAL MOTORS The universal motor operates on either dc or single-phase ac. Universal motors have fractional horsepower ratings and a high starting torque. Therefore, they are used in small appliances such as electric drills. As shown in figure 20-23, a universal motor is a dc series motor whose windings operate efficiently with either a dc or ac input.

(R20-17) By reversing the line connections of the universal motor in figure 20-23, what is the effect on the direction of rotation?

SPEED CONTROL USING SCRs Solid-state silicon-controlled rectifiers (SCRs) are used to regulate the motor speed to a specific torque requirement. The theory and applications of the SCR are explained in detail in Chapter 29.

The SCR has two stable states of operation. One stable state, or OFF state, has negligible current. The other stable state, or ON state, has a very high current. The current in the ON state is limited only by the resistance of the external circuit. The SCR acts as a switch to turn the power ON. A very small triggering current can be used to turn on a very large current.

In figure 20-24, an SCR is used to control an ac universal motor. The steps in the operation of this circuit are as follows:

1. The RC circuit determines when the SCR in the OFF state is turned to the ON state.
2. The time constant of the RC circuit is adjusted by the variable resistor (R).
3. At some time after the ac input voltage becomes positive, the gate terminal of the SCR becomes positive. The SCR is turned ON. It then conducts current from Terminal A (the anode) to Terminal K (the cathode). The motor continues to receive current for the rest of the positive half-cycle of input voltage.

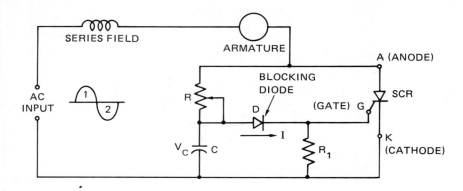

1. TORQUE PULSES FOR MOTOR OPERATION ARE PRODUCED ONLY DURING ALTERNATION 1 OF THE AC INPUT.
2. BLOCKING DIODE D ALLOWS CURRENT TO FLOW ONLY IN THE DIRECTION OF THE ARROW.
3. THE SCR FIRES (IS TURNED ON) ONLY DURING ALTERNATION 1, WHEN THE VOLTAGE ACROSS R_1 IS POSITIVE. ADJUSTMENT OF RESISTOR R VARIES THE POINT IN ALTERNATION 1 AT WHICH THE SCR FIRES.

Fig. 20-24 SCR motor control

4. When the input voltage becomes negative, the SCR no longer conducts and is turned OFF. In this state, no motor torque is produced.
5. The motor again receives current when the positive half-cycle of the input voltage charges the capacitor (C) to a positive value which is high enough to turn the SCR ON. This stage occurs when the voltage across R_1 is positive.

The internal components of an SCR motor controller are shown in figure 20-25.

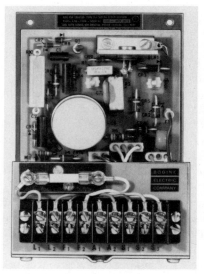

Fig. 20-25 Internal components of an SCR motor controller

**EXTENDED
STUDY TOPICS**

1. What is the purpose of a commutator in a dc motor?
2. The resistance in series with the shunt field of a dc shunt motor is increased. Describe the effect on the speed of the motor.
3. List the two methods used to reverse the direction of rotation of a dc shunt motor.
4. The polarity of the dc voltage supplied to the motor of figure 20-5 is reversed. Will the direction of rotation reverse?
5. Why does a dc motor draw more current when the load is increased?
6. Describe the effect on motor speed of a dc shunt motor running under no load, if its field opens.
7. Which type of dc motor has the highest starting torque?
8. Which type of dc motor has a low starting torque?
9. Calculate the speed of a two-pole ac motor at 60 Hz.
10. Calculate the rotor speed of a six-pole induction motor which has 5% slip at full load. The input frequency is 60 Hz.
11. What is the purpose of the centrifugal switch of a split-phase motor?
12. Explain the origin of the rotating magnetic field of a polyphase squirrel-cage induction motor.
13. How often are torque pulses produced in the motor control circuit of figure 20-24?

Chapter 21

Solar Cells, Fuel Cells, and Batteries

OBJECTIVES

After studying this chapter, the student will be able to

- explain how solar cells operate and list practical applications of solar cells.
- describe fuel cells in use at present and discuss their future development.
- discuss the theory and applications of various types of batteries.

BASIC CONCEPTS – SOLAR CELLS

A *photocell* is a two-terminal device which converts light energy into electrical energy. A *photovoltaic cell*, or a *solar cell*, is the name given to a self-generating photocell that produces a voltage when it is exposed to sunlight. Such a cell can supply a current at voltages up to approximately 0.6 V. Figure 21-1 shows the typical output characteristics of a solar cell. The solar cell operates in the constant current region when the output voltage is less than 0.4 V.

(R21-1)

The light intensity reaching a solar cell is 50 mW/cm². If the output voltage is 0.3 V, how can the output current be increased?

The maximum power delivered to an external load normally is only 12% of the total solar energy that strikes the solar cell. Some laboratory versions of solar cells have been able to deliver more than 15% of the solar energy input on the cell. Solar cells are connected in series to obtain higher voltages. For higher currents, they are connected in parallel. In addition, cells may be connected in series-parallel to meet specific voltage and current requirements. Manufacturers pack groups of cells in modules which are then used in arrays designed for particular applications. Figure 21-2 shows the output characteristic for a solar cell module when the incident sunlight has a value of 100 mW/cm².

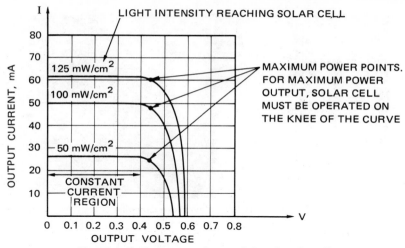

Fig. 21-1 Typical output characteristics of a solar cell

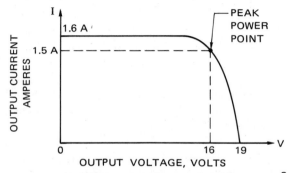

Fig. 21-2 Output characteristics for a solar cell module with 100 mW/cm^2 incident sunlight

(R21-2)   What is the peak power output of the solar cell module in figure 21-2?

(R21-3)   Define the following: a. photocell; b. solar cell; c. solar module; and d. solar array.

Solar Battery. A *solar battery* is a direct-current power source consisting of a solar array. It delivers useful amounts of power when sunlight shines on it. Applications of solar batteries include control devices, emergency telephone power supplies, space satellites, and portable radios. The cardlike solar battery shown in figure 21-3 can replace 4-V, 5-V, 6-V, or 9-V dry batteries in industrial applications. The solar cells in this battery are wired in series to obtain the correct output voltage. The same type of solar cells were used to power the instrumentation for the NASA Voyager spacecraft. When the incident sunlight has a value of 100 mW/cm^2, the solar battery converts this energy into 6.0 V with a current output of 25 mA. Figure 21-4 shows the mounting of silicon solar cells on a Telstar satellite.

Fig. 21-3 Cardlike solar battery

Fig. 21-4 Mounting silicon solar cells on Telstar satellite

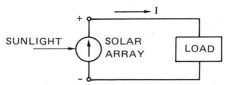

Fig. 21-5 Basic photovoltaic circuit

A basic photovoltaic circuit is shown in figure 21-5. If the sunlight is steady, the output is in the form of direct current. If the light varies in intensity, the output may be a pulsating unidirectional wave, or it may be an alternating-current wave superimposed on direct current.

ELECTRONIC ANALYSIS

A solar cell is a semiconductor diode. It is constructed so that light will reach the area of the junction of the materials that make up the diode. The theory and application of the semiconductor diode are covered in detail in Chapters 22 and 23.

Semiconductor Material. Silicon is commonly used as the semiconductor element in commercial solar cells. To produce excess negative or positive charges to carry electric currents, impurities are added to the silicon. For example, when phosphorus atoms are added to pure silicon, the material has excess electrons. The part of the silicon crystal with excess electrons is called *n-type silicon*. Excess positive charges are obtained by adding boron atoms to pure silicon. This part of the silicon crystal is called *p-type silicon*.

The junction between the *p*-type silicon wafer and the *n*-type silicon wafer is known as the *p-n* junction. It is in the area of the junction that the electric fields operate which give rise to the diode characteristics and the photovoltaic effect. When light strikes the solar cell, negative and positive charges flow inside the crystal. However, only negative charges (electrons) flow in the external circuit. Figure 21-6 shows the basic structure of a solar cell.

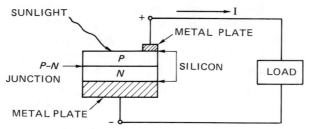

Fig. 21-6 Basic structure of a solar cell

(R21-4) Why is the metal plate connected to the *p-* section of the solar cell, figure 21-6, much smaller than the metal plate connected to the *n*-section?

SOLAR BATTERY CHARGER Storage batteries and a solar array are used together when electrical energy is required 24 hours a day year-round. A solar battery charging circuit is shown in figure 21-7. The blocking diode prevents the battery from discharging through the solar array when there is no sunshine. The shunt voltage regulator insures that battery life is not shortened by too great a charging voltage. The solar battery used should have ample energy storage capacity for those times when there is no sunlight.

Problem 1 The circuit of figure 21-7 is used in an earth satellite which operates on 12-V batteries. The batteries must supply a continuous current of 0.5 A to the load. Sunlight hits the solar array for 12 hours in every 24 hours. If the expected light intensity reaches the solar cells, the maximum power output is obtained when the cells operate at an output of 0.5 V and 60 mA. Calculate the number of solar cells required for the solar array. Determine how the cells are to be connected. Assume a voltage drop of 1 V across the blocking diode.

Solution The maximum output of the solar array is:

$$12 \text{ V} + 1 \text{ V} = 13 \text{ V}$$

$$\text{Number of cells connected in series} = \frac{\text{output voltage}}{\text{cell voltage}}$$

$$= \frac{13 \text{ V}}{0.5 \text{ V}}$$

$$= 26 \text{ cells}$$

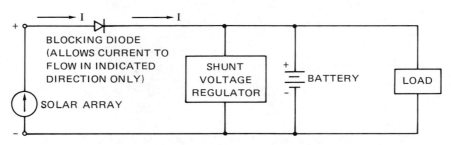

Fig. 21-7 Solar battery charging circuit

The charge taken from the batteries
over a 24-hour period   = (24 hr)(0.5 A)
                                                         = 12 ampere-hours

The solar cells must deliver 12 ampere-hours. However, the solar cells can deliver current only during the 12 hours in every 24 hours that they are illuminated.

The charging current from the solar cells $= \dfrac{12 \text{ ampere-hours}}{12 \text{ hours}}$
$= 1$ A

The total number of groups of cells in
parallel

$= \dfrac{\text{output current}}{\text{cell current}}$

$= \dfrac{1 \text{ A}}{60 \text{ mA}}$

$\cong 17$ groups in parallel

The total number of cells required is equal to:
(number of groups in parallel)(number of cells in series) = (17)(26) = 442 cells

(R21-5)   Determine the number and arrangement of the solar cells required for Problem 1, figure 21-7, if the maximum power output at the expected light intensity is 406 mV and 50 mA.

PRACTICAL APPLICATIONS   Figure 21-8 shows an artist's sketch of a practical application of a solar cell array. This installation was supplied by the NASA Lewis Research Center of Cleveland, Ohio under a cooperative agreement with the U.S. Agency for International Development (AID). The array supplies power to a water pump and grain grinder in the remote village of Tangaye, Upper Volta, in West Africa. At peak operation, the array can provide 1.8 kW of electricity. The solar panels

Fig. 21-8 Sunlight is converted to electricity to pump water and grind grain in the West African village of Tangaye

Fig. 21-9 RAMOS — solar powered weather stations

consist of silicon placed on a fiberglass base (substrate). This assembly is then enclosed (encapsulated) in a protective jacket of silicone. Each panel has eight solar modules. Each module operates at 10% efficiency and delivers 18.6 W when the light input is 100 mW/cm^2.

Solar cells are also being used to power six RAMOS (Remote Automatic Meteorological Observing System) for the National Weather Service, figure 21-9. The installations are at Strafford Shoals, New York; Clines Corners, New Mexico; South Point, Hawaii; Point Retreat, Alaska; Halfway Rock, Maine; and Loggerhead Key, Florida. Depending upon the location of each RAMOS, the power level of the solar cell array varies from 74 to 148 peak watts. The use of solar cells rather than conventional fuels saves up to $150 per year per site in fuel costs. An additional $3000 per year per site is saved in fuel transportation and maintenance costs.

Figure 21-10 shows a 185-W solar cell array that supplies energy to a refrigerator in the Papago Indian village of Sil Nakya, Arizona. This refrigerator is

Fig. 21-10 Photovoltaic powered refrigerator at Papago Indian village of Sil Nakya, Arizona

Fig. 21-11 Photovoltaic powered U.S. Forest Service lookout towers

used to keep medicines at, or below, room temperature to preserve their effectiveness. In addition, the refrigerator is used to store perishable foods which are required to improve the diet of all the people in the village.

U.S. Forest Service lookout towers also receive electrical power from 300-W solar cell arrays, figure 21-11. Each system consists of a 300-W solar array, 3 000 ampere-hours of battery storage, a battery charge controller, and instrumentation to indicate the state of the power system.

Another application of solar energy is shown in figure 21-12. Here, a solar cell array provides power to operate a duststorm warning sign in Arizona. The same photovoltaic power system also supplies electricity for lighting and radio

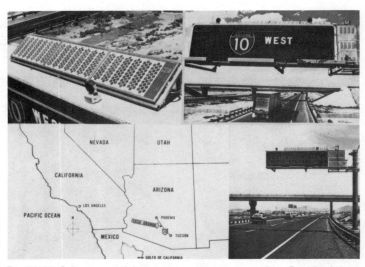

Fig. 21-12 Solar powered duststorm warning sign near Casa Grande, Arizona

Fig. 21-13 Solar power at a railroad crossing

communications. When all forty of these warning signs are solar powered, the savings in sign operating costs will be $12,000 a year.

Figure 21-13 shows a railroad crossing operated by a battery. The power source for charging the battery is a solar array with an output voltage of 13.8 V dc, 10.2 A (peak), and 166 W (peak). The six-cell lead-calcium battery has a rating of 12 V dc and 835 ampere-hours.

BASIC CONCEPTS OF FUEL CELLS   A fuel cell powerplant, figure 21-14, is a one-step electrochemical system in which electricity is generated directly from hydrocarbon fuels such as natural gas, gasoline, oil, or kerosene. The three basic parts of a fuel cell powerplant are:

  1. a fuel processor or reformer
  2. the fuel cell power section
  3. an inverter or power conditioner

(R21-6)   What is the basic difference between a fuel cell powerplant and a conventional powerplant?

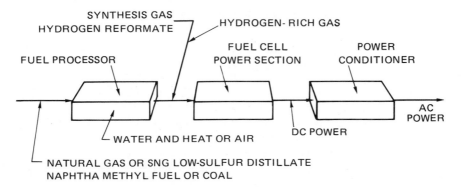

Fig. 21-14 Fuel cell powerplant

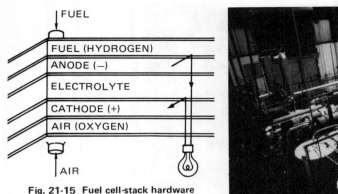

Fig. 21-15 Fuel cell-stack hardware

Fig. 21-16 Overall view of 1-MW fuel cell pilot plant

A chemical reaction in the *fuel processor*, or *reformer*, converts the hydrocarbon fuel in the presence of steam and a catalyst to hydrogen gas and carbon dioxide. The hydrogen formed in this part of the process serves as the fuel for the second basic element, the fuel cell itself.

The *fuel cell* works by means of electrochemical principles. The fuel cell-stack hardware, figure 21-15, consists of a fuel electrode (anode), an air electrode (cathode), and an ion conductor (electrolyte). The hydrogen fuel and oxygen from air are combined in the fuel cell to produce water and electricity. The process is continuous as long as fuel and air are supplied.

A single fuel cell generates approximately 1 V dc. Each square foot of electrode cross-sectional area can be expected to generate 100 W to 200 W of electricity. Fuel cells may be connected in series to obtain voltages of 100 V to 1 000 V, with power values ranging from kilowatts to megawatts.

The third basic element of the fuel cell powerplant is the *power conditioner*, or *inverter*. It changes the dc output of the fuel cell to ac at the frequency and voltage level required by the application.

PRACTICAL APPLICATIONS

A 1-MW fuel cell pilot plant is shown in figure 21-16. Hydrogen is obtained from naphtha in the fuel processor (in the rear of the plant at the right). The six cans in the foreground contain the fuel stacks which combine the hydrogen with air to produce water and a dc output. The reactor thermal control system is located at the left of the plant in the rear. This system provides air to the fuel cells and also acts as a heat exchanger.

A sketch of a 4.8-MW fuel cell demonstration plant planned by Consolidated Edison  of New York City is shown in figure 21-17.

A 26-MW fuel cell powerplant in the design stage is shown in figure 21-18.

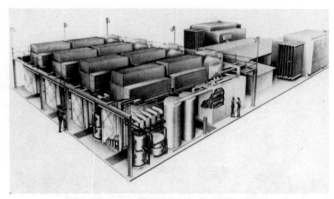

Fig. 21-17 4.8-MW fuel cell demonstration plant

Fig. 21-18 Proposed 26-MW fuel cell powerplant

RESEARCH   Progress is being made toward the objective of having molten carbonate fuel cell generators in commercial use by the mid 1980s. Such generators will be able to operate using a wide range of fuels, including coal. An increased electrical efficiency of approximately 50% will be provided by these fuel cells. Compare this value with earlier designs of systems which had an efficiency of only 38%. Carbonate fuel cell generators do not require precious metal catalysts. In addition, the fuel processing for this type of system is not as complex as the processing required for earlier fuel cells. Figure 21-19 shows an artist's rendering of a molten carbonate system using coal as the primary fuel.

BATTERIES   A cell is the basic unit of a battery. A battery consists of either a single or a combination of cells. A battery is an electrochemical system that converts chemical energy into electrical energy. A *primary battery* is a unit in which the chemical reactions cannot be reversed. When the chemical reactions are reversible, the device is called a *secondary battery*, a *rechargeable battery*, or a *storage battery*. Battery performance is defined by the maximum voltage and current it can supply, and by its ampere-hour rating. Some typical batteries are shown in figure 21-20.

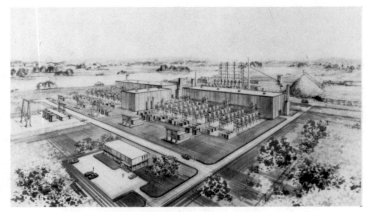

Fig. 21-19 Molten carbonate system for fuel cells

(A) Heavy duty lantern battery

(B) Heavy-duty industrial dry cell

(C) Transistor battery

(D) Camera battery

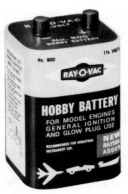

(E) Hobby battery

Fig. 21-20 Typical batteries

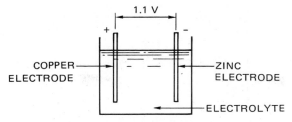

Fig. 21-21 Basic voltage cell

(R21-7) What is the difference between a primary battery and a secondary battery?

(R21-8) Is the automobile battery a primary or a secondary battery?

BASIC VOLTAGE CELL A basic voltage cell is shown in figure 21-21. The cell consists of one copper electrode and one zinc electrode placed in a dilute solution of sulfuric acid (the electrolyte). When the electrodes are placed in the electrolyte:

1. the surface of the zinc electrode dissolves in the sulfuric acid.
2. the zinc atoms react chemically with the acid to form the compound zinc sulphate.
3. the zinc electrode becomes negatively charged as it accumulates excess electrons.
4. hydrogen ions are released as zinc sulphate is formed. These ions travel to the copper electrode to acquire electrons. The copper electrode becomes positively charged.
5. the chemical action continues until the zinc electrode becomes so negative that it repels the negative sulphate molecules. At the same time, the positive charge on the copper electrode repels the positive hydrogen ions. The potential difference at the cell terminals reaches a value of 1.1 V.

External Circuit Connection. When an external circuit is connected to the terminals of the cell, the electrons can flow from the negative zinc electrode to the positive copper electrode. The zinc electrode loses negative charges as the copper electrode gains negative charges. That is, the copper electrode becomes less positive. The chemical action in the cell resumes as the electrolyte provides electrons to supply the current drawn from the cell.

The hydrogen released in the reaction forms bubbles which cling to the copper electrode. These bubbles reduce the ability of the cell to supply current because they reduce the active surface area of the copper electrode. This effect is called *polarization.* The effect of polarization can be overcome by the use of materials called *depolarizers.*

(R21-9) What is the direction of current (conventional) in the external circuit?

(R21-10) What happens to the zinc electrode as current is drawn from the cell?

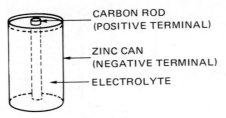

CARBON ROD
(POSITIVE TERMINAL)

ZINC CAN
(NEGATIVE TERMINAL)

ELECTROLYTE

Fig. 21-22 Basic zinc-carbon dry cell

DRY CELLS   The zinc-carbon cell, figure 21-22, is a primary cell. This type of cell is used in flashlights and other portable devices. The components of a zinc-carbon cell are the negative electrode (the zinc can), the positive electrode (the carbon rod), and the electrolyte (a moist paste). If the electrolyte dries out, the chemical reactions stop and the cell does not operate.

Chemical Reactions. The electrolyte paste is sal ammoniac (ammonium chloride) mixed with manganese peroxide and powdered graphite. The manganese peroxide acts as a depolarizer by combining with the oxygen at the positive electrode to yield manganese oxide and water. This reaction eliminates the polarizing effect of the hydrogen. At the same time, the electrolyte is kept moist.

Electrons accumulate on the zinc electrode which becomes negative. As zinc atoms combine with the electrolyte, electrons are removed from the carbon electrode, causing it to become positive. A potential difference of 1.5 V develops across the terminals of this cell.

LEAD-ACID   Lead plates immersed in an electrolyte of dilute sulfuric acid form the cells of a
BATTERY   lead-acid battery. The output of each cell is 2 V. Therefore, a 12-V automobile battery consists of six cells connected in series.

The electrode plates are made of lead-antimony alloy. The plates are provided with recesses which are filled with lead oxide. Each electrode consists of several plates connected in parallel and separated by porous rubber.

When the lead-acid cell is charged, the lead oxide on the positive plates changes to lead peroxide. On the negative plates, the lead oxide becomes spongy, or porous, lead.

When the battery is supplying a load, it is discharging. Atoms from the spongy lead on the negative plates combine with the sulphate molecules to form lead sulphate and hydrogen. Electrons remaining on the negative plates give the plates a negative potential. The hydrogen released in the reaction combines with the lead peroxide on the positive plate. As a result, electrons are removed from the plate and it has a positive potential.

When the hydrogen and lead peroxide combine at the positive plate, water and lead sulphate are produced. The water dilutes the electrolyte and the lead sulphate fills the pores of both plates. Thus, each cell becomes less efficient, and the battery output voltage decreases.

The battery can be recharged. This means that the cells are formed again and the electrolyte regains its strength. A 12-V automobile battery supplies a current of 250 A to the starter.

ELECTRICAL CHARACTER-ISTICS

The electrochemical reactions in the battery produce a potential difference across the terminals of a cell. The value of this potential difference depends upon the material of the electrodes only. It does *not* depend upon either the physical size of the electrodes, or the quantity of the electrolyte.

The amount of energy supplied by any cell, or battery, is given in terms of its rating in ampere-hours (Ah):

$$Ah = I \times t \qquad \qquad \text{Eq. 21.1}$$

where

Ah = rating of battery, in ampere-hours
I = current, in amperes
t = time, in hours.

Problem 2

How long can a 10-Ah battery supply a current of 1 ampere?

Solution

$Ah = I \times t$

$t = \dfrac{Ah}{I}$

$= \dfrac{10 \ Ah}{1 \ A}$

$= 10 \ h$

The potential difference across the cell terminals is called the *no-load output voltage*, or the *open-circuit output voltage*. The internal resistance of the cell causes the terminal voltage to drop when the cell supplies a load. The equivalent circuit of a cell supplying a load is given in figure 21-23. In this diagram, E_{oc} is the open-circuit, or no-load output voltage of the cell, and V_L is the load voltage.

$$V_L = E_{oc} - I_L \ R_{int} \qquad \qquad \text{Eq. 21.2}$$

$$I_L = \frac{E_{oc}}{R_{int} + R_L} \qquad \qquad \text{Eq. 21.3}$$

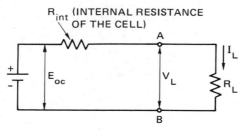

Fig. 21-23 Equivalent circuit of a cell supplying a load

Problem 3   A battery has an open-circuit output voltage of 6 V. The internal resistance of the battery is 0.3 Ω. Calculate the load voltage when the load current is 2 A.

Solution
$$\begin{aligned}
V_L &= E_{oc} - I_L R_{int} \\
&= 6\ V - (2\ A)(0.3\ \Omega) \\
&= 6\ V - 0.6\ V \\
&= 5.4\ V
\end{aligned}$$

(R21-11)   Calculate the value of R_L in Problem 3.

(R21-12)   Determine the value of the load current for Problem 3, if terminals A and B, figure 21-23, are short circuited.

When a number of cells are connected in series, or in parallel, the group is called a *battery*. The following rules apply:

1. Series-connected cells produce an output voltage equal to the sum of the individual voltages.

$$E_T = E_1 + E_2 + E_3 + \ldots \qquad \text{Eq. 21.4}$$

In this case, the cells may be different.

2. Series-connected cells can supply a maximum current equal to the maximum that can be taken from any one cell.

$$I_T = I_1 = I_2 = I_3 = \ldots \qquad \text{Eq. 21.5}$$

3. Parallel-connected like cells produce an output voltage equal to the terminal voltage of one cell.

$$E_T = E_1 = E_2 = E_3 = \ldots \qquad \text{Eq. 21.6}$$

4. Parallel-connected like cells can supply a maximum output current equal to the sum of the maximum currents from each cell.

$$I_T = I_1 + I_2 + I_3 + \ldots \qquad \text{Eq. 21.7}$$

MISCELLANEOUS BATTERIES   Figure 21-24 shows an alkaline manganese battery. This battery is a rechargeable, or secondary, battery. When compared to the carbon-zinc battery, it can sustain a constant terminal voltage of 1.5 V to supply heavy loads for a much longer time.

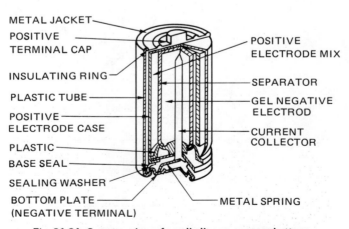

METAL JACKET

POSITIVE TERMINAL CAP

INSULATING RING

PLASTIC TUBE

POSITIVE ELECTRODE CASE

PLASTIC

BASE SEAL

SEALING WASHER

BOTTOM PLATE (NEGATIVE TERMINAL)

POSITIVE ELECTRODE MIX

SEPARATOR

GEL NEGATIVE ELECTROD

CURRENT COLLECTOR

METAL SPRING

Fig. 21-24 Cutaway view of an alkaline manganese battery

(A) Nickel-cadmium batteries

(B) DataSentry® nickel-cadmium batteries

(C) Rechargeable nickel-cadmium cells   (D) Nickel-cadmium storage battery

Fig. 21-25

Nickel-cadmium batteries are shown in figure 21-25. The output voltage of a nickel-cadmium cell is 1.3 V on open circuit and drops to 1.25 V under load. These batteries are used in electronic calculators, communications equipment, hearing aids, electric shavers, and movie cameras. For many applications, the storage battery in figure 21-25D is substituted for the lead-acid battery because of its many superior features. A battery charger is shown in figure 21-26.

The construction of a lithium battery is shown in figure 21-27. Coin size lithium batteries are shown in figure 21-28. These batteries are used in digital watches and calculators. The nominal voltage of a lithium battery is 3 V, and the ampere-hour rating is 90 mAh. The cells are hermetically sealed. They have a shelf life of five years or more.

A 9-V alkaline battery is shown in figure 21-29. This type of battery is designed for use in smoke detectors and other applications which require high reliability over long periods of time.

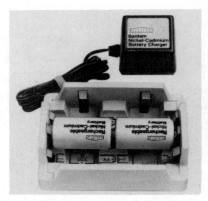

Fig. 21-26 Nickel-cadmium battery charger

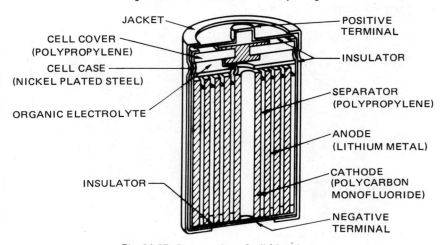

JACKET

POSITIVE
TERMINAL

CELL COVER
(POLYPROPYLENE)

INSULATOR

CELL CASE
(NICKEL PLATED STEEL)

SEPARATOR
(POLYPROPYLENE)

ORGANIC ELECTROLYTE

ANODE
(LITHIUM METAL)

CATHODE
(POLYCARBON
MONOFLUORIDE)

INSULATOR

NEGATIVE
TERMINAL

Fig. 21-27 Construction of a lithium battery

Fig. 21-28 Coin size long-life lithium battery

Fig. 21-29 9-V Alkaline battery

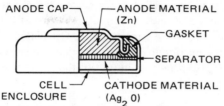

Fig. 21-30 Cutaway view of a silver oxide battery

A. Ultra-thin oxide watch cell

B. Silver oxide battery for a digital watch

Fig. 21-31

A cutaway view of a silver oxide battery is shown in figure 21-30. Silver oxide batteries, figure 21-31 are used in watches, such as low-drain LCD (liquid crystal display) watches. The cells have an output voltage of 1.5 V and an ampere-hour rating of 40 mAh. Advantages of the cells include a long storage life and the ability to operate over a wide temperature range.

EXTENDED STUDY TOPICS

1. A solar array with an area of 100 m^2 is covered with solar cells. Each cell is 2 cm $\times$ 2 cm. The average daytime level of illumination is 100 mW/cm^2. At this level, the solar cells have a maximum power output when the cell current equals 50 mA and the cell voltage equals 500 mV.
   a. Determine the number and arrangement of the solar cells to provide an output voltage of 120 V.
   b. If the sun shines for an average of twelve hours each day, calculate the value of kilowatt-hours generated by the solar array each day.
2. What are the basic components of a fuel cell powerplant?
3. List the three components of a fuel cell.
4. A battery consists of six cells connected in series. Each cell has an open-circuit terminal voltage of 1.5 V and an internal resistance of 0.1 Ω. Calculate: a. the open-circuit voltage of the battery, and b. the load voltage when $R_L = 5\ \Omega$.
5. A battery consisting of 1.5-V cells connected in parallel supplies current to a 0.3-Ω load resistor. If each cell can supply a maximum current of 1 A, determine the number of cells required. Neglect the internal resistance of each cell.

6. Eight voltage cells connected in parallel have an open-circuit terminal voltage if 1.5 V and an internal resistance of 0.1 Ω. Each cell is rated to supply a current of 1 A for ten hours. Calculate:
   a. the maximum load current that can be supplied.
   b. the ampere-hour output of the battery, and
   c. the load voltage.

Section 3
Electronics Fundamentals

| Chapter 22 | Semiconductor Theory |

OBJECTIVES After studying this chapter, the student will be able to

- discuss the concept of covalent bonding.

- explain the difference between intrinsic and extrinsic conduction in a semi-conductor.

- explain why and how a semiconductor is doped to obtain an *n*-type or a *p*-type crystal.

- analyze the action of charges at a *p-n* junction at room temperature.

ATOMIC PHYSICS The atom is the basic building block of all matter. (The student should review Chapter 2 at this time.) Figure 22-1 shows a nucleus and its orbiting electrons. The rotation of the electrons about the nucleus creates a centrifugal force which acts to move the electrons away from the nucleus. However, this centrifugal force is counteracted by the charge attraction between the orbiting electrons (negative charge) and the protons (positive charge) in the nucleus, figure 22-2. As a result of the action of two forces, the orbiting electrons keep their positions in the individual rings, or shells, of the atom.

(R22-1) Name the three basic particles of the atom.

(R22-2) Is a normal atom electrically neutral, positively charged, or negatively charged?

(R22-3) Define: a. positive ion, and b. negative ion.

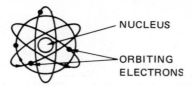

Fig. 22-1 Nucleus with its orbiting electrons

For each element, the electrons of all of the atoms are arranged in shells, or energy levels, in a definite pattern. Each shell has a limit as to the number of electrons it can hold. Table 22-1 lists the number of electrons needed to complete each shell. The table shows that shells M, N, O, and P are complete when the indicated number of electrons fill the shell. Except for the very simplest atoms, the highest or outermost level is never completely filled before electrons start appearing in the next higher level. The outer shell of an atom contains a maximum of eight electrons except for the atoms of those elements having electrons only in the K shell. Shells L, M, N, O, and P contain several rings or subshells.

Problem 1   A germanium atom contains 32 protons. How are the electrons distributed in the various shells?

Solution   From Table 22-1:
  K shell: 2 electrons
  L shell: 8 electrons
  M shell: 18 electrons
  N shell: $32 - (2 + 8 + 18) = 4$ electrons

(R22-4)   A silicon atom contains 14 protons. How are the electrons distributed in the various shells?

(R22-5)   What do germanium and silicon atoms have in common?

The electrical characteristics of an atom are determined by the electrons in the outer shell, or the valence electrons. This outer, or valence, shell may be partially or completely filled. The *valence* of an atom represents either the number of electrons in the outer ring, or the number of electrons required to complete the ring. Valence is an indication of how readily an atom will share elec-

NUCLEUS OF ATOM
ELECTRON IN ORBIT
CENTRIFUGAL FORCE DUE TO
ROTATION OF ELECTRON.
ELECTROSTATIC FORCE DUE TO THE
ATTRACTION BETWEEN THE NEGATIVE
ELECTRONS AND THE POSITIVE PROTONS
IN THE NUCLEUS.
THE ELECTROSTATIC FORCE AND
THE CENTRIFUGAL FORCE ARE EQUAL
AND OPPOSITE.
Fig. 22-2 Forces acting on orbiting electrons

TABLE 22-1 Number of Electrons Required
to Complete Shell

Shell Level	Electrons Required to Complete Shell
K	2
L	8
M	8 or 18
N	8, 18, or 32
O	8, 18, or 32
P	8 or 18
Q	8

NOTE: THE K AND L SHELLS ARE
NOT SHOWN FOR SILICON. THE
K, L, AND M SHELLS ARE NOT
SHOWN FOR GERMANIUM.

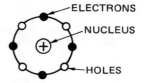

Fig. 22-3 Valence shell of silicon and germanium

trons with other atoms. The atoms of both germanium (Ge) and silicon (Si) have four electrons in a valence shell that can hold a maximum of eight electrons. Figure 22-3 shows that the valence shells of germanium and silicon have four electrons and four holes. A *hole* is defined as the lack of an electron in a shell where an electron could exist.

(R22-6) Are silicon and germanium atoms electrically neutral, positively charged, or negatively charged? Explain the reason for the answer.

(R22-7) What is the electrical charge of the holes shown in figure 22-3?

Each shell has an energy level. That is, a certain amount of energy must be applied to remove an electron from that shell. As the distance decreases between an electron in a shell and the nucleus of the atom, there is an increase in the forces that bind the electron to the shell. Since the electrons in the valence shell are the greatest distance from the nucleus, they require the least amount of energy to be removed from the atom. Energy levels are measured in electron volts (eV). An *electron volt* is the amount of energy required to move one electron through a potential difference of one volt.

CONDUCTORS, INSULATORS, AND SEMI- CONDUCTORS

Electrons are affected by forces due to other atoms when the atoms are close together, as in a solid. The energy levels that are occupied by the electrons then merge into bands of energy levels. As shown in the energy band diagram in figure 22-4, the electrons may exist in one of two distinct energy levels:
1. The valence band.
2. The conduction band.
An energy gap separates the two bands. No electrons can exist in this forbidden band.

Conduction Band. Electrons orbits in the conduction band are very large. Nuclear attraction has little effect on these electrons. As a result, only a small amount of energy is required to cause conduction band electrons to move within the material as free electrons.

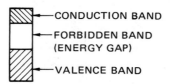

Fig. 22-4 Energy band diagram

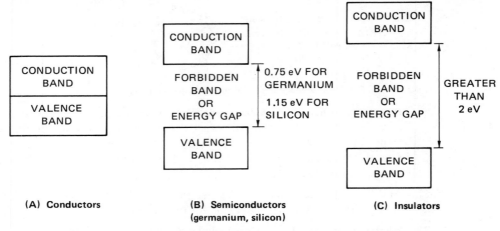

(A) Conductors

(B) Semiconductors
(germanium, silicon)

(C) Insulators

Fig. 22-5 Energy band diagrams for conductors, semiconductors, and insulators

Valence Band. Electrons in the valence band are usually orbiting their nucleus and are affected by nuclear attraction. This means that much larger amounts of energy must be applied before these electrons can be removed from the valence band.

Forbidden Band. For a particular element, the forbidden band may be nonexistent, small, or large. Figure 22-5 shows the energy band diagrams for conductors, semiconductors, and insulators.

There is no forbidden band in a conductor. In other words, the valence band energy level of a conductor is next to, or may overlap, the conduction band. Very little energy is required to move an electron from its atom into the conduction state. Even at room temperature, there may be enough energy to move the electrons. As a result, large numbers of electrons are made available for conduction.

At absolute zero ($-273°C$), the valence band of a semiconductor material is filled. There are no electrons in the conduction band. As shown in figure 22-5, semiconductors have a forbidden band. An electron may be in the valence band or in the conduction band, but it is never found in the band between these two levels. At room temperature, the available thermal energy causes some electrons to move from the valence band to the conduction band. When a potential is applied to the semiconductor, conduction occurs by two methods:

1. electron movement in the conduction band.
2. hole transfer in the valence band. (This concept is explained later in this chapter.)

In an insulator, there are no electrons in the conduction band. All of the electrons are located in the valence band. The forbidden band of an insulator is much larger than that of semiconductors. Conduction does not occur in an insulator when the applied voltage is less than the breakdown voltage of the insulator.

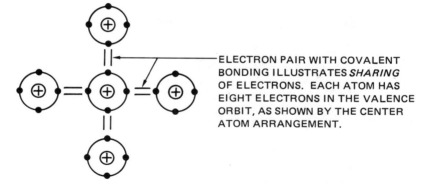

ELECTRON PAIR WITH COVALENT BONDING ILLUSTRATES *SHARING* OF ELECTRONS. EACH ATOM HAS EIGHT ELECTRONS IN THE VALENCE ORBIT, AS SHOWN BY THE CENTER ATOM ARRANGEMENT.

Fig. 22-6 Schematic diagram of covalent bonding of silicon and germanium

Both silicon and germanium are semiconductors. At room temperature, a silicon crystal has fewer free electrons than a germanium crystal. Because of this characteristic, silicon is the most commonly used semiconductor material.

(R22-8) Name the two possible energy bands in which electrons may exist.

(R22-9) Define the forbidden band.

COVALENT BONDING   Silicon and germanium atoms are able to share electrons with other atoms in an arrangement called *covalent bonding*. Each atom of these semiconductors has four electrons in its valence orbit. These electrons combine with the electrons of other atoms so that there are eight electrons in the valence orbit, figure 22-6. The electrons no longer belong to a single atom but are shared by adjacent atoms. Figure 22-7 is a detailed illustration of the sharing of electrons in covalent bonding.

(R22-10) Why does the core of the semiconductor crystal shown in figure 22-7 have a charge of +4?

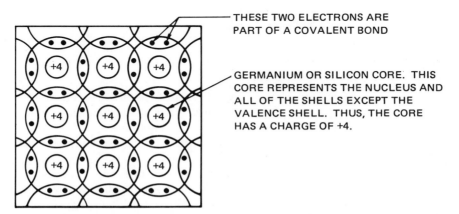

THESE TWO ELECTRONS ARE PART OF A COVALENT BOND

GERMANIUM OR SILICON CORE. THIS CORE REPRESENTS THE NUCLEUS AND ALL OF THE SHELLS EXCEPT THE VALENCE SHELL. THUS, THE CORE HAS A CHARGE OF +4.

Fig. 22-7 Schematic diagram of a semiconductor crystal showing covalent bonding

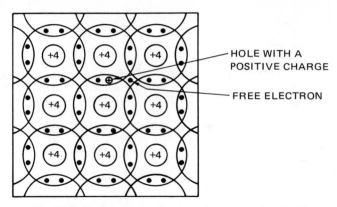

Fig. 22-8 Semiconductor showing a broken covalent bond

(R22-11)   Is the crystal of figure 22-7 neutral, positively charged, or negatively charged?

MOVEMENT OF CHARGES   An electron may gain enough energy at room temperature to leave the valence band and enter the conduction band where it is considered to be a free electron. When a covalent bond is broken, each electron released from the bond has gained energy equal to the value of the energy gap of that semiconductor. As shown in figure 22-5B, silicon has an energy gap of 1.1 eV and germanium has an energy gap of 0.72 eV. Silicon has a larger energy gap than germanium and therefore has fewer broken bonds. At room temperature, silicon has 1.5×10^{10} broken bonds per cubic centimeter and germanium has 2.5×10^{13} broken bonds per cubic centimeter.

Whenever a covalent bond is broken, an electron-hole pair is formed. The electron, or negative charge carrier, moves in a random fashion. At the broken bond, the absence of an electron leaves a hole with a positive charge, figure 22-8. In a pure semiconductor, the number of free electrons equals the number of holes. This type of semiconductor is known as an *intrinsic semiconductor.*

(R22-12)   Define an intrinsic semiconductor.

(R22-13)   How many holes per cubic centimeter are created in a germanium crystal at room temperature?

(R22-14)   How many electron-hole pairs per cubic centimeter are created in a silicon crystal at room temperature?

Figure 22-9 is the energy diagram of the hole current in a semiconductor. When the semiconductor material is subjected to thermal energy, such as room temperature heat energy, the following actions occur:

1. Electron A receives enough energy to leave the valence band and enter the conduction band.
2. A hole exists in the valence band at the former location of electron A.
3. The valence electron at B moves into the hole as a result of the covalent bonding force acting on it. The original hole disappears and a new hole appears at B.

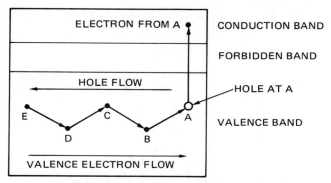

Fig. 22-9 Energy diagram of hole current in a semiconductor

4. The valence electron at C moves into the new hole at B. This hole disappears and a new hole appears at C.
5. The valence electron at D moves into the hole. The hole at C disappears and a new hole appears at D.
6. The valence electron at E moves into the hole. The hole at D disappears and a new hole appears at E.
7. In summary, the valence electrons move along the path EDCBA. The holes move through the valence band along the path ABCDE.

(R22-15)   Referring to the preceding analysis of hole current, what conclusion can be reached about the direction of valence electron flow (negative charges) and the direction of hole flow (positive charges)?

INTRINSIC CONDUCTION

Intrinsic conduction, figure 22-10, occurs in the following manner:
1. Electron A breaks away from atom 3 and becomes a free electron at room temperature.
2. A vacant position, or hole, is formed in the covalent structure of atom 3 because of the removal of the electron.
3. Atom 3 is a positive ion. It is assumed that the charge of the atom is concentrated at the hole. Thus, the hole represents a positive charge.
4. The entire crystal is still neutral because the positive charge of the hole is balanced by the negative charge of the free electron.
5. Free electron A is attracted to the positive terminal of the battery and is removed from the crystal. At the same time, however, the battery releases a free electron into the crystal at its negative terminal. The crystal remains neutral.
6. The removal of electron A from the covalent bond leaves a hole at B which attracts adjacent valence electrons in an effort to complete its valence ring.
7. This attraction is strong enough to break the covalent bond at atom 4. Electron C leaves atom 4 and moves to the hole in atom 3. The hole is now located in atom 4.

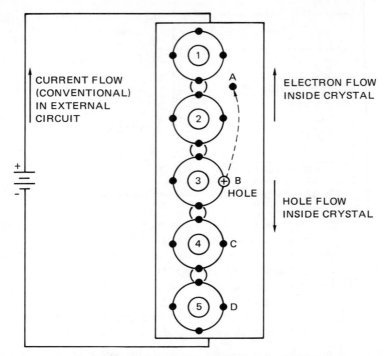

CURRENT FLOW
(CONVENTIONAL)
IN EXTERNAL
CIRCUIT

ELECTRON FLOW
INSIDE CRYSTAL

HOLE FLOW
INSIDE CRYSTAL

Fig. 22-10 Intrinsic conduction at room temperature

8. Electron D breaks away from its bond at atom 5 and moves to atom 4. The hole has now moved to atom 5.
9. An electron from the negative terminal of the battery enters the crystal and fills the hole at D in atom 5. The hole no longer exists.
10. The crystal has a negative charge due to the free electron in the semiconductor (see Step 5). The electron is drawn to the positive terminal of the battery and leaves the crystal.
11. The current in the external circuit is the *sum* of the hole current and the electron current inside the crystal.
12. Intrinsic conduction continues as more covalent bonds are broken at room temperature and more electron-hole pairs are formed.

(R22-16)   What charges flow in the external circuit of figure 22-10?

The drift velocity of holes in intrinsic conduction is nearly one-half the velocity of the electrons. Since the electrons are detached from their atoms, they have greater mobility than holes. Holes move more slowly as they must move in the valence band from atom to atom.

From time to time, a free electron may fall into a hole, causing the hole to disappear. This phenomenon is called *recombination.* It occurs continuously in semiconductors. However, new electron-hole pairs are produced by the action of heat energy. As a result, current continues to flow in the external circuit.

Lifetime is defined as the average time between the creation and the disappearance of an electron-hole pair. Lifetime varies from a few nanoseconds to several microseconds.

(R22-17)   Define: a. mobility; b. recombination; and c. lifetime.

SEMICONDUCTOR DOPING   *Doping* is a process in which impurity atoms are added to a pure semiconductor crystal to increase the number of free electrons, or the number of holes. Doping is required because an intrinsic semiconductor does not form enough electron-hole pairs to produce a usable current. Each impurity atom must have more than four electrons or less than four electrons in its valence orbit. In other words, the impurity atom must be nontetravalent. A doped cyrstal is called an *extrinsic semiconductor.*

The resistance of a doped (extrinsic) semiconductor is called its *bulk resistance.* A lightly doped semiconductor has a high bulk resistance. The bulk resistance of a semiconductor decreases as the doping increases because of the increased number of charge carriers.

(R22-18)   Define: a. doping; b. tetravalent atom; c. extrinsic semiconductor; and d. bulk resistance.

N-Type Semiconductors. An extrinsic semiconductor is created when controlled amounts of a pentavalent element are added to an intrinsic semiconductor in the molten state. (The atoms of a pentavalent element have five valence electrons.) When the doped semiconductor is cooled, the structure of the crystal lattice is reformed, causing a change in the characteristics of the semiconductor. Figure 22-11 shows how a pentavalent atom replaces a germanium or silicon atom in the lattice structure. Elements with pentavalent atoms include antimony, arsenic, and phosphorus.

Doping a semiconductor means the addition of one impurity atom to 10^6 germanium or silicon atoms. As shown in figure 22-11, four of the valence electrons of the pentavalent atom form covalent bonds with the neighboring valence electrons. The fifth valence electron of the pentavalent atom is a free electron. It does not form a covalent bond with any other electron.

This type of doped crystal is called an *n-type semiconductor.* Most of the conduction in this material takes place through *negative* charge carriers. The pentavalent impurity is called a *donor* because it donates an electron to the crystal.

(R22-19)   Is an *n*-type semiconductor negatively charged?

Figure 22-11 indicates that charge flow in the *n*-type crystal consists of the following:

1. electron flow due to the breaking up of covalent bonds by thermal agitation.
2. hole flow due to the breaking up of covalent bonds by thermal agitation.
3. electron flow due to donor atoms.

Thus, in an *n*-type semiconductor, the *negative charge carriers*, or electrons, are the *majority* carriers. The *positive charge carriers*, or holes, are the *minority*

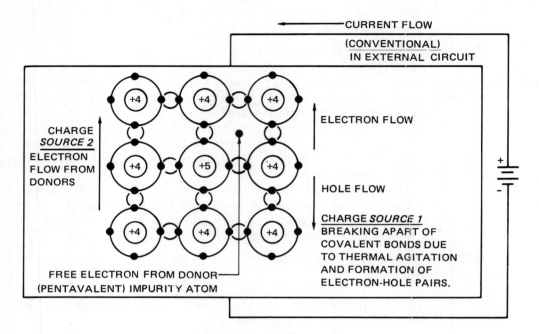

Fig. 22-11 *n*-type semiconductor

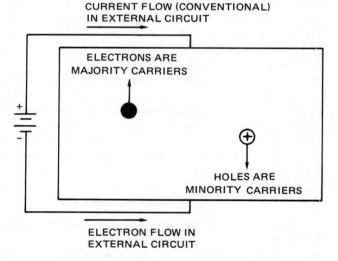

Fig. 22-12 Charge flow in an *n*-type semiconductor

carriers. Figure 22-12 shows both the flow of charges inside the crystal and the flow of electrons and conventional current in the external circuit.

(R22-20) What charges flow in the external circuit shown in figure 22-12?

(R22-21) How is the number of minority carriers affected by an increase in the temperature?

P-type Semiconductors. An extrinsic semiconductor is created when controlled amounts of a trivalent element are added to an intrinsic semiconductor in the molten state. The atoms of a trivalent element have three valence electrons. The characteristics of the semiconductor are altered when it is cooled and the structure of the crystal lattice is reformed. Figure 22-13 shows how a trivalent atom replaces a germanium or silicon atom in the lattice structure. Elements with trivalent atoms include aluminum, boron, gallium, and indium.

As shown in figure 22-13, the three valence electrons of the trivalent atom form covalent bonds with neighboring valence electrons. However, the impurity atom cannot fill all of the covalent bond requirements of adjacent atoms. As a result, a vacancy, or a hole, exists at one point in the structure of the crystal. For every impurity atom added to the intrinsic semiconductor material, there is one hole in the lattice structure. The holes are available for conduction in the same manner as the extra electrons in a pentavalent-doped semiconductor material.

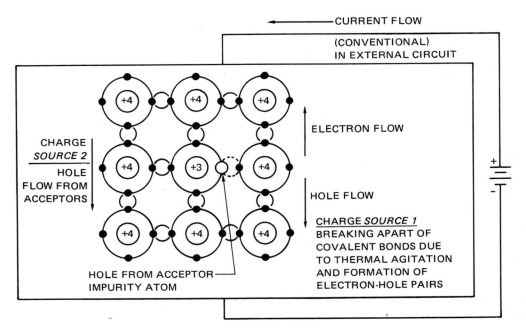

Fig. 22-13 *p*-type semiconductor

The crystal is called a *p*-type semiconductor because most of the conduction takes place by means of positive charge carriers, or holes. Since the trivalent impurities accept an additional electron, they are called *acceptor* materials.

(R22-22)   Is the *p*-type semiconductor positively charged?

The following types of charge flow inside a *p*-type crystal are shown in figure 22-13:

1. electron flow due to the breaking apart of covalent bonds by thermal agitation.
2. hole flow due to the breaking apart of covalent bonds by thermal agitation.
3. hole flow from acceptor atoms.

In a *p*-type semiconductor, the positive charge carriers, or holes, are the majority carriers. The minority carriers are the negative charge carriers, or the electrons. Figure 22-14 shows both the flow of charges in the crystal and the flow of electrons and conventional current in the external circuit.

(R22-23)   What charges flow in the external circuit of the *p*-type crystal shown in figure 22-14?

By comparing figures 22-12 and 22-14, it can be seen that the majority charge carriers in the *n*-type semiconductor and the majority charge carriers in the *p*-type semiconductor move in opposite directions. However, the direction of electron flow in the external circuit is the same in both cases.

The *p-n* Junction. A *p*-type semiconductor at room temperature is shown in figure 22-15. The doped holes of the impurity atoms are filled with electrons. Thus, holes with positive charges are created in the valence band of the intrinsic material. Each impurity atom has four electrons and three protons, resulting in a

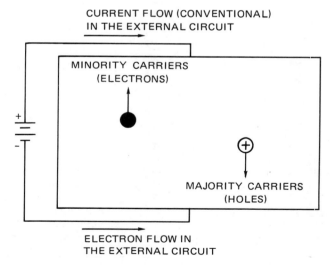

Fig. 22-14 Charge flow in a *p*-type semiconductor

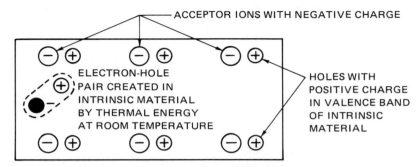

Fig. 22-15 *p*-type semiconductor at room temperature showing holes as the majority carriers

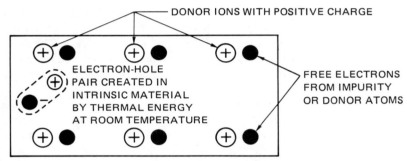

Fig. 22-16 *n*-type semiconductor at room temperature showing electrons as the majority carriers

negative acceptor ion. These negative ions are fixed in the crystal structure and cannot move in response to an external voltage source.

(R22-24) List the charges and carriers of figure 22-15.

An *n*-type semiconductor at room temperature is shown in figure 22-16. The doped electrons of the impurity atoms exist as free electrons in the conduction band. Each impurity atom has five protons and four electrons, resulting in a positive ion. The number of positive donor ions equals the number of donor electrons.

(R22-25) List the charges and carriers of figure 22-16.

p-type and *n*-type semiconductors alone are not very useful. These two materials must be joined together, figure 22-17, to form a junction to obtain several useful properties.

The *p-n* junction is not formed simply by placing the *p*-type and *n*-type materials in physical contact with each other. The intrinsic semiconductor material is doped while the crystal is being formed using one of the following processes:

    1. alloy junction
    2. diffusion process
    3. electrochemical etching and plating

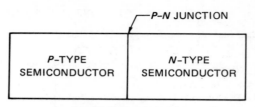

Fig. 22-17 *p-n* junction

4. epitaxial layer
5. grown junction

At the instant a *p-n* junction is formed, figure 22-18, the free electrons from the *n*-side cross the junction to fill some of the holes in the lattice structure on the *p*-side. As a result, positive ions are uncovered on the *n*-side and negative ions are uncovered on the *p*-side. The negative ions on the *p*-side form an area of negative charge. On the *n*-side, the positive ions form an area of positive charge. The areas where the ions exist are now depleted of majority carriers. The process of combination eliminates the negative electron charge carriers and the positive hole charge carriers. The area on each side of the junction is called a *depletion region*, figure 22-19.

The positive ions on the *n*-side of the junction repel the movement of positive holes from the *p*-side. The negative ions on the *p*-side of the junction repel the movement of negative electrons from the *n*-side. A barrier is established to the flow of majority carriers from either side of the junction to the other side. This opposition to the flow, or *barrier potential*, is represented by the battery in figure 22-20.

The barrier potential is a potential field force offered by the *p-n* junction to the flow of majority carriers. It has been shown that the barrier potential, or potential hill, at room temperature, is 0.7 V for silicon and 0.3 V for germanium.

The creation of a *p-n* junction led to the development of a device known as a *semiconductor diode*. This device is described in detail in Chapter 23. Examples of commercial semiconductor diodes are shown in figure 22-21.

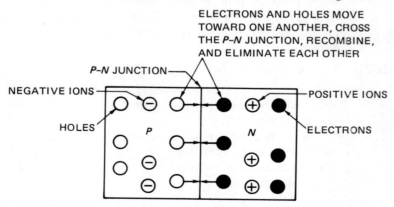

Fig. 22-18 Action at a *p-n* junction at the instant the junction is formed

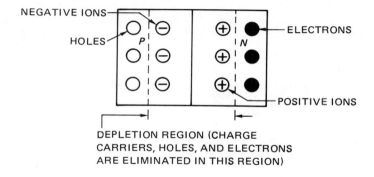

DEPLETION REGION (CHARGE
CARRIERS, HOLES, AND ELECTRONS
ARE ELIMINATED IN THIS REGION)

Fig. 22-19 The depletion region around a *p-n* junction

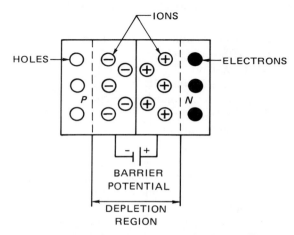

BARRIER
POTENTIAL

DEPLETION
REGION

Fig. 22-20 Formation of a barrier potential at equilibrium

Fig. 22-21A High-current metal oxide silicon diodes

**Fig. 22-21B Diodes with band markings to denote
n-type end of diode**

Fig. 22-21C Diodes with screw stud terminal

Fig. 22-21D Diode with cap end to denote *n*-type end of diode

(R22-26)   Define: a. barrier potential; b. potential hill; and c. depletion region.

EXTENDED STUDY TOPICS

1. How is an electron maintained in orbit?
2. Define the term hole.
3. What is a shell?
4. Define an electron volt.
5. Define the term conduction band.
6. Define valence band.
7. What is an electron-hole pair?
8. Do electrons and holes move at the same speed in a semiconductor? Explain the reasons for the answer.
9. Why is it necessary to dope a semiconductor?
10. What is the ratio of impurity atoms to intrinsic atoms in a doped semiconductor?
11. What type of impurities are used to make an *n*-type semiconductor?
12. What type of impurities are used to make a *p*-type semiconductor?
13. Is the direction of current flow in the external circuit of a *p*-type crystal the same or different from that of an *n*-type crystal?
14. What is the barrier potential of a. silicon, and b. germanium?
15. List the techniques used to form a *p-n* junction diode.

<table>
<tr><td>Chapter
23</td><td>Rectification and Power Supply Filters</td></tr>
</table>

OBJECTIVES   After studying this chapter, the student will be able to

- explain the operation of a diode with forward and reverse biasing.

- identify commercial diodes.

- use diode data sheets to solve problems.

- describe half-wave and full-wave rectifier circuits.

- discuss the operation of a filter using a capacitor input.

FORWARD BIASED DIODE   A diode is said to be *biased* when a dc voltage is applied across the diode. A diode is *forward biased* when the polarity of the dc voltage source is like that shown in figure 23-1.

(R23-1)   What conditions are to be met to forward bias a diode?

As shown in figure 23-1, the applied voltage causes holes on the *p*-side of the junction to be repelled from the positive bias terminal of the battery. These holes are driven toward the *p-n* junction. At the same time, electrons on the *n*-side are repelled from the negative bias terminal of the battery. The electrons also are driven toward the junction. The field due to the battery opposes the field due to the barrier potential. As a result, the width of the depletion region is reduced.

Charge carriers flow across the *p-n* junction when the forward bias voltage is greater than the barrier voltage. Electrons from the *n*-side cross the junction and

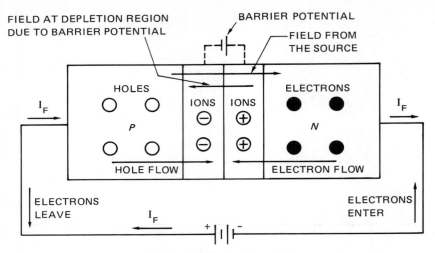

Fig. 23-1 Diode with forward bias

are attracted to the positive terminal of the battery. Holes from the *p*-side also cross the junction and are attracted toward the negative terminal of the battery. The conduction of the majority carriers gives rise to the current flow I_F.

Characteristics. Forward biased *p-n* junction diodes have the characteristics shown in figure 23-2. An analysis of this figure yields the following information about the diode.

1. There is little forward current until the barrier potential is overcome.
2. As the applied voltage nears the barrier potential, conduction band electrons and holes begin to cross the *p-n* junction in large numbers. The

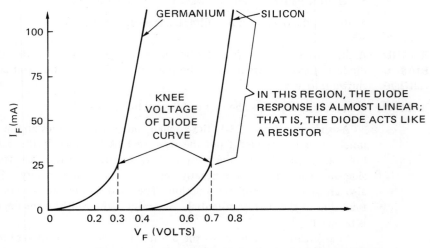

Fig. 23-2 Characteristics of *p-n* junction diodes with forward bias

voltage value at which the current begins to increase rapidly is known as the *knee voltage.*

3. In the region above the knee of the characteristic curve, the diode has an almost linear response and acts like a resistor.

4. The *p*-region and the *n*-region both have some resistance. The sum of these resistances is called the *bulk resistance* of the diode. The abbreviation for bulk resistance is r_B. For a forward biased diode, the bulk resistance is very low. The resistance range generally is from one ohm to 40 ohms. The value of r_B depends upon the doping of the semiconductor material and the size of the *p*- and *n*-regions.

(R23-2) What is the knee voltage of: a. a silicon diode, and b. a germanium diode?

(R23-3) List the conclusions that can be made by analyzing the forward characteristics of a *p-n* junction diode. Discuss: a. conduction, b. I_F, c. the resistance of the diode, and d. the response of the diode above the knee of the characteristic curve.

REVERSE BIASED DIODE

Figure 23-3 shows how the dc voltage source is connected to reverse bias the diode.

(R23-4) What requirements must be met to reverse bias a diode?

When the voltage is applied, electrons from the *n*-side are attracted to the positive bias terminal. At the same time, holes from the *p*-side are attracted to the negative bias terminal. The fields due to the barrier potential and the voltage source are in the same direction. Thus, the depletion region is widened. The barrier potential increases until it is equal to the supply voltage, V_R. At this equilibrium condition, majority carriers cannot flow across the junction. Because the depletion layer has almost no charge carriers, it acts like an insulator or dielectric. The *p*- and *n*-regions act like conductors. This combination has the effect of creating a parallel-plate capacitor.

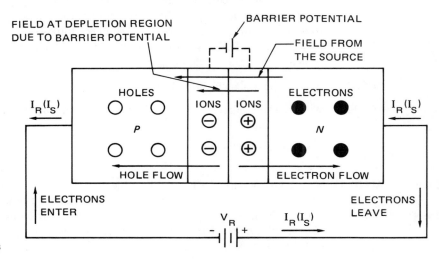

Fig. 23-3
Diode with reverse bias

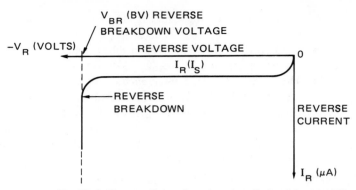

Fig. 23-4 Characteristics of *p-n* junction diode with reverse bias

Characteristics. The polarity of the applied voltage causes minority carriers to move across the junction of the reverse biased diode. Figure 23-4 shows the characteristics of a reverse biased *p-n* junction diode. When a very small reverse voltage (V_R) is applied to the diode, a reverse current (I_R) flows. After the available minority carriers cross the *p-n* junction, an increase in the bias voltage does not result in an increase in the current. This current is called the *reverse saturation current* (I_S). I_S for silicon diodes is less than 1 μA. For germanium diodes, I_S may be greater than 10 μA.

(R23-5) Define reverse saturation current.

(R23-6) Explain the origin of the minority carriers in a reverse biased diode.

As the reverse voltage increases, a point is reached where the minority carriers collide with atoms in the depletion region. The collisions produce enough energy to move electrons from the valence band to the conduction band. This effect is cumulative and causes an avalanche of carriers. As a result, there is a sharp increase in current, figure 23-4. The current surge occurs at the *avalanche breakdown voltage*, V_{BR} or BV. A resistor must be used in series with the diode to limit I_R. This resistor prevents I_R from reaching a value so high that the diode is destroyed. The combined forward and reverse characteristics of a diode are shown in the graph of figure 23-5.

(R23-7) What is the avalanche effect?

(R23-8) What does an analysis of figure 23-5 show about the current flow through a *p-n* junction diode?

DIODE SYMBOLS AND IDENTIFICATION The standard symbol for a diode is the arrowhead and bar shown in figure 23-6. The arrowhead shows the direction of conventional current flow when the diode is forward biased. For forward bias the *p*-side of the diode is the positive terminal, or the *anode*. The *n*-side of the diode is then the negative terminal and is called the *cathode*.

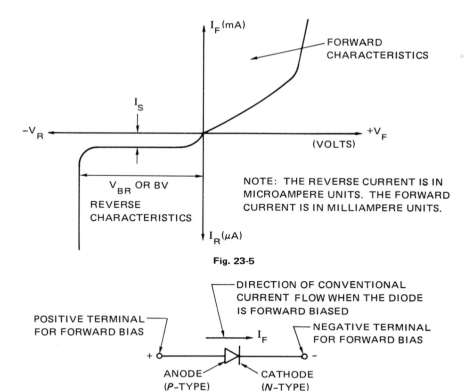

Fig. 23-5

DIRECTION OF CONVENTIONAL
CURRENT FLOW WHEN THE DIODE
IS FORWARD BIASED

POSITIVE TERMINAL
FOR FORWARD BIAS

NEGATIVE TERMINAL
FOR FORWARD BIAS

I_F

+ −

ANODE
(P-TYPE)

CATHODE
(N-TYPE)

Fig. 23-6 Diode symbol

(R23-9) Is the conventional direction of current flow in a forward biased diode in the direction of hole flow or electron flow?

Figure 23-7 shows various commercial diodes. Most diodes have a type number beginning with 1N. The numbers after 1N may be the serial number of the diode, or they may indicate the type number as read from the color bands. The color bands are read from the cathode end. For example, a diode with green, blue, and orange color bands is a type 1N563. The colors follow the first band of the color code for resistors. See Table 4-5 in Chapter 4.

(R23-10) Which of the three color bands of the type 1N563 diode is closest to the cathode end?

UNDERSTAND-
ING DIODE
DATA SHEETS

The data sheets provided by manufacturers for semiconductor devices, such as diodes, must be studied before the proper diode can be selected for an application. Typical data sheets are shown in figure 23-8.

Note at the top of the data sheet of figure 23-8A that a list is given of the type numbers of the fast switching diodes covered by this sheet. The mechanical data includes a diagram of the package shape and the dimensions of the device.

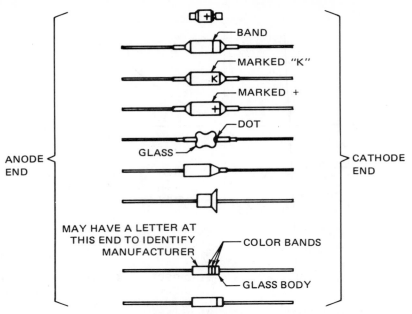

Fig. 23-7 Commercial diodes

The absolute maximum ratings at 25°C are then tabulated. Values greater than these will damage the diode.

Figure 23-8B lists electrical characteristics of the diode at 25°C. The parameters listed on the data sheets are defined as follows:

V_{RM}: *Peak reverse voltage*, or *peak inverse voltage*. This value is the absolute maximum reverse biasing that can be applied across the diode. For the data sheet shown in figure 23-8A, V_{RM} is called the *working peak reverse voltage*.

V_{BR}: *Reverse breakdown voltage*. This value is the minimum reverse voltage at which the device may break down.

I_F: *Steady-state forward current*. This value is the maximum current that may be passed continuously through the diode. Generally, I_F is determined at 25°C and is derated if the diode is to operate at a higher temperature. In the data sheet shown in figure 23-8A, I_F is called the *average rectified forward current*. (Some manufacturers label this current I_O.)

I_{FM} (surge): *Peak surge current*. This value is the amount of current that may be passed through the diode for one second at 25°C. The surge current is much higher than I_F. I_{FM} (surge) flows briefly when the circuit is first switched on.

I_R: *Static reverse current*. This value is the reverse saturation current (I_s) for a specified bias voltage and device temperature, figure 23-8B. I_R is very low because the reverse resistance of a diode ranges from 5 MΩ to 1 000 MΩ.

FAST SWITCHING DIODES

- ● **Rugged Double-Plug Construction**

Electrical Equivalents

1N914 . . . 1N4148 . . . 1N4531
1N914A . . . 1N4446
1N914B . . . 1N4448
1N916 . . . 1N4149
1N916A . . . 1N4447
1N916B . . . 1N4449

mechanical data

Double-plug construction affords integral positive contacts by means of a thermal compression bond. Moisture-free stability is ensured through hermetic sealing. The coefficients of thermal expansion of the glass case and the dumet plugs are closely matched to allow extreme temperature excursions. Hot-solder-dipped leads are standard.

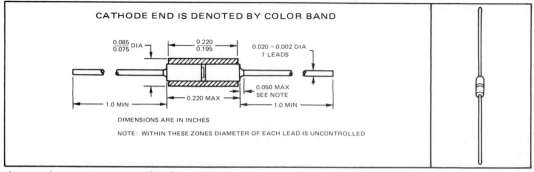

absolute maximum ratings at specified free-air temperature

		1N914 1N914A 1N914B	1N915	1N916 1N916A 1N916B	1N917	UNIT
Working Peak Reverse Voltage from −65°C to 150°C		75*	50*	75*	30*	V
Average Rectified Forward Current (See Note 1)	at (or below) 25°C	75*	75*	75*	50*	mA
	at 150°C	10*	10*	10*	10*	
Peak Surge Current, 1 Second at 25°C (See Note 2)		500*	500	500*	300	mA
Continuous Power Dissipation at (or below) 25°C (See Note 3)		250*	250	250*	250	mW
Operating Free-Air Temperature Range		−65 to 175				°C
Storage Temperature Range		−65 to 200*				°C
Lead Temperature 1/16 Inch from Case for 10 Seconds		300				°C

NOTES: 1. These values may be applied continuously under a single-phase 60-Hz half-sine-wave operation with resistive load.
2. These values apply for a one-second square-wave pulse with the devices at nonoperating thermal equilibrium immediatly prior to the surge.
3. Derate linearly to 175° free-air temperature at the rate of 1.67 mW/°C.

Fig. 23-8A. Diode data sheet, including mechanical data and absolute maximum ratings

1N914 SERIES AND 1N915

***electrical characteristics at 25° free-air temperature (unless otherwise noted)**

PARAMETER		TEST CONDITIONS	1N914		1N914A		1N914B		1N915		UNIT
			MIN	MAX	MIN	MAX	MIN	MAX	MIN	MAX	
$V_{(BR)}$	Reverse Breakdown Voltage	$I_R = 100\ \mu A$	100		100		100		65		V
I_R	Static Reverse Current	$V_R = 10\ V$								25	nA
		$V_R = 20\ V$		25		25		25			nA
		$V_R = 20\ V,\ T_A = 100°C$						3		5	μA
		$V_R = 20\ V,\ T_A = 150°C$		50		50		50			μA
		$V_R = 50\ V$								5	μA
		$V_R = 75\ V$		5		5		5			μA
V_F	Static Forward Voltage	$I_F = 5\ mA$					0.62	0.72	0.6	0.73	V
		$I_F = 10\ mA$		1							V
		$I_F = 20\ mA$ (See Note 4)				1					V
		$I_F = 50\ mA$ (See Note 4)								1	V
		$I_F = 100\ mA$ (See Note 4)						1			V
C_T	Total Capacitance	$V_R = 0$, f = 1 MHz		4		4		4		4	pF

1N916 SERIES AND 1N917

***electrical characteristics at 25°C free-air temperature (unless otherwise noted)**

PARAMETER		TEST CONDITIONS	1N916		1N916A		1N916B		1N917		UNIT
			MIN	MAX	MIN	MAX	MIN	MAX	MIN	MAX	
$V_{(BR)}$	Reverse Breakdown Voltage	$I_R = 100\ \mu A$	100		100		100		40		V
I_R	Static Reverse Current	$V_R = 10\ V$								50	nA
		$V_R = 20\ V$		25		25		25			nA
		$V_R = 20\ V,\ T_A = 100°C$						3		25	μA
		$V_R = 20\ V,\ T_A = 150°C$		50		50		50			μA
		$V_R = 75\ V$		5		5		5			μA
V_F	Static Forward Voltage	$I_F = 0.25\ mA$								0.64	V
		$I_F = 1.5\ mA$								0.74	V
		$I_F = 3.5\ mA$								0.83	V
		$I_F = 5\ mA$					0.63	0.73			V
		$I_F = 10\ mA$ (See Note 4)		1						1	V
		$I_F = 20\ mA$ (See Note 4)				1					V
		$I_F = 30\ mA$ (See Note 4)						1			V
C_T	Total Capacitance	$V_R = 0$, f = 1 MHz		2		2		2		2.5	pF

NOTE 4: These parameters must be measured using pulse techniques, $t_w = 300\ \mu s$, duty cycle < 2%.

Fig. 23-8B. Diode data sheet showing electrical characteristics at 25°C

V_F: *Static forward voltage drop.* This value is the maximum forward drop at 25°C determined at specific values of I_F, figure 23-8B.

P: *Continuous power dissipation* (25°C). This value is the maximum power that the diode can dissipate continuously in free air without damaging the diode. This rating must be downgraded at higher temperatures.

(R23-11) Use figure 23-8 to determine the following parameters for a 1N914 diode at 25°C: a. V_{RM}, b. V_{BR}, c. I_F, d. I_{FM} (surge), e. I_R, f. V_F, and g. P

HALF-WAVE RECTIFIER A half-wave rectifier circuit is shown in figure 23-9A. This circuit converts the ac input voltage of figure 23-9B into the pulsating unidirectional voltage of figure 23-9C. It is assumed that an ideal diode is used in the circuit. Thus, the peak value of the output pulses is equal to the peak value of the input pulses. In addition, the period of the output wave is equal to the period of the input wave. The dc component of the output wave is found using the following equation:

$$V_{dc} = \frac{V_m}{\pi} = 0.318 \, V_m \qquad \text{Eq. 23.1}$$

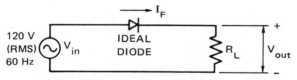

(A) Half-Wave Rectifier Circuit

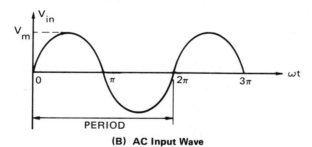

(B) AC Input Wave

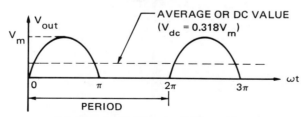

(C) Pulsating Unidirectional Output Wave

Fig. 23-9 The half-wave rectifier

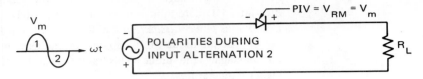

Fig. 23-10 Peak inverse voltage (PIV) of a diode

Problem 1   The input voltage of a half-wave rectifier circuit is e = 170 sin 377 t. Find: a. the dc component of the output voltage, and b. the fundamental frequency of the output voltage.

Solution   a. V_m = 170 V

$$V_{dc} = \frac{V_m}{\pi}$$

$$= \frac{170 \text{ V}}{3.141\ 6}$$

$$= 54.1 \text{ V}$$

b. ω = 377 = $2\pi f$

$$f = \frac{377 \text{ radians per second}}{(2)(3.141\ 6) \text{ radians per hertz}}$$

$$= 60 \text{ Hz}$$

(R23-12)   Assume R_L is 1 000 Ω in problem 1.
Find: a. I_m and b. I_{dc}.

When the diode is not conducting, figures 23-9C and 23-10, the applied voltage across the terminals of the diode is at its peak value. This maximum voltage is called the peak inverse voltage (PIV), or the peak reverse voltage. It represents the maximum voltage the diode must withstand during the reverse part of the cycle. In equation form:

$$PIV = V_m \qquad\qquad \text{Eq. 23.2}$$

(R23-13)   Determine the peak inverse voltage (PIV) for Problem 1.

Figure 23-11 shows high-voltage, high-current silicon rectifiers used in industrial and military applications.

Fig. 23-11 High-voltage, high-current silicon rectifiers

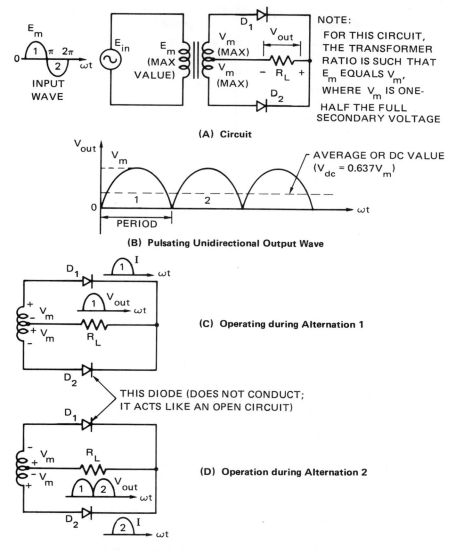

(A) Circuit

NOTE:
FOR THIS CIRCUIT,
THE TRANSFORMER
RATIO IS SUCH THAT
E_m EQUALS V_m,
WHERE V_m IS ONE-
HALF THE FULL
SECONDARY VOLTAGE

(B) Pulsating Unidirectional Output Wave

AVERAGE OR DC VALUE
($V_{dc} = 0.637V_m$)

(C) Operating during Alternation 1

THIS DIODE (DOES NOT CONDUCT;
IT ACTS LIKE AN OPEN CIRCUIT)

(D) Operation during Alternation 2

Fig. 23-12 Operation of a full-wave rectifier circuit with a center-tapped transformer

FULL-WAVE RECTIFIER WITH CENTER-TAPPED TRANSFORMER

A full-wave rectifier circuit with a center-tapped transformer is shown in figure 23-12. The operation of this circuit occurs as follows:

1. During input alternation 1, or the positive half-cycle, D_1 is forward biased and D_2 is reverse biased.
2. A pulsating unidirectional current flows through D_1 and the load (R_L). A pulsating unidirectional voltage drop appears across the load.
3. There is no current through D_2. This diode acts like an open switch (assuming that the diodes are ideal, and that I_R is zero at all times).

4. During input alternation 2, or the negative half-cycle, D_2 is forward biased and D_1 is reverse biased.

5. A pulsating unidirectional current flows through D_2 and the load. The direction of the current is the same as in step 2. A pulsating unidirectional voltage drop appears across the load (R_L).

6. There is no current through D_1.

The average or dc value of the full-wave output is determined from the following equation:

$$V_{dc} = \frac{2\,V_m}{\pi} = 0.637\,V_m \qquad\qquad \text{Eq. 23.3}$$

The fundamental period of the output wave is one-half the period of the input wave. Therefore, the fundamental frequency of the output wave is twice the frequency of the input wave, as shown in the following equation:

$$f_{out} = 2\,f_{in} \qquad\qquad \text{Eq. 23.4}$$

Problem 2 A full-wave rectifier circuit with a center-tapped transformer has an input voltage e = 170 sin 377 t. Find: a. the dc component of the output voltage, and b. the frequency of the output voltage.

Solution a. V_m = 170 V (Note: V_m = one-half the full secondary voltage)

$$V_{dc} = \frac{2\,V_m}{\pi}$$
$$= \frac{(2)(170\text{ V})}{3.141\ 6}$$
$$= 108.2\text{ V}$$

b. ω = 377 = $2\pi f$ (for the input wave)

$$f = \frac{377 \text{ radians per second}}{(2)(3.141\ 6) \text{ radians per hertz}}$$
$$= 60\text{ Hz}$$

$$f_{out} = 2\,f_{in}$$
$$= (2)(60\text{ Hz})$$
$$= 120\text{ Hz}$$

(R23-14) Assume R_L = 1 000 Ω for the circuit in Problem 2.
Find: a. I_m and b. I_{dc}.

(R23-15) What does an analysis of Review Questions (R23-12) and (R23-14) show?

Figure 23-13 shows the circuit at the instant the secondary voltage reaches its maximum value, or V_m. As shown in the following equation, the reverse voltage across the nonconducting diode (D_2) is 2 V_m.

$$PIV = 2\,V_m \qquad\qquad \text{Eq. 23.5}$$

(R23-16) Determine the PIV for the circuit in Problem 2.

BRIDGE RECTIFIER Figure 23-14 shows a bridge rectifier circuit having a full-wave output. The current flow in the circuit during alternations 1 and 2 is also shown.

The operation of this circuit is explained in the following steps:

1. During input alternation 1 (the positive half-cycle), diodes D_2 and D_3 are forward biased and diodes D_1 and D_4 are reversed biased.

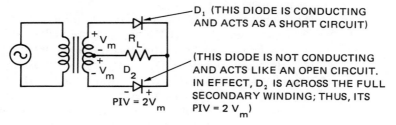

Fig. 23-13 Peak inverse voltage of a diode in a full-wave rectifier circuit with a center-tap transformer

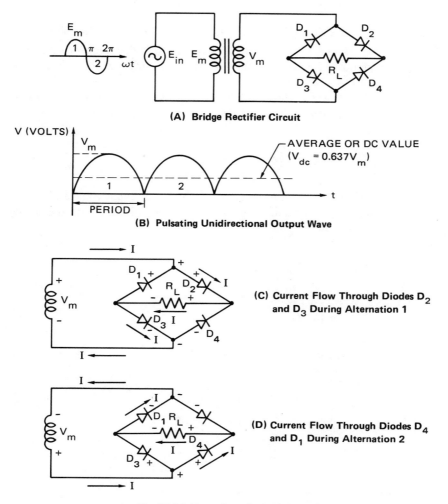

(A) Bridge Rectifier Circuit

(B) Pulsating Unidirectional Output Wave

(C) Current Flow Through Diodes D_2 and D_3 During Alternation 1

(D) Current Flow Through Diodes D_4 and D_1 During Alternation 2

Fig. 23-14 Operation of a bridge rectifier

2. A pulsating unidirectional current flows through D_2, the load R_L, and D_3. A pulsating unidirectional voltage drop appears across the load R_L.

3. There is no current through D_1 and D_4. These diodes act like open switches.

4. During input alternation 2 (the negative half-cycle) diodes D_1 and D_4 are forward biased and diodes D_2 and D_3 are reverse biased.

5. A pulsating unidirectional current flows through D_4, the load R_L, and D_1. This current has the same direction as the current of step 2. A pulsating unidirectional voltage drop appears across the load R_L.

6. There is no current through D_2 and D_3.

The average value, or the dc value, of the full-wave output is given by the following equation:

$$V_{dc} = \frac{2\,V_m}{\pi} = 0.637\,V_m \qquad\qquad \text{Eq. 23.6}$$

(R23-17)   What does an analysis of Equation 23.6 and Equation 23.3 show?

The frequency of the output wave is twice the frequency of the input wave. Equation 23.4 for the full-wave rectifier with a center-tapped transformer applies also to the bridge rectifier.

Figure 23-15 shows the bridge rectifier circuit during the positive half-cycle of the input alternation. Diodes D_2 and D_3 are conducting; they act like closed switches. Diodes D_1 and D_4 are reverse biased, and are connected directly across V_m.

Therefore, for each diode,

$$PIV = V_m \qquad\qquad \text{Eq. 23.7}$$

Problem 3   A bridge rectifier circuit, figure 23-14, has an input voltage of e = 240 sin 377 t. Find: a. the dc component of the output voltage and b. the frequency of the output voltage.

Solution   a. V_m = 240 V

$$V_{dc} = \frac{2\,V_m}{\pi}$$

$$= \frac{(2)(240\ \text{V})}{3.141\ 6}$$

$$= 153\ \text{V}$$

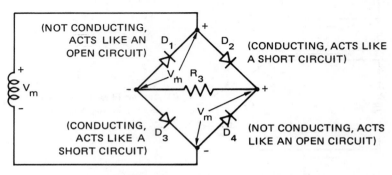

Fig. 23-15 PIV of a diode in a bridge rectifier circuit

Alternate solution:

$$V_{dc} = 0.637\ V_m$$
$$= (0.637)(240\ V)$$
$$= 153\ V$$

b. $f_{out} = 2\ f_{in}$
$$= 2(60\ Hz)$$
$$= 120\ Hz$$

(R23-18)   Calculate the peak inverse voltage for each diode of the circuit in Problem 3.

Figure 23-16 shows some commercial bridge rectifiers. The rectifiers in figure 23-16A have PIV ratings in the range from 50 V to 600 V, and an output current of 17.5 A. Subminiature bridge rectifiers, figure 23-16B, have PIV ratings in the range from 50 V to 1 000 V, and an output current of 1.0 A. The high-current bridge rectifiers, figure 23-16C, are used in power supplies, ac to dc converters, and motor control circuits for both military and industrial appli-

(A) Single-phase bridge rectifiers in aluminum cases

(B) Subminiature silicon bridge rectifiers

(C) High-current bridge rectifiers for military and industrial application

Fig. 23-16 Commercial bridge rectifiers

SURGE LIMITING RESISTOR USED TO LIMIT THE
SURGE CURRENT THROUGH THE DIODE WHEN
THE POWER SUPPLY IS FIRST SWITCHED ON.

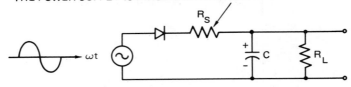

Fig. 23-17 Half-wave rectifier circuit with a capacitor-input filter

cations. These rectifiers can be used for single-phase operation with PIV ratings in the range from 50 V to 600 V, and an output current of 150 A.

CAPACITOR-
INPUT FILTER

Half-wave and full-wave unidirectional pulsations can be converted into a constant dc voltage by filtering or smoothing out the ac variations. Figure 23-17 shows a half-wave rectifier circuit with a capacitor-input filter.

The operation of this circuit is described in the following steps.

1. The capacitor is charged to the peak input voltage when the diode is forward biased.
2. When the diode is reverse biased, the capacitor discharges partially through the load R_L. The capacitor discharges just past the positive peak. At this point, the capacitor voltage V_m is greater than the decreasing supply voltage.
3. The diode is again forward biased near the peaks of the input voltage because the capacitor retains some positive charge. The diode then passes a current pulse to the capacitor, replacing the charge lost to the load.
4. As a result, the output voltage is a dc voltage with a superimposed ripple waveform. For the circuit of figure 23-17, the ripple voltage frequency is the same as the input frequency. Figure 23-18 shows the curve of the charge and discharge cycle for a capacitive filter. The decrease in the voltage before the capacitor is recharged is called the *ripple voltage.*
5. The amount of capacitor discharge between the voltage peaks is controlled by the RC time constant of the filter capacitor C and the load resistor R_L.

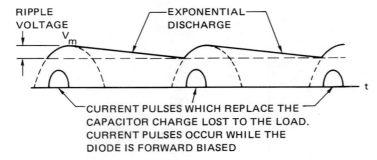

Fig. 23-18 Curve of the charge and discharge cycle for a capacitor-input filter in a half-wave rectifier circuit

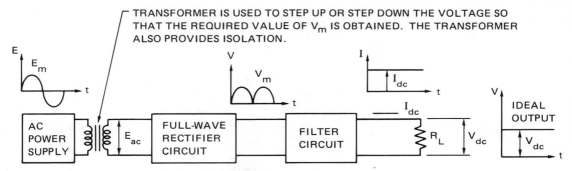

TRANSFORMER IS USED TO STEP UP OR STEP DOWN THE VOLTAGE SO THAT THE REQUIRED VALUE OF V_m IS OBTAINED. THE TRANSFORMER ALSO PROVIDES ISOLATION.

Fig. 23-19 Block diagram of an unregulated power supply

The capacitor-input filter circuit is also called a *peak rectifier*, or a *peak detector*. The time constant of the capacitor-input filter, $R_L C$, is greater than the period T of the input voltage. As a result, the capacitor loses only a small part of its charge before it is recharged.

(R23-19) Theoretically, what is the maximum dc voltage that can be obtained with a capacitor-input filter?

(R23-20) What is the effect on the ripple voltage when the load resistor, R_L, is decreased in value?

UNREGULATED POWER SUPPLIES

Figure 23-19 is a block diagram of an unregulated power supply which changes an ac voltage to a dc voltage. The value of V_{dc} changes as R_L is varied, or as E_{ac} changes. Two commercial unregulated power supplies are shown in figure 23-20.

(A) Castor configuration   (B) Two-wheel configuration

Fig. 23-20 Commercial unregulated power supplies

Fig. 23-21 Multiunit electrolytic capacitor

A multiunit electrolytic capacitor with three ratings is shown in figure 23-21. This type of capacitor is used in the complex filter circuits of power supplies. Multiunit capacitors can be used in dc circuits only because their polarities must be observed at all times. Electrolytic capacitors require less space than equivalent paper capacitors.

LABORATORY EXERCISE 23-1 CHARACTERISTICS AND OPERATION OF RECTIFIER CIRCUITS

PURPOSE In completing this exercise, the student will

- determine experimentally the characteristics and operation of half-wave and full-wave rectifier circuits.
- determine experimentally the operation of a capacitor-input filter.

EQUIPMENT AND MATERIALS
1 Oscilloscope, dual trace
1 Transformer, center-tapped, 120 V/60 V (120 V/12.6 V)
2 Diodes, 1N67A
1 Resistor, 10 kΩ, 2 W
1 Resistor, 1 kΩ, 2 W
1 Capacitor, 100 μF
1 Capacitor, 10 μF

PROCEDURE
1. Construct the circuit shown in figure 23-22.
   a. Energize the ac power supply.
   b. Display both the voltage V_S and the power supply ac voltage on the oscilloscope at the same time.
   c. Measure the peak value of each voltage.

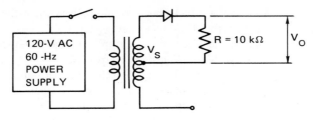

Fig. 23-22

d. Measure the voltage V_0 using the oscilloscope with dc coupling.
e. With the oscilloscope connected across R, connect the 10-μF capacitor across R. Observe and record the waveform of V_0.
f. Record the peak-to-peak amplitude of the ripple voltage.
g. Replace the 10-μF capacitor with the 100-μF capacitor. Observe and record the waveform of V_0.
h. Record the peak-to-peak amplitude of the ripple voltage.
i. Remove the 10-kΩ resistor and replace it with the 1-kΩ resistor. Observe and record the waveform of V_0.

2. Construct the circuit shown in figure 23-23.
   a. Energize the ac power supply.
   b. Display both the voltage V_S and the power supply ac voltage on the oscilloscope at the same time.
   c. Measure the peak value of each voltage.
   d. Measure the voltage V_0 with the oscilloscope.
   e. Connect the oscilloscope to display V_0 and observe the waveforms for the following output load combinations. (Note: in each case R and C must be connected in parallel.)
      (1) R = 10 kΩ, C = 10 μF
      (2) R = 10 kΩ, C = 100 μF
      (3) R = 1 kΩ, C = 100 μF
   Sketch the voltage waveforms for V_0 for each combination. Record the amplitude of the ripple voltage.

3. a. Explain the operation of the half-wave rectifier circuit in figure 23-22.
   b. Explain the effect of connecting a capacitor across the 10-kΩ resistor in figure 23-22.

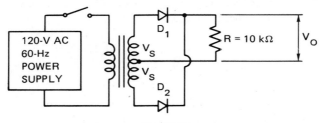

Fig. 23-23

c. What is the effect on the ripple voltage when the 10-μF capacitor is replaced with a 100-μF capacitor (step 1.g.)?

d. What is the effect on the ripple voltage when the 10-kΩ resistor is replaced with a 1-kΩ resistor (step 1.i.)?

e. Explain the operation of the full-wave rectifier circuit in figure 23-23.

f. What is the effect on the ripple voltage when the resistors and capacitors are used in different combinations in the full-wave rectifier circuit?

g. What are the advantages and disadvantages of the half-wave and full-wave rectifier circuits?

h. Discuss the operation of the capacitor-input filter.

EXTENDED STUDY TOPICS

1. What is the barrier potential for germanium and for silicon?
2. What is the procedure for biasing a diode?
3. What is the bulk resistance of a diode? What are typical values in a forward biased diode?
4. The following data is given for a half-wave rectifier circuit: e (input) = 50 sin 377 t and R_L = 1 kΩ. Assume that the diode has a forward voltage drop V_F of 1 V.
   Find: a. the maximum value of the output load voltage.
   b. the peak value of load current.
5. If a half-wave rectifier circuit has an input frequency of 1 000 Hz, what is the frequency of the output wave?
6. Calculate the peak inverse voltage for the circuit in topic 4.
7. In a full-wave rectifier circuit with a center-tapped transformer, V_m (one-half of the secondary voltage) = 200 V and R_L = 2 000 Ω.
   Find: a. the dc output voltage.
   b. the dc output current.
   c. the minimum PIV rating of each diode.
8. A full-wave bridge rectifier operates under the following conditions: the input voltage e equals 100 sin 377 t; the transformer is a 1 : 2 step-up transformer; and R_L equals 1 000 Ω.
   Find: a. the dc output voltage.
   b. the dc output current.
   c. the minimum PIV rating of each diode.
9. Define ripple voltage in a capacitor-input filter.
10. Compare the responses of a capacitor-input filter with a full-wave voltage input and a capacitor-input filter with a half-wave voltage input.
11. What is a peak detector?

<table>
<tr><td>

Chapter

24

</td><td>

Bipolar Junction Transistors and Configurations

</td></tr>
</table>

OBJECTIVES   After studying this chapter, the student will be able to

- explain the operation of NPN and PNP bipolar transistors.
- discuss dc currents and dc voltages in transistor circuits.
- explain the characteristics of the three transistor configurations.
- interpret transistor data sheets.
- construct the dc load line.

BIPOLAR TRANSISTOR THEORY   The bipolar junction transistor (BJT) is a doped silicon or germanium semiconductor crystal. It contains alternate regions of n-type and p-type material. The BJT is a three-terminal device consisting of two p-n junction diodes placed back to back. Figure 24-1 shows the arrangement of layers and the circuit symbols for NPN and PNP bipolar junction transistors.

Junction transistors normally are identified by a type number, such as 2N718. The "2N" indicates that the transistor has two junctions. The following observations can be made about the NPN transistor of figure 24-1.

1. The emitter is heavily doped so that it can emit or inject electrons into the base.
2. The base is very thin and is lightly doped. Most of the electrons injected by the emitter pass through the base to the collector.
3. The collector is neither as heavily doped as the emitter nor as lightly doped as the base. The collector receives or collects the electrons from the

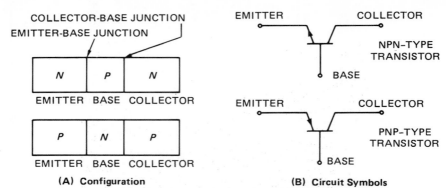

(A) Configuration   (B) Circuit Symbols

Fig. 24-1 NPN and PNP bipolar junction transistors

base. This section controls the flow of electrons to the external load circuit. The collector must dissipate more heat than the emitter and base. Therefore, it is the largest region in the junction transistor.

4. The transistor has two junctions. One junction is between the emitter and base regions and the other junction is between the collector and base regions.

The PNP transistor is the complement of the NPN transistor. That is, its construction, currents, and voltages are opposite to those of the NPN transistor.

A matched pair of NPN dual transistors is shown in the photomicrograph of figure 24-2. Such a device is used in differential amplifiers. Two types of transistor packages are shown in figure 24-3.

(R24-1)   What is a bipolar junction transistor?

(R24-2)   Does an NPN transistor consist of one, two, or three separate crystals?

(R24-3)   What is the number of terminals in a bipolar junction transistor?

(R24-4)   Name the three regions of a BJT.

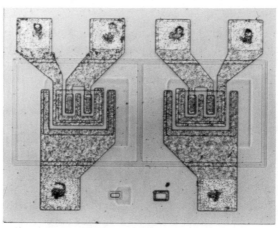

Fig. 24-2 Photomicrograph of NPN dual transistors

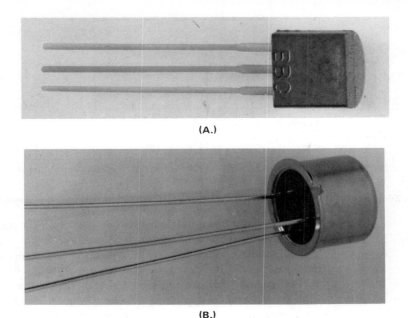

(A.)

(B.)

Fig. 24-3 Case styles for bipolar junction transistors

DC TRANSISTOR CURRENTS

The conventional direction of current flow in a transistor is shown by the direction of the arrow on the emitter terminal. (Refer to figure 24-1.) When the arrow points *in*, the device is a PNP transistor. An NPN transistor is indicated if the arrow does *not* point *in* (that is, it points *out*).

The flow of majority charge carriers in an NPN transistor is shown in figure 24-4. The following observations can be made by studying the figure.

1. Electrons are the majority carriers in the *n*-type emitter.

2. The junction between the base and emitter regions is forward biased by the voltage V_{EE}. As a result, the electrons in the emitter cross the emitter-base junction and enter the base.

3. The electrons from the emitter enter the *p*-type base and diffuse through the base. Some of these electrons travel to the base terminal and leave the region as the base current, I_B. Most of the electrons reach the collector-base junction.

4. A bias voltage, V_{CC}, is applied between the collector and base terminals. The polarity of this voltage is such that the collector-base junction is reverse biased for the majority carriers in both the base and collector. However, the electrons that were injected into the base region are minority carriers there. The base collector junction is "forward biased" for those carriers; in other words, they "fall" to the collector.

5. The electrons reaching the collector-base junction are drawn to the collector terminal by the positive potential there.

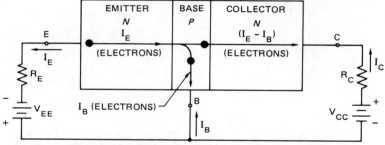

NOTE: THE CONVENTIONAL CURRENT FLOW DIRECTION IS USED TO SHOW
THE EXTERNAL OR TERMINAL CURRENTS I_E, I_B AND I_C.

Fig. 24-4 Flow of majority charge carriers (electrons) from the emitter in an NPN transistor

6. Most of the electrons that start in the emitter reach the collector to form the conventional collector current, I_C.

7. The reverse bias on the collector-base junction causes another type of current flow across this junction. This current consists of minority charge carriers in both the base and the collector. That is, the current is due to electrons from the *p*-type base and holes from the *n*-type collector. The current due to the flow of minority charges across the collector-base junction is very small. It is called the *collector-base reverse saturation current* or the *collector cutoff current*. It is usually labeled I_{CBO}, or I_{CO}. This reverse saturation current is very sensitive to temperature. For each 10°C rise in temperature, I_{CBO} nearly doubles. I_{CBO} is independent of all other currents. Figure 24-5 indicates that I_{CBO} continues to flow when the emitter is open and the collector-base junction is reverse biased.

(R24-5) Define I_{CBO}.

(R24-6) What effect does temperature have on the reverse saturation current?

The voltage bias required for the normal operation of an NPN transistor is shown in figure 24-6. The actual directions of the three external terminal currents are also shown. It can be seen that the emitter is negative with respect to the base. In other words, the emitter-base diode is forward biased. The collector is positive with respect to the base. Therefore, the collector-base diode is

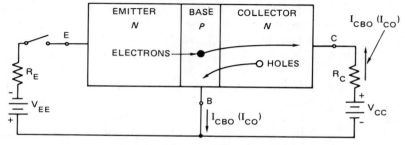

Fig. 24-5 Analysis of I_{CBO} (I_{CO})

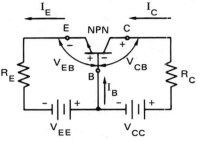

Fig. 24-6 Terminal or external currents for the normal operation of an NPN transistor. (The emitter-base junction is forward biased and the collector-base junction is reverse biased)

reverse biased. The base current, I_B, is the part of I_E entering the base terminal, minus, I_{CBO}. The collector current, I_C, is the part of I_E flowing across the CB junction, plus I_{CBO}. Terminal currents can be measured with very sensitive instruments, such as the picoammeter shown in figure 24-7.

(R24-7) What are the bias requirements for normal transistor operation?

A transistor current equation derived from figure 24-6 defines transistor action:

$$I_E = I_B + I_C \qquad\qquad \text{Eq. 24.1}$$

where

I_B = base current
I_C = collector current
I_E = emitter current

When I_{CBO} is considered, the collector current may be expressed as follows:

$$I_C = \alpha I_E + I_{CBO} \qquad\qquad \text{Eq. 24.2}$$

where

I_C = collector current
I_{CBO} = reverse saturation current
I_E = emitter current
α (alpha) = the portion of I_E collected at the CB junction.

Fig. 24-7 Picoammeter used to measure transistor currents

Equation 24.2 can be solved for α.

$$\alpha = \frac{I_C - I_{CBO}}{I_E} \qquad \text{Eq. 24.3}$$

When the reverse saturation current (I_{CBO}) is small compared to the collector current, Equation 24.3 becomes:

$$\alpha \cong \frac{I_C}{I_E} \qquad \text{Eq. 24.4}$$

Values of α range from 0.90 to 0.999. At the upper end of this range, most of the emitter current becomes the collector current. The parameter α is also known as the dc short-circuit current gain for the common-base configuration. It is written as α_{DC} or h_{FB}.

(R24-8) Write the current equation defining transistor action.

(R24-9) Write the equation for I_C using the reverse saturation current.

(R24-10) Does the circuit of figure 24-6 have more or less current in the collector than in the emitter? (Neglect I_{CBO}.)

I_B can be solved in terms of I_E by combining Equations 24.1 and 24.2 as follows:

$$I_E = I_B + I_C$$
$$I_C = I_E - I_B$$
$$I_E - I_B = \alpha I_E + I_{CBO}$$
$$I_B = (1 - \alpha)\, I_E - I_{CBO} \qquad \text{Eq. 24.5}$$

Problem 1 A transistor has the following values: $\alpha = 0.98$; $I_{CBO} = 5\ \mu A$; and $I_E = 5.25$ mA. Calculate a. I_B and b. I_C.

Solution a. $I_B = (1 - \alpha)\, I_E - I_{CBO}$
$I_B = (1 - 0.98)(5.25 \times 10^{-3})\,A - (5.0 \times 10^{-6})\,A$
$ = (0.02)(5.25 \times 10^{-3})\,A - (5.0 \times 10^{-6})\,A$
$ = (105 \times 10^{-6})\,A - (5 \times 10^{-6})\,A$
$ = 100\ \mu A$

b. $I_C = I_E - I_B$
$ = (5.25 \times 10^{-3})\,A - (100 \times 10^{-6})\,A$
$ = (5.25 \times 10^{-3})\,A - (0.10 \times 10^{-3})\,A$
$ = (5.15 \times 10^{-3})\,A$
$ = 5.15$ mA

(R24-11) Solve Problem 1 for I_B and I_C neglecting I_{CBO}.

DC TRANSISTOR VOLTAGES Figure 24-8 shows the required bias and supply voltage polarities for both NPN and PNP transistors. For the NPN transistor:

1. The base is positive with respect to the emitter; that is, the emitter is negative with respect to the base. The base-emitter junction is forward biased.
2. The collector is more positive than the base; that is, the base is negative with respect to the collector. The base-collector junction is reverse biased.

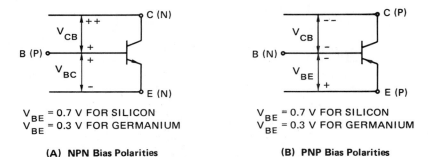

V_{BE} = 0.7 V FOR SILICON
V_{BE} = 0.3 V FOR GERMANIUM

V_{BE} = 0.7 V FOR SILICON
V_{BE} = 0.3 V FOR GERMANIUM

(A) NPN Bias Polarities

(B) PNP Bias Polarities

Fig. 24-8 Bias voltage polarities for NPN and PNP transistors

For the PNP transistor:

1. The base is negative with respect to the emitter; that is, the emitter is positive with respect to the base. The base-emitter junction is forward biased.
2. The collector is more negative than the base. Therefore, the base is positive with respect to the collector. The base-collector junction is reverse biased.

As shown in figure 24-8, typical base-emitter voltages for both NPN and PNP transistors are 0.7 V for silicon and 0.3 V for germanium. Transistor voltages may be measured using the very sensitive digital multimeter shown in figure 24-9.

(R24-12) What is the base-emitter voltage, V_{EB}, of an NPN silicon transistor?

(R24-13) What is the base-emitter voltage, V_{EB}, of a PNP silicon transistor?

MODES OF OPERATION Transistors may be connected to give any one of three different modes of operation or configurations. One of the three transistor leads is common to both the input and output circuits for each of these modes. Each mode is known by the name of the common terminal. For example, when the base is common, the circuit operates in the common-base, or CB, mode. When the emitter is common, the circuit operates in the common-emitter, or CE, mode. When the collector is

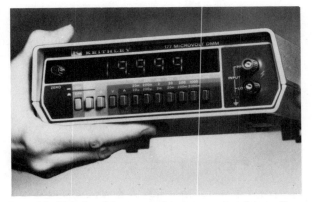

Fig. 24-9 Digital multimeter used to measure transistor voltages

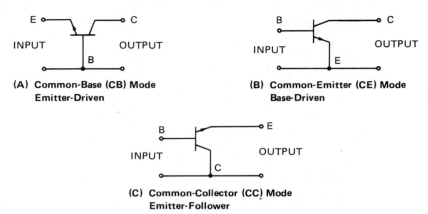

(A) **Common-Base (CB) Mode**
Emitter-Driven

(B) **Common-Emitter (CE) Mode**
Base-Driven

(C) **Common-Collector (CC) Mode**
Emitter-Follower

Fig. 24-10 The three modes, or configurations, for an NPN transistor

common, the circuit operates in the common-collector, or CC, mode. Figure 24-10 illustrates the three modes, or configurations, for an NPN transistor.

(R24-14) List the three different modes of operation or configurations in which a transistor may be connected.

Common-base Characteristics. Figure 24-11 shows a CB configuration using an NPN transistor. The input is applied between the emitter and base and the output is taken between the collector and base. Figure 24-12A is a graph of the emitter current versus the emitter-base voltage for various constant values of V_{CB}. This graph is known as the input characteristics for the transistor. The emitter-base junction is forward biased; therefore, these input characteristics resemble those of a forward-biased diode. Changes in the collector-to-base voltage, V_{CB}, have a slight effect on I_E. Note that I_E increases slightly for the same value of V_{EB} as V_{CB} is increased.

The output characteristics, figure 24-12B, are a graph of the collector current versus the collector-base voltage for various constant values of I_E. For normal operation, the collector-base junction is reverse biased. The value at which the collector current becomes constant (saturates) is determined by the emitter current. The collector current value is independent of the collector-base voltage.

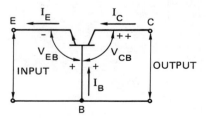

Fig. 24-11 CB configuration using an NPN transistor (see figure 24-6 for the complete circuit with resistors and power supplies)

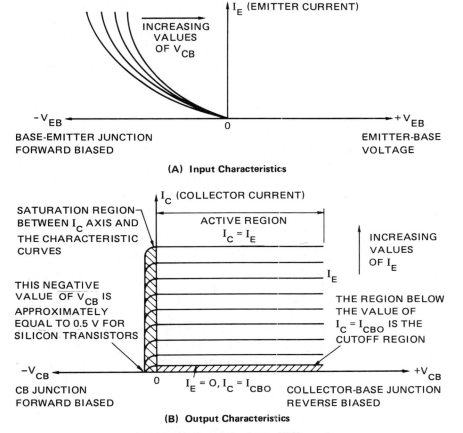

I_E (EMITTER CURRENT)

INCREASING VALUES OF V_{CB}

$-V_{EB}$ 0 $+V_{EB}$

BASE-EMITTER JUNCTION FORWARD BIASED

EMITTER-BASE VOLTAGE

(A) Input Characteristics

I_C (COLLECTOR CURRENT)

SATURATION REGION BETWEEN I_C AXIS AND THE CHARACTERISTIC CURVES

ACTIVE REGION $I_C = I_E$

INCREASING VALUES OF I_E

THIS NEGATIVE VALUE OF V_{CB} IS APPROXIMATELY EQUAL TO 0.5 V FOR SILICON TRANSISTORS

I_E

THE REGION BELOW THE VALUE OF $I_C = I_{CBO}$ IS THE CUTOFF REGION

$-V_{CB}$ 0 $I_E = O, I_C = I_{CBO}$ $+V_{CB}$

CB JUNCTION FORWARD BIASED

COLLECTOR-BASE JUNCTION REVERSE BIASED

(B) Output Characteristics

Fig. 24-12 CB characteristics for an NPN transistor

As shown in figure 24-12B, there is still a collector current when V_{CB} is reduced to zero. The current exists because of a barrier potential at the CB junction. The direction of this potential aids the flow of the collector current. The student should recall from previous study that the barrier potential is positive on the n-type collector side of the CB junction. The potential is negative on the p-type base side of the CB junction. Therefore, to reduce I_C to zero, the CB junction must be forward biased. For silicon transistors, the value of the bias voltage (V_{CB}) is approximately 0.5 V.

Figure 24-12B also shows that I_C equals I_{CBO} when I_E is zero. The region of operation below I_{CBO} is called the *cutoff region*. I_{CBO} is very small, in the order of nanoamperes (nA) in silicon transistors and microamperes (μA) in germanium transistors. Therefore, when the CB transistor configuration is inserted into a circuit, the output voltage of the transistor, V_{CB}, is nearly equal to the supply voltage V_{CC}. The voltage drop across the collector resistor (R_C) can be neglected because of the very low value of I_{CBO}.

The *active region* is the region in which I_C is nearly equal to I_E, figure 24-12B.

Increases in the emitter current (I_E) cause the collector current (I_C) to reach a value beyond which it no longer increases for constant values of V_{CC} and R_C. This region is called the *saturation region* and it lies between the I_C axis and the characteristic curves, figure 24-12B. In this region, the voltage V_{CB} is negligible and I_C depends upon the supply voltage V_{CC} and the collector resistor R_C. The value of the collector current in the saturation region is determined as follows:

$$I_{C(sat)} = \frac{V_{CC}}{R_C}$$

Eq. 24.6

where

$I_{C(sat)}$ = collector saturation current
V_{CC} = collector supply voltage
R_C = collector resistor

(R24-15) Define the active region for an NPN transistor operating in the CB mode.

(R24-16) Define the saturation region for an NPN transistor operating in the CB mode.

(R24-17) Define the cutoff region for an NPN transistor operating in the CB mode.

(R24-18) Calculate $I_{C(sat)}$ for an NPN transistor operating in the CB mode if: V_{CC} = 10 V, R_C = 5 kΩ, V_{EE} = 1.5 V, and R_E = 1 kΩ.

Common-emitter Characteristics. A CE configuration using an NPN transistor is shown in figure 24-13. The input is applied between the base and emitter and the output is taken between the collector and emitter. The graph of the base current versus the base-emitter voltage for various constant values of V_{CE} forms the input characteristics, figure 24-14A. This graph shows that V_{CE} does not control the base current. These input characteristics are similar to those of a forward-biased diode.

The output characteristics, figure 24-14B, are a graph of the collector current versus the collector-emitter voltage for various values of I_B. For all values of V_{CE} greater than 0.2 V to 0.5 V, I_C saturates, or reaches a constant value that is independent of V_{CE}. This value of the collector current is controlled by the magnitude of the base current only.

The dc short-circuit current gain in the CE configureation is determined as follows:

$$\beta = \frac{I_C}{I_B}$$

Eq. 24.7

where

β (beta) = dc short-circuit current gain in the CE mode
I_C = collector current
I_B = base current

It should be noted that manufacturers of semiconductor devices also refer to β as h_{FE}. When α is known, β can be found.

$$\beta = \frac{\alpha}{1 - \alpha}$$

Eq. 24.8

NOTE: $V_{CE} > V_{BE}$ SO THAT V_{CB} WILL HAVE A
POSITIVE VALUE (REVERSE BIAS ON CB JUNCTION)

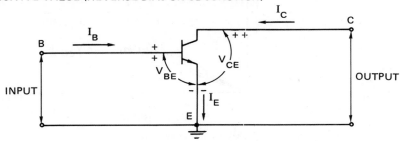

(A) CE Configuration Using An NPN Transistor

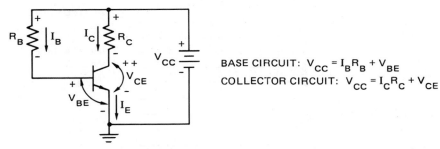

BASE CIRCUIT: $V_{CC} = I_B R_B + V_{BE}$
COLLECTOR CIRCUIT: $V_{CC} = I_C R_C + V_{CE}$

**(B) CE Configuration Using An NPN Transistor,
With Bias Voltages and Resistors**

Fig. 24-13

Similarly, when β is known , α can be found:

$$\alpha = \frac{\beta}{\beta + 1}$$

Eq. 24.9

Problem 2   For the transistor whose characteristics are shown in figure 24-14, $I_C = 5$ mA and $I_B = 50$ μA. a. Calculate the β of the transistor. b. Calculate the α of the transistor if it is connected in the CB mode.

Solution   a. $\beta = \dfrac{I_C}{I_B}$

$= \dfrac{(5 \times 10^{-3})\text{A}}{(50 \times 10^{-6})\text{A}}$

$= 100$

b. $\alpha = \dfrac{\beta}{\beta + 1}$

$= \dfrac{100}{100 + 1}$

$= 0.99$

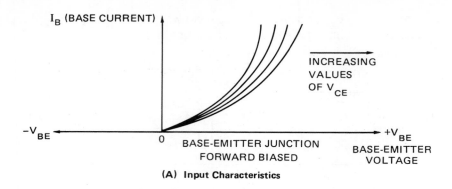

(A) Input Characteristics

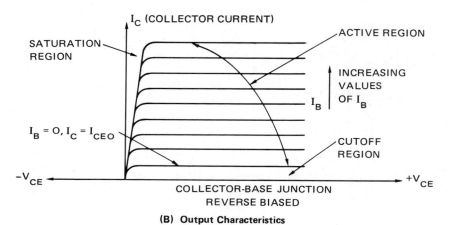

(B) Output Characteristics

Fig. 24-14 CE characteristics for an NPN transistor

(R24-19)   What conclusion can be reached from the calculations in Problem 2 concerning the current gain in the CE mode and the current gain in the CB mode?

(R24-20)   What is the value of β if the α of a transistor is 0.98?

Normal transistor operation in the CE mode requires that (1) the base-emitter junction be forward biased (that is, V_{BE} must be positive), and (2) the collector-emitter junction be reverse biased (that is, V_{CE} must be positive).

To maintain the reverse bias on the collector, V_{CE} must be greater than V_{BE}. In other words, for the common emitter mode, the collector must be more positive than the base.

(R24-21)   For normal bias operation in the CE mode, is V_{CB} positive or negative?

When the base is open-circuited ($I_B = 0$), the collector current I_C is the reverse saturation current in the CE mode. For this case, I_C is known as I_{CEO} and is related to the reverse saturation current in the CB mode according to the following equation:

$$I_{CEO} = (\beta + 1) \, I_{CBO} \qquad\qquad \text{Eq. 24.10}$$

I_C can be determined when either I_{CBO} or I_{CEO} is known:

$$I_C = \beta I_B + (\beta + 1) I_{CBO} \qquad \text{Eq. 24.11}$$

$$I_C = \beta I_B + I_{CEO} \qquad \text{Eq. 24.12}$$

I_{CEO} is much larger than I_{CBO}. However, I_{CEO} generally is very small when compared to βI_B. This means that I_{CEO} can be neglected and Equation 24.12 reduces to Equation 24.7.

Therefore,

$$I_C \cong \beta I_B$$

The emitter current is expressed as:

$$I_E = (\beta + 1) I_B + I_{CEO} \qquad \text{Eq. 24.13}$$

When I_{CEO} is neglected, Equation 24.13 reduces to:

$$I_E = (\beta + 1) I_B \qquad \text{Eq. 24.14}$$

Since β is very much greater than 1, Equation 24.14 can be written as:

$$I_E \cong \beta I_B \qquad \text{Eq. 24.15}$$

Since $I_C \cong \beta I_B$, the following approximation is true:

$$I_C \cong I_E \qquad \text{Eq. 24.16}$$

Problem 3   For a transistor connected in the CE mode, I_B = 0.5 mA and I_C = 100 mA. Calculate: a. β, and b. I_E (neglect I_{CEO}).

Solution   a. β $= \dfrac{I_C}{I_B}$

$$= \frac{(100 \times 10^{-3})\ A}{(0.5 \times 10^{-3})\ A}$$

$$= 200$$

b. I_E $= (\beta + 1) I_B$

$$= (200 + 1)(0.5\ mA)$$

$$= 100.5\ mA$$

COMMON COL-LECTOR CHAR-ACTERISTICS   A common collector (CC) configuration using an NPN transistor is shown in figure 24-15. The input is applied between the base and the collector and the output is taken between the emitter and the collector. Another way in which the common collector configuration can be drawn is given in figure 24-16. In this case, note that the output voltage in the CC mode is always equal to the input voltage, minus the voltage drop between the base and the emitter. Therefore, the output voltage is nearly equal to the input voltage. In other words, the

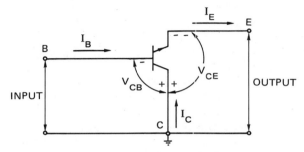

Fig. 24-15 CC configuration using an NPN transistor

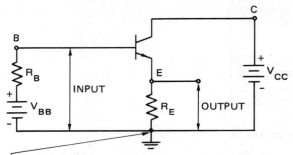

THIS TERMINAL IS EQUIVALENT TO THE COLLECTOR
TERMINAL SINCE THE INTERNAL RESISTANCE OF THE
BATTERY OR POWER SUPPLY (V$_{CC}$) IS NEGLIGIBLE.
THEREFORE, THE OUTPUT IS DIRECTLY ACROSS THE
COLLECTOR-EMITTER TERMINALS. POINT C AND
GROUND ARE AT THE SAME AC POTENTIAL.

Fig. 24-16   Alternate method of showing a CC configuration using an NPN transistor

output voltage closely follows the input voltage. This characteristic of the CC mode is the reason why it is also known as an emitter-follower.

The characteristic curves of the CC mode are identical to the characteristic curves of the CE mode. The equations developed in this Chapter for the CE mode apply also to the CC mode.

(R24-22)   Why is the CC configuration called an emitter-follower?

UNDERSTANDING TRANSISTOR DATA SHEETS   Manufacturers' data sheets must be studied carefully before the proper transistor can be selected for a particular application. Typical data sheets are shown in figure 24-17.

The information given in figure 24-17A includes: 1. the type number of the device, such as 2N718; 2. a descriptive title and a list of the major applications for which the transistor is designed; 3. drawings showing mechanical data such as the package shape, the dimensions, and the identification of the collector, base, and emitter leads; and 4. the absolute maximum voltage and current values which the transistor can withstand without breaking down at 25°C. Actual operating conditions must not approach these maximum ratings if transistor breakdown is to be avoided. When the transistor is operated at temperatures greater than 25°C, the maximum ratings must be adjusted downward.

Figure 24-17B lists the electrical characteristics of the transistor at 25°C. The following parameters are commonly listed on manufacturers' data sheets.

BV_{CBO}:   Collector-base breakdown voltage; this is the breakdown voltage for a reverse-biased collector-base junction.

BV_{CEO}:   Collector-emitter breakdown voltage; this is the collector-to-emitter breakdown voltage with the base open circuited.

BV_{EBO}:   Emitter-base breakdown voltage. The emitter-base junction is reverse biased.

Highly Reliable, Versatile Devices Designed for Amplifier, Switching and Oscillator Applications from <0.1 ma to >150 ma, dc to 30 mc

- High Voltage   • Low Leakage
- Useful h_{FE} Over Wide Current Range

*mechanical data

Device types 2N717, 2N718, 2N718A, 2N730, 2N731, and 2N956 are in JEDEC TO-18 packages.
Device types 2N696, 2N697, 2N1420, 2N1507, 2N1613, and 2N1711 are in JEDEC TO-5 packages.

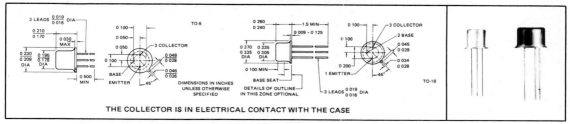

THE COLLECTOR IS IN ELECTRICAL CONTACT WITH THE CASE

*absolute maximum ratings at 25°C free-air temperature (unless otherwise noted)

	2N696 2N697	2N717 2N718	2N718A	2N730 2N731	2N956	2N1420 2N1507	2N1613	2N1711	UNIT
Collector-Base Voltage	60	60	75	60	75	60	75	75	v
Collector-Emitter Voltage (See Note 1)	40	40	50	40	50	30	50	50	v
Collector-Emitter Voltage (See Note 2)			32						v
Emitter-Base Voltage	5	5	7	5	7	5	7	7	v
Collector Current				1.0		1.0		1.0	a
Total Device Dissipation at (or below) 25°C Free-Air Temperature (See Note Indicated in Parenthesis)	0.6 † (3)	0.4 †† (5)	0.5 (7)	0.5 †† (9)	0.5 (7)	0.6 † (3)	0.8 (10)	0.8 (10)	w
Total Device Dissipation at (or below) 25°C Case Temperature (See Note Indicated in Parenthesis)	2.0 † (4)	1.5 †† (6)	1.8 (8)	1.5 †† (6)	1.8 (8)	2.0 † (4)	3.0 (11)	3.0 (11)	w
Total Device Dissipation at 100°C Cast Temperature	1.0 †	0.75 ††	1.0	0.75 ††	1.0	1.0 †	1.7	1.7	w
Operating Collector Junction Temperature	175†	175††	200	175††	200	175†	200	200	°C
Storage Temperature Range	−65°C to 300°C								

NOTES 1. This value applies when the base-emitter resistance (R_{BE}) is equal to or less than 10 ohms.
2. This value applies when the base-emitter diode is open-circuited.
3. Derate linearly to 175°C free-air temperature at the rate of 4.0 mw/C°.
4. Derate linearly to 175°C case temperature at the rate of 13.3 mw/C°.
5. Derate linearly to 175°C free-air temperature at the rate of 2.67 mw/C°.
6. Derate linearly to 175°C case temperature at the rate of 10.0 mw/C°.
7. Derate linearly to 200°C free-air temperature at the rate of 2.86 mw/C°.
8. Derate linearly to 200°C case temperature at the rate of 10.3 mw/C°.
9. Derate linearly to 175°C free-air temperature at the rate of 3.33 mw/C°.
10. Derate linearly to 200°C free-air temperature at the rate of 4.56 mw/C°.
11. Derate linearly to 200°C case temperature at the rate of 17.2 mw/C°.

*Indicates JEDEC registered data

† Texas Instruments guarantees its types 2N696, 2N697, 2N1420, and 2N1507 to be capable of the same dissipation as registered and shown for types 2N1613 and 2N1711 with appropriate derating factors shown in Notes 10 and 11. See derating curves, page 8.

†† Texas Instruments guarantees its types 2N717, 2N718, 2N730, and 2N731 to be capable of the same dissipation as registered and shown for types 2N718A and 2N956 with appropriate derating factors shown in Notes 7 and 8. See derating curves, page 8.

(A) Absolute maximum ratings at 25°C

Fig. 24-17 Transistor data sheet

*electrical characteristics at 25°C free-air temperature (unless otherwise noted)

PARAMETER		TEST CONDITIONS	TO-18 → 2N718A / TO-5 → 2N1613		2N1420		2N1507		2N956 / 2N1711		UNIT
			MIN	MAX	MIN	MAX	MIN	MAX	MIN	MAX	
BV_{CBO}	Collector-Base Breakdown Voltage	$I_C = 100\ \mu a,\ I_E = 0$	75		60		60		75		v
BV_{CEO}	Collector-Emitter Breakdown Voltage	$I_C = 30\ ma,\ I_B = 0$, (See Note 12)					25				v
BV_{CER}	Collector-Emitter Breakdown Voltage	$I_C = 100\ ma,\ R_{BE} = 10\ \Omega$, (See Note 12)	50		30		30		50		v
BV_{EBO}	Emitter-Base Breakdown Voltage	$I_C = 100\ \mu a,\ I_C = 0$	7						7		v
		$V_{CB} = 30\ v,\ I_E = 0$				1.0		1.0			μa
		$V_{CB} = 30\ v,\ I_E = 0,\ T_A = 150°C$				100		50			μa
I_{CBO}	Collector Cutoff Current	$V_{CB} = 60\ v,\ I_E = 0$		0.010						0.010	μa
		$V_{CB} = 60\ v,\ I_E = 0\ T_A = 150°C$		10						10	μa
I_{CER}	Collector Cutoff Current	$V_{CE} = 20\ v,\ R_{BE} = 100\ k\Omega$						10			μa
I_{EBO}	Emitter Cutoff Current	$V_{EB} = 5\ v,\ I_C = 0$		0.01				100		0.005	μa
h_{FE}	Static Forward Current Transfer Ratio	$V_{CE} = 10\ v,\ I_C = 10\ \mu a$							20		
		$V_{CE} = 10\ v,\ I_C = 100\ \mu a$	20						35		
		$V_{CE}\ 10\ v,\ I_C = 10\ ma$, (See Note 12)	35						75		
		$V_{CE} = 10\ v,\ I_C = 10\ ma,\ T_A = -55°C$, (See Note 12)	20						35		
		$V_{CE} = 10\ v,\ I_C = 150\ ma$, (See Note 12)	40	120	100	300	100	300	100	300	
		$V_{CE} = 10\ v,\ I_C = 500\ ma$, (See Note 12)	20						40		
V_{BE}	Base-Emitter Voltage	$I_B = 15\ ma,\ I_C = 150\ ma$, (See Note 12)		1.3		1.3		1.3		1.3	v
$V_{CE(salt)}$	Collector-Emitter Saturation Voltage	$I_B = 15\ ma,\ I_C = 150\ ma$, (See Note 12)		1.5		1.5		1.5		1.5	v
h_{ib}	Small-Signal Common-Base Input Impedance	$V_{CB} = 5\ v,\ I_C = 1\ ma,\ f = 1\ kc$	24	34					24	34	ohm
		$V_{CB} = 10\ v,\ I_C = 5\ ma,\ f = 1\ kc$	4	8					4	8	ohm
h_{rb}	Small-Signal Common-Base Reverse Voltage Transfer Ratio	$V_{CB} = 5\ v,\ I_C = 1\ ma,\ f = 1\ kc$		3×10^{-4}						5×10^{-4}	
		$V_{CB} = 10\ v,\ I_C = 5\ ma,\ f = kc$		3×10^{-4}						5×10^{-4}	
h_{ob}	Small-Signal Common-Base Output Admittance	$V_{CB} = 5\ v,\ 1\ ma,\ f = 1\ kc$	0.1	0.5					0.1	0.5	μmho
		$V_{CB} = 10\ v,\ I_C = 5\ ma,\ f = 1\ kc$	0.1	1.0					0.1	1.0	μmho
h_{fe}	Small-Signal Common-Emitter Forward Current Transfer Ratio	$V_{CE} = 5\ v,\ I_C = 1\ ma,\ f = 1\ kc$	30	100					50	200	
		$V_{CE} = 10\ v,\ I_C = 5\ ma,\ f = 1\ kc$	35	150					70	300	
h_{fe}	Small-Signal Common-Emitter Forward Current Transfer Ratio	$V_{CE} = 10\ v,\ I_C = 50\ ma,\ f = 20\ mc$	3.0		2.5		2.5		3.5		
C_{ob}	Common-Base Open-Circuit Output Capacitance	$V_{CB} = 10\ v,\ I_E = 0,\ f = 1\ mc$		25		35		35		25	pf
C_{ib}	Common Base Open-Circuit Input Capacitance	$V_{EB} = 0.5\ v,\ I_C = 0\ f = 1\ mc$		80						80	pf

See operating and switching characteristics for types 2N718A, 2N956, 2N1613, and 2N1711 on page 4.

NOTE 12: These parameters must be measured using pulse techniques. PW ≤ 300 μsec, Duty Cycle ≤ 2%. Pulse width must be such that halving or doubling does not cause a change greater than the required accuracy of the measurement.

*Indicates JEDEC registered data.

(B) Electrical Characteristics at 25°C

Fig. 24-17 Transistor data sheet (cont.)

V_{BE}:	Base-emitter voltage is the dc voltage drop across a forward-biased base-emitter junction.
$V_{CE(sat)}$:	Collector-emitter saturation voltage
I_{CBO} (or I_{CO}):	Collector cutoff current.
h_{FE}:	Static forward current transfer ratio. In other words, h_{FE} is the ratio of the dc collector current to the dc base current in the CE mode or β.

(R24-23) Refer to figure 24-17 and determine the following parameters for the 2N718A transistor at 25°C: a. BV_{CBO}, b. I_{CBO}, c. h_{FE}, d. V_{BE}, and e. $V_{CE(sat)}$.

DC LOAD LINE AND QUIESCENT POINT

A common-emitter amplifier circuit is shown in figure 24-18. An equation can be written relating the circuit conditions in the output circuit. This expression, which is known as the dc load line equation, is written as follows:

$$V_{CC} = I_C R_L + V_{CE} \qquad \text{Eq. 24.17}$$

The graph of this equation is a straight line. Two points on this graph can be obtained as follows:

- Point No. 1 (the horizontal intercept): when $I_C = 0$, then $V_{CC} = V_{CE} = 20$ V

- Point No. 2 (the vertical intercept): when $V_{CE} = 0$, then $I_{C(sat)} = \dfrac{V_{CC}}{R_L}$

  $= \dfrac{20 \text{ V}}{(10 \times 10^3) \, \Omega} = 2$ mA

These points are located on the output characteristic curves of the transistor in figure 24-19. I_B has a value of 20 μA. This value represents the dc bias current, or the base quiescent current, when there is no ac signal applied to the circuit. It is called I_{BQ}. The intersection of the dc load line and I_{BQ} establishes the quiescent point Q. At the quiescent point, the dc collector-to-emitter voltage, V_{CEQ}, is 12.2 V and the dc collector current, I_{CQ}, is 0.78 mA. The load line shown in figure 24-19 is based on the conditions of $V_{CC} = 20$ V and R_L (R_C) = 10 kΩ. If either V_{CC} or R_L is changed, a new dc load line must be drawn.

(R24-24) If I_B equals 40 μA in figure 24-18, determine a. I_{BQ}, b. V_{CEQ}, and c. I_{CQ} using the CE output characteristics of figure 24-19.

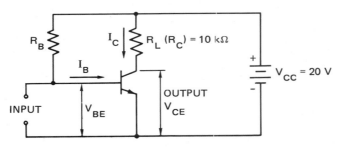

Fig. 24-18 Common-emitter amplifier circuit

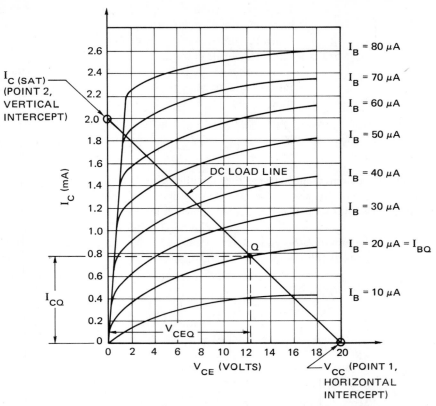

Fig. 24-19 CE output characteristics and plot of the dc load line

LABORATORY EXERCISE 24-1 THE COMMON-EMITTER TRANSISTOR CONFIGURATION

PURPOSE

In completing this exersise, the student will

- determine experimentally the β of the transistor in the CE mode.

- observe the changes in β as the operating values of I_B, I_C, and V_{CE} vary.

EQUIPMENT AND MATERIALS

1 Dc power supply
1 VOM
1 Decade resistance box
1 Transistor, 2N718
1 Resistor, 1 kΩ, 2 W
1 Resistor, 5.1 kΩ, 2 W

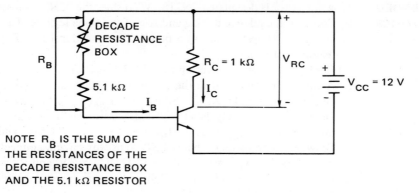

NOTE R_B IS THE SUM OF
THE RESISTANCES OF THE
DECADE RESISTANCE BOX
AND THE 5.1 kΩ RESISTOR

Fig. 24-20

PROCEDURE

1. Construct the circuit shown in figure 24-20.
   a. Turn on the dc power supply and adjust it to +12 volts.
   b. Adjust the decade resistance box, while measuring the voltage across R_C, such that $V_{RC} = 1$ V.
   c. Measure V_{BE}.
   d. Record V_{RC} and V_{BE} in Table 24-1. Repeat steps 1.a. through 1.d for values of $V_{RC} = 2, 4, 8$, and 10 V respectively.

2. Answer the following questions and perform the calculations indicated.
   a. Calculate the current I_B using the values of V_{CC} and R_B and each value of V_{BE}. Record these values in Table 24-1.
   b. Calculate the collector current I_C using the value of R_C. Record these values in Table 24-1.
   c. Calculate and record in Table 24-1 the value of β for each set of readings obtained from steps 2.a. and 2.b.
   d. What does β represent?
   e. Plot a graph of β versus I_{CQ} (the quiescent collector current).
   f. Does β change as the operating point changes?

TABLE 24-1

Measured		Calculated		
V_{RC}(V)	V_{BE} (V)	I_C (mA)	I_B (µA)	β
1				
2				
4				
8				
10				

EXTENDED
STUDY TOPICS

1. A transistor is designated 2N718. What does the "2N" represent?
2. Compare the relative doping and size of the three regions of a PNP transistor.
3. Name the largest region of a bipolar junction transistor. Explain why this region must be large.
4. Describe the transistor action for a PNP transistor.
5. A transistor connected in the common-base configuration has the following currents: $I_C = 5.0$ mA, $I_B = 50\ \mu$A, and $I_{CBO} = 5\ \mu$A. Find: a. I_E and b. α.
6. Draw and completely label a PNP transistor circuit connected in the CB mode. Assume $V_{CC} = 21$ V, $R_C = 10$ kΩ, $V_{EE} = 4.5$ V, and $R_E = 2$ kΩ.
7. Draw and completely label a PNP transistor circuit connected in the CE mode. Assume $V_{CC} = 15$ V, $R_C = 5$ kΩ, $V_{BB} = 6$ V, and $R_B = 15$ kΩ.
8. Draw and completely label a PNP transistor circuit connected in the CC mode. Assume $V_{EE} = 31$ V, $R_E = 20$ kΩ, $V_{BB} = 30$ V, and $R_B = 1$ MΩ.
9. Define: a. base-driven circuit, b. emitter-driven circuit, and c. emitter-follower circuit.
10. What is the difference between α and β for a transistor?
11. For the circuit shown in figure 24-18, $V_{CC} = 15$ V, $R_L = 10$ kΩ, and $I_B = 20\ \mu$A. The CE output characteristics for the transistor are shown in figure 24-19.

    a. Plot the dc load line.

    b. Find: I_{BQ}, I_{CQ}, and V_{CEQ}.

<table>
<tr><td>

Chapter

25
</td><td>

Electronic Amplifiers
</td></tr>
</table>

OBJECTIVES

After studying this chapter, the student will be able to

- explain the power dissipation limits of a power transistor.
- apply decibel notation.
- analyze currents and voltages in an amplifier.
- determine graphically voltage gain, current gain, and power gain.
- calculate input impedance, output impedance, voltage gain, current gain, and power gain.
- discuss the characteristics of amplifiers.

POWER TRANSISTORS

The power dissipated in the collector region of a power transistor connected in the CE mode is given by the following equation:

$$P_D = V_{CE} \, I_C \qquad \qquad \text{Eq. 25.1}$$

where

P_D = power dissipation, in watts
V_{CE} = collector-to-emitter voltage, in volts
I_C = collector current, in amperes

A specification sheet for a power transistor is shown in figure 25-1. Figure 25-2 is the graph of Equation 25.1 for this power transistor. The figure shows the permissible region of operation for the transistor. When operating the power transistor, the collector voltage and collector current values must remain in the region of the transistor characteristics. That is, the values must fall below or to

Product Preview

The RF Line

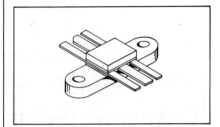

20 W − 870 MHz

RF POWER TRANSISTOR

NPN SILICON

NPN SILICON RF POWER TRANSISTOR

. . . designed for 12.5 volt UHF large-signal, common-base amplifier applications in industrial and commercial FM equipment operating in the range of 806 – 947 MHz.

- Specified 12.5 Volt, 870 MHz Characteristics:
    Output Power = 20 Watts
    Minimum Gain = 6.0 dB
    Efficiency = 50%

- Series Equivalent Large-Signal Characterization

- Internally Matched Input for Broadband Operation

- 100% Tested for Load Mismatch Stress at All Phase Angles with 20:1 VSWR @ 16 Volt Supply and 50% RF Overdrive

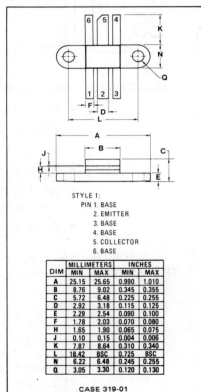

STYLE 1:
PIN 1. BASE
     2. EMITTER
     3. BASE
     4. BASE
     5. COLLECTOR
     6. BASE

MAXIMUM RATINGS

Rating	Symbol	Value	Unit
Collector-Emitter Voltage	V_{CEO}	16	Vdc
Collector-Base Voltage	V_{CBO}	36	Vdc
Emitter-Base Voltage	V_{EBO}	4.0	Vdc
Collector Current — Continuous	I_C	3.5	Adc
Total Device Dissipation @ T_C = 25°C (1) Derate Above 25°C	P_D	60 0.46	Watts W/°C
Storage Temperature Range	T_{stg}	−65 to +200	°C

THERMAL CHARACTERISTICS

Characteristic	Symbol	Max	Unit
Thermal Resistance, Junction to Case	$R_{\theta JC}$	2.2	°C/W

(1) These devices are designed for RF operation. The total device dissipation rating applies only when the devices are operated as RF amplifiers.

DIM	MILLIMETERS		INCHES	
	MIN	MAX	MIN	MAX
A	25.15	25.65	0.990	1.010
B	8.76	9.02	0.345	0.355
C	5.72	6.48	0.225	0.255
D	2.92	3.18	0.115	0.125
E	2.29	2.54	0.090	0.100
F	1.78	2.03	0.070	0.080
H	1.65	1.90	0.065	0.075
J	0.10	0.15	0.004	0.006
K	7.87	8.64	0.310	0.340
L	18.42	BSC	0.725	BSC
N	6.22	6.48	0.245	0.255
Q	3.05	3.30	0.120	0.130

CASE 319-01

This is advance information and specifications are subject to change without notice.   ©MOTOROLA INC., 1977   NP-76

Fig. 25-1 Power transistor specification sheet

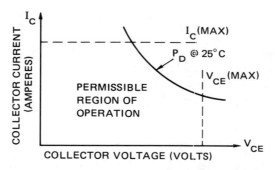

Fig. 25-2 Graph of equation 25.1 showing the permissible region of operation for a power transistor

the left of the maximum power dissipation curve. A design safety factor of 25% may be used. In other words, the transistor will be operated at a value 25% less than its maximum power rating.

(R25-1) For the power transistor whose specifications are given in figure 25-1, find: a. I_C(max), b. V_{CE}(max), and c. P_D.

Problem 1 For the transistor whose specifications are given in figure 25-1, calculate the operating value of V_{CE} if $I_C = 3A$. Use a design safety factor of 25%.

Solution

$$\text{Operating } P_D = (0.75) P_D$$
$$= (0.75)(60 \text{ W})$$
$$= 45 \text{ W}$$

$$P_D = V_{CE} I_C$$

$$V_{CE} = \frac{P_D}{I_C}$$
$$= \frac{45 \text{ W}}{3 \text{ A}}$$
$$= 15 \text{ V}$$

Power transistors are shown in figure 25-3. Figure 25-4 shows an assortment of power amplifiers, and an internal view of a power amplifier.

DECIBELS *Power gain* (G) is defined as the ratio of the ac output power to the ac input power:

$$G = \frac{P_2}{P_1} \qquad \text{Eq. 25.2}$$

where

G = power gain
P_2 = output power (P_o), in watts
P_1 = input power (P_{in}), in watts

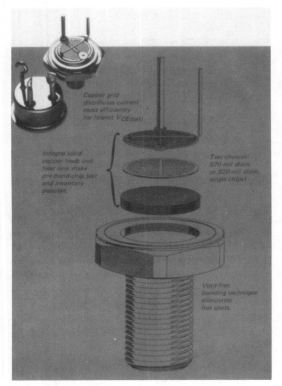

Fig. 25-3A Construction of a power transistor

Fig. 25-3B Power transistor

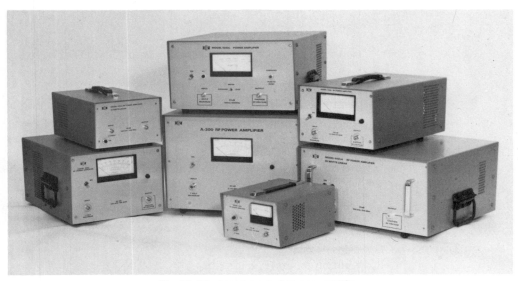

Fig. 25-4A Assortment of power amplifiers

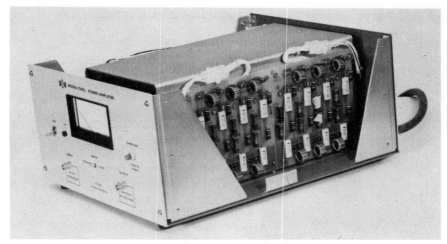

Fig. 25-4B Internal view of a power amplifier

Problem 2 The input power to an amplifier is 0.5 W. Calculate the gain of the amplifier if the output power is 20 W.

Solution $G = \dfrac{P_2}{P_1}$

$= \dfrac{20 \text{ W}}{0.5 \text{ W}}$

$= 40$

Figure 25-5 shows two cascaded power amplifiers. The overall power gain of two or more cascaded stages is the product of the power gain for each stage.

$$G_T = G_1 G_2 \qquad\qquad \text{Eq. 25.3}$$

Problem 3 For the cascaded amplifiers in figure 25-5, $G_1 = 30$ and $G_2 = 40$. Find the overall power gain, G_T.

Solution $G_T = G_1 G_2$

$= (30)(40)$

$= 1\ 200$

Bel power gain is defined as the common logarithm (log to the base 10) of the power gain. (The abbreviation for the bel unit is B.)

$$\text{Gain, in bels (B)} = \log G \qquad\qquad \text{Eq. 25.4}$$

Problem 4 An amplifier has a power gain of 1 000. Calculate the bel power gain.

Solution $\text{Gain} = \log G$

$= \log 1\ 000$

$= 3 \text{ B}$

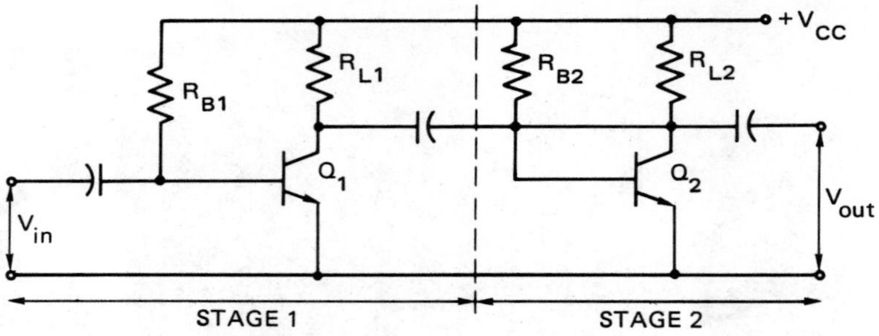

Fig. 25-5 Two-stage cascaded power amplifiers

(R25-2) Find the bel power gain of an amplifier having a power gain of 100.

When the bel power gain is known, the power gain can be determined as follows:

$$G = \text{antilog gain (B)} \qquad \text{Eq. 25.5}$$

Problem 5 Calculate the power gain of an amplifier having a bel power gain of 0.3 B.
Solution G = antilog gain (B)
= antilog 0.3 B = $10^{(0.3)}$
= 2

(R25-3) Find the power gain of an amplifier having a bel power gain of 4 B.

The bel is an inconveniently large unit of measurement for power. Therefore, the *decibel* (dB), or one-tenth of a bel, is commonly used to state power ratios.

$$\text{Gain (dB)} = 10 \log G \qquad \text{Eq. 25.6}$$

Problem 6 An amplifier has a gain of 2. Calculate its gain in decibels (dB).
Solution   Gain (dB) = 10 log G
= 10 log 2
= 10(0.3)
= 3 dB

Problem 6 shows that a power gain of 2 is a 3-dB increase in power. Problem 7 shows the procedure to be followed when the amplifier output is decreased.

Problem 7 The output of an amplifier decreases to one-half of its former output. Find the change in decibels.

Solution   Gain (dB) = $10 \log \dfrac{P_2}{P_1}$
= 10 log 1/2

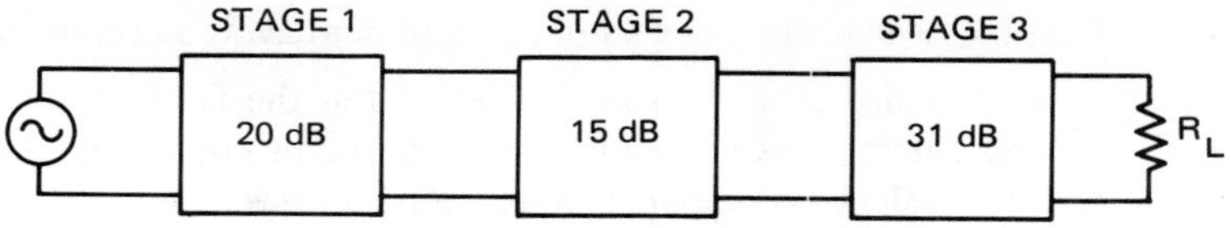

Fig. 25-6

When the ratio $\frac{P_2}{P_1}$ is less than 1, the terms are inverted and the gain becomes negative.

Therefore,

$$\text{Gain (dB)} = -10 \log \frac{2}{1}$$
$$= -10 \log 2$$
$$= (-10)(0.3)$$
$$= -3 \text{ dB}$$

(R25-4) The output of an amplifier changes from 2 W to 0.5 W. Find the dB change.

When amplifiers are cascaded, the total power gain in decibels is the algebraic sum of the individual power gains.

$$dB_T = dB_1 + dB_2 + dB_3 + \ldots + dB_n \qquad \text{Eq. 25.7}$$

where dB_n is the power gain of the last of "n" amplifiers cascaded together.

Problem 8 Figure 25-6 shows three cascaded stages. Calculate the total decibel power gain.
Solution $dB_T = 20 \text{ dB} + 15 \text{ dB} + 31 \text{ dB}$
$= 66 \text{ dB}$

(R25-5) For the amplifier stages shown in figure 25-6, the output of stage 3 drops to 20 dB. Calculate the total decibel power gain of these cascaded amplifiers.

Recall that power is proportional to the square of the voltage or to the square of the current. This relationship means that voltage and current ratios can be expressed in decibel units as follows:

$$\text{Gain (dB)} = 20 \log \frac{V_2}{V_1} = 20 \log A_v \qquad \text{Eq. 25.8}$$

where A_v = voltage gain

$$\text{Gain (dB)} = 20 \log \frac{I_2}{I_1} = 20 \log A_i \qquad \text{Eq. 25.9}$$

where A_i = current gain

Problem 9 Calculate the decibel voltage gain of an amplifier having a voltage gain of 100.
Solution $\text{Gain (dB)} = 20 \log A_v$
$= 20 \log 100$
$= (20)(2)$
$= 40 \text{ dB}$

(R25-6) Find the decibel current gain of an amplifier having a current gain of 10.

A reference level for sound is defined as the faintest sound that the human ear can detect. The value of 0 dB is assigned to this level. Some typical sounds and their values as compared to the reference level are:

15 dB the rustle of leaves in a gentle breeze

40 dB the noise in a typical home

100 dB sound noise in a subway

A change in acoustical power must be no less than 1 dB if the human ear is to detect the change.

The reference level for electronic instruments is 1 mW. Indicating meters calibrated to read decibel values have scales labeled in dBm.

$$\text{Gain (dBm)} = 10 \log \frac{P_o}{1 \text{ mW}} = 10 \log Gm \qquad \text{Eq. 25.10}$$

where

P_o = output power, in watts

Gm = gain with the input reference equal to 1 mW

Problem 10 The output of an amplifier equals 0.1 W. Express the output in dBm units.

Solution
$$\text{Gain (dBm)} = 10 \frac{P_o}{1 \text{ mW}}$$
$$= 10 \log \frac{0.1 \text{ W}}{1 \text{ mW}}$$
$$= 10 \log 100$$
$$= (10)(2)$$
$$= 20 \text{ dBm}$$

(R25-7) Express an output of 0.5 mW in dBm units.

ANALYSIS OF CURRENTS AND VOLTAGES A CE amplifier circuit is shown in figure 25-7A. The dc and ac equivalents of this circuit are shown in figures 25-7B and 25-7C. For the frequency being considered, the capacitors are short circuits to the ac currents and open circuits to the dc currents. A study of these figures results in the following conclusions:

1. I_{BQ} and I_{CQ} are the dc quiescent currents, or the currents which flow without the application of ac.
2. e_s is the sinusoidal input voltage.
3. i_s is the sinusoidal input current. A small percentage of i_s bypasses the transistor and returns to the source through R_B. R_B has a very large value. Thus, this bypass current is negligible and it is assumed that all of the ac input current (i_s) enters the base of the transistor.
4. i_B is the total current entering the base of the transistor. This current is the sum of I_{BQ} and the input sinusoidal current, i_s.

$$I_B(\text{max}) = I_{BQ} + \sqrt{2} I_s \qquad \text{Eq. 25.11}$$
$$I_B(\text{min}) = I_{BQ} - \sqrt{2} I_s \qquad \text{Eq. 25.12}$$

where I_s is the rms value of the sinusoidal input current.

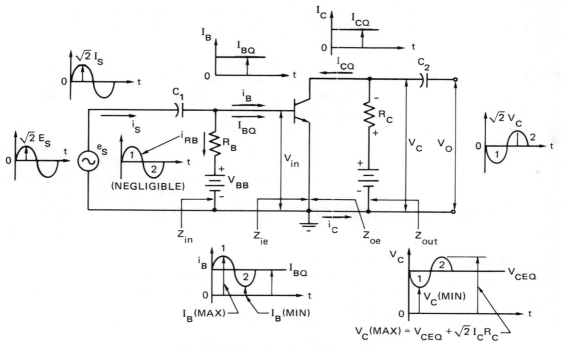

Fig. 25-7A Complete CE amplifier circuit

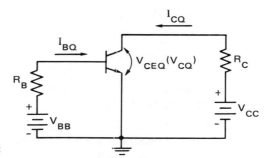

Fig. 25-7B Dc equivalent circuit

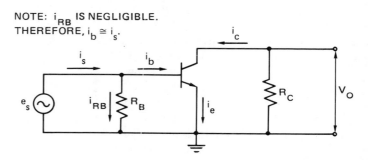

Fig. 25-7C Ac equivalent circuit

5. i_C is the total current entering the collector of the transistor. This current is the sum of I_{CQ} and the sinusoidal collector current, i_c.

$$I_C(max) = I_{CQ} + \sqrt{2}\,I_c \qquad\qquad \text{Eq. 25.13}$$

$$I_C(min) = I_{CQ} - \sqrt{2}\,I_c \qquad\qquad \text{Eq. 25.14}$$

where I_C is the rms value of the sinusoidal collector current.

6. V_C is the total voltage across the CE terminals. V_C is the sum of V_{CEQ} and the ac voltage drop across R_C.

$$V_C(max) = V_{CEQ} + \sqrt{2}\,I_C\,R_C \qquad\qquad \text{Eq. 25.15}$$

$$V_C(min) = V_{CEQ} - \sqrt{2}\,I_C\,R_C \qquad\qquad \text{Eq. 25.16}$$

7. During alternation No. 1 of the input voltage, i_c increases as e_s increases. The ac voltage drop across R_C becomes more negative. Therefore, the output voltage, V_o, is $180°$ out of phase with the input voltage.

(R25-8) What is the phase relationship between the output voltage and the input voltage for a CE amplifier?

Problem 11 The circuit of figure 25-7A has the following values: $I_{BQ} = 15\ \mu A$, $E_s = 5$ mV, $i_s = 10\ \mu A$ peak-to-peak, $V_{CC} = 15$ V, and $R_C = 3$ kΩ. Neglect the ac current drawn by R_B. Graphically solve for a. the voltage gain, b. the current gain, and c. the power gain.

Solution Figure 25-8 shows the load line drawn on the transistor characteristics.

The cutoff point, or the horizontal intercept, occurs at 15 V.

The saturation point, or the vertical intercept, occurs at 5 mA.

$$I_C(sat) = \frac{V_{CC}}{R_C}$$

$$= \frac{15\text{ V}}{3 \times 10^3\ \Omega}$$

$$= 5\text{ mA}$$

The intersection of the load line and the curve for $I_{BQ} = 15\ \mu A$ locates the quiescent point, Q. At this point, $V_{CQ} = 7.3$ V and $I_{CQ} = 2.6$ mA.

As the input signal current (i_s) varies, the output of the transistor will vary over the operating, or active, portion of the load line between points X and Y.

At point X (alternation 1):
$$I_B(max) = I_{BQ} + \sqrt{2}\,I_s$$
$$= 15\ \mu A + 5\ \mu A$$
$$= 20\ \mu A$$

$I_C(max)$ is read as 3.1 mA
$V_C(min)$ is read as 5.7 V

At Point Y (alternation 2):
$$I_B(min) = I_{BQ} - \sqrt{2}\,I_s$$
$$= 15\ \mu A - 5\ \mu A$$
$$= 10\ \mu A$$

$I_C(min)$ is read as 2.1 mA
$V_C(max)$ is read as 8.9 V

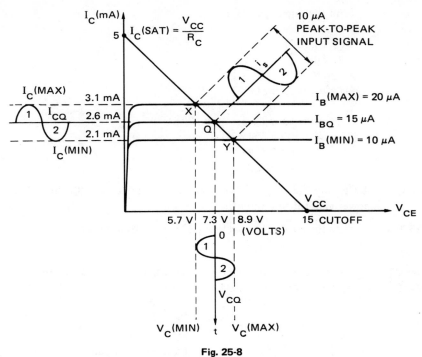

Fig. 25-8

a. Find the voltage gain, A_v.

Peak-to-peak collector voltage swing $= V_C(max) - V_C(min)$
$= 8.9\ V - 5.7\ V$
$= 3.2\ V$ (peak-to-peak)

$$V_C(\text{rms}) = \frac{(0.707)(3.2\ V)}{2}$$
$= 1.13\ V$ (ac rms value of the output voltage)

$$A_v = \frac{\text{output voltage}}{\text{input voltage}}$$
$$= \frac{V_C}{E_S}$$
$$= \frac{1.13\ V}{5 \times 10^{-3}\ V}$$
$$= 226$$

b. Find the current gain, A_i

Peak-to-peak collector current swing $= I_C(max) - I_C(min)$
$= 3.1\ mA - 2.1\ mA$
$= 1.0\ mA$

$$A_i = \frac{\text{output current}}{\text{input current}}$$
$$= \frac{1.0 \times 10^{-3}\ A}{10 \times 10^{-6}\ A}$$
$$= 100$$

c. Find the power gain, G.

$$G = A_v \, A_i \qquad\qquad \text{Eq. 25.17}$$

$= (226)(100)$

$= 22\,600$

$= 226 \times 10^4$

Gain dB $= 10 \log G$

$\qquad\qquad = 10 \log [(2.26)(10^4)]$

$\qquad\qquad = 10(\log 2.26 + \log 10^4)$

$\qquad\qquad = 10(0.354 + 4)$

$\qquad\qquad = 10(4.354)$

$\qquad\qquad = 43.54 \text{ dB}$

(R25-9)   A CE amplifier has a voltage gain of 100 and a current gain of 10. Calculate the power gain for this amplifier.

(R25-10)   Express the answer to question (R25-9) in dB notation.

INPUT Z_{ie} is the input impedance of a transistor connected in the CE mode. Z_{ie} is
IMPEDANCE   defined by the following equation:

$$Z_{ie} = \frac{\Delta V_{BE}}{\Delta I_B} \qquad\qquad \text{Eq. 25.18}$$

where

$\qquad Z_{ie} \qquad$ = input impedance of a transistor connected in the CE mode.

$\qquad \Delta V_{BE}$ = change in the base-to-emitter voltage

$\qquad \Delta I_B \qquad$ = change in the base current

Problem 12   A change of 2 mV occurs in V_{BE} when the base current of a transistor changes by 1 μA. Find the input impedance Z_{ie}.

Solution $Z_{ie} \quad = \dfrac{\Delta V_{BE}}{\Delta I_B}$

$\qquad\qquad = \dfrac{(2 \times 10^{-3}) \text{ V}}{(1 \times 10^{-6}) \text{ A}}$

$\qquad\qquad = 1.0 \text{ k}\Omega$

For the transistor used in the CE amplifier circuit shown in figure 25-7A, the input impedance of the amplifier is called Z_{in}. The value of Z_{in} is equal to the parallel combination of Z_{ie} and R_B.

Problem 13   For a CE amplifier, $R_B = 15$ kΩ and $Z_{ie} = 1$ kΩ. Find the input impedance of the amplifier.

Solution $Z_{in} \quad = Z_{ie} \parallel R_B$

$\qquad\qquad = \dfrac{(1\,000 \; \Omega)(15\,000 \; \Omega)}{(1\,000 \; \Omega) + (15\,000 \; \Omega)}$

$\qquad\qquad = 937.5 \; \Omega$

OUTPUT IMPEDANCE The output impedance of a transistor connected in the CE mode is called Z_{oe}. Z_{oe} is defined as follows:

$$Z_{oe} = \frac{\Delta V_{CE}}{\Delta I_C}$$

Eq. 25.19

where

Z_{oe} = output impedance of a transistor connected in the CE mode.

ΔV_{CE} = change in the collector-to-emitter voltage.

ΔI_C = change in the collector current.

Problem 14 For a transistor connected in the CE mode, V_{CE} changes by 15 V when the collector current changes by 15 μA. Find the output impedance Z_{oe} for this transistor.

Solution
$$Z_{oe} = \frac{\Delta V_{CE}}{\Delta I_C}$$
$$= \frac{15 \text{ V}}{(15 \times 10^{-6}) \text{ A}}$$
$$= 1 \text{ M}\Omega$$

The output impedance, Z_{out}, of the CE amplifier shown in figure 26-7A is equal to the parallel combination of Z_{oe} and R_C.

Problem 15 Find the output impedance of a CE amplifier with R_C = 10 kΩ and Z_{oe} = 1 MΩ.

Solution
$$Z_{out} = Z_{oe} \| R_C$$
$$= \frac{(1 \times 10^6 \text{ }\Omega)(10 \times 10^3 \text{ }\Omega)}{(1 \times 10^6 \text{ }\Omega) + (10 \times 10^3 \text{ }\Omega)}$$
$$= 9.9 \text{ k}\Omega$$

VOLTAGE GAIN The voltage gain, A_v, for the CE amplifier shown in figure 25-7A is expressed as follows:

$$A_v = \frac{\Delta V_o}{\Delta V_{in}} = \frac{\Delta V_{CE}}{\Delta V_{BE}}$$

Eq. 25.20

$$A_v = \frac{\beta R_C}{Z_{ie}}$$

Eq. 25.21

Problem 16 Find the voltage gain of a CE amplifier if β = 50, R_C = 10 kΩ, and Z_{ie} = 1.0 kΩ.

Solution
$$A_v = \frac{\beta R_C}{Z_{ie}}$$
$$= \frac{(50)(10 \times 10^3 \text{ }\Omega)}{(1.0 \times 10^3 \text{ }\Omega)}$$
$$= 500$$

(R25-11) A change of 2 V occurs in V_{CE} when V_{BE} of a CE amplifier changes by 2 mV. Find the voltage gain of the amplifier.

CURRENT GAIN   The current gain of a transistor is equal to β.

$$A_i = \frac{\Delta I_C}{\Delta I_B} = \beta \qquad \text{Eq. 25.22}$$

(R25-12)   A transistor has a β equal to 50. What is the current gain of the transistor when it is connected to the CE mode?

The circuit shown in figure 25-7A is reexamined to determine what happens to the ac signal current, I_S. Note that I_S follows two paths; that is, part of I_S flows through R_B and part flows into the base of the transistor. The current component that flows through R_B bypasses the transistor and is not amplified. The part of I_S that enters the base, i_b, is amplified. The following equation gives the value of the *circuit* current gain based on the factors just described.

$$A_i' = \frac{\Delta I_C}{\Delta I_S} = \frac{\beta R_B}{R_B + Z_{ie}} \qquad \text{Eq. 25.23}$$

where A_i' = the circuit current gain (*not* the transistor current gain)

Problem 17   Find the circuit current gain of a CE amplifier if $\beta = 50$, $Z_{ie} = 1.0 \text{ k}\Omega$, and $R_B = 30 \text{ k}\Omega$.

Solution
$$\begin{aligned} A_i' &= \frac{\beta R_B}{R_B + Z_{ie}} \\ &= \frac{(50)(30 \times 10^3 \ \Omega)}{(30 \times 10^3 \ \Omega) + (1.0 \times 10^3 \ \Omega)} \\ &= 48.4 \end{aligned}$$

(R25-13)   A CE amplifier has a circuit current gain of 100 and an input signal current (I_S) equal to 2 mA. Calculate the value of the collector current, I_C.

POWER GAIN   The power gain of a transistor is expressed as follows:

$$A_p = G = A_v A_i = \frac{\text{Ac power output}}{\text{Ac power input}} \qquad \text{Eq. 25.24}$$

where A_p, or G = power gain of the transistor

To calculate the power gain of the *circuit*, the current gain of the *circuit*, A_i, must be used (Equation 25.23):

$$A_p' = G' = A_v A_i' \qquad \text{Eq. 25.25}$$

where A_p', or G' = the *circuit* power gain

Problem 18   For a CE amplifier, $\beta = 50$ and $A_v = 300$. Calculate the power gain of the transistor.

Solution
$$\begin{aligned} A_p &= G = A_v A_i \\ &= A_v \beta \\ &= (300)(50) \\ &= 15\ 000 \\ &= 15 \times 10^3 \end{aligned}$$

(R25-14)   The circuit current gain of the CE amplifier of Problem 18 = 48. Calculate the power gain of the circuit.

SUMMARY OF AMPLIFIER CHARACTERISTICS

An amplifier circuit using a transistor connected in the CE mode has the following characteristics:

1. Best power gain.
2. Good voltage gain, with 180° phase inversion.
3. Good current gain.
4. Moderate input and output impedances.
5. Most frequently used amplifier circuit.

An amplifier circuit using a transistor connected in the CB mode has the following characteristics:

1. Low power gain.
2. Best voltage gain, with no phase inversion.
3. Current gain is less than one.
4. Lowest input impedance, but high output impedance.
5. Used in high-frequency tuned circuits and oscillators.

An amplifier circuit using a transistor connected in the CC mode has the following characteristics:

1. Low power gain.
2. Voltage gain less than one, with no phase inversion.
3. Good current gain (equal to CE mode).
4. Highest input impedance, but low output impedance.
5. Used for impedance matching.

LABORATORY EXERCISE 25-1 THE CE AMPLIFIER

PURPOSE

In completing this exercise, the student will

- become familiar with the characteristics of a CE amplifier.

- observe experimentally the waveforms at various points in the circuit.

- determine experimentally the voltage gain of the amplifier.

EQUIPMENT AND MATERIALS

1 Audio generator
1 Dc power supply
1 Transistor, 2N3904
1 Resistor, 47 Ω, 2 W
2 Resistors, 4.7 kΩ, 2 W
4 Resistors, 10 kΩ, 2 W
2 Capacitors, 1 μF
1 Capacitor , 100 μF
1 Oscilloscope

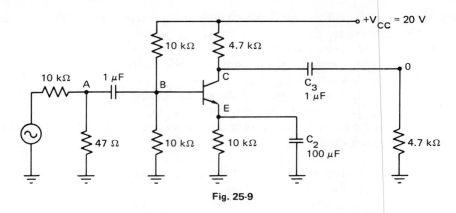

Fig. 25-9

PROCEDURE

1. Construct the circuit shown in figure 25-9.
    a. Turn on the dc power supply and adjust it to +20 V.
    b. Adjust the audio generator to 1 kHz. Set the output level so that a sinusoidal wave of 20 mV peak-to-peak is observed across the 47-Ω resistor. The oscilloscope must be connected between Point A and ground.
    c. Using the oscilloscope, measure the dc and ac voltages at Points A, B, C, E, and O, each with respect to ground. Record these voltage values.
2. Answer the following questions and perform the calculations required.
    a. Using the measured voltages, calculate the voltage gain from the base to the collector.
    b. Using the measured voltages, calculate the voltage gain from point A to Point O.
    c. What is the phase relationship between the voltage at Point A and the voltage at Point O?
    d. Is there any dc voltage at Point A? Explain the answer.
    e. What is the reason for connecting the 100-μF capacitor between the emitter and ground?

EXTENDED STUDY TOPICS

1. An amplifier has a power gain of 10 000. Find its bel power gain.
2. Calculate the gain of an amplifier having a bel power gain of 4.3 B.
3. Calculate the decibel gain of an amplifier having a power gain of 16.
4. The output of an amplifier changes from 24 W to 3 W. Find the dB change.
5. Three cascaded stages have the following gains: Stage 1, 20 dB; Stage 2, 15 dB; and Stage 3, −10 dB. Find the total dB power gain.
6. Find the dB voltage gain of an amplifier having a voltage gain of 200.
7. Find the power output of an amplifier whose output measures −6 dBm.
8. Calculate the current gain of a CE amplifier having a power gain of 40 dB, if its voltage gain is 200.
9. A CE amplifier has the following parameters: Z_{ie} = 2 kΩ, R_B = 250 kΩ, R_C = 5 kΩ, β = 75, and Z_{oe} = 1 MΩ. Calculate: a. the input impedance, b. the output impedance, c. A_v, d. A_i, e. A_i', f. G, and g. G'.
10. Express G and G' determined in topic 9 in dB notation.

Chapter 26

Field Effect Transistors and Configurations

OBJECTIVES

After studying this chapter, the student will be able to

- explain the operation of JFETs.
- describe the characteristics of the three JFET configurations.
- interpret the JFET data sheets.
- describe the construction and operation of MOSFETs.

THE JUNCTION FIELD EFFECT TRANSISTOR (JFET)

The junction field effect transistor is a unipolar transistor which can function using majority carriers only. The name of this device is frequently abbreviated to FET or JFET. The JFET is a voltage operated device. This means that the value of the input impedance approaches infinity. An n-channel JFET less bias voltage is shown in figure 26-1.

The majority carriers, or conduction-band electrons, flow through the n-type channel in the JFET. The ends of the n-channel are called the *drain* and *source*. When the drain and source terminals can be interchanged with no change in the operation of the device, the JFET is said to be symmetrical. The JFET is asymmetrical (nonsymmetrical) when the drain and source terminals are *not* interchangeable because the performance of the device would be affected.

As shown in figure 26-1, two p-regions are embedded in the sides of the n-channel of the JFET. Each of these p-regions is called a *gate*. The device is called a *dual-gate* JFET when the manufacturer connects a separate external lead to each gate. When the two gates are connected internally by the manufacturer, the device is called a *single-gate FET*. Diagrams of the two types of JFETs are

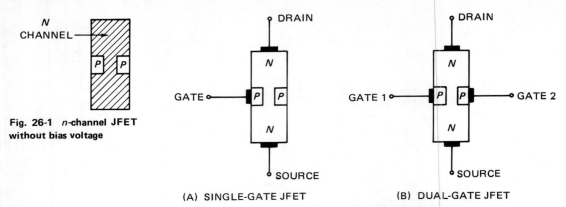

Fig. 26-1 *n*-channel JFET
without bias voltage

(A) SINGLE-GATE JFET   (B) DUAL-GATE JFET

Fig. 26-2 Two types of JFETs

shown in figure 26-2. A "black box" symbolic representation of a single-gate JFET is given in figure 26-3.

(R26-1)   What is Z_{in} for the ideal JFET shown in figure 26-3?

(R26-2)   Name the three terminals of a JFET.

(R26-3)   Why is a JFET called a unipolar transistor?

(R26-4)   Classify a BJT and a JFET as a voltage-operated device or a current-operated device.

A photomicrograph of a dual JFET is shown in figure 26-4. A JFET is shown in figure 26-5.

BIASING THE JFET   The polarities required to bias an *n*-channel JFET are shown in figure 26-6. The gate is reverse biased. The gate voltage is called V_{GS} when the source is used as a reference. The drain is reverse biased also. The drain voltage is called V_{DS} when the source is used as a reference.

(R26-5)   Compare the biasing polarities of the drain and gate of a JFET with the biasing polarities of the collector and base of a BJT.

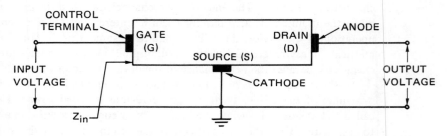

Fig. 26-3 "Black box" symbolic representation of a single-gate JFET

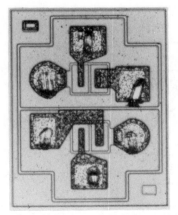

Fig. 26-5 JFET

Fig. 26-4 Photomicrograph of a dual JFET

As shown in figure 26-7, depletion layers or regions are formed around each *p-n* junction. These depletion layers are the reason why this type of device is known as a field effect transistor.

A study of figures 26-6 and 26-7 leads to the following conclusions:

1. The gate current is zero since the gate is reverse biased.
2. The current flows from the source to the drain through the narrow channel between the depletion layers. It was shown in Chapters 22 and 23 that the depletion layers do not contain majority conduction carriers.
3. The size of the depletion layers determines the width of the conducting channel.
4. The depletion layers are closer to each other when the gate voltage, V_{GS}, becomes more negative.
5. The gate voltage controls the current I_D between the drain and the source. As V_{GS} becomes more negative, I_D decreases.

(R26-6) Why is a JFET called a field effect transistor?

(R26-7) If V_{GS} becomes less negative, will I_D increase or decrease?

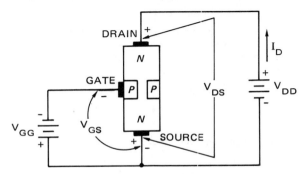

Fig. 26-6 Biasing an *n*-channel JFET

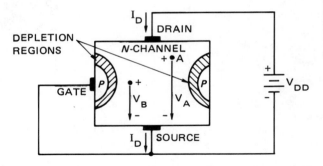

Note: In a single-gate JFET both P-regions have the same polarity since they are connected internally. The N-channel acts like a resistor due to its resistive characteristics. In the section of the channel between point A and the source, the drain current (I_D) causes a voltage drop. This voltage drop biases the gate with respect to that section of the channel. The gate regions are negative with respect to point A by the voltage V_A. The depletion regions penetrate into the channel at point A by an amount proportional to the negative bias V_A. The voltage drop V_B is less than V_A. Therefore, the depletion regions are narrower and penetrate the channel less at point B. This difference in voltage drops along the channel varies the gate-source bias and is responsible for the shape of the depletion regions penetrating the N-channel.

(A) Gate bias due to drain current creating internal voltage drops along the channel

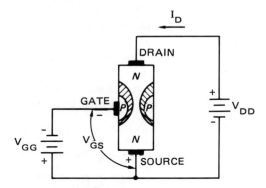

Note: An increase in the negative gate bias widens the depletion layers. As a result, they penetrate deeper into the N-channel. As the negative gate bias increases further, the depletion layers penetrate deeper and pinch off the current I_D

(B) Large negative gate bias

Fig. 26-7 Depletion layers around the p-n junctions of a JFET

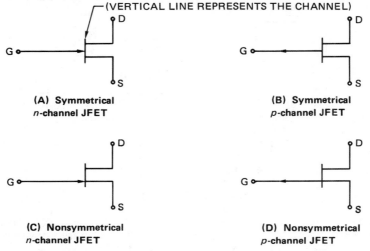

Fig. 26-8 Schematic symbols for *n*-channel and *p*-channel JFETs

SCHEMATIC
REPRESEN-
TATION

Figure 26-8 gives the schematic symbols for *n*-channel JFETs and *p*-channel JFETs. The *p*-channel JFET complements the *n*-channel JFET. The rest of this chapter describes *n*-channel JFETs only.

DRAIN CHARAC-
TERISTICS FOR
SHORTED-GATE
OPERATION

When the gate is connected directly to the source, the JFET has no external bias, and V_{GS} = 0. This condition is called *shorted-gate operation*, figure 26-9A. Figure 26-9B is a graph of the drain current versus the drain voltage during shorted-gate operation. The following observations can be made by studying this graph.

1. The curve is very similar to the collector curve of a BJT.
2. When V_{DS} = 0, I_D = 0.
3. When V_{DS} increases slightly, a small value of I_D flows. This small current causes a voltage drop along the channel. As a result, the gate-channel junctions are reverse biased by a small amount. There is very little penetration of the depletion regions into the channel and the channel resistance is only slightly affected.
4. As V_{DS} increases, the drain current shows an almost linear increase in Region A. This region is called the *channel ohmic region.*
5. The drain current levels off in Region B. Several names are given to this region, including the *normal*, or *active, operating* region, and the *pinchoff,* or *saturation*, region. The JFET acts like a constant current source in Region B.
6. The normal operating region for the JFET lies in the area between V_{DS} = V_P and V_{DS} = V_{DS}(max):
$$V_P < V_{DS} < V_{DS} \text{ (max)} \qquad \text{Eq. 26.1}$$
7. The drain current becomes constant above the value of V_{DS} known as the *pinchoff voltage*, V_p. When V_{DS} equals V_p, the channel becomes narrow and the depletion layers almost touch, figure 26-9C.

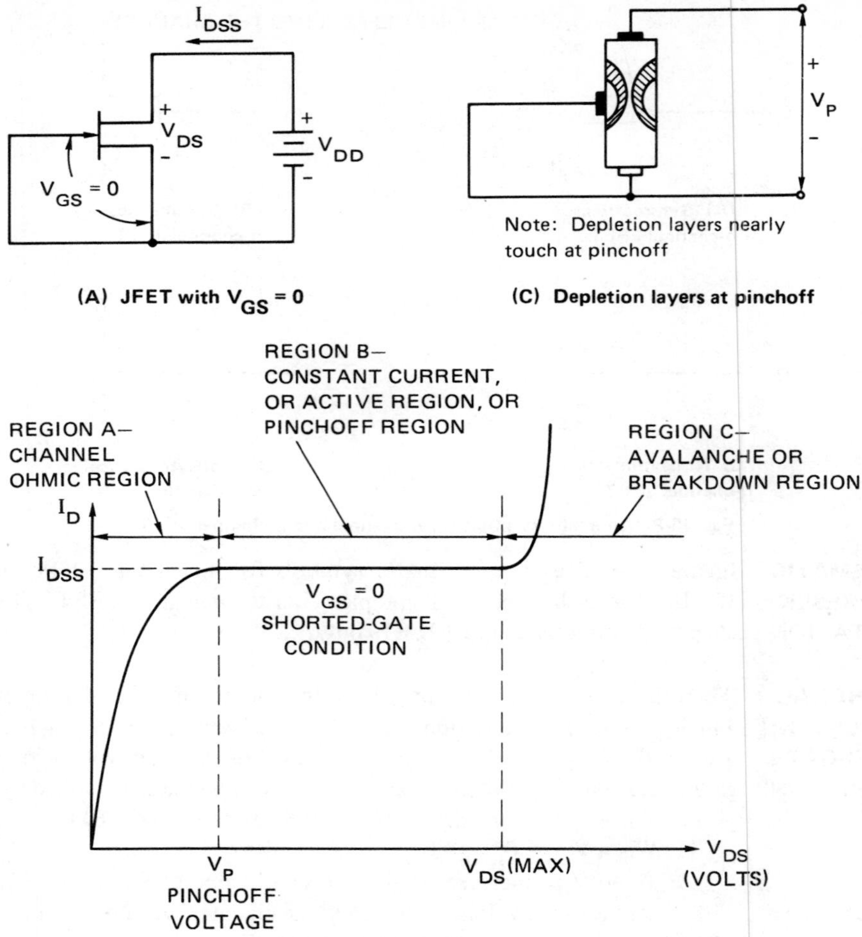

(A) JFET with $V_{GS} = 0$

(C) Depletion layers at pinchoff

Note: Depletion layers nearly touch at pinchoff

REGION A—
CHANNEL
OHMIC REGION

REGION B—
CONSTANT CURRENT,
OR ACTIVE REGION, OR
PINCHOFF REGION

REGION C—
AVALANCHE OR
BREAKDOWN REGION

$V_{GS} = 0$
SHORTED-GATE
CONDITION

V_P
PINCHOFF
VOLTAGE

V_{DS}(MAX)

V_{DS}
(VOLTS)

(B) Graph of I_D versus V_{DS}

Fig. 26-9 Shorted-gate operation

8. I_D in Region B is called the drain-to-source current in the shorted-gate condition, or I_{DSS}. Since I_{DSS} is measured in the shorted-gate condition, it is the maximum drain current that can be obtained for the normal operation of a JFET.

9. As V_{DS} increases to V_{DS}(max), the gate-channel junction breaks down and the drain current increases very rapidly. The JFET may be destroyed in this region which is known as the *avalanche*, or *breakdown, region*.

(R26-8) Assume that $V_{DS} = V_P$. a. Describe the status of the depletion layers of the JFET. b. What is the value of the drain current?

(R26-9) What is the maximum possible value of the drain current?

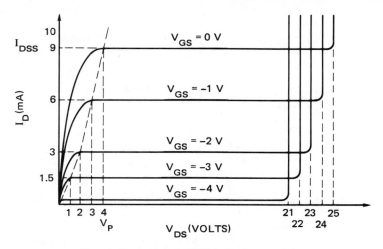

Fig. 26-10 *n*-channel JFET drain characteristics

(R26-10) For what values of V_{DS} does the JFET operate in the constant current region?

DRAIN CHARAC-
TERISTICS WITH
EXTERNAL BIAS

A family of I_D versus V_{DS} curves for a JFET can be obtained by using several values of negative external bias voltage. Figure 26-10 shows a typical family of characteristic curves. It can be seen that the value of V_{DS} for breakdown is reduced as the negative gate bias voltage is increased.

The dashed line shown in figure 26-10 is drawn through the points at which I_D saturates for each level of the gate bias voltage. When $V_{GS} = 0$, I_D saturates when $I_{DSS} = 9$ mA. At $V_{GS} = 0$, the pinchoff voltage, V_P, equals 4 V and the breakdown voltage equals 25 V. Therefore, the active region for this JFET is:

$$4 \text{ V} < V_{DS} < 25 \text{ V}$$

The bottom curve of figure 26-10 is important because it shows the value of V_{GS} at which the drain current is nearly zero. This voltage is called the *gate-source cutoff voltage* and is designated as V_{GS}(off).

When $V_{GS} = V_{GS}$(off), the depletion layers touch causing the drain current to be almost zero. Recall that V_P is the value of drain voltage that pinches off I_D for shorted-gate operation. Therefore,

$$V_P = |V_{GS}\text{ (off)}| \qquad \text{Eq. 26.2}$$

Manufacturers' data sheets normally list the value of V_{GS}(off), but not V_P. For the JFET whose characteristics are shown in figure 26-10, the data sheet lists a value of –4 V for V_{GS}(off). This means that $V_P = +4$ V.

(R26-11) Using figure 26-10, find the normal range of a. the drain voltage, b. the gate voltage, and the drain current.

TRANSFER CHAR-
ACTERISTICS

The JFET transfer characteristics can be determined experimentally by maintaining a constant value of V_{DS} and varying V_{GS}. Another way of obtaining these characteristics is to use the output characteristics of figure 26-10. Corres-

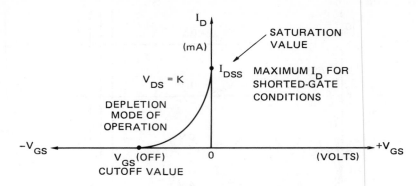

Fig. 26-11 Transconductance curve for a typical JFET

ponding values of V_{GS} and I_D can be read from these curves for a constant value of V_{DS}. The rssulting graph of I_D versus V_{GS} relates the output current to the input voltage and is called a *transconductance curve.*

Figure 26-11 shows the transconductance curve for a typical JFET. This curve is part of a parabola whose equation is:

$$I_D = I_{DSS} \left[1 - \frac{V_{GS}}{V_{GS}(\text{off})} \right]^2 \qquad \text{Eq. 26.3}$$

For a JFET, the output current varies as the square of the input voltage. Therefore, JFETs are known as square-law devices.

Problem 1   A JFET has the following characteristics: I_{DSS} = 16 mA and $V_{GS}(\text{off})$ = –8 V. Calculate I_D when V_{GS} = –2 V.

Solution $$\begin{aligned} I_D &= I_{DSS} \left[1 - \frac{V_{GS}}{V_{GS}(\text{off})} \right]^2 \\ &= 16 \text{ mA} \left[1 - \frac{(-2 \text{ V})}{(-8 \text{ V})} \right]^2 \\ &= 16 \text{ mA} \left[1 - \frac{1}{4} \right]^2 \\ &= 16 \text{ mA} \left(\frac{3}{4} \right)^2 \\ &= 9 \text{ mA} \end{aligned}$$

(R26-12)   Find I_D for the JFET of Problem 1, when V_{GS} = –6 V.

The transconductance, g_m, of a JFET is expressed by the following equation:

$$g_m = \frac{\Delta I_D}{\Delta V_{GS}} \text{ for a constant } V_{DS} \qquad \text{Eq. 26.4}$$

where

g_m = transconductance (The value of this parameter formerly was expressed in μmhos. The accepted SI unit for transconductance is the microsiemens, μS.)

ΔI_D = change in the drain current

ΔV_{GS} = corresponding change in the gate voltage

Problem 2   A change in the gate voltage of 0.1 V produces a change in the drain current of 0.4 mA. V_{DS} is a constant 15 V. Find g_m.

Solution $g_m = \dfrac{\Delta I_D}{\Delta V_{GS}}$

$= \dfrac{0.4\ mA}{0.1\ V}$

$= 4(10^{-3})$ S

$= 4\ 000\ \mu S$

Some manufacturers refer to the transconductance as the *forward transfer admittance*, or the *transadmittance*. In this case, the symbol Y_{fs} is used rather than g_m.

The drain resistance (r_d) is the ac resistance between the drain and the source terminals when the JFET is operating in the pinchoff region. The drain resistance is expressed as:

$$r_d = \frac{\Delta V_{DS}}{\Delta I_D}\ \text{for a constant}\ V_{GS} \qquad\qquad \text{Eq. 26.5}$$

Problem 3   A change in the drain-to-source voltage of 3 V produces a change in the drain current of 0.03 mA. V_{GS} is a constant –2 V. Find the ac drain resistance.

Solution $r_d = \dfrac{\Delta V_{DS}}{\Delta I_D}$

$= \dfrac{3\ V}{0.03\ mA}$

$= 100\ k\Omega$

(R26-13)   Calculate r_d if a 3-V change in the drain-to-source voltage produces a 0.3-mA change in the drain current.

Some manufacturers list a value for the *output admittance*, or the *output conductance*, Y_{os}, rather than r_d. When Y_{os} is given, the following equation shows how r_d is calculated:

$$r_d = \frac{1}{Y_{os}} \qquad\qquad \text{Eq. 26.6}$$

where Y_{os} = the output admittance, or the output conductance.

UNDERSTANDING JFET DATA SHEETS   The data sheets for a JFET (figure 26-4) are shown in figure 26-12.

(R26-14)   Refer to the data sheets of figure 26-12 and find:
a. the maximum drain-to-source voltage.
b. the maximum power dissipation.
c. the maximum g_m for typical operation.
d. the maximum I_{DSS}.
e. the maximum operating gate current.
f. the maximum pinchoff voltage.

ULTRA LOW NOISE . . . . $e_n = 8nV/\sqrt{Hz}$ TYP.

LOW LEAKAGE . . . . $I_G = 50pA$ max.

LOW DRIFT . . . . $\left|\dfrac{\Delta V_{GS_{1-2}}}{\Delta T}\right| = 5\mu V/°C$ max.

LOW OFFSET VOLTAGE . . . . $|V_{GS_{1-2}}| = 5mV$ max.

LINEAR TEMPERATURE TRACKING   TDN = $\pm1\mu V/°C$

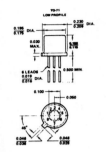

NOTES:
All dimensions in inches.
Leads are gold-plated KOVAR.
Package weight is 0.60 grams.
Lead Configuration on page 6.

ABSOLUTE MAXIMUM RATINGS (Note 1)
@ 25°C (unless otherwise noted)

Maximum Temperatures
Storage Temperature	-65° to	+150°C
Operating Junction Temperature		+150°C
Lead Temperature (Soldering, 10 second time limit)		+300°C

Maximum Power Dissipation
Device Dissipation @ Free Air-Total	400mW

Maximum Voltage and Current for Each Transistor
$-V_{GSS}$	Gate to Drain or Source Voltage	40V
$-V_{DSO}$	Drain to Source Voltage	40V
$-I_{G(f)}$	Forward Current	50mA

ELECTRICAL CHARACTERISTICS @ 25°C (unless otherwise noted)

SYMBOL	CHARACTERISTICS	MP 840	MP 841	MP 842	UNITS	CONDITIONS
$\left|\dfrac{\Delta V_{GS_{1-2}}}{\Delta T}\right|$ max.	Drift vs Temperature	5	10	40	$\mu V/°C$	$V_{DG} = 20V$, $I_D = 200\mu A$ $T_A = -55°C$ to $+25°C$ to $+125°C$
$|V_{GS_{1-2}}|$ max.	Offset Voltage, +25°C	5	5	25	mV	
TDN typ	Temp Drift Nonlinearity	±1	±1	±1	$\mu V/°C$	$V_{DG} = 20V$, $I_D = 200\mu A$
TDN max.		±3	±3	±3	$\mu V/°C$	$T_A = -55°C$ to $+25°C$ to $+125°C$

Notes and Additional Electrical Characteristics on next page.

Fig. 26-12 Data sheets for a JFET

SYMBOL	CHARACTERISTICS	MIN.	TYP.	MAX.	UNITS	CONDITIONS
Y_{fss}	Transconductance Full Conduction	1000		4000	μmho	V_{DG} = 20V, V_{GS} = 0, f = 1kHz
Y_{fs}	Typical Operation	500		1000	μmho	
$\left\|\dfrac{Y_{fs_{1-2}}}{Y_{fs}}\right\|$	Mismatch	–	0.6	3	%	V_{DG} = 20V, I_D = 200μA
I_{DSS}	Drain Current Full Conduction	0.5	2	5	mA	V_{DG} = 20V, V_{GS} = 0
$\left\|\dfrac{I_{DSS_{1-2}}}{I_{DSS}}\right\|$	Mismatch at Full Conduction		1	5	%	
$-I_G$	Gate Current Operating	–	20	50	pA	V_{DG} = 20V, I_D = 200μA
$-I_G$	High Temperature	–	–	50	nA	V_{DG} = 20V, I_D = 200μA, T_A = +125°C
$-I_G$	Reduced V_{DG}	–	5	–	pA	V_{DG} = 10V, I_D = 200μA
I_G (f) D*	Forward Current	–	–	50	mA	Any Condition
$-I_{GSS}$	At Full Conduction	–	–	100	pA	V_{DG} = 20V, V_{DS} = 0
Y_{oss}	Output Conductance Full Conduction	–	–	10	μmho	V_{DG} = 20V, V_{GS} = 0
Y_{os}	Operating	–	0.1	1	μmho	V_{DG} = 20V, I_D = 200μA
$\|Y_{os_{1-2}}\|$	Differential	–	0.01	0.1	μmho	
CMR	Common Mode Rejection $-20 \log \left\|\dfrac{\Delta V_{GS_{1-2}}}{\Delta V_{DS}}\right\|$	–	100	–	dB	ΔV_{DS} = 10 to 20V, I_D = 200μA
CMR		–	75	–	dB	ΔV_{DS} = 5 to 10V, I_D = 200μA
V_{GS} (off) or V_P	Gate Voltage Pinchoff Voltage	1	2	4.5	V	V_{DS} = 20V, I_D = 1nA
V_{GS}	Operating Range	0.5	–	4	V	V_{DS} = 20V, I_D = 200μA
BV_{GSS}	Breakdown Voltage	40	–	–	V	V_{DS} = 0, I_D = 1nA
V_{GSS} D*	To Source or Drain	–	–	40	V	Any Condition
V_{GGO}	Gate-to-Gate Breakdown	40	–		V	I_G = 1nA, I_D = 0, I_S = 0
V_{DSO} D*	Drain-Source Voltage	–	–	40	V	Any Condition
NF	Noise Figure	–	–	0.5	dB	V_{DS} = 20V, V_{GS} = 0, R_G = 10MΩ f = 100Hz, NBW = 6Hz
e_n	Voltage	–		15	nV/$\sqrt{\text{Hz}}$	V_{DS} = 20V, I_D = 200μA, f = 10Hz NBW = 1Hz
e_n	Voltage			10	nV/$\sqrt{\text{Hz}}$	V_{DS} = 20V, I_D = 200μA, f = 1kHz NBW = 1Hz
C_{iss}	Capacitance Input	–	–	25	pF	V_{DS} = 20V, V_{GS} = 0, f = 1MHz
C_{rss}	Reverse Transfer	–	–	5	pF	
C_{dd}	Drain to Drain	–	0.1	–	pF	V_{DG} = 20V, I_D = 200μA
T_S D*	Temperature Storage	-65	–	+150	°C	Any Condition
T_J D*	Junction	–	–	+150	°C	Any Condition
T_L D*	Lead	–	–	+300	°C	10 sec. max.-1/16" or more from case
P_D D*	Dissipation - both sides	–	–	400	mW	T_A = +25°C, Derate 3.3mW/°C

*Note: These ratings are limiting values above which the serviceability of any semiconductor may be impaired.

Ⓡ Applied MATERIALS TECHNOLOGY, INC.

Fig. 26-12 Data sheets for a JFET (cont'd)

g. the maximum $V_{GS}(\text{off})$.

h. the maximum r_d for full conduction.

BASIC JFET CIRCUITS

The JFET can be connected in three basic modes, or configurations, as follows: 1. common source, 2. common drain, or 3. common gate. The common-source mode is commonly used because of its voltage amplification feature and high input impedance.

(R26-15) Name the two advantages of the common-source JFET mode.

The Common-source Configuration. A common-source (CS) JFET amplifier is shown in figure 26-13. This circuit is the JFET equivalent of the BJT common-emitter circuit.

Problem 4
Solution

If $I_D = 1.5$ mA in figure 25-13, find the value of V_{DS}.

V_{DS} = dc voltage from the drain to ground

$V_{DD} = V_{DS} + I_D R_L$ (This expression is obtained by applying Kirchhoff's Voltage Law in the output circuit.)

$$\begin{aligned}
V_{DS} &= V_{DD} - I_D R_L \\
&= 20 \text{ V} - (1.5 \text{ mA})(10 \text{ k}\Omega) \\
&= 20 \text{ V} - 15 \text{ V} \\
&= 5 \text{ V}
\end{aligned}$$

The ac voltage gain is the ratio of the output ac voltage to the input ac voltage. The ac voltage gain for the CS JFET mode is given by the following equation:

$$A_V \cong -g_m R_L \qquad \text{Eq. 26.7}$$

(R26-16) What is indicated by the minus sign in Equation 26.7?

Problem 5 The CS amplifier of figure 25-13 has a $g_m = 3\,000\ \mu S$. Calculate the ac voltage gain.

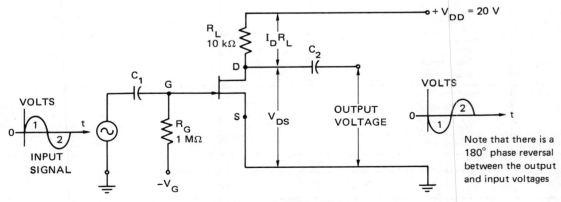

Note that there is a 180° phase reversal between the output and input voltages

Fig. 26-13 CS JFET amplifier

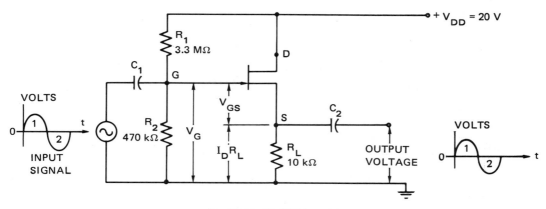

Fig. 26-14 CD JFET amplifier

Solution $A_V \cong -g_m R_L$
$\cong -(3\ 000 \times 10^{-6}\ S)(10 \times 10^3\ \Omega)$
$\cong -30$

(R26-17) A CS amplifier uses a JFET with a g_m of 4 000 μS. If $R_L = 7\ k\Omega$, calculate A_V.

The Common-drain Configuration. A common-drain (CD) JFET amplifier is shown in figure 26-14. This circuit is the JFET equivalent of the BJT common-collector circuit. The amplifier has a voltage gain of nearly one. There is no phase shift between the input and the output. Since the output voltage changes follow the input voltage changes approximately, this amplifier is called a *source follower*.

Problem 6 If $I_D = 1$ mA and $V_{GS} = -2$ V in figure 26-14, find the value of V_G.
Solution V_G = dc voltage from the gate-to-ground
$V_G = V_{GS} + I_D R_L$ (This expression is obtained by applying Kirchhoff's Voltage Law in the input circuit)
$V_G = (-2\ V) + (1 \times 10^{-3}\ A)(10\ k\Omega)$
$= -2\ V + 10\ V$
$= 8\ V$

(R26-18) Is the voltage gain of a CD amplifier equal to 1, greater than 1, or less than 1?

The Common-gate Configuration. A common-gate (CG) JFET amplifier is shown in figure 26-15. This circuit is the JFET equivalent of the BJT common-base circuit. Voltage is amplified in this circuit and there is no phase shift between the input and output. A major disadvantage of this circuit is that it has a very low input impedance.

Problem 7 Find the value of V_D in figure 26-15, if $I_D = 1$ mA.
Solution V_D = dc voltage from drain to ground.
$V_{DD} = V_D + I_D R_L$ (This expression is obtained by applying Kirchhoff's Voltage Law in the output circuit)

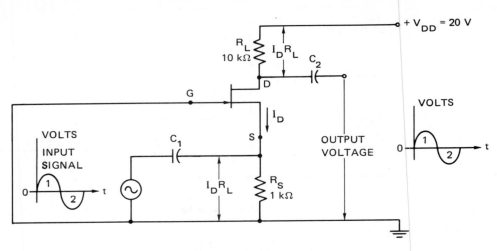

Fig. 26-15 CG JFET amplifier

$$V_D = V_{DD} - I_D R_L$$
$$= 20\,V - (1 \times 10^{-3}\,A)(10\,k\Omega)$$
$$= 20\,V - 10\,V$$
$$= 10\,V$$

The ac voltage gain for the CG mode is expressed as follows:

$$A_V \cong g_m R_L \qquad\qquad\qquad\text{Eq. 26.8}$$

(R26-19)   Compare Equation 26.8 with Equation 26.7.

Problem 8   The CG amplifier of figure 26-15 has a $g_m = 3\,000\,\mu S$. Calculate the ac voltage gain.

Solution $A_V \cong g_m R_L$
$$\cong (3\,000 \times 10^{-6}\,S)(10 \times 10^3\,\Omega)$$
$$\cong 30$$

(R26-20)   A CG amplifier uses a JFET with a g_m of $4\,000\,\mu S$. Calculate A_V if $R_L = 6.8$ $k\Omega$.

THE INSULATED GATE FET OR MOSFET   Figure 26-16 shows the structure of an insulated gate FET (IGFET) or a metal oxide semiconductor FET (MOSFET). The biasing polarities for an IGFET with a created channel are shown in figure 26-17. The conductivity of the created channel is increased (enhanced) by the positive bias of the gate. As a result, the full name of this device is "enhancement mode IGFET."

Two symbols for the n-channel IGFET are shown in figure 26-18. Note that the gate does not contact the channel directly. Figure 26-18A shows the source and substrate connected internally. Figure 26-18B shows the substrate connection brought out independently of the source. The line representing the channel is shown in three sections. This means that the channel does not exist until a positive gate voltage is applied.

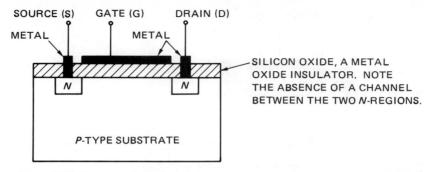

Fig. 26-16 Structure of an insulated gate FET or metal oxide semiconductor FET (MOSFET)

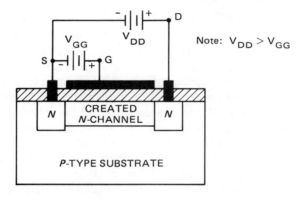

Note: $V_{DD} > V_{GG}$

THE *N*-CHANNEL IS CREATED AS FOLLOWS:

1. Negative charge carriers are induced in the *p*-type substrate when V_{GG} is applied to make the gate positive with respect to the source and the substrate.
2. The induced charge carriers are electrons which are attracted to the positive voltage on the metal plate. However, the silicon oxide insulator prevents the electrons from traveling to the gate terminal. A capacitor, in effect, is created by the gate, silicon oxide dielectric, and the *n*-channel. Therefore, the electrons accumulate on the surface of the substrate, just below the metal plate, and create an *n*-channel between the two *n*-regions.
3. The *n*-channel extends between the drain and the source.

Fig. 26-17 Biasing polarities for *n*-channel enhancement mode IGFET

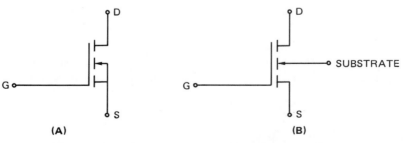

Fig. 26-18 Symbols for *n*-channel enhancement mode IGFET

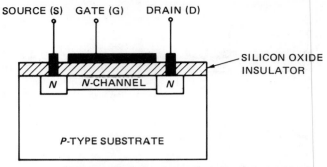

Fig. 26-19 *n*-channel depletion-enhancement mode IGFET

(R26-21)   What do the letters IGFET represent?

The structure of an *n*-channel depletion-enhancement mode IGFET is shown in figure 26-19. Basically, this structure is identical to the enhancement mode IGFET shown in figure 26-16. However, a lightly doped *n*-type channel exists between the heavily doped drain and the source. If the drain is made positive with respect to the source, the drain current will flow even when the gate voltage is zero. As a result, the device operates in the enhancement mode. If the gate is made negative with respect to the source, positive charge carriers will be induced in the *n*-type channel. These carriers will combine with free electrons, causing an increase in the channel resistance and a decrease in the drain current. The device operates in the depletion mode because the channel is depleted of free electrons. The symbols for the depletion-enhancement mode IGFET are similar to those for the enhancement mode device. However, note in figure 26-20 that the line indicating the channel is now solid. (Compare this figure with figure 26-18 for the enhancement mode IGFET.)

A MOSFET is shown in figure 26-21. The drain characteristics of an *n*-channel enhancement mode IGFET are illustrated in figure 26-22. The drain characteristics for an *n*-channel depletion-enhancement mode IGFET are shown in figure 26-23.

(R26-22)   Refer to figure 26-23. If V_{DS} = 8 V, how does operation in the enhancement mode differ from operation in the depletion mode with regard to the resulting drain current?

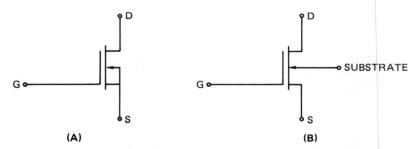

Fig. 26-20 Symbols for the *n*-channel depletion-enhancement mode IGFET

Fig. 26-21 MOSFET

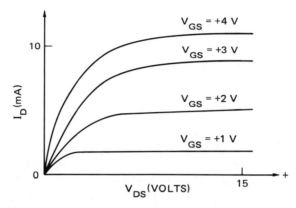

Fig. 26-22 Drain characteristics for an *n*-channel enhancement mode IGFET

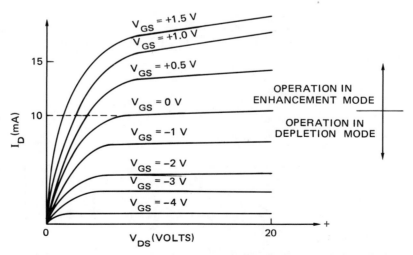

Fig. 26-23 Drain characteristics for an *n*-channel depletion-enhancement mode IGFET

LABORATORY EXERCISE 26-1 CHARACTERIS- TICS AND OPERA- TION OF JFET AMPLIFIERS

PURPOSE   By completing this exercise, the student will study experimentally the character- istics and operation of JFET amplifiers.

EQUIPMENT AND MATERIALS
1 Dc power supply
1 Audio generator
1 JFET, MPF 102
1 Resistor, 1.2 kΩ, 2 W
1 Resistor, 2.2 kΩ, 2 W
1 Resistor, 5.6 kΩ, 2 W
1 Resistor, 220 kΩ, 2 W
1 Capacitor, 100 μF
1 Oscilloscope

PROCEDURE
1. Construct the circuit shown in figure 26-24.
   a. Adjust the audio generator to 1 kHz.
   b. Using the oscilloscope, set the output voltage of the audio generator to measure 0.1 V peak-to-peak across the 220-kΩ resistor.
   c. Measure the output voltage using the oscilloscope.
2. Construct the circuit shown in figure 26-25.
   a. Adjust the audio generator to 1 kHz.
   b. Using the oscilloscope, set the output of the audio generator to measure 1 V peak-to-peak across the 220-kΩ resistor.
   c. Measure the output voltage using the oscilloscope.

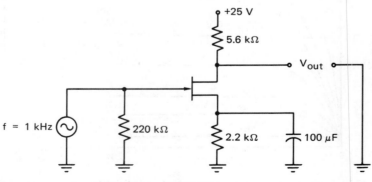

Fig. 26-24

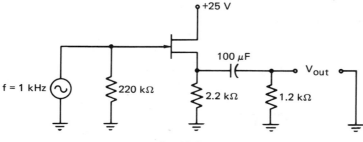

Fig. 26-25

3. Answer the following questions, and perform the calculations requested for the circuit in figure 26-24.
   a. Calculate the voltage gain from the readings taken in the laboratory.
   b. Calculate the voltage gain from the values of the components. The JFET used in the circuit has a minimum g_m of 2 000 μS and a maximum g_m of 7 500 μS. (Assume g_m = 5 000 μS.)
   c. In what mode is this circuit operating?
4. Answer the following questions, and perform the calculations requested for the circuit in figure 26-25.
   a. Answer questions 3.a., 3.b., and 3.c.
   b. What is the difference between the circuit of figure 26-25 and the circuit of figure 26-24?

EXTENDED STUDY TOPICS

1. Define: a. symmetrical JFET and b. nonsymmetrical JFET.
2. Compare the three terminals of a JFET with the corresponding terminals of a BJT.
3. Describe the conditions which result when $V_{DS} = V_p$ in a JFET operating in the shorted-gate condition.
4. Why are JFETs called square-law devices?
5. A JFET has the following characteristics: I_{DSS} = 20 mA, $V_{GS}(off)$ = –8 V. Calculate the drain current when V_{GS} = –4 V.
6. The JFET operates in the (enhancement) (depletion) mode. (Select the correct word to complete the statement.)
7. If a JFET has a drain resistance of 10 kΩ, when ΔV_{DS} = 3 V, find the change in the drain current.
8. Compare the JFET configurations with the corresponding BJT modes.
9. A CF JFET amplifier is designed to have a voltage gain of 50. Calculate the value of R_L required if g_m = 2 000 μS.
10. A CG JFET amplifier is designed to have a voltage gain of 60. Calculate the value of R_L required if g_m = 3 000 μS.
11. a. An n-channel depletion-enhancement mode IGFET has a value of V_{GS} = +1 V. In what mode is the device operating?
    b. In what mode does the IGFET operate when V_{GS} is changed to –1 V?

Section 4
Applied Electronics Technology

Chapter 27

Communication Circuits

OBJECTIVES

After studying this chapter, the student will be able to

- state basic feedback concepts.
- explain the characteristics and the operation of sinusoidal oscillators.
- describe the Hartley and Colpitts oscillators.
- analyze an AM wave.
- discuss the principles of AM transmission.
- explain the principles that govern the operation of the superheterodyne AM radio receiver.
- explain the concepts of frequency modulation.
- describe the flow of video and sound signals through a television receiver.

FEEDBACK CONCEPTS

Figure 27-1 shows an amplifier with feedback and a gain of A_f. The derivation of the gain formula for this amplifier is also shown in the figure. The gain is:

$$A_f = \frac{A}{1 - A\beta} \qquad \text{Eq. 27.1}$$

where

$\quad$ A $\quad$ = open-loop amplifier gain with switch S open

$$\quad\quad = \frac{V_O}{V_{in}}$$

$\quad$ β $\quad$ = the fraction of the output voltage fed back to the input when switch S is closed. β is also known as the *feedback factor*.

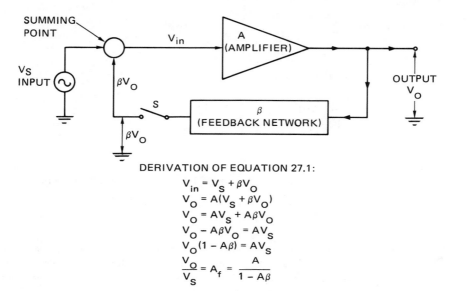

SUMMING POINT

V_S INPUT

βV_O

βV_O

V_{in}

A (AMPLIFIER)

S

β (FEEDBACK NETWORK)

OUTPUT V_O

DERIVATION OF EQUATION 27.1:

$$V_{in} = V_S + \beta V_O$$
$$V_O = A(V_S + \beta V_O)$$
$$V_O = AV_S + A\beta V_O$$
$$V_O - A\beta V_O = AV_S$$
$$V_O(1 - A\beta) = AV_S$$
$$\frac{V_O}{V_S} = A_f = \frac{A}{1 - A\beta}$$

Eq. 27.1

Fig. 27-1 An amplifier with feedback and a gain of A_f

$$A_f = \text{closed-loop amplifier gain with switch S closed} = \frac{V_o}{V_s}$$

There are three operating modes for the amplifier in figure 27-1.

1. $\beta A < 0$ (BA is negative). In this condition, A_f is less than A. Therefore, the circuit is a negative feedback amplifier. (Applications of negative feedback amplifiers are given in Chapters 28 and 33.)

2. $0 < \beta A < 1$. In this condition, A_f is greater than A. The circuit is a positive feedback amplifier.

3. $\beta A = +1$. In this condition, A_f approaches infinity. As a result, the circuit goes into sustained oscillation. This condition is known as the *Barkhausen criterion for oscillation.*

   Note: According to Equation 27.1, A_f has a mathematical value of infinity when $\beta A = +1$. However, this does not happen electrically and the circuit goes into sustained oscillation. The infinite gain means that the oscillator can supply its own signal and achieve self-sustained operation. The amplifier amplifies the signal input, V_s. A portion of the output voltage, βV_o, is fed back to the amplifier. This feedback voltage has the same amplitude and phase as the input signal, V_s. Therefore, the input signal can be removed and the circuit continues to oscillate. The output voltage V_o is a sinusoidal waveform.

(R27-1) Define: a. open-loop amplifier, and b. closed-loop amplifier.

(R27-2) Define: a. negative feedback amplifier, and b. positive feedback amplifier.

Problem 1 The single-stage CE amplifier shown in figure 27-1 is connected to a feedback network. Find A_f when $A = -100$ and $\beta = \frac{1}{10}$.

Solution $A_f = \dfrac{A}{1 - A\beta}$

$\qquad\qquad = \dfrac{-100}{1 - (-100)\left(\dfrac{1}{10}\right)}$

$\qquad\qquad = \dfrac{-100}{1 - (-10)}$

$\qquad\qquad = \dfrac{-100}{11}$

$\qquad\qquad = -9.1$

(R27-3) Is the amplifier shown in figure 27-1 a negative feedback amplifier or a positive feedback amplifier?

SINUSOIDAL OSCILLATORS A sinusoidal oscillator provides an ac sinusoidal output with no ac input. The input for this oscillator is the dc power supply voltage used to bias the transistors. A sinusoidal oscillator amplifier has positive, or regenerative, feedback. To sustain oscillation, two conditions must be met.

1. The feedback must be positive. That is, the feedback energy and the input energy must be in phase, figure 27-2.
2. The feedback must be large enough to overcome the internal losses in the circuit and prevent the oscillations from dying away.

The diagram of figure 27-3 shows a system containing a resonant circuit to provide oscillation. The constants of the LC resonant, or tank, circuit determine the frequency of oscillation. An adjustable radio-frequency oscillator coil is shown in figure 27-4. This coil is used in the oscillator circuit of broadcast band receivers. The transistors shown in figure 27-5 are used in the amplifier section of these receivers.

(R27-4) List the three basic parts of a sinusoidal oscillator system.

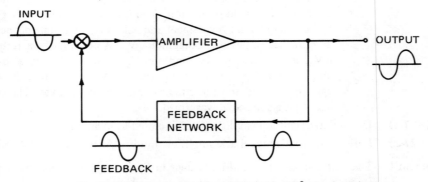

Fig. 27-2 Positive feedback amplifier (with 180° phase shift)

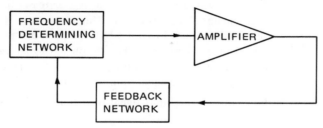

Fig. 27-3 Sinusoidal oscillator system

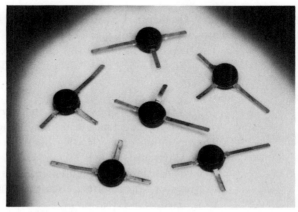

Fig. 27-5   Transistors used in amplifier section of broadcast band receivers

Fig. 27-4 Radio-frequency oscillator coil

STARTING THE SINUSOIDAL OSCILLATOR

The sinusoidal oscillator depends upon the *noise voltages* generated in the resistors in the circuit for its starting voltage because it has no external input. Each resistor acts like a small voltage source due to the random motion of electrons. A band of sinusoidal frequencies is produced with an upper value of more than 10^{12} Hz. The following sequence of actions occurs in starting the oscillator.

1. When the dc power supply is turned on, the only signals in the system are the low-amplitude noise voltages.
2. The noise voltages are amplified and appear at the output terminals.
3. The amplified noise is returned to the feedback network and the resonant tank circuit.
4. The amplified noise is filtered. One sinusoidal frequency is more favored than other frequencies and enters the amplifier with the correct phase for positive feedback.
5. The loop gain $A\beta$ is greater than unity, causing oscillations to build up at the frequency fed back.

6. When the desired level is reached, $A\beta$ decreases to unity, and the output signal remains at a constant amplitude.

Note: $A\beta$ is reduced to unity by the action of a nonlinear resistor placed in the feedback network. When the output signal reaches the proper value, the nonlinear resistor causes β to decrease. Thus, the oscillator automatically makes $A\beta$ equal to unity after the oscillations build up.

(R27-5) What is the source of the input signal to the sinusoidal oscillator system shown in figure 27-3?

(R27-6) Why is an oscillator called a self-excited amplifier?

OSCILLATOR CIRCUITS

Hartley Oscillator. Circuits for both the Hartley oscillator and the Colpitts oscillator are shown in figure 27-6.

The frequency of the Hartley circuit is expressed as:

$$f_o \cong \frac{1}{2\pi \sqrt{LC}}$$ Eq. 27.2

where

f_o = frequency of oscillation, in hertz
L = total inductance of the coil, in henries

The operation of the Hartley oscillator circuit, figure 27-6A, is described in the following steps:

1. When a resonant circuit is used as the feedback element, in place of a resistor, the resulting circuit oscillates at the resonant frequency determined from Equation 27.2.
2. Components R_L, R_B, and C_C function as they do in any CE amplifier.
3. Components L_1, L_2 and C form the parallel resonant circuit, or tank circuit. This circuit is the frequency determining network. L_1 and L_2 are

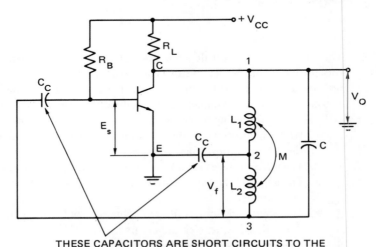

THESE CAPACITORS ARE SHORT CIRCUITS TO THE
AC POTENTIAL AT THE FREQUENCY OF OSCILLATION

Fig. 27-6A Hartley oscillator circuit

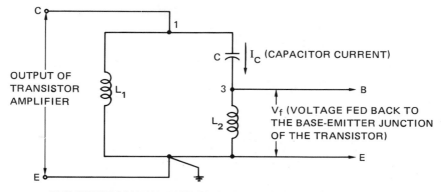

THE SERIES COMBINATION OF C AND L_2 IS DESIGNED SO THAT I_C LEADS THE VOLTAGE V_{12} ACROSS THE TANK CIRCUIT BY APPROXIMATELY 90°. I_C CAUSES A VOLTAGE DROP ACROSS L_2. THIS VOLTAGE IS IN PHASE WITH E_S AND IS THE FEEDBACK VOLTAGE, V_f.

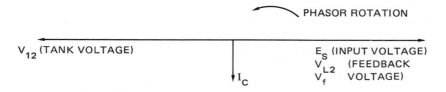

Fig. 27-6B Phasor analysis of Hartley oscillator circuit

mutually coupled so that the total inductance L is used in Equation 27.2, where $L = L_1 + L_2 + 2M$.

4. Feedback is provided by the inductive voltage divider. The voltage across L_2 is fed back to the input.

5. The phasor diagram of figure 27-6B shows that the CE amplifier provides a 180° phase shift to the tank circuit.

6. The load for the transistor consists of L_1 in parallel with the series combination of L_2 and C. Note that terminal 2 of the tank circuit is at the ac ground potential.

7. An additional 180° phase shift is provided by the resonant circuit. As a result, the feedback voltage, V_f, is in phase with the input voltage, E_s. The student must be aware that the ac input voltage E_s is a self-generated voltage. E_s is not supplied externally.

8. The amplifier performs one 180° phase inversion and the resonant circuit performs the additional 180° phase inversion. The result of these inversions is positive feedback. In other words, the feedback voltage V_f is in phase with the self-generated input voltage E_s.

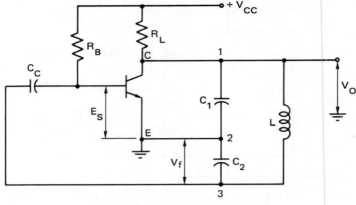

Fig. 27-6C Colpitts oscillator circuit

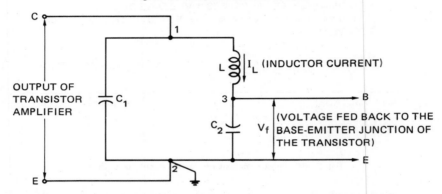

THE SERIES COMBINATION OF L AND C_2 IS DESIGNED SO THAT I_L LAGS
THE VOLTAGE V_{12} ACROSS THE TANK CIRCUIT BY APPROXIMATELY 90°.
I_L CAUSES A VOLTAGE DROP ACROSS C_2. THIS VOLTAGE IS IN PHASE
WITH E_S AND IS THE FEEDBACK VOLTAGE V_f.

Fig. 27-6D Phasor analysis of Colpitts oscillator circuit

The operation of the Colpitts oscillator circuit, figure 27-6C, is described in
the following steps:

1. The resonant frequency for this circuit is also determined from Equation
   27.2.
2. Components R_L, R_B, and C_C function as they do in any CE amplifier.

3. Components L, C_1, and C_2 form the parallel resonant circuit, or tank circuit. This circuit is the frequency determining network. The series equivalent (C) of capacitors C_1 and C_2 is used in Equation 27.2, where

$$C = \frac{C_1 C_2}{C_1 + C_2}.$$

4. Feedback is provided by the capacitive voltage divider. The voltage across C_2 is fed back to the input.

5. The phasor diagram of figure 27-6D shows that the CE amplifier provides a $180°$ phase shift to the tank circuit.

6. Since terminal 2 of the tank circuit is at the ac ground potential, the load for the transistor is C_1 in parallel with the series combination of L and C_2.

7. An additional $180°$ phase shift is provided by the resonant circuit. As a result, the feedback voltage, V_f, is in phase with the input voltage, E_s. As in the Hartley oscillator, the ac input voltage E_s is a self-generated voltage and is not supplied by an external source.

8. The amplifier performs one $180°$ phase inversion and the resonant circuit performs the additional $180°$ phase inversion, resulting in positive feedback. This means that the feedback voltage V_f is in phase with the self-generated input voltage E_s.

Problem 2 An 800-μH coil and a 100-pF capacitor are used in the Hartley oscillator circuit, figure 27-6A. Calculate the frequency of oscillation.

Solution $f_o \cong \dfrac{1}{2\pi \sqrt{LC}}$

$\cong \dfrac{1}{2\pi \sqrt{(800 \times 10^{-6}\ \text{H})(100 \times 10^{-12}\ \text{F})}}$

$\cong 563\ \text{kHz}$

(R27-7) How is feedback obtained in the Hartley oscillator?

(R27-8) How is feedback obtained in the Colpitts oscillator?

AMPLITUDE MODULATION Modulation is a process in which one wave shape is changed or modified by means of another wave shape. (This means that modulation translates a band of frequencies to some other portion of the frequency spectrum.) In communications applications, amplitude modulation (AM) is a method in which information contained in the modulating signal is impressed upon a high-frequency carrier wave, figure 27-7A. As a result, the amplitude of the carrier varies with the modulating sound signal. The three waveshapes of figure 27-7B show the effect of amplitude modulation.

The equation of the *unmodulated* carrier wave is:

$$v_c = A \sin 2\pi f_c t \qquad \text{Eq. 27.3}$$

where

A = peak value of the unmodulated carrier

f_c = carrier frequency

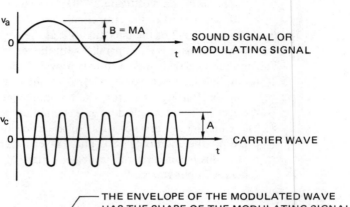

Fig. 27-7A Signal flow diagram

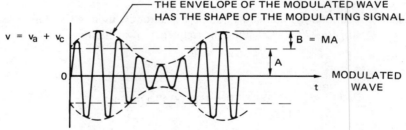

Fig. 27-7B Waveshapes

The equation of the modulating signal, which is a single sound tone, is:

$$v_a = B \sin 2\pi f_a t \qquad \text{Eq. 27.4}$$

where

B = peak value of the modulating signal

f_a = frequency of the sound tone

The modulation factor (M) is the peak value of the modulating signal divided by the peak value of the unmodulated carrier wave. Expressed as a percentage, the modulation factor is:

$$M = \frac{B}{A} \times 100\% \qquad \text{Eq. 27.5}$$

The modulation factor can vary from 0% to 100% with no distortion. The most desirable modulation is 100%. However, there is distortion when the modulation increases beyond 100%. Figure 27-8 shows three degrees of modulation. The dashed curve in each case is called the *envelope* of the waveform. As shown in the figure, there is an upper envelope and a lower envelope.

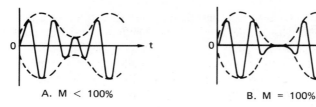

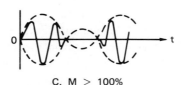

Fig. 27-8 Three degrees of amplitude modulation

The modulated wave has three frequency components.

1. The first component is the frequency f_c (carrier frequency).
2. The second component is the sum of f_c and f_a (modulating signal frequency) and is called the *upper sideband.*
3. The third component is the frequency at $f_c - f_a$ and is known as the *lower sideband.*

The three components of the modulated wave are shown in figure 27-9.

The percentage modulation can be determined from an oscilloscope display, figure 27-10, using the following equation:

$$M = \frac{MAX_{p\text{-}p} - MIN_{p\text{-}p}}{MAX_{p\text{-}p} + MIN_{p\text{-}p}} \times 100\% \qquad \text{Eq. 27.6}$$

A mathematical explanation of amplitude modulation is presented next to help the student understand the process. The carrier signal given by Equation 27.3 is also represented by the following equation:

$$v_c = A \cos 2\pi f_c\, t$$

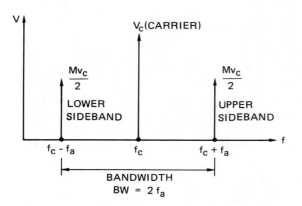

Fig. 27-9 Three components of an AM wave on a voltage versus frequency spectrum

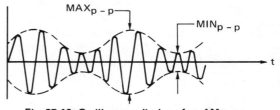

Fig. 27-10 Oscilloscope display of an AM wave

The phase angle is assumed to be zero since it has no effect on amplitude modulation. The cosine form of the equation is used because it is easier to solve mathematically.

In amplitude modulation, the modulating signal controls the magnitude of the carrier wave, A. As shown in figure 27-7B, a modulating envelope is produced on the carrier waveform. This modulation envelope follows the modulating signal waveform. In other words, when the envelope increases in the positive direction, it also increases in the negative direction.

The amplitude of the modulation envelope is a fraction M of the amplitude of the unmodulated wave. Based on this definition for modulation, Equation 27.4 for the modulating signal becomes:

$$v_a = MA \cos 2\pi f_a t$$

When a carrier is amplitude modulated by a sine-wave signal, the amplitude of the carrier contains the sinusoidal variations expressed by:

$$(1 + M \cos \omega_a t) A$$

The instantaneous voltage of the resultant modulated wave is:

$$v = (1 + M \cos \omega_a t) A \cos \omega_c t \qquad \text{(See figure 27-7B.)}$$

This expression can be expanded to yield:

$$v = A \cos \omega_c T + MA \cos \omega_c t \cos \omega_a t$$

The following trigonometric expansion formula can be used to simplify the preceding equation:

$$\cos x \cos y = 1/2 \cos (x + y) + 1/2 \cos (x - y)$$

Substituting the expansion formula for the product of the two cosine waves in the formula for v gives:

$$v = A \cos 2\pi f_c t + \frac{MA}{2} \cos 2\pi (f_c + f_a) t + \frac{MA}{2} \cos 2\pi (f_c - f_a) t$$

This derivation shows that the equation of an amplitude modulated wave contains three terms. That is, amplitude modulation does not change the original carrier wave, but adds two other frequency components to it. The first term is identical to Equation 27.3 for the unmodulated wave. The frequency of the second term is $(f_c + f_a)$ and the frequency of the third term is $(f_c - f_a)$. For example, when the carrier frequency is 5 000 Hz and the modulating signal has a frequency of 200 Hz, the frequencies of the three terms in the modulated carrier are 5 000 Hz, 5 200 Hz and 4 800 Hz. The frequency at $(f_c + f_a)$, or 5 200 Hz, is called the upper sideband. The frequency at $(f_c - f_a)$, or 4 800 Hz, is called the lower sideband. The audio signal f_a has a frequency of 200 Hz. However, the total bandwidth is from 4 800 Hz to 5 200 Hz, or 400 Hz. Thus, it can be concluded that the bandwidth in amplitude modulation is equal to twice the frequency of the modulating signal, f_a.

Problem 3   A 100-kHz carrier wave is modulated by a 5-kHz audio tone. Find: a. the frequencies that make up the modulated wave, and b. the bandwidth of the modulated wave.

Solution   a. The frequency content of the modulated wave is f_c, $f_c + f_a$, and $f_c - f_a$.

$$f_c = 100 \text{ kHz}$$
$$f_c + f_a = (100 + 5) \text{ kHz} = 105 \text{ kHz}$$
$$f_c - f_a = (100 - 5) \text{ kHz} = 95 \text{ kHz}$$

b. BW $= 2 f_a$
$$= 2(5 \text{ kHz})$$
$$= 10 \text{ kHz}$$

(R27-9) Determine the percentage modulation of the wave in Problem 3 if the maximum peak-to-peak height of the modulated wave on the oscilloscope is 6 centimeters, and the minimum peak-to-peak height is 2 centimeters.

The power versus frequency spectrum for an AM wave is shown in figure 27-11. The total power is:

$$P_T = \frac{M^2 \, P_c}{4} + \frac{M^2 \, P_c}{4} + P_c \qquad \text{Eq. 27.7}$$

$$P_T = P_c \left(1 + \frac{M^2}{2} \right) \qquad \text{Eq. 27.8}$$

The two equations for P_T show that the power content of the carrier, P_c, is independent of the percentage modulation. The two sidebands contain equal amounts of power, as shown in the following equation:

$$P_{lsb} = P_{usb} = \frac{P_T - P_c}{2} \qquad \text{Eq. 27.9}$$

where

P_{lsb} = power of lower sideband
P_{usb} = power of upper sideband

Problem 4   An AM transmitter delivers a total power of 120 W to an antenna. The percentage modulation is 100%. Determine the power contained at the carrier frequency, and the power contained in each of the sidebands.

Solution $P_T = P_c \left(1 + \dfrac{M^2}{2} \right)$

$120 \text{ W} = P_c \left(1 + \dfrac{1^2}{2} \right)$

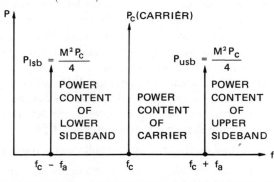

Fig. 27-11 Power versus frequency spectrum of an AM wave

$$120 \text{ W} = P_c \, (1.5)$$

$$P_c = \frac{120 \text{ W}}{1.5}$$

$$= 80 \text{ W}$$

$$P_{lsb} = P_{usb} = \frac{P_T - P_c}{2}$$

$$= \frac{120 \text{ W} - 80 \text{ W}}{2}$$

$$= 20 \text{ W}$$

An analysis of problem 4 leads to the following conclusions:

1. The waveform at the carrier frequency carries no information. Therefore, two-thirds of the power radiated by the antenna contains no useful, or audio, information.
2. The two sidebands contain the same audio information. As a result, the total useful radiated power is equal to one-third of the total radiated power. Since the wave is 100% modulated, this useful radiated power is the maximum amount that can be obtained with AM.

(R27-10) What is the total useful radiated power transmitted by the antenna described in Problem 4?

(R27-11) Solve Problem 4 if the percentage modulation is 50%.

(R27-12) Analyze question (R27-11) and state the conclusion that can be made regarding the radiation of useful power, or power containing audio information.

AM BROADCAST RADIO TRANSMISSION

Figure 27-12 illustrates the essential components of an AM broadcast radio transmitter.

1. The radio frequency (RF) oscillator generates a sinusoidal RF carrier signal in the AM broadcast band. This frequency is assigned and is in the range from 550 kHz to 1 600 kHz.
2. An audio frequency signal is produced by the sound waves striking the microphone. This signal is in the range from 20 Hz to 20 000 Hz.
3. The audio amplifier increases the signal to a level of several volts.
4. The audio signal modulates the RF carrier in the AM modulator.
5. The modulated RF signal is then power amplified. The amplifier circuits used are tuned to the RF frequency band in the RF power amplifier.
6. The AM radio frequency signal is sent from the antenna of the transmitter as electromagnetic radiation.

Figure 27-13 shows an audio amplifier which is designed for an industrial communication system. An AM transmitter designed for an industrial communication system, figure 27-14, operates in the 300 Hz to 30 kHz range. The AM receiver for this system is shown in figure 27-15.

(R27-13) At what speed do electromagnetic radio waves travel through space?

(R27-14) Does the radio broadcast band or the industrial communication band contain a frequency of 1 kHz?

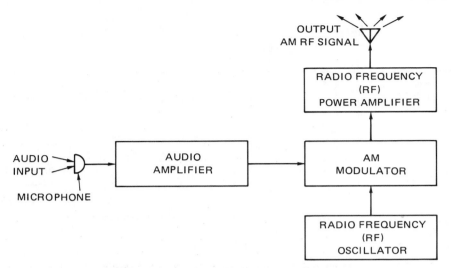

Fig. 27-12 Components of an AM broadcast radio transmitter

Fig. 27-13 Plug-in audio amplifier module for an industrial communication system

Fig. 27-14 Plug-in AM transmitter module for an industrial communication system

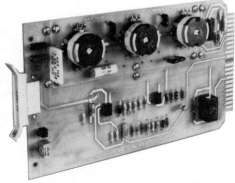

Fig. 27-15 Plug-in AM receiver module for an industrial communication system

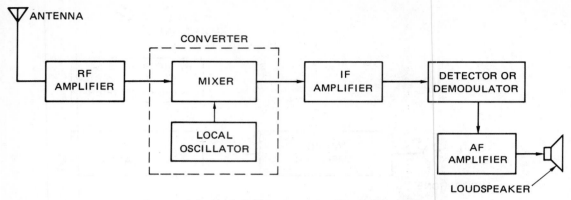

Fig. 27-16 Block diagram of a superheterodyne AM receiver

THE SUPER-HETERODYNE AM RECEIVER

The block diagram of a superheterodyne AM receiver, also known as a "superhet" is shown in figure 27-16. This receiver is used in modern radio receivers. A heterodyne, or signal beating principle converts the incoming RF signal to a carrier having the same audio modulation as the original carrier, but at a fixed, lower frequency. The intermediate frequency (IF) is 455 kHz.

The operation of a superheterodyne receiver is described in the following steps:

1. The AM signal is picked up by the antenna and is connected to the RF amplifier. This amplifier is tuned to the carrier frequency of the station to be received. The bandwidth of the RF amplifier is wide enough to pass the carrier signal and both sideband signals. The selectivity of the RF amplifier must be high enough to reject, or attenuate, all signals of adjacent radio stations.

2. The AM signal from the RF amplifier is centered around the carrier frequency. The converter changes the AM signal so that it is centered around an intermediate frequency (IF). This change in the signal is made by a local oscillator which is tuned in such a manner so that its output is always 455 kHz higher than the incoming carrier frequency. The tuned circuits in the RF amplifier and the local oscillator track each other. The difference in frequency between the local oscillator and all other stations tuned in the broadcast band is always 455 kHz. This feature makes possible the use of very selective high-gain tuned IF stages. These stages are designed to operate at one fixed frequency. Therefore, wide bandwidth amplification is not required. (This feature is both difficult to design and expensive to build.)

3. The signal is decoded in the detector, or demodulator. The detector removes the IF carrier. The output of the detector contains only the original audio modulating frequencies. Figure 27-17 shows a diode detector circuit, including wave shapes at various points in the circuit.

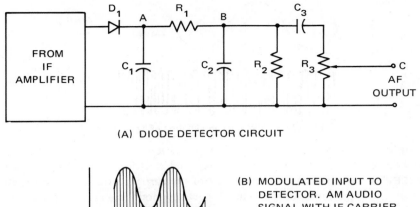

(A) DIODE DETECTOR CIRCUIT

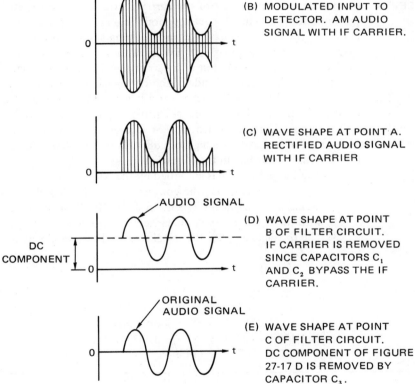

(B) MODULATED INPUT TO DETECTOR. AM AUDIO SIGNAL WITH IF CARRIER.

(C) WAVE SHAPE AT POINT A. RECTIFIED AUDIO SIGNAL WITH IF CARRIER

AUDIO SIGNAL

(D) WAVE SHAPE AT POINT B OF FILTER CIRCUIT. IF CARRIER IS REMOVED SINCE CAPACITORS C_1 AND C_2 BYPASS THE IF CARRIER.

ORIGINAL AUDIO SIGNAL

(E) WAVE SHAPE AT POINT C OF FILTER CIRCUIT. DC COMPONENT OF FIGURE 27-17 D IS REMOVED BY CAPACITOR C_3.

Fig. 27-17 Diode detector circuit with wave shapes at various points in the circuit.

4. The audio signal from the detector is increased in the AF amplifier. Finally, it is sent to the loudspeaker where it is reproduced as a sound wave.

(R27-15) What two functions are performed by the detector?

(R27-16) What two functions are performed by the converter stage?

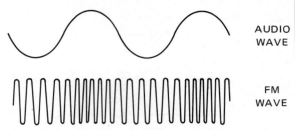

Fig. 27-18 Frequency modulation of a carrier signal

FREQUENCY MODULATION (FM) CONCEPTS

Examples of noise in AM transmission and reception include lightning and radiation from fluorescent devices and electrical machinery. Frequency modulation (FM) varies the frequency of the carrier at the audio signal rate while keeping the amplitude constant. Frequency modulation eliminates much of the unwanted noise. At the same, it allows the broadcasting station to use higher modulating frequencies than is possible with AM operation.

The frequency modulation of a carrier signal is shown in figure 27-18. Note in the figure that the amplitude of the FM signal does not vary. In addition, the frequency of the signal is higher for a higher audio signal amplitude. In other words, the carrier frequency changes as the amplitude of the audio signal changes.

The block diagram of a basic FM receiver is shown in figure 27-19. This receiver receives signals ranging from 88 MHz to 108 MHz. The operation of the FM receiver is described in the following steps.

1. The FM radio wave is amplified by the RF amplifier.
2. The FM signal enters the converter where it is combined with the local oscillator signal in the mixer. A 10.7-MHz IF signal leaves the converter with the audio information carried as frequency variations.
3. The IF stages amplify the signal to a level of several volts.

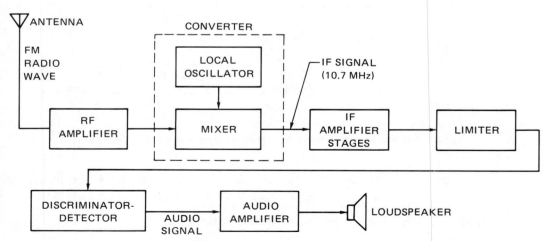

Fig. 27-19 Block diagram of a basic FM receiver

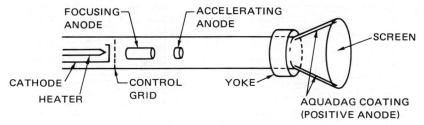

Fig. 27-20 TV picture tube (CRT)

4. The limiter circuit prevents any amplitude variations.
5. The discriminator-detector circuits convert the FM frequency variations to their original audio frequency variations. The voltage output of the detector is changed by the frequency variation of the FM signal.
6. The audio signal is increased by the AF amplifier. Then, the signal is sent to the loudspeaker where it is reproduced as a sound wave.

(R27-17) What are the two characteristics of an FM wave?

(R27-18) What is: a. the range of the FM band, and b. the IF in an FM receiver?

(R27-19) What is the purpose of the limiter circuit in an FM receiver?

TELEVISION CONCEPTS

Television is a combination of an AM system and an FM system. The video information is transmitted by the AM system. The FM system transmits the sound portion. A single television channel requires a bandwidth of 6 MHz. Compare this value with a bandwidth of 10 kHz for an AM radio broadcast station.

The picture tube, or cathode-ray tube (CRT), of a television is shown in the the schematic drawing in figure 27-20. The CRT consists of three sections:

1. An electron gun is located in the neck of the tube. It generates the electron beam by thermionic emission.
2. The yoke is made up of a set of electromagnetic coils. The vertical deflection coils move the electron beam up or down. The horizontal deflection coils move the electron beam to the right or left.
3. The phosphor-coated screen lights up wherever the electron beam strikes it.

A television image, like motion pictures, consists of a series of still pictures flashed on a screen at a rapid rate. The rate is fast enough to give the image the appearance of continuous motion. When thirty pictures per second are flashed on the screen, the resulting motion seems to be continuous with no flicker.

To obtain a continuous picture, a scene is scanned one line at a time for 525 lines per frame. A method of scanning known as *interlaced scanning* is shown in figure 27-21. A complete picture, or one frame, is divided into two fields of 262.5 lines each. The odd-numbered scan lines are shown during the first field, and the even-numbered scan lines are shown during the second field. Therefore, one frame, or 525 lines, is made up of two fields containing 262.5 lines each.

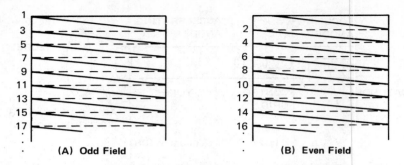

Fig. 27-21 Interlaced scanning (the odd field plus the even field make up one frame)

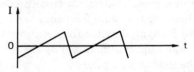

Fig. 27-22 Sawtooth current wave supplied to horizontal deflection coil

The sawtooth current wave, figure 27-22, is supplied to the horizontal deflection coils. This wave causes the beam of electrons to move across the screen of the CRT in a linear fashion, and then return to the starting side of the screen. The frequency of this wave is 15 750 Hz, or 525 lines times 30 frames per second. To prevent retrace lines from appearing on the screen, the beam is returned to the starting side of the screen. Before it can retrace, however, it is blanked off by a cutoff pulse applied to the control grid of the electron gun.

The current waveform for the vertical deflection coils is similar in shape to the waveform for the horizontal deflection coils, but at a frequency of 60 Hz. To prevent retrace lines from appearing on the screen, the beam is cut off during the vertical retrace period.

Figure 27-23 is a block diagram of a typical TV receiver. The sequence of operation for this receiver is as follows:

1. The sound and picture portions of the TV signal are received by the antenna. They next pass through the preselector and video amplifier.
2. The two signals enter the converter. The sound IF is 41.25 MHz, and the picture IF is 45.75 MHz.
3. The signals enter the video detector and mixer. The picture signal is demodulated and sent to the video amplifier. The video detector also acts as a mixer for the sound portion of the IF signal. That is, it heterodynes, or beats, the sound portion of the IF signal with the carrier frequency of the picture portion of the IF signal. The sound signal is shifted to a new IF with a center frequency of 4.5 MHz.
4. The new sound IF is separated and carried as an FM signal through the audio IF, the FM demodulator, the AF amplifier, and finally to the loudspeaker.

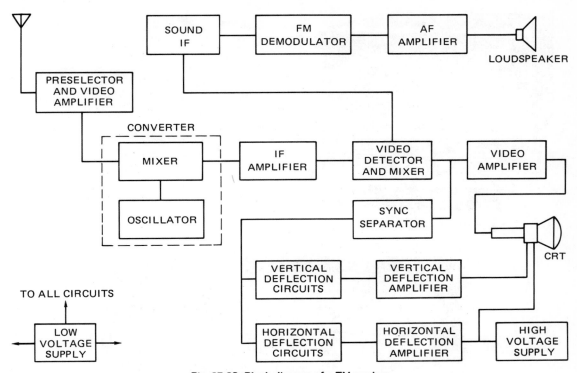

Fig. 27-23 Block diagram of a TV receiver

5. The signals needed to synchronize the receiver deflection circuits with the camera deflection circuits are removed by the sync separator and sent to these deflection circuits.

6. A low-voltage power supply provides the various stages with direct current.

7. A high-voltage power supply provides the high voltage needed by the picture tube.

(R27-20)   Which system (AM or FM) is used to transmit a. the video signal, and b. the sound signal?

(R27-21)   Name the three sections of a picture tube.

(R27-22)   How many lines are scanned in a TV picture for a. each frame, and b. each field?

Some of the instruments used in testing and analyzing communication circuits are shown in figure 27-24. Figure 27-24A is a frequency counter. Figure 27-24B shows a function generator with AM/FM modulation. This generator is used as a signal source for testing amplifiers, receivers, filter circuits, and logic circuits. The instrument shown in figure 27-24C measures distortion. It is used by technicians in the design and development of audio, broadcast and high-fidelity equipment, and amplifiers.

Fig. 27-24A Frequency counter

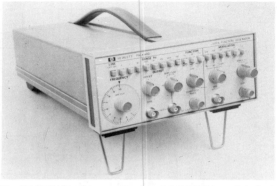

Fig. 27-24B Function generator with AM/FM modulation

Fig. 27-24C Distortion measurement set

LABORATORY EXERCISE 27-1 CHARACTERISTICS AND OPERATION OF A SINUSOIDAL OSCILLATOR

PURPOSE In completing this exercise, the student will become familiar with the characteristics and operation of a sinusoidal oscillator.

EQUIPMENT AND MATERIALS
1 Dc power supply
1 Transistor, 2N3904
1 Resistor, 4.7 kΩ, 2 W
1 Resistor, 10 kΩ, 2 W
1 Resistor, 22 kΩ, 2 W
1 Resistor, 47 kΩ, 2 W

1 Capacitor, 0.001 μF
1 Capacitor, 0.01 μF
1 Capacitor, 0.1 μF
1 Coil, 100 μH
1 Oscilloscope
1 Potentiometer, 100 kΩ, 5 W

PROCEDURE
1. Construct the circuit shown in figure 27-25.
   a. Energize the dc power supply.
   b. Adjust the 100-kΩ potentiometer to obtain a sinusoidal output voltage, V_{out}.
   c. Observe the output waveform, V_{out}, with the oscilloscope. Measure and record the following: 1.) period T, 2.) frequency f_o, and 3.) V_{out} (peak-to-peak).
   d. Connect a 10-kΩ resistor across the output terminal. Measure and record V_{out} (peak-to-peak).
2. Answer the following questions, and perform the calculations requested:
   a. List the components that make up the resonant tank circuit.
   b. Describe the input to this oscillator.
   c. Calculate the resonant frequency of the oscillator. Compare this value with the value measured in step 1.c.
   d. What inductor value is required to double the resonant frequency?
   e. How much ac power is lost in the 10-kΩ resistor in step 1.d.?

EXTENDED STUDY TOPICS
1. What is the Barkhausen criterion for sustained oscillation in a sinusoidal oscillator?
2. How does a Colpitts oscillator differ from a Hartley oscillator?
3. A 500-μH coil and two 100-pF capacitors are used in the Colpitts oscillator circuit of figure 27-6C. Calculate the frequency of oscillation.

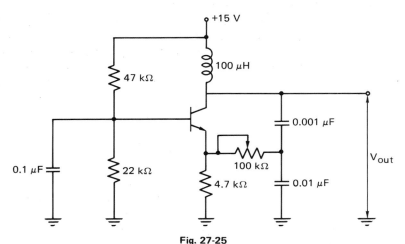

Fig. 27-25

4. A 150-kHz carrier wave is modulated by a 5-kHz audio tone and displayed on an oscilloscope screen. The maximum peak-to-peak height of the modulated wave is 8 cm and the minimum peak-to-peak height is 2 cm. The transmitter is operated at a total output power of 100 kW.

   a. What is the frequency content of the modulated wave?

   b. Determine the bandwidth of the modulated wave.

   c. Determine the percentage modulation.

   d. Find the power content of the carrier.

   e. How much power is contained in each of the sidebands?

   f. How much power is radiated by the carrier if no modulating signal is impressed on the carrier?

5. Can the human hear radio frequency signal? Explain.

6. What is the main advantage of the superheterodyne radio principle?

7. List the two advantages of FM.

8. What is the function of the discriminator-detector circuits in an FM receiver?

9. Explain how the sawtooth current wave is used in the horizontal deflection coils of a television CRT.

10. Calculate the frequency of the sawtooth current wave in Topic 9.

11. What action prevents the appearance of horizontal retrace lines on a TV screen?

Chapter 28

Operational Amplifiers

OBJECTIVES   After studying this chapter, the student will be able to

- describe the characteristics of an operational amplifier.

- design an inverting amplifier.

- design a noninverting amplifier.

- discuss the use of operational amplifiers in industry.

INTRODUCTION   An *operational amplifier* is a direct-coupled, high gain, integrated circuit voltage amplifier which is fabricated on a single chip of silicon. This amplifier has an open loop gain of 10^5 to 10^9. The schematic diagram of an operational amplifier, commonly known as an *op amp*, is shown in figure 28-1. In its original applications, the operational amplifier performed mathematical functions as an integral part of an analog computer. Table 28-1 lists a number of current op amp applications. In these applications, the external components will determine the overall response characteristics of an op amp circuit.

Commercial op amps are shown in figure 28-2 and sockets for op amps are shown in figure 28-3. Figure 28-4 is a photomicrograph of a type 741 op amp. The equivalent circuit of the 741 op amp is given on the data sheet shown in figure 28-5. The connection diagrams for various op amp package types are also shown in figure 28-5.

(R28-1)   What is the number of transistors in the equivalent circuit of a 741 op amp?

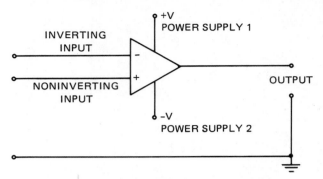

Fig. 28-1 Schematic of an operational amplifier

TABLE 28-1

Applications of Op Amps	
Active filters	Laboratory power supplies
Analog instrumentation	Peak detection circuits
Bridge amplifier	Process controllers
Circuit isolation	Servo controllers
Current amplification	Signal generation (such as sine wave,
Current-to-voltage conversion	sawtooth, square wave)
Differential amplifier	Signal multiplication
Differentiation	Summation of signals
Discrimination	Voltage amplification
Impedance matching	Voltage or current comparison
Integration	Voltage or current limiting
Instrumentation amplifier	

Fig. 28-2 Commercial op amps

Fig. 28-3 Op amp sockets

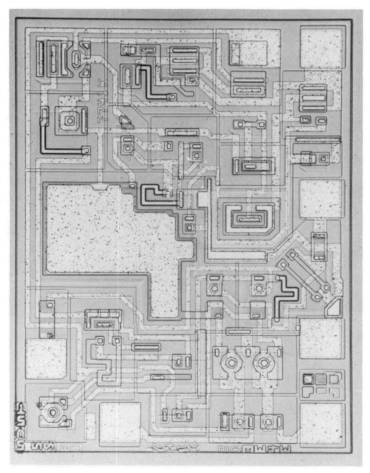

Fig. 28-4 Photomicrograph of a type 741 op amp

GENERAL DESCRIPTION — The μA741 is a high performance monolithic Operational Amplifier constructed using the Fairchild Planar* epitaxial process. It is intended for a wide range of analog applications. High common mode voltage range and absence of "latch-up" tendencies make the μA741 ideal for use as a voltage follower. The high gain and wide range of operating voltage provides superior performance in integrator, summing amplifier, and general feedback applications.

- NO FREQUENCY COMPENSATION REQUIRED
- SHORT CIRCUIT PROTECTION
- OFFSET VOLTAGE NULL CAPABILITY
- LARGE COMMON-MODE AND DIFFERENTIAL VOLTAGE RANGES
- LOW POWER CONSUMPTION
- NO LATCH UP

ABSOLUTE MAXIMUM RATINGS

Supply Voltage	
Military (741)	±22 V
Commercial (741C)	±18 V
Internal Power Dissipation (Note 1)	
Metal Can	500 mW
DIP	670 mW
Mini DIP	310 mW
Flatpak	570 mW
Differential Input Voltage	±30 V
Input Voltage (Note 2)	±15 V
Storage Temperature Range	
Metal Can, DIP, and Flatpak	$-65°$C to $+150°$C
Mini DIP	$-55°$C to $+125°$C
Operating Temperature Range	
Military (741)	$-55°$C to $+125°$C
Commercial (741C)	$0°$C to $+70°$C
Lead Temperature (Soldering)	
Metal Can, DIP, and Flatpak (60 seconds)	$300°$C
Mini DIP (10 seconds)	$260°$C
Output Short Circuit Duration (Note 3)	Indefinite

EQUIVALENT CIRCUIT

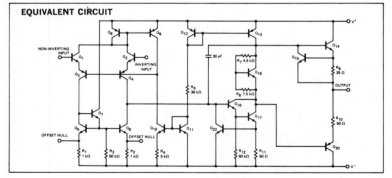

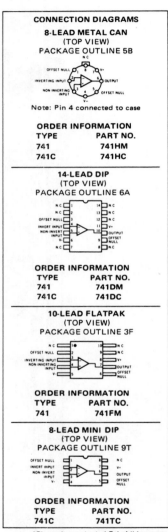

CONNECTION DIAGRAMS

8-LEAD METAL CAN
(TOP VIEW)
PACKAGE OUTLINE 5B

Note: Pin 4 connected to case

ORDER INFORMATION

TYPE	PART NO.
741	741HM
741C	741HC

14-LEAD DIP
(TOP VIEW)
PACKAGE OUTLINE 6A

ORDER INFORMATION

TYPE	PART NO.
741	741DM
741C	741DC

10-LEAD FLATPAK
(TOP VIEW)
PACKAGE OUTLINE 3F

ORDER INFORMATION

TYPE	PART NO.
741	741FM

8-LEAD MINI DIP
(TOP VIEW)
PACKAGE OUTLINE 9T

ORDER INFORMATION

TYPE	PART NO.
741C	741TC

*Planar is a patented Fairchild process.

Fig. 28-5 Data sheet for a type 741 op amp

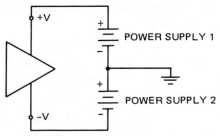

Fig. 28-6 Power supply connections

POWER SUPPLY
CONNECTIONS

The output terminal of the op amp must be able to supply positive or negative output voltages. Ground potential is the zero voltage reference level for all input, output, and power supply voltages. The op amp requires both positive (+) and negative (–) supply voltages, as shown in figure 28-6.

(R28-2) Why is it necessary to ground the center tap of the two power supplies shown in figure 28-6?

INPUT-OUTPUT
TERMINAL
CONNECTIONS

Op amps have two input terminals and one output terminal. The power supply connections are shown in figure 28-6. It is common practice not to show the power supply connections on the schematic, figure 28-7. The student must realize that the op amp will function properly only when the power supplies are connected, even when the connections are not shown on the schematic.

An algebraic sign identifies each input terminal. A *minus* (–) sign identifies the *inverting input terminal.* Any voltage applied to this terminal appears at the output terminal with a reversed polarity. This means that the output voltage, V_o, is 180° out of phase with the input voltage.

A *plus* (+) sign identifies the *noninverting terminal.* Any voltage applied to this input terminal appears at the output terminal with the same polarity. Thus, V_o is in phase with the input voltage.

A voltage signal of any polarity or phase angle may be applied to either, or both, input terminals. The (+) or (–) terminal markings indicate if the output voltage is in phase, or 180° out of phase, with the input signal.

(R28-3) What is the phase angle of V_o if the input voltage to the minus (–) terminal has a phase angle of –30°?

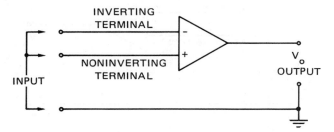

Fig. 28-7 Input-output terminal connections

(R28-4)   What is the phase angle of V_o if the input voltage to the plus (+) terminal has a phase angle of $-30°$?

FREQUENCY RESPONSE   The op amp will amplify signals over a wide frequency range. Since the op amp is also a dc amplifier, it can be used to amplify unchanging signals and very slowly changing signals, as well as rapidly varying signals.

(R28-5)   Can an op amp be used to amplify the output voltage of a battery?

(R28-6)   Does an op amp have a low-frequency limit in its circuit response?

DIFFERENTIAL AMPLIFIER CHARACTERISTICS   The op amp is a *differential amplifier.* That is, it responds to the difference in the voltages applied to the two input terminals. If the same signal voltage is applied to each of the input terminals, the two amplified output signals are equal and 180° out of phase. As a result, the output signals cancel each other and the output voltage, V_o, is zero. Figure 28-8 shows an op amp with two input voltages.

The symbol A is used to represent the open loop voltage gain of the op amp. V_n is the input voltage supplied between the inverting input terminal and ground. It is the voltage V_n to ground. The amplifying action of the op amp causes V_n to undergo a 180° phase reversal. V_p is the input voltage supplied between the noninverting input terminal and ground. It is the voltage V_p to ground. There is no phase angle change in V_p as a result of the amplifying action of the op amp.

$(V_p - V_n)$ is the differential input voltage which is amplified by the operational amplifier. V_o is the output voltage and is expressed by the following equation:

$$V_o = A(V_p - V_n) \qquad\qquad \text{Eq. 28.1}$$

An op amp amplifies only the differential input voltage. That is, it is not affected by, or it rejects, the signals which are common to both inputs, such as noise signals. This characteristic is called *common-mode rejection.*

(R28-7)   Which type of op amp amplifier is used to reject common-mode noise?

(R28-8)   Must the noise in each input lead have exactly the same frequency and amplitude to be rejected?

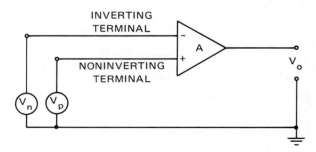

Fig. 28-8 Op amp with two input voltages

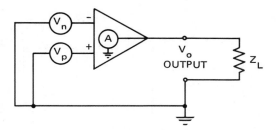

Fig. 28-9 The ideal op amp

SINGLE INPUT OP AMP

A single input op amp corresponds to an op amp with one input signal grounded. If V_p = 0, the op amp is an inverting amplifier, and the following equation expresses V_o:

$$V_o = -AV_n \qquad \text{Eq. 28.2}$$

If V_n = 0, the op amp is a noninverting amplifier. In this case, V_o is given by:

$$V_o = AV_p \qquad \text{Eq. 28.3}$$

THE IDEAL OP AMP

Figure 28-9 shows an ideal op amp with the following characteristics:

Voltage gain (A) = ∞
Bandwidth (BW) = ∞
Input impedance (Z_i) = ∞
Output impedance (Z_o) = 0

The infinite input impedance of an ideal op amp means that it does not draw any current from the input source. In other words, the op amp does not appear to be a load to the input source.

(R28-9) What is the value of the current drawn from the input circuits if V_n = +0.2 V and V_p = +0.1 V for the op amp shown in figure 28-9?

The output impedance of an ideal op amp is zero. This means that the ideal op amp appears to be a constant voltage source to the load. Thus, the voltage output of the op amp is not affected by the output load.

(R28-10) For the ideal op amp in figure 28-9, V_o is 2 V when Z_L = 100 Ω. What is the value of V_o when Z_L = 1 000 Ω?

THE PRACTICAL OPEN LOOP OP AMP

Figure 28-10 is a schematic of an input-output amplifier showing a practical open loop op amp. For this circuit, the input impedance is Z_i, the output impedance is Z_o, and the gain is A.

(R28-11) As Z_L varies, what is the effect of Z_o on V_o?

NEGATIVE FEEDBACK

The infinite gain characteristic of an ideal op amp means that any voltage input causes the output to saturate. As a result, the output voltage rises to its maximum value, or the value of the power supply voltage. The op amp can saturate in either the positive or the negative direction.

(R28-12) Refer to figure 28-11 and determine V_o when V_p = –2V.

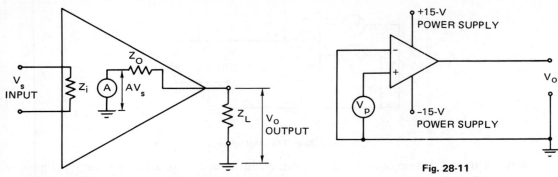

Fig. 28-10 Practical input-output op amp amplifier

Fig. 28-11

An op amp with negative feedback, figure 28-12, can overcome the problem of saturation. For the circuit shown, resistor R_1 is the input component and resistor R_F is the feedback component. When R_F is connected as shown, negative feedback results. The output voltage, V_o, is fed back to the input circuit and is 180° out of phase with the input voltage. Negative feedback affects this circuit in the following ways: it 1. decreases gain, 2. improves gain sensitivity, 3. prevents immediate saturation, and 4. increases bandwidth. For a practical op amp, the closed loop gain is less than 500 and the output voltage, V_o, is ±10 V.

The circuit in figure 28-12 is analyzed as follows:

1. Current I_1 flows through the input component, R_1, when V_s is applied to the circuit.
2. A small voltage, V_i, develops across the minus (−) and positive (+) input terminals of the op amp.
3. The high-gain op amp amplifies this differential voltage, V_i.
4. The amplified voltage, V_o, is 180° out of phase with V_i and V_s.
5. V_o sends a feedback current, I_F, through the feedback component, R_F.

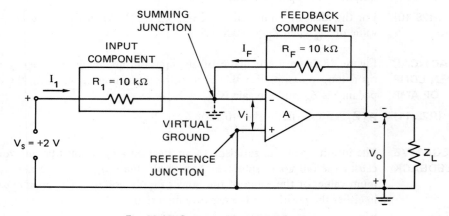

Fig. 28-12 Op amp with negative feedback

6. I_1 and I_F are equal in magnitude, but $180°$ out of phase with each other. The algebraic sum of I_1 and I_F is zero at the negative input terminal. Thus, this terminal is called the *summing junction*.

(R28-13) What prevents the internal flow of current in the op amp between the minus (–) and the positive (+) input terminals?

V_i is usually less than 1 mV. Thus, it is assumed to be zero for calculation purposes. Since V_i is approximately zero, the inverting input terminal is virtually at the same potential as the noninverting terminal, which is connected to ground. Therefore, the summing junction in this circuit is called *virtual ground*. There is no current between the summing junction and the positive (+) input terminal since the input impedance of the op amp is almost infinite. The positive (+) input terminal is called the *reference junction*. The two input terminals are nearly at the same voltage level, or, for this circuit, the zero ground reference level.

The fact that $V_i = 0$, leads to the concept that a virtual short circuit, or virtual ground, exists at the input to the op amp. It is assumed that no current flows into the op amp, because Z_i is almost infinite. Figure 28-13 shows this concept.

Before op amp circuits can be designed, the student must understand two design assumptions, or criteria, resulting from the previous analysis:

Criteria 1: Current does not flow into, or out of, any op amp input terminal.

Criteria 2: There is no difference in potential between the two input terminals of an op amp. Both the summing junction and the reference junction are always at tht same voltage level. This condition is true in all op amp circuit configurations.

(R28-14) What circuit requirement is necessary to prevent saturation?

(R28-15) Where is the reference junction?

(R28-16) Which terminal is the summing junction?

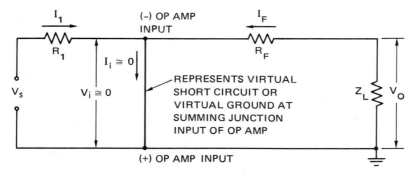

Fig. 28-13 Virtual short circuit or virtual ground at input to the op amp

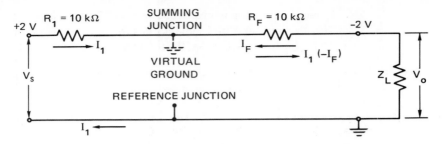

Fig. 28-14 Equivalent circuit of figure 28-12

NEGATIVE FEEDBACK CIRCUIT ANALYSIS

The equivalent circuit for the circuit of figure 28-12 is shown in figure 28-14. There is no current between the summing junction, or virtual ground, and the reference junction. I_1 flows from the +2-V terminal of V_s through R_1 to the summing junction. The source voltage is connected to the inverting input terminal of the op amp. Therefore, V_o is 180° out of phase with V_s, resulting in a negative voltage at the nongrounded terminal of Z_L. The output voltage, V_o, sends a current, I_F, through R_F in the direction shown. The algebraic sum of I_1 and I_F is zero:

$$I_F = I_1 \underline{/180°}$$
$$I_1 + I_F = I_1 \underline{/0°} + I_1 \underline{/180°} = 0$$
$$I_1 = -I_F$$

Since $I_1 = -I_F$, I_1 flows through R_1 and R_F as shown. The voltage drop across R_1 always equals the input voltage, V_s, since the summing junction is at the virtual ground potential. For the amplifier shown in figure 28-12, $R_1 = R_F$. Therefore, V_o must be –2 V to establish the summing junction connection at virtual ground potential. This amplifier is an inverting amplifier with an output voltage equal and opposite to the input voltage.

Problem 1   Calculate the gain of the amplifier shown in figure 28-12.

Solution $A = \dfrac{\text{output voltage}}{\text{input voltage}}$

$ = \dfrac{V_o}{V_s}$

$ = \dfrac{-2\text{ V}}{+2\text{ V}}$

$ = -1$

The input current, I_1, is independent of the feedback component, R_F, as determined from the following equation:

$$I_1 = \frac{V_s}{R_1} \qquad\qquad \text{Eq. 28.4}$$

The input of this amplifier is electrically isolated from the output. Therefore, the circuit of figure 28-14 can be simplified even more to become the equivalent circuit shown in figure 28-15. The current flow through the input component controls the current flow through the feedback component. A voltage, V_o, be-

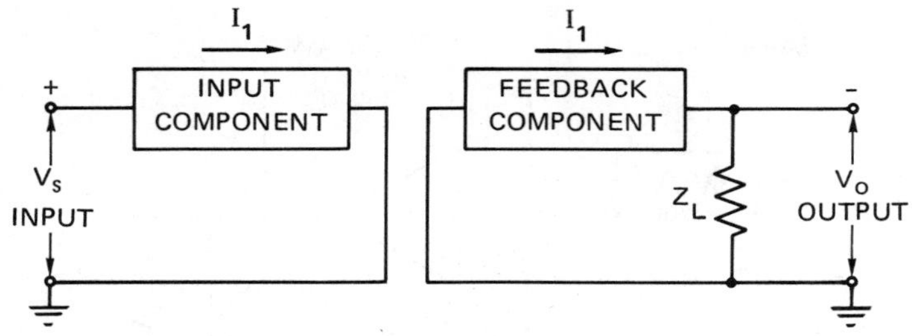

Fig. 28-15 Simplified equivalent circuit of figure 28-14

tween the output terminal and ground is equivalent to reading the voltage drop across R_F, because the summing junction is at virtual ground potential. I_F is given by the following equation:

$$I_F = \frac{V_o}{R_F}$$ Eq. 28.5

(R28-17) What is the value of the input current, I_1, in figure 28-14?

(R28-18) If V_s in figure 28-14 is 2 V $\underline{/45°}$, what is the value of V_o?

(R28-19) For the circuit in figure 28-14, does the value of I_F change when the value of Z_L is changed?

DESIGN OF AN INVERTING AMPLIFIER

If an inverting amplifier is to have a gain greater than one, R_F must be greater than R_1. The circuit of figure 28-16 shows an amplifier with $R_F = 500 \text{ k}\Omega$ and $R_1 = 100 \text{ k}\Omega$. Figure 28-17 is the equivalent circuit of this amplifier.

The equation for the gain of this amplifier is derived as follows:

$$A = \frac{\text{output voltage}}{\text{input voltage}}$$

$$= \frac{V_o}{V_s}$$

but,

$$V_o = I_F R_F$$

and,

$$V_s = I_1 R_1$$

Therefore,

$$A = \frac{I_F R_F}{I_1 R_1}$$

Since,

$$I_1 = -I_F,$$

$$A = \frac{I_F R_F}{-I_F R_1}$$

$$= -\frac{R_F}{R_1}$$ Eq. 28.6

Equation 28.6 shows that the gain of an inverting amplifier is negative. The gain is equal to the ratio of the values of the feedback resistor and the input resistor.

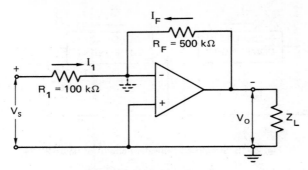

Fig. 28-16 Inverting amplifier

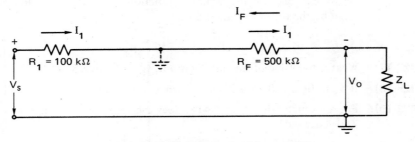

Fig. 28-17 Equivalent circuit of figure 28-16

This means that the desired voltage gain can be obtained by varying the ratio of the two resistors. Note that the load impedance, Z_L, does not appear in Equation 28.6. Therefore, the gain of the circuit is independent of Z_L.

(R28-20) Calculate the gain of the inverting amplifier shown in figure 28-16.

(R28-21) For an inverting amplifier, $V_s = -2$ V, $R_1 = 100$ kΩ, and A = 2.
a. Find the value of V_o.
b. Find the value of R_F.

DESIGN OF A NONINVERTING AMPLIFIER

A noninverting amplifier is shown in figure 28-18. The summing junction voltage equals the reference junction voltage. The value of the feedback component, R_F, must fulfill the conditions stated in Criterion 2, page 479.

The summing junction voltage is +1 V. The voltage drop across R_1, V_1, is 1 V. The equivalent circuit for this amplifier is shown in figure 28-19. Note that the current through R_F is the same as the current through R_1, or $I_F = I_1$. A voltage, V_o, between the output terminal and ground is equivalent to the voltage across R_1 and R_F. The voltage drop across R_F is ten times that of the voltage drop across R_1, since $R_F = 10 R_1$. The gain of this amplifier is found as follows:

$$V_F = 10 V_1$$
$$= (10)(1 \text{ V})$$
$$= 10 \text{ V}$$

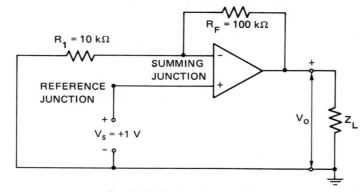

Fig. 28-18 Noninverting amplifier

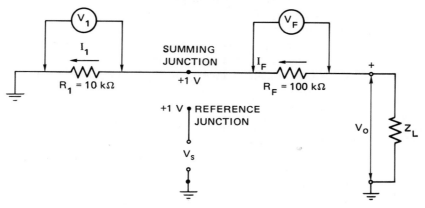

Fig. 28-19 Equivalent circuit of figure 28-18

But,

$$V_o = V_1 + V_F$$
$$= 1\,V + 10\,V$$
$$= 11\,V$$

Therefore,

$$A = \frac{V_o}{V_s}$$
$$= \frac{11\,V}{1\,V}$$
$$= 11$$

The voltage gain of this circuit is 11.

The equation for the gain of the circuit is derived as follows:

$$A = \frac{V_o}{V_s}$$
$$V_o = V_1 + V_F$$
$$= I_1 R_1 + I_F R_F$$

But,

$$I_1 = I_F$$

Therefore,

$$V_o = I_1 R_1 + I_1 R_F$$
$$= I_1(R_1 + R_F)$$
$$V_s = I_1 R_1 \text{ (a voltage equivalent to the input voltage}$$
is always dropped across the input
resistor)

$$A = \frac{I_1(R_1 + R_F)}{I_1 R_1}$$

$$= \frac{R_1 + R_F}{R_1} = 1 + \frac{R_F}{R_1} \qquad \text{Eq. 28.7}$$

Equation 28.7 shows that the gain of a noninverting amplifier is positive, and always equals one plus the ratio of the value of the feedback resistor to the input resistor. The desired voltage gain is obtained by varying the ratio of these resistors. Note that the load impedance, Z_L, does not appear in Equation 28.7. Therefore, the gain of the circuit is independent of Z_L. The circuit of figure 28-19 can be simplified even more to the equivalent circuit shown in figure 28-20. Current flow through the input component controls the current through the feedback component.

(R28-22)   A noninverting amplifier has the following specifications: $R_1 = 10 \text{ k}\Omega$, $R_F = 10 \text{ k}\Omega$, $V_s = +1 \text{ V}$, and $Z_L = 1 \text{ k}\Omega$.
a. Calculate the gain.
b. Find the value of V_o.
c. What is the effect of Z_L on the gain of this amplifier?

OP AMP
APPLICATIONS   Figure 28-21 shows a summing amplifier with three inputs. Each input is connected through a separate resistor. Each input adds a voltage term to the output which is obtained in the same manner as in the basic inverting amplifier. Any number of resistors and corresponding voltage inputs may be used. The following equation is used to determine the output voltage:

$$V_o = -\left[\left(\frac{R_F}{R_1}\right)V_1 + \left(\frac{R_F}{R_2}\right)V_2 + \left(\frac{R_F}{R_3}\right)V_3\right] \qquad \text{Eq. 28.8}$$

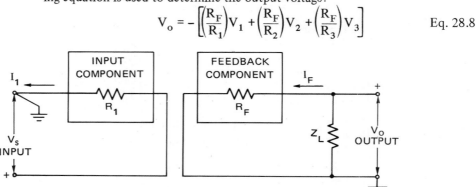

Fig. 28-20 Simplified equivalent circuit of figure 28-19

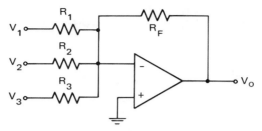

Fig. 28-21 Summing op amp with three inputs

Op Amp Integrator. An *op amp integrator* is a circuit that sums, or integrates, the input signal for a period of time. As shown in the integrator circuit of figure 28-22, the input component is a resistor and the feedback component is a capacitor. The charging current of the capacitor is equal to the current through the input resistor. The output voltage, V_o, equals the voltage across the capacitor. V_o is proportional to the negative of the integral (sum) of the input voltage. The op amp integrator is used in analog computers and industrial process controllers.

Op Amp Differentiator. An *op amp differentiator* is a circuit that responds to the changes, or differences, in the input signal. As shown in the differentiator circuit of figure 28-23, the input component is a capacitor and the feedback component is a resistor. The capacitor does not pass direct current. Therefore, current through the input capacitor and the feedback resistor occurs only with changes in the input signal. The output of the op amp differentiator equals the voltage drop across the feedback resistor. V_o is proportional to the negative of the derivative of the input signal. This circuit is used in instrumentation control to determine the rate of change of voltages.

Op Amp Sinusoidal Oscillator. Figure 28-24 is the circuit diagram of a sinusoidal oscillator, using an op amp. The frequency of oscillation of this circuit is given by the following equation:

$$f = \frac{1}{2\pi\, RC} \qquad\qquad \text{Eq. 28.9}$$

The potentiometer shown in the circuit is varied slightly until an undistorted sine wave appears at the output.

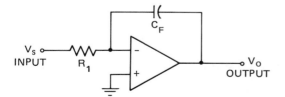

Fig. 28-22 Op amp integrator

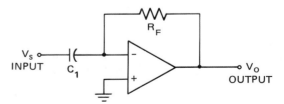

Fig. 28-23 Op amp differentiator

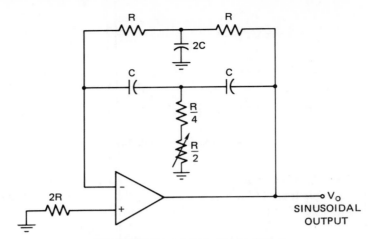

Fig. 28-24 Op amp sinusoidal oscillator

LABORATORY
EXERCISE 28-1
THE OP AMP
AMPLIFIER

PURPOSE In completing this exercise, the student will study the characterisrics of an op amp amplifier.

**EQUIPMENT
AND MATERIALS**
2 Dc power supplies, 15 V
1 Op amp, type 741C
2 Resistors, 1 kΩ, 2 W
1 Resistor, 10 kΩ, 2 W
1 Resistor, 100 kΩ, 2 W
2 Capacitors, 1 μF
1 Sine-wave generator
1 Oscilloscope
1 Ac millivoltmeter

PROCEDURE
1. Construct the circuit shown in figure 28-25. Use the 10-kΩ resistor as the feedback resistor, R.
   a. Connect the oscilloscope to display the voltage V_o.
   b. Adjust the sine-wave generator so that V_o is equal to 20 V peak-to-peak, at 1 kHz.
   c. Measure the voltage of the input signal at pin 3 of the op amp using the ac millivoltmeter.
   d. Change the feedback resistor to 100 kΩ. Repeat steps 1.a. through 1.c.
2. Answer the following questions, and perform the calculations requested:
   a. What is the maximum possible swing in the output voltage of this amplifier?
   b. What is the phase angle between V_o and V_{in}?

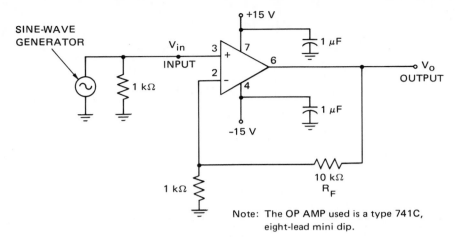

Note: The OP AMP used is a type 741C,
eight-lead mini dip.

Fig. 28-25

c. What type of amplifier does this circuit represent?
d. Calculate the voltage gain of this circuit from the values of the circuit components, using values of 10 kΩ and 100 kΩ for R_F.
e. Calculate the voltage gain of this circuit using the measured values of steps 1.b., 1.c., and 1.d.
f. Compare the results of step 2.d. with the results of step 2.e.

EXTENDED STUDY TOPICS

1. What are the characteristics of an op amp which make it an ideal input-output device?
2. What is the significance of the algebraic terminal markings inside the op amp symbol?
3. What are the two basic design criteria used in designing op amp circuits?
4. Explain the meaning of virtual ground.
5. Does an actual short circuit exist between the input terminals of an inverting amplifier?
6. What is the significance of the high input impedance of an op amp amplifier?
7. What does the low output impedance of an op amp amplifier indicate?
8. An inverting amplifier operates under the following conditions: R_1 = 50 kΩ, R_F = 0.5 MΩ, V_s = –1 V. Find V_o.
9. Calculate the output voltage of a noninverting amplifier when V_s = 2 V, R_F = 500 kΩ, and R_1 = 100 kΩ.
10. Calculate the output voltage of an op amp summing amplifier with the following inputs: R_1 = 500 kΩ, R_2 = 250 kΩ, R_3 = 1 MΩ, R_F = 1 MΩ, V_1 = +4 V, V_2 = –3 V, and V_3 = +6 V.
11. Design an op amp circuit with a voltage gain equal to –100 using a 10-kΩ input resistor. Draw a schematic diagram and label it completely.

Chapter 29

Regulating Semiconductors and Regulated Power Supplies

OBJECTIVES   After studying this chapter, the student will be able to

- describe the characteristics of a Zener diode.
- use the data sheets for a Zener diode to design a voltage regulating circuit.
- discuss the characteristics and applications of the SCR, DIAC, and TRIAC.
- explain the methods of firing and extinguishing an SCR.
- describe the principles and uses of commercial regulated power supplies.

ZENER DIODE CONCEPTS   The silicon voltage regulator diode, or Zener diode, is designed to operate in the breakdown region with reverse bias. When it is operated in this region, it provides a nearly constant voltage. However, when the Zener diode is operated with forward bias, it behaves much like the forward biased rectifier junction diodes described in Chapter 23. The manufacturer varies the amount of doping used to produce Zener diodes capable of operating at specific Zener breakdown voltages.

(R29-1)   What is the difference between Zener diodes and rectifier junction diodes?

ZENER AND AVALANCHE BREAKDOWN   Two effects can cause breakdown at a reverse biased *p-n* junction. *Zener breakdown* and *avalanche breakdown* can occur independently of one another or they can occur at the same time.

Zener breakdown occurs in a typical Zener diode when the reverse bias is less than 5 V. When the diode is heavily doped, the depletion layer at the *p-n* junction is very narrow. As the electric field across the depletion layer reaches a value of 300 000 volts per centimeter, large numbers of electrons in the depletion region break their covalent bonds. As a result, a large reverse current flows. This characteristic is called Zener breakdown, or high-field emission.

Avalanche breakdown occurs in the same typical Zener diode when the reverse bias is greater than 5 V. The reverse saturation current (I_S) across the reverse biased *p-n* junction consists of minority charge carriers (refer to Chapter 23). The velocity and energy content of the minority charge carriers increase as the reverse bias is increased. When the high-energy charge carriers strike atoms in the depletion region, other charge carriers break away from their atoms. This effect is called *ionization by collision.* The newly created charge carriers are accelerated to a high-energy state and are at a high speed. They strike other atoms, causing more ionization. The number of charge carriers increases, or avalanches, rapidly and avalanche breakdown results.

(R29-2) What value of electric field strength is necessary to cause Zener breakdown?

(R29-3) What are the two effects that cause breakdown at a reverse biased *p-n* junction?

ZENER DIODE CHARACTERISTIC CURVE AND PARAMETERS The symbol and the equivalent circuits for the Zener diode are shown in figure 29-1. Figure 29-2 is the characteristic curve for a typical Zener diode. The data sheet given in figure 29-3 covers a group of Zener diodes having nominal Zener voltages ranging from 6.8 V to 200 V. Some of the important Zener parameters indicated in figures 29-2 and 29-3 are as follows:

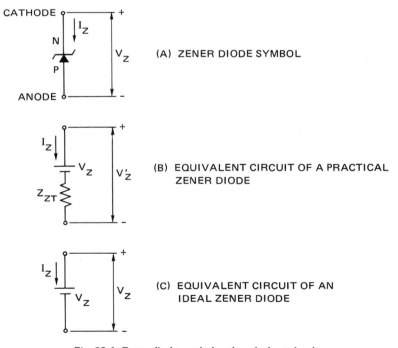

Fig. 29-1 Zener diode symbol and equivalent circuits

1. V_Z is the Zener breakdown voltage. The suffix "B" of the number identifying the type of diode indicates that the Zener diode has a voltage tolerance of ±5%. The suffix "A" indicates that the voltage tolerance is ±10%. A ±20% tolerance is indicated by the absence of a suffix.
2. I_{ZT} is the test current at which V_Z is measured.
3. I_{ZK} is the Zener current near the knee of the characteristic. This value is the minimum Zener current required to sustain breakdown.
4. I_{ZM} is the maximum Zener current. This value is limited by the maximum power dissipation.
5. P_{ZM} is the maximum power dissipation rating of the Zener diode. The Zener diode will not be destroyed as long as P_Z is less than P_{ZM}. Zener diodes are available with power ratings ranging from 1/4 W to more than 50 W. The relationship between I_{ZM} and P_{ZM} is expressed by the following equation:

$$I_{ZM} = \frac{P_{ZM}}{V_Z} \qquad \text{Eq. 29.1}$$

6. Z_{ZT} is the Zener dynamic impedance. The value of this parameter can be determined by referring to figure 29-2 and calculating the reciprocal of the slope of the characteristic at the test current I_{ZT}.

$$Z_{ZT} = \frac{\Delta V_Z}{\Delta I_Z} \qquad \text{Eq. 29.2}$$

It is desirable for Z_{ZT} to be very small. When ΔV_Z approaches zero, Z_{ZT} also approaches zero.

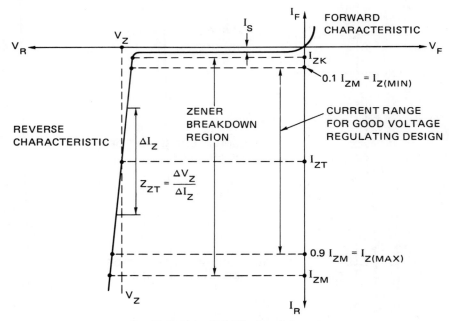

Fig. 29-2 Zener diode characteristic curve

FEATURES

- ZENER VOLTAGE RANGE: 6.8V TO 200V
- 1N3016B THROUGH 1N3051B HAVE JAN, JANTX, and JANTXV QUALIFICATION
- S1N3016B THROUGH S1N3051B ALSO AVAILABLE

*MAXIMUM RATINGS

Junction and Storage Temperatures: $-65°C$ to $+175°C$

DC Power Dissipation: 1 Watt

Derating: 6.67 mW/°C above 25°C

Forward Voltage @ 200 mA: 1.5 Volts

*ELECTRICAL CHARACTERISTICS @ 25°C

JEDEC TYPE NUMBER	NOMINAL ZENER VOLTAGE V_Z @ I_Z (Note 2)	ZENER TEST CURRENT I_Z	MAXIMUM ZENER IMPEDANCE (Note 3)			MAXIMUM ZENER CURRENT I_{ZM} (Note 4)	TYPICAL TEMP. COEFF. OF ZENER VOLTAGE
			Z_{ZT} @ I_Z	Z_{ZK} @ I_{ZK}			
	VOLTS	mA	OHMS	OHMS	mA	mA	%/°C
1N3016B	6.8	37	3.5	700	1.0	140	.040
1N3017B	7.5	34	4.0	700	.5	125	.045
1N3018B	8.2	31	4.5	700	.5	115	.048
1N3019B	9.1	28	5	700	.5	105	.050
1N3020B	10	25	7	700	.25	95	.055
1N3021B	11	23	8	700	.25	85	.060
1N3022B	12	21	9	700	.25	80	.065
1N3023B	13	19	10	700	.25	74	.065
1N3024B	15	17	14	700	.25	63	.070
1N3025B	16	15.5	16	700	.25	60	.070
1N3026B	18	14	20	750	.25	52	.075
1N3027B	20	12.5	22	750	.25	47	.075
1N3028B	22	11.5	23	750	.25	43	.080
1N3029B	24	10.5	25	750	.25	40	.080
1N3030B	27	9.5	35	750	.25	34	.085
1N3031B	30	8.5	40	1000	.25	31	.085
1N3032B	33	7.5	45	1000	.25	28	.085
1N3033B	36	7.0	50	1000	.25	26	.085
1N3034B	39	6.5	60	1000	.25	23	.090
1N3035B	43	6.0	70	1500	.25	21	.090
1N3036B	47	5.5	80	1500	.25	19	.090
1N3037B	51	5.0	95	1500	.25	18	.090
1N3038B	56	4.5	110	2000	.25	17	.090
1N3039B	62	4.0	125	2000	.25	15	.090
1N3040B	68	3.7	150	2000	.25	14	.090
1N3041B	75	3.3	175	2000	.25	12	.090
1N3042B	82	3.0	200	3000	.25	11	.090
1N3043B	91	2.8	250	3000	.25	10	.090
1N3044B	100	2.5	350	3000	.25	9.0	.090
1N3045B	110	2.3	450	4000	.25	8.3	.095
1N3046B	120	2.0	550	4500	.25	8.0	.095
1N3047B	130	1.9	700	5000	.25	6.9	.095
1N3048B	150	1.7	1000	6000	.25	5.7	.095
1N3049B	160	1.6	1100	6500	.25	5.4	.095
1N3050B	180	1.4	1200	7000	.25	4.9	.095
1N3051B	200	1.2	1500	8000	.25	4.6	.100

*JEDEC Registered Data

Fig. 29-3 Data sheet for Zener diodes

SILICON
1 WATT
ZENER DIODES

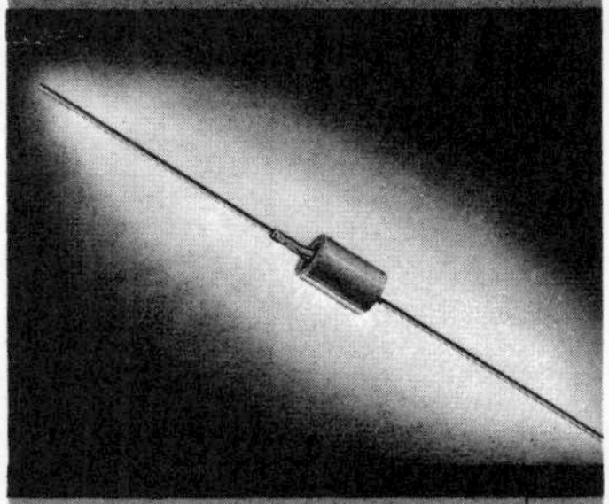

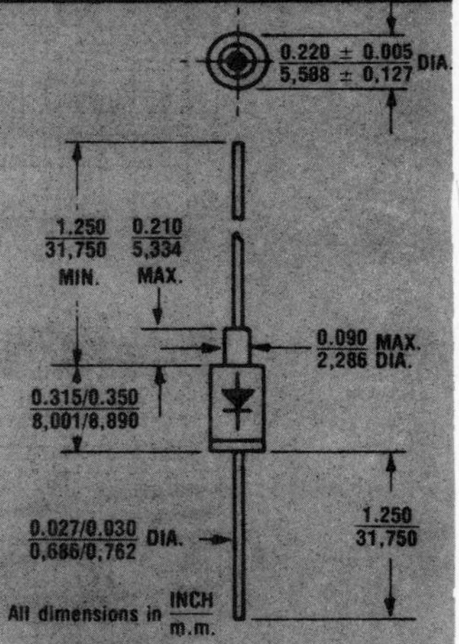

$\frac{0.220 \pm 0.005}{5,588 \pm 0,127}$ DIA.

$\frac{1.250}{31,750}$ MIN. $\frac{0.210}{5,334}$ MAX.

$\frac{0.090}{2,286}$ MAX. DIA.

$\frac{0.315/0.350}{8,001/8,890}$

$\frac{0.027/0.030}{0,666/0,762}$ DIA.

$\frac{1.250}{31,750}$

All dimensions in $\frac{INCH}{m.m.}$

FIGURE 1

MECHANICAL CHARACTERISTICS

CASE: DO-13, welded, hermetically sealed metal and glass.

FINISH: All external surfaces are corrosion resistant and leads solderable.

THERMAL RESISTANCE: 100° C/W (Typical) junction to ambient

POLARITY: Cathode connected case.

WEIGHT: 1.4 grams.

MOUNTING POSITION: Any.

73

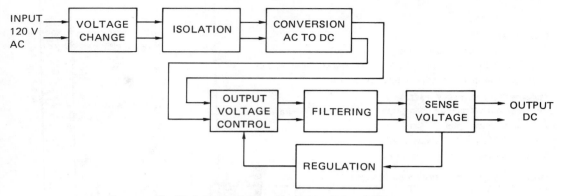

Fig. 29-4 Block diagram of a regulated power supply

7. Z_{ZK} is the Zener impedance near the knee of the characteristic. Z_{ZK} is calculated at the value of I_Z equal to I_{ZK}. Z_{ZK} is much larger than Z_{ZT}.

(R29-4) Refer to the data sheet of figure 29-3 and determine the following electrical characteristics for the 1N3031B Zener diode: a. V_Z, b. voltage tolerance, c. I_{ZT}, d. Z_{ZT}, e. Z_{ZK}, f. I_{ZK}, g. I_{ZM}, and h. P_{ZM}.

REGULATED POWER SUPPLIES   The dc power supply is a primary component in any communication, instrumentation, computer, or electronic system. The power supply converts the 120-V (rms), 60-Hz ac supply voltage into the required dc voltage. A typical power supply contains the following circuits:

1. A transformer is included to step up or step down the ac line voltage to meet the requirements of the dc output voltage. The transformer also provides isolation between the ac input section and the dc output section.
2. The rectifier circuit converts the ac input voltage into a unidirectional output voltage.
3. The filter circuit removes or minimizes the ripple.
4. A regulator circuit maintains the dc output voltage at a constant value for varying loads.

A block diagram of a regulated power supply is shown in figure 29-4. Normally, changes in the current requirements of the load circuits cause the output voltage to change over an unacceptable range. To prevent changes in the output voltages, its level is sensed and this information is fed back through regulation circuits to an output voltage control circuit. This circuit holds the output voltage constant for changes in the load current.

Figure 29-5 illustrates a three-terminal integrated circuit (IC) voltage regulator. This regulator combines Zener diodes and transistors on a single chip. It is easy to use, offers high reliability, and keeps the output voltage constant.

An assortment of commercial power supplies is shown in figure 29-6. Figure 29-7 shows two different models of regulated power supplies. Figure 29-8A

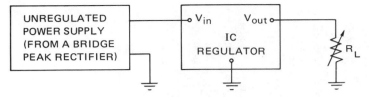

Fig. 29-5 Three-terminal integrated circuit (IC) voltage regulator

Fig. 29-6A Power supply for laboratory and test equipment

Fig. 29-6B High-efficiency power supply

Fig. 29-6C High-power switching power supply

Fig. 29-6D Internal view of power supply shown in figure 29-6C

shows a constant current power supply. A power supply used for microprocessor applications is shown in figure 29-8B.

An inverter, or an uninterruptible power system, protects critical ac loads against both power line disturbances and loss of commercial ac power. Inverters supply ac power from the energy stored in the batteries of the system whenever the commercial ac power is unavailable. Such systems are used to supply power to computers and computer-based equipment such as data acquisition, process control, communications, instrumentation, and security systems. An uninter-

Fig. 29-7A Regulated power supply

Fig. 29-7B Regulated power supply

Fig. 29-8A   Constant current
power supply

Fig. 29-8B Power supply used for microprocessor applications

ruptible power system is shown in figure 29-9A. The features of this system are shown in schematic form in figure 29-9B. The operation of the SCR used in this power system is discussed later in this chapter.

(R29-5)   What is an inverter?

(R29-6)   List the four major circuits in a regulated power supply.

(R29-7)   Which element of a regulated power supply is responsible for increasing safety and isolating the ac power supply ground from the chassis ground of the power supply?

(R29-8)   What are the two primary functions of a regulated power supply?

Fig. 29-9A Uninterruptible power system

OPERATION

A Topaz UPS consists of an isolation transformer, a rectifier, a battery charger, a battery, a static inverter and a transfer switch. The block diagram illustrates system operation.

When the ac mains power is normal, power is passed through the isolation transformer, where the extremely low interwinding capacitance eliminates high-speed, high-amplitude line transients. The output of the transformer is fed to the rectifier where ac power is converted to dc. The static inverter converts the dc input back into ac to drive the critical load.

Ac mains power is also converted to regulated dc by the battery charger, maintaining the battery in a fully charged state. The regulated battery voltage is isolated from the rectifier output/inverter by a power SCR.

If the ac mains voltage falls more than 15% below nominal, the SCR turns on, connecting the battery to the inverter input. The stored energy at the rectifier is sufficient to provide power to the inverter during turn-on of the SCR so that no discontinuity of power is seen by the load. The inverter will continue to supply stable, transient-free power to the critical load until the system battery is discharged.

When ac mains power is restored, the inverter is again automatically supplied with dc power from the rectifier. The SCR is turned off and the battery is automatically recharged to ensure power to the critical load during the next ac mains outage.

Should the monitor sense loss of ac voltage at the inverter output, a transfer switch automatically transfers the load to the ac mains.

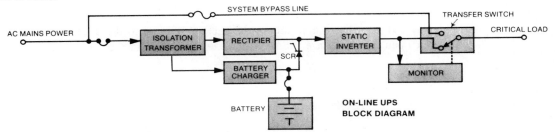

Fig. 29-9B Operation of an uninterruptible power supply (UPS)

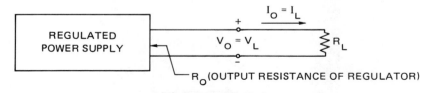

(A) No-Load (Open Circuit) Operation

(B) Full-Load Operation

Fig. 29-10 Regulated power supply operation

BASIC REGULATOR CONCEPTS

A *regulator* is any circuit that maintains a rated output voltage under all conditions. The conditions include two extremes:

1. operation at no load, or an open circuit.
2. operation at a full-load output current.

These two conditions are illustrated in the diagram of the regulated power supply, figure 29-10. In this figure:

V_O = output voltage
I_O = output current
V_{OC} = open circuit voltage (no load)
I_L = load current
V_L = load voltage (with R_L as a load)

An ideal regulator maintains the output voltage V_O at V_{OC} and supplies any current, I_L within a specified range. A practical regulator has the characteristics shown in figure 29-11. The output voltage under load (V_L) is lower than the open circuit, or no-load, voltage (V_{OC}). The percentage regulation is given by the following equation:

$$\% \text{ regulation} = \frac{V_{OC} - V_L}{V_L} \times 100\% \qquad \text{Eq. 29.3}$$

Figure 29-10B shows that $V_L = I_L R_L$. Thus, Equation 29.3 can be rewritten as follows:

$$\% \text{ regulation} = \frac{V_{OC} - V_L}{I_L R_L} \times 100\% \qquad \text{Eq. 29.4}$$

The output resistance of the regulator (R_O) is equal to the difference between the open circuit voltage and the voltage under load divided by the current I_L.

$$R_O = \frac{V_{OC} - V_L}{I_L} \qquad \text{Eq. 29.5}$$

Rewriting Equation 29.4 gives:

$$\% \text{ regulation} = \frac{R_O}{R_L} \times 100\% \qquad \text{Eq. 29.6}$$

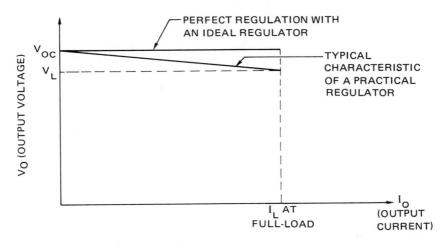

Fig. 29-11 Voltage characteristics of a power supply

The previous analysis leads to the following conclusions concerning regulation:
1. Voltage regulation is a measure of the change in the load voltage due to the load current.
2. The percentage regulation becomes lower (regulation improves) as the output resistance of the regulator decreases.
3. A good regulator has a low-output resistance.
4. A perfect regulator has no output resistance.

Problem 1   In figure 29-10, V_{OC} = 25 V, V_L = 24 V, R_L = 200 Ω. Calculate a. % regulation and b. R_O.

Solution   a. % regulation $= \dfrac{V_{OC} - V_L}{V_L} \times 100\%$

$$= \frac{25\ V - 24\ V}{24\ V} \times 100\%$$

$$= 4.17\%$$

b. $I_L = \dfrac{V_L}{R_L}$

$$= \frac{24\ V}{200\ \Omega}$$

$$= 0.12\ A$$

$R_O = \dfrac{V_{OC} - V_L}{I_L}$

$$= \frac{25\ V - 24\ V}{0.12\ A}$$

$$= 8.33\ \Omega$$

(R29-9)   Define voltage regulation.

(R29-10)   What is the effect of a decrease in the output resistance of a regulator on the percentage regulation?

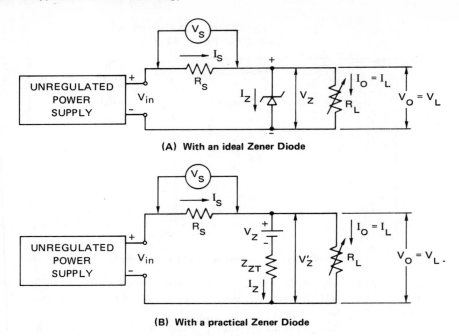

(A) With an ideal Zener Diode

(B) With a practical Zener Diode

Fig. 29-12 Zener diode as a voltage regulator

(R29-11)   What is the output resistance of a perfect regulator?

(R29-12)   Calculate the output resistance of the regulator described in Problem 1, using Equation 29.6.

THE ZENER DIODE AS A VOLTAGE REGULATOR

Figure 29-12 shows a Zener diode used as a voltage regulator. The following equations apply to the circuits shown:

$$V_O = V_Z \qquad \text{Eq. 29.7}$$

$$I_S = \frac{V_{in} - V_O}{R_S} \qquad \text{Eq. 29.8}$$

$$I_Z = I_S - I_L \qquad \text{Eq. 29.9}$$

$$I_L = \frac{V_O}{R_L} \qquad \text{Eq. 29.10}$$

$$V_S = V_{in} - V_Z \qquad \text{Eq. 29.11}$$

$$V_S = I_S R_S \qquad \text{Eq. 29.12}$$

Figure 29-12 represents the most basic and inexpensive form of a voltage regulator. The unregulated dc voltage is V_{in}. R_S is the series dropping resistor. The regulated output is taken across the Zener diode. The unregulated dc voltage at the input reverse biases the Zener diode. The dc input voltage must be larger than the Zener voltage of the diode. When the Zener diode is considered to be ideal, figure 29-12A, the output voltage of the regulator is equal to the

Zener voltage, V_Z. For the practical Zener diode, the output voltage is equal to V'_Z, figure 29-22B. V'_Z is given by the following equation:

$$V'_Z = V_Z + I_Z Z_{ZT} \qquad \text{Eq. 29.13}$$

The series dropping resistor, R_S, must take up the excess voltage supplied by the unregulated power supply. In addition, any fluctuations in V_{in} are absorbed by a voltage drop across this resistor, as shown by Equation 29.11. Changes in the load current (I_L) are offset by changes in the Zener current (I_Z). The current that is not drawn by the load is drawn by the Zener diode.

When the load is not connected so that R_L equals infinity, I_L equals zero and all of the current passes through the Zener diode. As a result,

$$I_Z = I_S = I_{Z(max)} \qquad \text{Eq. 29.14}$$

The Zener diode must be able to pass all of the current I_S. I_Z, or $I_{Z(max)}$, can be equal to or less than I_{ZM}, which is the maximum Zener current determined by the maximum power dissipation. For good design, it is advisable to limit $I_{Z(max)}$ to $0.9\, I_{ZM}$, as shown in figure 29-2.

To operate in the voltage regulating region, the Zener diode must pass current for all expected source voltages and all required load currents. The worst case occurs when the Zener current drops to a minimum for a minimum source voltage and a maximum load current. This minimum current must be equal to or greater than I_{ZK}, which is the minimum Zener current required to sustain breakdown. It is recommended that $I_{Z(min)}$ be equal to or greater than $0.1\, I_{ZM}$, as shown in figure 29-2.

The value of the maximum allowable series limiting resistor is obtained by solving Equations 29.7 through 29.12 simultaneously for R_S:

$$R_{S(max)} = \frac{V_{in(min)} - V_O}{I_{L(max)}} \qquad \text{Eq. 29.15}$$

Unless R_S is less than $R_{S(max)}$, the Zener regulator will not regulate for low source voltages and high load currents. In addition, R_S must have the proper wattage rating to enable it to dissipate the required power.

(R29-13)   Write an equation for $I_{L(max)}$ in terms of I_S and I_Z.

(R29-14)   Write an equation for $I_{L(min)}$ in terms of I_S and I_Z.

(R29-15)   Write an equation for a. $R_{L(max)}$ and b. $R_{L(min)}$ in terms of V_Z and I_L.

(R29-16)   Calculate the range of Zener current in a voltage regulator circuit using a 1N3022B Zener diode.

(R29-17)   What is the maximum value of I_S in a voltage regulator circuit using a 1N3022B Zener diode?

Problem 2   The following values are given for the circuit shown in figure 29-12A: $V_{in} = 20$ V, and $V_Z = 14$ V.
a. Find V_S.
b. Is the current I_S constant under these conditions?
c. List the two functions of I_S.

Solution   a. $V_S = V_{in} - V_Z$
$= 20\ V - 14\ V$
$= 6\ V$

b. The current I_S is constant and is given by Equation 29.8:
$$I_S = \frac{V_{in} - V_O}{R_S}$$
(Note that $V_O = V_Z$.)

c. (1) I_S must be large enough to supply the load current I_L.
(2) I_S must keep the Zener diode "turned on" since $I_S = I_Z + I_L$ (see Equation 29.9)

Problem 3   The following values are assumed for the circuit shown in figure 29-12A: $V_{in} = 20\ V$, $V_Z = 14\ V$, $I_{Z(min)} = 10\ mA$, and I_L varies between 20 mA and 50 mA. Calculate a. R_S, b. the power rating of the dropping resistor, R_S, and c. the Zener power rating, P_{ZM}.

Solution   a. At a maximum load current:
$I_S = I_{Z(min)} + I_{L(max)}$
$= 10\ mA + 50\ mA$
$= 60\ mA$, or 0.06 A
$V_S = V_{in} - V_Z$
$= 20\ V - 14\ V$
$= 6\ V$
$$R_S = \frac{V_S}{I_S}$$
$$= \frac{6\ V}{0.06\ A}$$
$= 100\ \Omega$

b. $P_{R_S} = V_S I_S$
$= (6\ V)(0.06\ A)$
$= 0.36\ W$
Therefore, a 1/2-W resistor is used for R_S.

c. $I_{Z(max)} = I_S - I_{L(min)}$
$= 60\ mA - 20\ mA$
$= 40\ mA$
$P_{ZM} = V_Z I_Z$
$= (14\ V)(40\ mA)$
$= 0.56\ W$
A 1-W Zener diode is used.

Problem 4   The following values are assumed for the circuit of figure 29-12B: $R_S = 200\ \Omega$, $V_Z = 100\ V$, $Z_{ZT} = 20\ \Omega$, and $I_{ZM} = 100\ mA$.

a. What variation in the unregulated power supply voltage, V_{in}, causes a 1% variation in V_O when $I_L = 50\ mA$?

b. With V_{in} constant at 112.2 V, calculate the variation in V_L for a change in I_L from 50 mA to 10 mA.

Solution a. $I_{Z(min)} = 0.01 \, I_{ZM}$
$= (0.1)(100 \text{ mA})$
$= 10 \text{ mA, or } 0.01 \text{ A}$

$I_S = I_{Z(min)} + I_L$
$= 10 \text{ mA} + 50 \text{ mA}$
$= 60 \text{ mA, or } 0.06 \text{ A}$

$V_O = V'_Z = V_Z + I_Z \, Z_{ZT}$
$= 100 \text{ V} + (0.01 \text{ A})(20 \, \Omega)$
$= 100 \text{ V} + 0.2 \text{ V}$
$= 100.2 \text{ V}$

$V_{in} = V_S + V'_Z$
$= I_S R_S + V'_Z$
$= (0.06 \text{ A})(200 \, \Omega) + 100.2 \text{ V}$
$= 12 \text{ V} + 100.2 \text{ V}$
$= 112.2 \text{ V}$

V_O increases by 1%, or 1 V.
That is,

V_O increases to 101.2 V and V'_Z also becomes 101.2 V.

$I_Z = \dfrac{V'_Z - V_Z}{Z_{ZT}}$

$= \dfrac{101.2 \text{ V} - 100 \text{ V}}{20 \, \Omega}$

$= 0.06 \text{ A}$

$V_{in} = V'_Z + V_S$
$= V'_Z + (I_Z + I_L) R_S$
$= 101.2 \text{ V} + (0.06 \text{ A} + 0.05 \text{ A})(200 \, \Omega)$
$= 101.2 \text{ V} + 22 \text{ V}$
$= 123.2 \text{ V}$

A variation in V_{in}, or the unregulated power supply voltage, from 112.2 V to 123.2 V (a change of 11 V) produces a change in V_O of only 1 V when $I_L = 50$ mA.

b. From part a of this problem:
$V_{in} = 112.2 \text{ V}, V_O = V'_Z = 100.2 \text{ V}, I_L = 50 \text{ mA, and } I_Z = 10 \text{ mA}.$

Kirchhoff's Voltage Law is applied to the left-hand loop to obtain the following expression:

$V_{in} - V_S - V'_Z = 0$
$V_{in} - (I_L + I_Z) R_S - V_Z - I_Z Z_{ZT} = 0$
$V_{in} - I_L R_S - I_Z R_S - V_Z - I_Z Z_{ZT} = 0$
$V_{in} - I_L R_S - V_Z = I_Z R_S - I_Z Z_{ZT}$

$I_Z = \dfrac{V_{in} - I_L R_S - V_Z}{R_S + Z_{ZT}}$

As I_L varies from 50 mA to 10 mA:

$$I_Z = \frac{112.2 \text{ V} - (0.01 \text{ A})(200 \text{ }\Omega) - 100 \text{ V}}{200 \text{ }\Omega + 20 \text{ }\Omega}$$

$$\cong 0.046 \text{ A}$$

$$V_O = V'_Z = V_Z + I_Z Z_{ZT}$$
$$= 100 \text{ V} + (0.046 \text{ A})(20 \text{ }\Omega)$$
$$\cong 100.9 \text{ V}$$

In part a, V_O = 100.2 V
In part b, V_O changes to 100.9 V.

A 40-mA variation in the load current from 50 mA to 10 mA produces a change in the output voltage (V_O) of only 0.7 V (100.9 V – 100.2 V = 0.7 V).

Problem 5   Design a Zener voltage regulator circuit to supply 10 V from a 20-V dc source. Select a diode from those listed on the data sheet shown in figure 29-3. Calculate the minimum value of R_L required to achieve good regulation. (Refer to figure 29-12A.)

Solution   The Zener diode selected is a type 1N3020B with a nominal V_Z of 10 V and I_{ZM} of 95 mA.

$$V_S = V_{in} - V_Z$$
$$= 20 \text{ V} - 10 \text{ V}$$
$$= 10 \text{ V}$$

$$I_{Z(max)} = 0.9 \text{ } I_{ZM}$$
$$= (0.9)(95 \text{ mA})$$
$$= 85.5 \text{ mA}$$

$$I_S = I_{Z(max)} = 85.5 \text{ mA}$$

$$R_S = \frac{V_S}{I_S}$$
$$= \frac{10 \text{ V}}{85.5 \text{ mA}}$$
$$\cong 117 \text{ }\Omega$$

$$\text{Power } (R_S) = \frac{V_S^2}{R_S}$$
$$= \frac{(10 \text{ V})^2}{(117 \text{ }\Omega)^2}$$
$$= 0.855 \text{ W}$$
$$\cong 1 \text{ W}$$

$$I_{Z(min)} = 0.1 \text{ } I_{ZM}$$
$$= (0.1)(95 \text{ mA})$$
$$= 9.5 \text{ mA}$$

$$I_{L(max)} = I_S - I_{Z(min)}$$
$$= 85.5 \text{ mA} - 9.5 \text{ mA}$$
$$= 76 \text{ mA}$$

$$R_{L(min)} = \frac{V_Z}{I_{L(max)}}$$
$$= \frac{10 \text{ V}}{76 \text{ mA}}$$
$$\cong 132 \ \Omega$$

(R29-18)   Calculate the value of $R_{L(max)}$ for the circuit of Problem 5.

(R29-19)   In Problem 5, determine the range (where applicable) of: a. I_L, b. I_Z, c. I_S, d. V_O, e. R_L, and f. R_S.

(R29-20)   List the three conditions which must be satisfied when designing a Zener voltage regulator.

SERIES TRANSISTOR REGULATOR   The Zener diode regulator, figure 29-12, has the following disadvantages:
1. limited power handling capability.
2. limited regulation and ripple reduction since Z_{ZT} is not zero.

To overcome these limitations, a Zener diode is used at low currents as the reference element and a power transistor is used as the high-current control element. This arrangement is shown in the series transistor regulator of figure 29-13.

This type of regulator provides regulation over a wide range of loads. The output voltage of this circuit, V_O, is the Zener voltage minus the base emitter voltage. That is,

$$V_O = V_Z - 0.7 \text{ V (ideal Zener diode)} \qquad \text{Eq. 29.16}$$
$$V_O = V'_Z - 0.07 \text{ V (practical Zener diode)} \qquad \text{Eq. 29.17}$$

The unregulated dc input voltage must be greater than the desired output voltage by at least 1 V. The collector-to-emitter voltage is the difference between the unregulated input voltage and the regulated output voltage:

$$V_{CE} = V_{in} - V_O \qquad \text{Eq. 29.18}$$

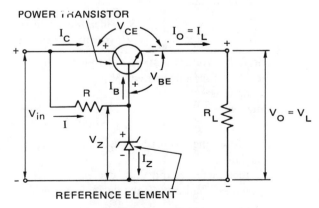

Fig. 29-13 Series transistor regulator

The series regulator operates as follows:

1. The current supplied to the load, I_O, is essentially the same as the collector current of the transistor, I_C.
2. The collector current is essentially equal to βI_B.
3. The transistor base current, I_B, changes as R_L changes.
4. The actual amount of change in the output voltage is very small. That is, it depends upon the amount of change in V_{BE} resulting in the desired change in I_B.
5. The collector current then changes to provide the output current required to maintain a constant output voltage.

(R29-21) What are the two disadvantages of the Zener diode regulator shown in figure 29-12?

(R29-22) Explain the use of the Zener diode and the transistor in the circuit shown in figure 29-13.

SERIES REGULATOR WITH TRANSISTOR FEEDBACK
The performance of the regulator circuit in figure 29-13 can be improved by sensing the output voltage and forcing the series transistor to adjust to the load. This operation is the basic principle of feedback, and is shown in figure 29-14. Feedback is used to regulate the output voltage which is expressed by the following equation:

$$V_O = V_{in} - V_{CE1} \qquad \text{Eq. 29.19}$$

When V_{CE1} is forced to decrease by the same amount as the unregulated input V_{in}, a constant output, V_O, is maintained. The circuit of figure 29-14 operates as follows:

1. Assume that V_O decreases because of a decrease in V_{in}.
2. The decrease in V_O is sensed across the resistor, R_4. In other words, the feedback voltage, V_f, is decreased.

$$V_f = V_{BE2} + V_Z \text{ (ideal Zener diode)} \qquad \text{Eq. 29.20}$$
$$V_f = V_{BE2} + V'_Z \text{ (practical Zener diode)} \qquad \text{Eq. 29.21}$$

3. The Zener voltage is fixed. Therefore, the decrease in the feedback voltage, V_f, causes a decrease in V_{BE2}.
4. I_{C2} decreases.
5. However,

$$I_{B1} = I - I_{C2}. \qquad \text{Eq. 29.22}$$

Therefore, the decrease in I_{C2} causes I_{B1} to increase since I remains constant.

6. I_{C1} then increases and V_{CE1} decreases.
7. I_O increases and cancels the decrease in the output voltage, V_O.

(R29-23) How is regulation achieved in the circuit shown in figure 29-14?

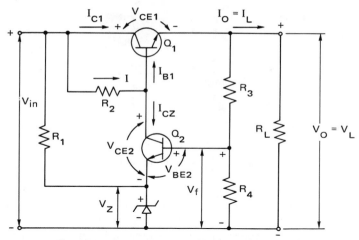

Fig. 29-14 Series regulator with transistor feedback

BASIC SHUNT REGULATOR

A basic shunt regulator is shown in figure 29-15. The regulated output voltage is the sum of the Zener and base emitter voltages, as shown in the following equations:

$$V_O = V_Z + V_{BE} \text{ (ideal Zener diode)} \qquad \text{Eq. 29.23}$$
$$V_O = V'_Z + V_{BE} \text{ (practical Zener diode)} \qquad \text{Eq. 29.24}$$

The shunt regulator operates in the following manner:

1. Assume that R_L decreases.
2. V_O then decreases.
3. V_{BE} decreases, according to Equations 29.23 and 29.24, since V_Z and V'_Z are constant.
4. I_B and I_C decrease.
5. Since the input current I is nearly constant, I' is also constant. (I_B is negligible compared to I or I'.)
6. The output current (I_O) increases because of the decrease in I_C:

$$I' = I_C + I_O \qquad \text{Eq. 29.25}$$

7. The increase in the output current I_O tends to keep the output voltage V_O constant.

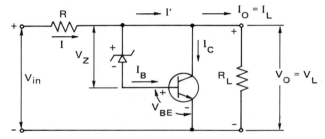

Fig. 29-15 Basic shunt regulator

(A) Stud Package   (B) Plastic Package

Fig. 29-16 SCR configurations

SILICON CONTROLLED RECTIFIER (SCR)

The silicon controlled rectifier (SCR) is a solid-state device whose base material is silicon. It belongs to a group of semiconductor devices known as thyristors. These devices have a control mechanism. The SCR has two stable states of operation. One stable mode is called the OFF state and has negligible current. The other stable mode, called the ON state, has a high current which is limited by the resistance of the external circuit.

The SCR combines selected features of both the diode and the transistor. SCRs can be used to switch power on or off. They can control a power value as high as 10 MW, with individual ratings as high as 2000 A at 1800 V. Because the SCR has no moving parts, it is far superior to a mechanical switching device.

The SCR has a large power gain. It is an extremely fast switch and can be used with devices having a frequency range as high as 50 kHz. Typical high-frequency applications include induction heating equipment and ultrasonic cleaning. Some SCR configurations are illustrated in figure 29-16.

To operate as a power switch, the SCR requires:
1. A source of power.
2. A load to be supplied with power.
3. A trigger circuit.

The schematic symbol for the SCR and a block diagram of a typical SCR circuit are shown in figure 29-17. To establish forward conduction, the anode must be positive with respect to the cathode. However, to turn on the SCR, a positive pulse of sufficient magnitude must be applied to the gate to establish the gate current I_G. The SCR remains on even after the trigger pulse has ceased. This means that the trigger circuit no longer controls the load current I_L. The SCR can be switched off by one or more of the following methods:
1. Remove the power source.
2. Reduce the value of the power source voltage so that the current through the SCR drops below the holding current value (I_H).
3. Reverse the polarity of the power supply. (An ac power supply illustrates what happens when the polarity is reversed.)
   Note: The student must remember that the SCR is *not* turned off by removing the trigger pulse.

(R29-24)   List the three basic requirements for an SCR circuit.

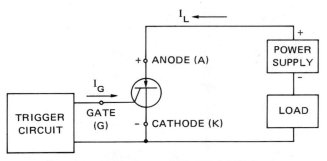

Fig. 29-17 Block diagram of a typical SCR circuit

(R29-5) What is the function of the trigger circuit?

SCR Construction and Theory of Operation. The block diagram of figure 29-18A shows that the SCR consists of four layers of semiconductor material arranged in the following order — PNPN. There are three *p-n* junctions between the anode and the cathode. Each junction has the same characteristics as any diode junction. The two-transistor equivalent circuits, figures 29-18B and 29-18C are used to explain the operation of an SCR.

If the SCR is biased using the polarities shown in figure 29-18A, junctions J_1 and J_3 are forward biased. Junction J_2 is reverse biased to prevent current across the junction. When the gate voltage is zero, and V_{AK} is less than the forward breakover voltage, no current can flow through the SCR. According to figure 29-18C:

1. The collector of Q_2 is connected to the base of Q_1.
2. The collector of Q_1 is connected to the base of Q_2.
3. The emitter of Q_2 is the cathode terminal.
4. The emitter of Q_1 is the anode terminal.
5. The junction of the collector of Q_1 and the base of Q_2 is the gate terminal.

When V_G is applied as a positive gate-to-cathode trigger voltage, for the schematic shown in 29-18C, the following sequence of operation occurs:

1. The emitter junction of Q_2 is forward biased.
2. Gate current I_G flows.
3. Base current I_{B2} flows.
4. Q_2 switches on.
5. Collector current I_{C2} flows.
6. Q_1 switches on since I_{B1} is the same as I_{C2}.
7. Collector current I_{C1} flows.
8. I_{C1} increases and causes a corresponding increase in I_{B2}. The base current I_{B2} is maintained by I_{C1}.
9. The increase in I_{B2} results in a further increase in I_{C2}. A regenerative increase in the collector current of each transistor results. That is, each collector supplies the base current for the other transistor.
10. When V_G, the trigger voltage, is removed, $I_G = 0$.

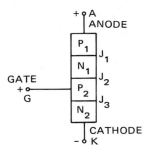

(A) SCR block diagram

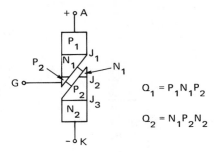

(B) Block diagram of the two-transistor
equivalent of an SCR

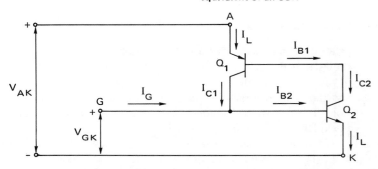

(C) Two-transistor equivalent used on an SCR schematic diagram

Fig. 29-18 Analyzing the SCR by means of a two-transistor equivalent

11. Both Q_1 and Q_2 remain on. The currents I_{C2} (I_{B1}) and I_{C1} (I_{B2}) continue to be supplied by the voltage V_{AK}. *Latching* is the term applied to the ability of the SCR to remain on when the triggering current is removed.

(R29-26)   Does the SCR turn on in the schematic shown in figure 29-18C if $V_G = 0$ and V_{AK} is less than the forward breakover voltage?

SCR Characteristics and Ratings. The SCR characteristics are shown in figure 29-19. The following currents and voltages are shown on the characteristics:

1. The *forward breakover* voltage is the voltage at which the SCR switches from a nonconducting state to a conducting state with no triggering gate signal. Manufacturers identify this voltage as $V_{(BR)F}$, V_{BO}, or $V_{F(BO)}$. Normally, this voltage is not exceeded.

2. The *holding current*, I_H, or the latching current, is the current value below which the SCR switches from on to off. I_H is the minumum level of I_F that maintains SCR conduction.

3. The *forward* and *reverse blocking regions* correspond to the open circuit condition for the SCR. It is in these regions that the flow of current is blocked from the anode to the cathode.

4. The *reverse breakdown voltage* is equivalent to the Zener or avalanche region of the two-layer semiconductor diode. Normally, this voltage is not exceeded.

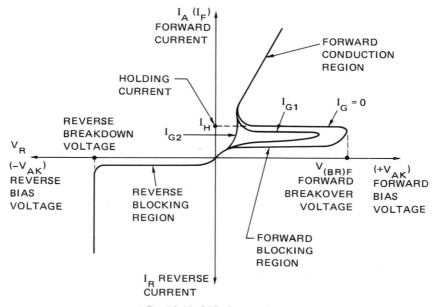

Fig. 29-19 SCR characteristics

The characteristics shown in figure 29-19 are similar to those of the semiconductor diode. However, one difference is the horizontal offshoot that occurs before the SCR enters the forward conduction region. When $I_G = 0$, V_F must reach the largest breakover voltage before the SCR can enter the conduction region. Unlike the diode, the breakover point of the SCR can be changed by increasing the gate current. When the gate current is increased to I_{G1}, as shown in figure 29-19, the value of V_F required for conduction is reduced. When the gate current is increased to I_{G2}, the SCR fires at very low voltage values. In this case, the characteristics begin to resemble those of a basic *p-n* junction diode.

(R29-27) Is a single SCR able to pass current in both directions?

(R29-28) How can the breakover point of the SCR be changed?

SCR Power Control. A single SCR conducts only on that half-cycle of ac voltage when the proper polarity is applied to the anode and cathode. The point when the positive gate pulse is applied determines the level of the output. *Phase control* is the most common form of power control. In this mode of operation, the SCR is held in the off state. As a result, all current flow is blocked in the circuit except for a small leakage current. The SCR remains off for a portion of the positive half-cycle until it is triggered or fired into conduction. The control circuitry determines when the SCR is fired.

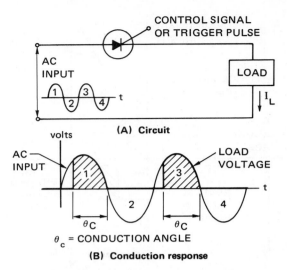

(A) Circuit

(B) Conduction response

Fig. 29-20 Power control using a single SCR

A single SCR connected in series with the load is shown in figure 29-20. The SCR controls the power delivered to the load in the following manner:

1. The control or trigger pulse signal keeps the SCR in the off state for the first part of the positive half-cycle of the ac input.
2. The SCR fires at the desired time and causes the current I_L. This current depends upon the applied voltage and the load.
3. The load voltage during the conduction angle, θ_C, is nearly equal to the applied ac voltage. (The voltage drop across the SCR, approximately 1 V, is neglected.)
4. Once the positive gate signal applied by the control circuit fires the SCR, the gate circuit loses control. The SCR continues to fire as long as its anode is positive with respect to the cathode.
5. The SCR turns off at the end of the positive half-cycle. (The SCR actually turns off when I_L falls below the holding current, I_H. This condition occurs when the input is still slightly positive.)
6. The SCR cannot be fired again until the gate signal voltage becomes positive during alternation 3 of the input voltage.

The previous description of the operation of the SCR leads to the following conclusions:

1. A single SCR controls only one half-cycle.
2. The amount of power delivered to the load is proportional to the length of time that the SCR is conducting.
3. The output power can be varied from nearly 0% of the available input power to almost 50% of the available input power. The SCR may be on for the full half-cycle, in which case the conduction angle is equal to 180°, or it may be off for the full half-cycle, with a conduction angle equal to 0°.

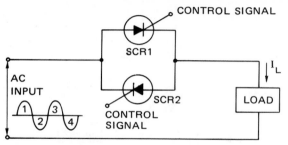

CONTROL SIGNAL

SCR1

AC INPUT

SCR2

CONTROL SIGNAL

I_L

LOAD

(A) Circuit with two SCRs connected in an inverse parallel arrangement

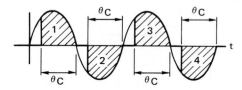

(B) Conduction response

Fig. 29-21 Power control using two SCRs

The circuit shown in figure 29-21 uses two SCRs connected in an inverse parallel arrangement. That is, the anode of one SCR is connected to the cathode of the other SCR. This circuit controls both the negative and the positive half-cycles of the ac input.

This circuit operates in the following manner:

1. During the positive half-cycle of the ac input voltage (alternations 1 and 3), the anode of SCR2 is negative and SCR2 is off. At the same time, SCR1 is fired into conduction. (The method of firing was explained for the single SCR circuit shown in figure 29-20.)

2. During the negative half-cycle of the ac input voltage (alternations 2 and 4), the anode of SCR1 is negative and SCR1 is off. At the same time, SCR2 is fired into conduction by an appropriate signal to its gate.

3. SCR1 controls the positive half-cycle and SCR2 controls the negative half-cycle. As a result, this circuit provides full-wave power control.

The circuit shown in figure 29-22 represents an alternate method of obtaining full-wave power control. The operation of this circuit is as follows:

1. The ac input is full-wave rectified by the bridge circuit consisting of diodes D_1, D_2, D_3, and D_4.

2. By controlling the conduction angle of the SCR, the power delivered to the load can be varied from 0% to 100% of the available input power.

Figure 29-23 shows a basic circuit that can be used to control the conduction angle, θ_C, of an SCR. This circuit operates in the following manner:

1. During alternation 1 of the input voltage, the capacitor is charged by way of R_1.

2. When the capacitor voltage reaches the gate firing value for the SCR, the SCR is turned on.

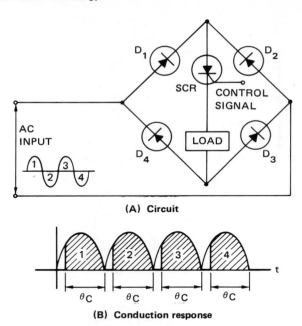

(A) Circuit

(B) Conduction response

Fig. 29-22 Full-wave bridge with single-SCR control

3. The SCR continues to conduct until the input voltage goes to zero in alternation 1.
4. When the SCR is on, C_1 is discharged. The student should observe the short circuit across C_1, while the SCR is on.
5. The cycle is repeated when the input becomes positive again, during alternation 3.
6. The capacitor C_1 controls the conduction angle. The rate at which C_1 charges determines when its voltage is high enough to fire the SCR.
7. The conduction angle can be controlled by adjusting R_1.

(R29-29)   What is the most common form of power control?

(R29-30)   Define the conduction angle.

(R29-31)   For the circuit shown in figure 29-20, what is the maximum output power delivered to the load?

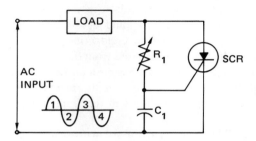

Fig. 29-23 Basic circuit used to control the conduction angle, θ_C, of an SCR

(R29-32)   For the circuit shown in figure 29-21, what is the maximum output power delivered to the load?

(R29-33)   What is the maximum output power which the circuit of figure 29-22 can deliver to the load?

SCR Applications. Common applications of SCRs include:
Relay controls
Time-delay circuits
Regulated power supplies
Static switches
Motor controls
Choppers
Inverters
Cycloconverters
Battery chargers
Heater controls
Phase controls
Controlled bridge rectifiers
Spot and seam welding control equipment
Lighting controls

An SCR isolation trigger transformer is shown in figure 29-24. Industrial and commercial applications of this transformer include motor speed controls, lighting controls, and heater controls. Power controllers designed for 100% conduction are shown in figure 29-25. These controllers are used in laboratory furnaces, laboratory ovens, plastic extruders, plastic molding machinery, food packaging machinery, and resistance heating devices. Power controllers designed for phase angle firing are illustrated in figure 29-26. These controllers are used in applications requiring sophisticated and precise controls, such as vacuum furnaces, heat treating furnaces, infrared heating devices, radio frequency heating equipment, large ovens, fan motors, pump motors, variable resistance loads, and silicon carbide heating elements.

Fig. 29-24 SCR isolation trigger transformer

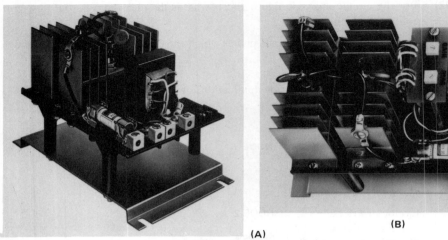

(A)

(B)

Fig. 29-25 Power controller for zero crossover firing (100% conduction)

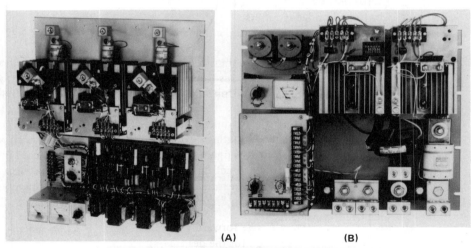

(A)   (B)

Fig. 29-26 Power controller for phase angle firing

THE DIAC   The DIAC is a two-terminal, solid-state, bidirectional thyristor. One type of DIAC is a parallel inverse combination of five semiconductor layers, figure 29-27A. Figure 29-27B shows the various symbols that represent the DIAC. These symbols indicate that the current can flow in either direction. Note that the device is shown with two emitters (two anodes) and no base connection.

DIACs are manufactured in either one of two forms:

1. a five-layer device called a *bidirectional diode thyristor.*
2. a three-layer device called a *bidirectional trigger diode.*

The characteristic curves for both of these devices are similar to the curves shown in figure 29-27C. These curves may be used to analyze each type of DIAC as follows:

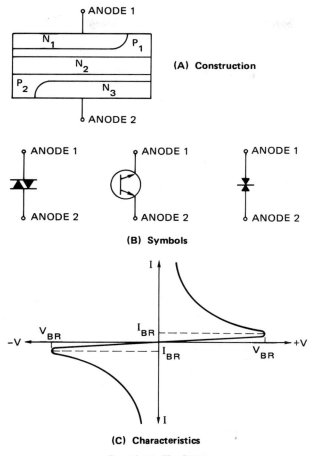

(A) Construction

(B) Symbols

(C) Characteristics

Fig. 29-27 The DIAC

1. Current flow in the forward and reverse directions is similar. This characteristic accounts for the designation of the device as a diode alternating current or DIAC.
2. The DIAC does not conduct appreciable current in either direction until its breakover voltage, V_{BR}, is reached.
3. The breakdown voltages are nearly equal in either direction.
4. Once the breakover voltage is reached, in either direction, there is a negligible resistance. The voltage drop across the DIAC changes rapidly to a much lower value. As a result, the current flow increases.

The DIAC is used as a triggering device, as for example, in figure 29-28, where it is used to fire an SCR. The DIAC in this circuit operates in the following manner:

1. During the positive half-cycle of the ac input, the capacitor C_1 charges through R_1.

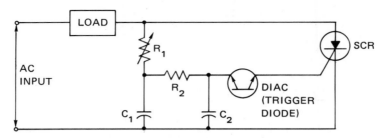

Fig. 29-28 A DIAC used as a triggering device to fire an SCR

2. The voltage across C_1 causes C_2 to charge through R_2.
3. The SCR remains off as long as the voltage across C_2 is less than the breakover voltage of the DIAC.
4. When the voltage across C_2 reaches the breakover voltage of the DIAC, the DIAC fires, or conducts. A high-current gate pulse is supplied which then fires the SCR.
5. Capacitors C_1 and C_2 are then discharged by way of the DIAC and the gate of the SCR.
6. The conduction angle is controlled by the point at which the voltage across C_2 reaches the firing potential of the DIAC. This angle can be adjusted by the variable resistor R_1. When R_1 is decreased, the capacitors charge more quickly and the firing times of the DIAC and SCR are decreased. This means that decreasing R_1 increases the conduction angle.

The circuit shown in figure 29-28 provides half-wave control. If the inverse parallel connection of two SCRs is used, as shown in figure 29-21A, full-wave control is obtained.

(R29-34) Compare a *p-n* junction diode and a DIAC.

(R29-35) List the two types of DIACs.

(R29-36) If the resistor R_1 in the circuit of figure 29-28 is increased in value, what is the affect upon the conduction angle of the SCR?

THE TRIAC   The TRIAC is similar to the DIAC with one exception. The TRIAC has a gate terminal that is used to control the conditions at which this bilateral device is turned on in either direction. The device is essentially a pair of SCRs with a common gate for the two anodes. The TRIAC can be viewed as two SCRs connected back-to-back, or in an inverse parallel arrangement. The device has three terminals, figure 29-29A. The main terminal 1 is labeled MT1 and the other main terminal is labeled MT2. By convention, the forward direction of the TRIAC is assumed to be from MT2 to MT1. When MT2 is positive with respect to MT1, the TRIAC is forward biased. For the reverse direction, the TRIAC is reverse biased when MT1 is positive with respect to MT2. The gate

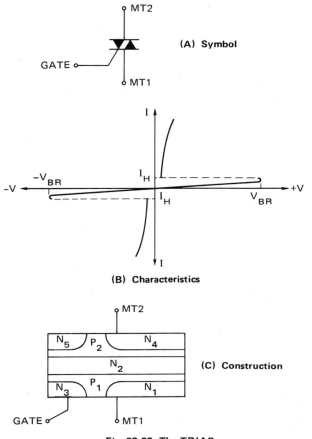

(A) Symbol

(B) Characteristics

(C) Construction

Fig. 29-29 The TRIAC

current can control the action of the TRIAC in either direction. The mode of control is similar to that described for the SCR. The characteristic curves are shown in figure 29-29B. The basic construction of the TRIAC is shown in figure 29-29C.

Figure 29-30A shows a TRIAC used in a light intensity control circuit (light dimmer). In this circuit, the TRIAC controls the ac power to the load by switching on or off during the positive and negative alternations of the ac input. The circuit fires in both the positive and negative alternations since both the DIAC and TRIAC conduct in either direction. Figure 29-30B shows the waveform of the current I_L through the lighting load. The student should refer to the previous explanation of the circuit shown in figure 29-28.

(R29-37) What is the purpose of the variable resistor, R_1, in figure 29-30A?

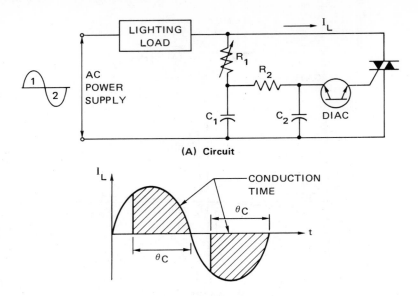

(A) Circuit

(B) Waveform of current I_L through a lighting load

Fig. 29-30 TRIAC used in the light intensity control (light dimmer) of a lighting load

LABORATORY EXERCISE 29-1 ZENER DIODE CHARACTERISTICS

PURPOSE   In completing this exercise, the student will

- determine the volt-ampere characteristics of a Zener diode with reverse biasing.
- observe the use of a Zener diode as a voltage regulator.

EQUIPMENT AND MATERIALS

1 Power supply, dc
1 Digital multimeter
1 Analog multimeter
1 Zener diode, 1N3019B
1 Resistor, 3.3 kΩ, 2 W
1 Resistor, 500 Ω, 5 W
1 Switch, SPST

PROCEDURE

1. Connect the circuit shown in figure 29-31.
   a. Refer to the data sheet shown in figure 29-3 and determine the following electrical characteristics of the Zener diode: a. P_{ZM}, b. V_Z, and c. I_{ZM}.

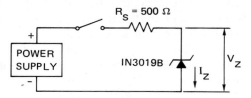

Fig. 29-31

 b. Calculate the range of voltage regulating current for the diode, that is, $0.1\,I_{ZM}$ to $0.9\,I_{ZM}$.

 c. Set the output of the dc power supply so that V_Z measures 1.0 V. Measure I_Z.

 d. Repeat step 1.c. for V_Z values at one-volt intervals between 2 V and 9.1 V. CAUTION: Do not exceed the maximum rating of the diode.

2. Connect the circuit shown in figure 29-32.

 a. Slowly increase the output of the dc power supply until I_Z measures 28 mA.

 b. Measure V_{in}, V_O, and I_S.

 c. Determine and record the range of variation of V_{in} over which V_O remains within ±5% of its value in step 2.b.

 d. Measure V_{in}, V_O, and I_S when $I_Z = 0.1\,I_{ZM}$.

 e. Measure V_{in}, V_O, and I_S when $I_Z = 0.9\,I_{ZM}$.

3. Answer the following questions and perform the calculations requested:

 a. Plot a graph of I_Z versus V_Z using the data in step 2.

 b. Compare the graph plotted in step 3.a. with the expected reverse characteristics of the Zener diode.

 c. In step 2.c., does the voltage V_O remain constant within ±5% of its value in step 2.b.?

 d. What is the significance of the data obtained in step 2.b.?

 e. Describe how the Zener diode circuit of figure 29-32 operates as a voltage regulator. Use the data from step 2 in this explanation.

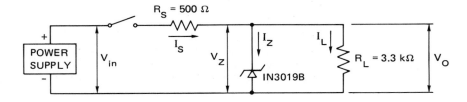

Fig. 29-32

EXTENDED STUDY TOPICS

1. A 1N3022B Zener diode is used in a voltage regulator circuit with $R_S = 180\ \Omega$ and $V_{in} = 25$ V. R_L ranges from 200 Ω to infinity. Calculate: a. I_S, b. $I_{L(min)}$, c. $I_{L(max)}$, d. $I_{Z(min)}$, and e. $I_{Z(max)}$.

2. Is the circuit of Topic 1 operating in the voltage regulator region?

3. A 1N3025B Zener diode is used in a voltage regulator circuit. If I_S measures 75 mA at the instant I_L drops to 5 mA, what happens?

4. A 1N3016B Zener diode is used in a voltage regulator circuit under the following conditions: V_{in} = 15 V to 20 V, I_L = 20 mA to 100 mA, and V_Z = 6.8 V. Calculate the value of R_S needed to hold V_O constant under all conditions of changes in V_{in}, and changes in I_L.

5. If R_S in Topic 4 is set equal to 100 Ω, describe the effect on the circuit.

6. A voltage regulator using a 1N3016B Zener diode operates under the following conditions: V_{in} = 15 V to 20 V, and R_S = 70 Ω. Find: a. the $I_{Z(min)}$ rating of the diode, b. the $I_{Z(max)}$ rating of the diode, c. I_S when V_{in} = 15 V and 20 V, d. $R_{L(min)}$ when V_{in} = 15 V and 20 V, e. $R_{L(max)}$ when V_{in} = 20 V, and f. I_Z when V_{in} = 15 V and R_L is set at $R_{L(max)}$.

7. List the three methods which can be used to switch an SCR to the off state.

8. Describe the meaning of latching as applied to an SCR.

9. Why is the voltage drop across an SCR small during conduction?

10. Compare an SCR, a DIAC, and a TRIAC.

Chapter 30

Digital Logic Fundamentals

OBJECTIVES
After studying this chapter, the student will be able to

- define the difference between analog and digital signals and devices.
- use the decimal, binary, octal, and hexadecimal system notations.
- discuss the BCD (binary coded decimal) system.
- explain the use of diodes and transistors as switches.
- discuss digital logic gates and the basic principles of Boolean algebra.
- use the NAND and NOR gates as universal logic elements.
- apply the EXCLUSIVE-OR and EXCLUSIVE-NOR functions.
- discuss the organization and format of computer words.

ANALOG AND DIGITAL SIGNALS AND DEVICES
An analog signal is a smooth, continuously varying signal, such as the height of a column of mercury in a liquid thermometer. The height of the mercury column varies as the temperature varies. Therefore, the indication is said to be representative of the temperature, or *analogous* to the temperature. An analog device has the ability to manipulate, or measure, analog signals.

A digital signal is discrete, or discontinuous. For example, the time displayed by a digital clock is a digital signal. A digital device can process or count digital signals. Such a device can determine the presence, or absence, of a signal; whether the signal is true or false; and if there is a punched hole in a computer card. Numerical values are represented by a finite number of digits which can be stored, displayed, counted, or processed.

(R30-1)   Classify the following devices as analog devices or digital devices: liquid-filled thermometer, adding machine, hand calculator, and slide rule.

(R30-2)   How does an analog device differ from a digital device?

DECIMAL NUMBER SYSTEM   A number system can be formed by selecting a set of symbols to represent numerical values. The number of symbols used is called the *base* or *radix* of that number system.

   The decimal number system is formed by selecting the numerical symbols 0 through 9 (that is, 0, 1, 2, 3, 4, 5, 6, 7, 8, 9). Each symbol indicates a different numerical quantity; the smallest quantity is 0 and the largest quantity is 9. Since ten different symbols are used, the base or radix of the decimal number system is ten.

   To express a quantity greater than nine, two or more symbols must be used. Each symbol, called a *digit*, acquires a value determined by its position with relation to the other digits of the quantity being expressed. This type of number system is known as a positional system. A *decimal number* is the sum of the products of the digits and their respective positional weights. Table 30-1 illustrates the positional weights for several digit locations. The table shows that each position has a value that is a power of ten. Positions with positive powers of ten are located to the left of the units, or reference, digit. Positions with negative powers of ten are located to the right of the reference digit; that is, they are located to the right of the decimal point.

Problem 1   Write the decimal number 6543_{10} in positional notation.

Solution   The number 10 written to the right and slightly below the number 6543 indicates that the number is a decimal number having a base of ten. As stated previously, a decimal number is the sum of the products of the digits and their respective positional weights.

$$6543_{10} = (6 \times 10^3) + (5 \times 10^2) + (4 \times 10^1) + (3 \times 10^0)$$
$$= 6000 + 500 + 40 + 3$$
$$= 6543_{10}$$

(R30-3)   What is the radix or base of the decimal number system?

(R30-4)   How many symbols are used in the decimal number system?

(R30-5)   Write the decimal number 385.23_{10} in positional notation.

TABLE 30-1   Decimal Number System: Digit Locations and Positional Weights

10^3	10^2	10^1	10^0	$\bullet$	10^{-1}	10^{-2}	10^{-3}
1000	100	10	1 (Reference)	$\bullet$ (Decimal point)	1/10	1/100	1/1000

TABLE 30-2 Binary Number System: Bit Locations and Positional Weights

←	2^3	2^2	2^1	2^0	●	2^{-1}	2^{-2}	2^{-3}	→
←	8	4	2	1 (Reference)	● (Binary point)	1/2	1/4	1/8	→

BINARY NUMBER SYSTEM

Digital electronic devices and circuits operate in one of two states: ON or OFF, CLOSED or OPEN, TRUE or FALSE, or HIGH or LOW. All of these conditions can be represented by the two symbols used in the binary number system, 0 or 1. Because only two symbols are used to form all binary numbers, the base of the binary number system is 2.

Binary numbers are formed by grouping together several 0s and 1s. The digits in a binary number are called *bits*. This term is a contraction of the words "binary digit."

Each bit of a binary number acquires a value determined by its position relative to the other bits of that number. A *binary number* is the sum of the products of the bits and their respective positional weights. Table 30-2 gives the positional weights for several bit locations. The table shows that the value of each position is a power of the base 2. Positions having positive powers are located to the left of the unit, or reference, digit. Positions having negative powers are located to the right of the reference digit. That is, these positions are located to the right of the binary point.

The *binary point*, which separates the integer part of the binary number from the fractional part, corresponds to the decimal point in the decimal (base 10) system. The bit on the extreme left in a binary number is called the *most significant bit*, abbreviated MSB. This bit has the greatest relative value of all of the bits in the binary number. The bit at the extreme right in a binary number is called the *least significant bit*, abbreviated LSB. The bit has the smallest relative value of all of the bits in the binary number.

Problem 2
Solution

Find the value of the MSB and LSB in the binary number 1000.101_2.

The bit on the extreme left, or the most significant bit (MSB), is 1. It is the fourth bit to the left of the binary point. From Table 30-2, the positional weight of this bit is equal to 2^3 or 8. Therefore, MSB = (1×8) = 8.

The bit at the extreme right, or the least significant bit (LSB), is 1. It is the third bit to the right of the binary point. From Table 30-2, the positional weight of this bit is equal to 2^{-3}, or 1/8. Therefore, LSB = $(1 \times \frac{1}{8})$ = $\frac{1}{8}$.

Numbers can be expressed in the decimal number system or in the binary number system. The decimal numbers 0 through 15 and their equivalent binary numbers are listed in Table 30-3. The binary number 101 is read as "one-zero-one", and *not* as "one hundred and one."

TABLE 30-3 Equivalent Decimal and Binary Numbers

Decimal Number	Binary Number
0	0
1	1
2	10
3	11
4	100
5	101
6	110
7	111
8	1000
9	1001
10	1010
11	1011
12	1100
13	1101
14	1110
15	1111

(R30-6)   What is a bit?

(R30-7)   Find the value of the MSB and LSB in the binary number 1010_2.

(R30-8)   What is the decimal equivalent of the binary number 1010_2?

BINARY TO DECIMAL CONVERSION   The decimal equivalent of a binary number can be found as follows:
1. Each bit is multiplied by its positional weight.
2. The products are added to obtain the decimal number.

Problem 3 Solution   Find the decimal equivalent of the binary number 10101_2.

$$10101_2 = (1 \times 2^4) + (0 \times 2^3) + (1 \times 2^2) + (0 \times 2^1) + (1 \times 2^0)$$
$$= 16 + 0 + 4 + 0 + 1$$
$$= 21_{10}$$

(R30-9)   Calculate the decimal equivalent of 1011_2.

(R30-10)   Find the decimal equivalent of 101010_2.

DECIMAL TO BINARY CONVERSION   A method of converting a decimal number to the binary number requires the successive division of the decimal number by the base 2. The resulting quotient of a division is divided again to obtain another quotient. The process is continued until a zero quotient is obtained. The remainders at each division are the binary digits, in order from the LSB to the MSB. This method is known as the *double dabble process*.

Problem 4   Find the binary equivalent of the decimal number 13_{10}.

REMAINDERS

Solution

$13 \div 2 = 6$	1 (LSB)
$6 \div 2 = 3$	0
$3 \div 2 = 1$	1
$1 \div 2 = 0$	1 (MSB)

Therefore, $13_{10} = 1101_2$.

The student must remember that the MSB is the bit at the extreme left and the LSB is the bit at the extreme right of the binary number.

(R30-11)   Find the binary equivalent of the decimal number 24_{10}.

BINARY ADDITION   There are four basic rules for adding binary digits:

Rule 1: $0 + 0 =$ 0 (zero plus 0 equal 0)
Rule 2: $0 + 1 =$ 1 (zero plus 1 equals 1)
Rule 3: $1 + 0 =$ 1 (one plus 0 equals 1)
Rule 4: $1 + 1 = 10$ (one plus 1 equals 0 with a carryover of 1 to the next higher column)

Note that three of the addition rules yield a single binary digit. According to rule 4, the addition of two 1s yields the binary number 10.

Problem 5   Add 10111_2 and 00111_2.
Solution

```
 10111
+00111
 11110  Answer
```

The steps in the addition of the two binary numbers are as follows:
1. Start with the LSB: $1 + 1 = 10$. The zero is recorded as the answer in that position, and a 1 is carried to the next column.
2. In the next place, $1 + 1 = 10$ plus the carry of $1 = 11$. The 1 is recorded as the answer in that position, and a 1 is carried to the third position.
3. $1 + 1 = 10$, plus the carry of $1 = 11$. The 1 is recorded as the value in that position, and a 1 is carried to the fourth position.
4. $0 + 0 = 0$, plus the carry of $1 = 1$. There is no carry to the next position.
5. In the MSB position, $1 + 0 = 1$.

(R30-12)   Add 101101_2 and 110110_2.

OCTAL NUMBER SYSTEM   The base of the octal number system is 8. The symbols in this system are 0, 1, 2, 3, 4, 5, 6, 7. The basic octal symbols are written in digit notation to yield a useful number. Each digit acquires a value determined by its position with relation to the other digits of the number. An *octal number* is the sum of the

TABLE 30-4 Octal Number System: Digit Locations and Positional Weights

8^3	8^2	8^1	8^0	•	8^{-1}	8^{-2}	8^{-3}
512	64	8	1	•	1/8	1/64	1/512
			(Reference)	(Octal point)			

products of the digits and their respective positional weights. Table 30-4 gives the positional weights for several digit locations. The table shows that the value of each position is expressed as a power of the base 8. The positions of positive powers are located to the left of the unit, or reference, digit. The positions of negative powers are located to the right of the reference digit. That is, these positions are located to the right of the octal point.

(R30-13)   What is the base of the octal number system?

(R30-14)   How many digits (symbols) are used in the octal number system?

OCTAL TO DECIMAL CONVERSION

The decimal equivalent of an octal number is found in the following manner:
1. Each digit is multiplied by its positional weight.
2. The products are added to obtain the decimal number.

Problem 6   Find the decimal equivalent of the octal number 2371_8.

Solution
$$2371_8 = (2 \times 8^3) + (3 \times 8^2) + (7 \times 8^1) + (1 \times 8^0)$$
$$= (2 \times 512) + (3 \times 64) + (7 \times 8) + (1 \times 1)$$
$$= 1024 + 192 + 56 + 1$$
$$= 1273_{10}$$

(R30-15)   Convert 1024_8 to its decimal equivalent.

DECIMAL TO OCTAL CONVERSION

A decimal number is converted to its equivalent octal number by successively dividing by eight. The resulting quotient of each division is divided again to obtain another quotient. The process is continued until a zero quotient results. The remainders at each step are the octal digits, in order from the least significant digit (LSD) to the most significant digit (MSD). This method is similar to the one used to convert a decimal number to a binary number.

Problem 7   Find the octal number equivalent of the decimal number 359_{10}.

REMAINDERS

Solution
$$359 \div 8 = 44 \qquad 7 \text{ (LSD)}$$
$$44 \div 8 = 5 \qquad 4$$
$$5 \div 8 = 0 \qquad 5 \text{ (MSD)}$$

Therefore, $359_{10} = 547_8$

(R30-16)   Convert the decimal number 435_{10} to octal notation.

TABLE 30-5 Equivalent Binary Numbers and Octal Digits

Binary Number	Octal Digit
000	0
001	1
010	2
011	3
100	4
101	5
110	6
111	7

CONVERSIONS BETWEEN THE OCTAL AND BINARY SYSTEMS

Octal numbers provide a more efficient way of writing digital information or instructions for computers and other digital equipment. A multibit binary number can be expressed by a few octal digits.

Table 30-5 lists three-bit binary numbers and their equivalent octal digits. This table can be used to convert numbers between the two systems. Since the bases 2 and 8 are related ($2^3 = 8$), it is a simple matter to convert between binary and octal numbers. The exponent "3" indicates that a three-bit binary number is a one-digit octal number and vice versa.

A binary number is converted to its octal equivalent in the following manner:

1. Start at the binary point of the binary number and group the bits by threes in both directions.
2. Each group of three bits is replaced by its equivalent octal digit to yield the octal number.
3. Zero may be added to the left of the MSB or to the right of the LSB to obtain a group of three bits.

An octal number is converted to its binary equivalent in the following manner:

1. Each octal digit is replaced by its equivalent binary group of three bits.
2. The procedure for converting a binary number to its octal equivalent is followed in reverse.

Problem 8

Convert 001111000101_2 to octal notation.

Solution

The binary number is divided into groups of three bits starting from the right. Each group of three bits is replaced by the appropriate octal digit listed in Table 30-5.

001 111 000 101

Binary 101 = octal 5 (LSD)
Binary 000 = octal 0
Binary 111 = octal 7
Binary 001 = octal 1 (MSD)

Therefore, the conversion is: $001111000101_2 = 1705_8$

Problem 9 Convert 10111011.1111_2 to octal notation.

Solution Starting at the binary point of the binary number, the bits are grouped in units of three in both directions.

10 111 011 . 111 1

Zeros are added to the left of the MSB and to the right of the LSB to obtain groups of three bits.

010 111 011 . 111 100

Each group of three bits is replaced by the appropriate octal digit listed in Table 30-5.

Binary 100 = octal 4 (LSD)
Binary 111 = octal 7
Binary 011 = octal 3
Binary 111 = octal 7
Binary 010 = octal 2 (MSD)

Therefore, the conversion is: $10111011.1111_2 = 273.74_8$

Problem 10 Convert 47_8 to binary notation.

Solution Octal 7 = binary 111 (LSB)
Octal 4 = binary 100 (MSB)

Therefore, the conversion is: $47_8 = 100111_2$

(R30-17) Convert 170_8 to binary notation.

(R30-18) Convert 10110000011_2 to octal notation.

HEXADECIMAL NUMBER SYSTEM

Hexadecimal numbers are used to describe computer operation and programming. The hexadecimal number system is a base 16 system consisting of ten digits (0, 1, 2, 3, 4, 5, 6, 7, 8, 9) and six letters (A, B, C, D, E, F). Table 30-6 lists the equivalents for decimal, binary, octal, and hexadecimal numbers.

The basic hexadecimal symbols are written in positional notation to yield a useful number. Each digit of a number acquires a value determined by its position with relation to the reference point in the number. A hexadecimal number is the sum of the products of the digits and their respective positional weights. Table 30-7 shows the positional weights of several digit locations in the hexadecimal system. The table shows that the value of each position is a power of 16. Positions having positive powers are located to the left of the unit, or reference, digit. Positions having negative powers are located to the right of the reference digit. That is, the negative positions are located to the right of the hexadecimal point.

(R30-19) What is the base of the hexadecimal number system?

(R30-20) How many digits (symbols) are used in the hexadecimal number system?

TABLE 30-6 Number System Equivalents

Decimal Number	Binary Number	Octal Number	Hexadecimal Digit
0	0000	0	0
1	0001	1	1
2	0010	2	2
3	0011	3	3
4	0100	4	4
5	0101	5	5
6	0110	6	6
7	0111	7	7
8	1000	10	8
9	1001	11	9
10	1010	12	A
11	1011	13	B
12	1100	14	C
13	1101	15	D
14	1110	16	E
15	1111	17	F

CONVERSIONS BETWEEN THE HEXADECIMAL AND BINARY SYSTEMS

The same techniques used to convert between the octal and binary systems can be used to convert between the hexadecimal and binary systems. It is a simple matter to convert between binary numbers and hexadecimal numbers because the bases 2 and 16 are related ($2^4 = 16$). To make the conversion to a hexadecimal number, the binary bits are grouped in fours.

Problem 11

Convert 10110011.1101_2 to hexadecimal notation.

Solution

Starting at the binary point of the binary number, the bits are grouped in fours in both directions.

1011 0011 . 1101

Each group of four bits is replaced by the appropriate hexadecimal digit, as given in Table 30-6.

Binary 1101 = hexadecimal D (LSD)
Binary 0011 = hexadecimal 3
Binary 1011 = hexadecimal B (MSD)

Therefore, the conversion is: $10110011.1101_2 = B3.D_{16}$

TABLE 30-7 Hexadecimal Number System: Digit Locations and Positional Weights

16^3	16^2	16^1	16^0	•	16^{-1}	16^{-2}	16^{-3}
4096	256	16	1 (Reference)	• (Hex. point)	1/16	1/256	1/4096

Problem 12   Convert $2FC4_{16}$ to binary notation.

Solution   Hexadecimal 4 = binary 0100 (LSB)
Hexadecimal C = binary 1100
Hexadecimal F = binary 1111
Hexadecimal 2 = binary 0010 (MSB)

Therefore, the conversion is: $2FC4_{16} = 0010111111000100_2 = 10111111000100_2$

(R30-21)   Convert 1010101101_2 to hexadecimal notation.

(R30-22)   Convert F3.2A to binary notation.

BCD CODE   Codes are used in digital equipment to represent numerical information. The use of a code simplifies and reduces the digital circuitry needed to process information. The reliability of the digital system is greatly improved when the digital switching circuitry is simplified.

One of the most widely used codes in digital circuitry is the *binary coded decimal* (BCD). The BCD code is also known as the "8421" code. BCD coding is used in the following applications:

1. Binary codes sent to a digital computer from a numeric input device.
2. Binary codes from the digital computer to a numeric output device.
3. Displays associated with analog-to-digital converters.
4. Electronic hand-held calculators.

In the BCD code, each of the ten symbols of the decimal system is represented by four binary bits. The positional weights of the four binary bits are $2^3, 2^2, 2^1, 2^0$ or 8, 4, 2, 1 respectively.

The main advantage of this code is the ease of converting between the 8421 code numbers and decimal numbers. Table 30-8 lists the equivalent four-bit BCD code for each of the ten decimal symbols. Any decimal number can be expressed in the 8421 code by replacing each decimal digit with its appropriate four-bit code. The BCD code for a decimal number of more than one digit is obtained by replacing each digit by its four-bit BCD code. The BCD representation of 23_{10} is 00100011_{BCD}, since the BCD code for 2 is 0010 and the BCD

TABLE 30-8   BCD (8421)
Code Equivalents

BCD (8421)	Decimal
0000	0
0001	1
0010	2
0011	3
0100	4
0101	5
0110	6
0111	7
1000	8
1001	9

code for 3 is 0011. The student must remember that the BCD code for a decimal number larger than 9 is *not* the actual binary equivalent. (The true binary equivalent of 23_{10} is 10111_2.) The following examples illustrate conversions between decimal notation and BCD notation.

Problem 13   Convert 321_{10} into the 8421 code.
Solution   Decimal 1 = BCD 0001 (LSD)
Decimal 2 = BCD 0010
Decimal 3 = BCD 0011 (MSD)

Therefore, the conversion is: $321_{10} = 001100100001_{BCD}$

Problem 14   Convert 1100001100001_{BCD} to decimal notation.
Solution   The bits are grouped by fours as follows:

0001 1000 0110 0001

BCD 0001 = decimal 1 (LSD)
BCD 0110 = decimal 6
BCD 1000 = decimal 8
BCD 0001 = decimal 1 (MSD)

Therefore, the conversion is: $1100001100001_{BCD} = 1861_{10}$

(R30-23)   Convert 150_{10} into the 8421 code.

(R30-24)   Convert 100101110100_{BCD} to decimal notation.

ORGANIZATION AND FORMAT OF COMPUTER WORDS

Binary numbers are used in digital computer applications to represent data, codes, and instructions. This information is processed by the electronic circuitry of the computer. Information processed by digital electronic systems is in the form of binary bits in groups of fixed length. Each group of bits is called a *word*. The number of bits in a word is known as its *word length*. Computers and other digital equipment use words of various lengths. For example, typical word lengths are 4 bits, 8 bits, 12 bits, and 16 bits. Table 30-6 shows the sixteen 4-bit binary words. When using word notation, each word must be written with the same number of bits. For example, the 4-bit word for 3 is written as 0011. It cannot be written as 11 because significant 0s in a word cannot be omitted.

In some computer applications, a word is processed in pieces. An eight-bit group is called a *byte* and a four-bit group is called a *nibble*.

The manner in which information is organized in a word is called its *format*. Figure 30-1 illustrates two 8-bit word formats.

When the information consists of data to be processed, such as letters, numbers, or symbols, the package of information is called a *data word*. The format used for the data word is shown in figure 30-1A. A data word is used to indicate a signed number. The 8-bit word shown in figure 30-1A is divided into two parts. The first part is the MSB and is known as the *sign bit*. When the sign bit is 0, the

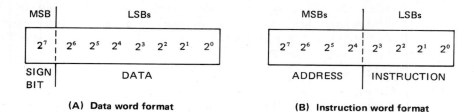

(A) Data word format   (B) Instruction word format

Fig. 30-1 Word formats

number is positive; when the sign bit is 1, the number is negative. The remaining bits of the data word identify the value of the numerical operation.

When the computer is being told what to do, the package of information is called an *instruction word*. The format for an instruction word is shown in figure 30-1B. The 8-bit word is divided into two 4-bit nibbles. The four LSBs are a binary code used to indicate an instruction. The MSBs represent a 4-bit *address*. In other words, this address is a code which selects the device to receive the instruction.

The two binary digits (0 and 1) are used in the binary number system to count more than two events. When all of the possible combinations of these two digits are used, four events can be counted or represented, Table 30-9. As shown in the equation, the total number of possible combinations (4) is equal to the base of the binary number system (2) raised to a power equal to the available number of digits (2).

$$N = 2^n \qquad\qquad \text{Eq. 30.1}$$

TABLE 30-9   All Possible Combinations of Binary
Digits

Two Digits	Three Digits	Four Digits
00	000	0000
01	001	0001
10	010	0010
11	011	0011
	100	0100
	101	0101
	110	0110
	111	0111
		1000
		1001
		1010
		1011
		1100
		1101
		1110
		1111

where
$\qquad$ N = total number of possible combinations
$\qquad$ 2 = base of binary number system
$\qquad$ n = number of binary digits

Table 30-9 also shows that for three binary digits, there are 2^3 or 8 possible combinations of digits. In addition, for four binary digits, there are 2^4 or 16 possible combinations of digits.

(R30-25) Define a. word, b. byte, and c. nibble.

(R30-26) In the binary word 1111001100001111, determine: a. the number of bits, b. the number of bytes, and c. the number of nibbles.

(R30-27) What is the sign and value of the data word 10010101?

(R30-28) How many pieces of discrete information can be represented by a binary word consisting of 16 bits?

THE DIODE AS A SWITCH
The diode can be described as a one-way device which simulates a switch. The "switch" is *closed* when the diode is forward biased and *open* when the diode is reverse biased. A switch has the following characteristics:
1. Zero voltage drop when closed.
2. Zero current when open.

Figure 30-2 shows the characteristics of a switch and an ideal diode. In many digital logic applications, a diode is assumed to have the characteristics shown.

(R30-29) What are the two characteristics of an ideal diode?

THE TRANSISTOR AS A SWITCH
The common emitter transistor circuit shown in figure 30-3 functions as a switch. The characteristic curves, figure 30-3D, are an idealized representation of the operation of the transistor.

The circuit shown in figure 30-3A operates as an amplifier under the following quiescent conditions:
1. The input voltage or controlling voltage is V_{BEQ}.
2. The input current is I_{BQ}.
3. The collector current is I_{CQ}.

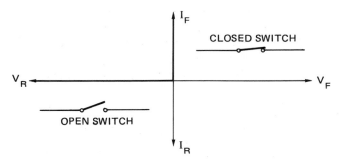

Fig. 30-2 Characteristics of a switch and an ideal diode.

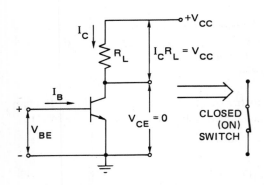

(A) Common emitter circuit operating
under quiescent conditions

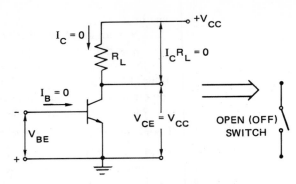

(B) Ideal transistor switch operation with
switch in off or open position

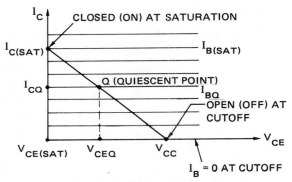

(C) Ideal transistor switch operation
with switch in on closed position

(D) Ideal characteristic curves

Fig. 30-3 The transistor as a switch

4. The collector-to-emitter voltage is given by the following equation:

$$V_{CQ} = V_{CEQ} = V_{CC} - I_{CQ}R_L \qquad \text{Eq. 30.2}$$

5. As shown in figure 30-3D , the transistor does not operate as a switch when it operates at the quiescent point because it is in the active region of operation.

Figure 30-3B represents the operation of the circuit as an ideal switch in the OFF (OPEN or CUTOFF) position. In this mode of operation:

1. V_{BE} may be zero or the transistor may be reverse biased as shown.
2. The input current $I_B = 0$.
3. The collector current $I_C = 0$.
4. Substituting values in Equation 30.2 gives:

$$V_{CE} = V_{CC} - I_C R_L = V_{CC} - (0)(R_L)$$
$$V_{CE} = V_{CC}$$

5. The transistor operates as a switch in the OFF (OPEN or CUTOFF) position. As shown in figure 30-3D, the collector-to-emitter voltage, V_{CE}, equals the supply voltage, V_{CC}.

Figure 30-3C represents the operaton of the circuit as an ideal switch in the ON (CLOSED or SATURATION) position. In this mode of operation:

1. The transistor input junction is forward biased. In other words, V_{BE} is positive.
2. The input current, I_B, is made large enough to ensure that $I_C R_L$ is equal to the supply voltage, V_{CC}.
3. Substituting values in Equation 30.2 gives:
$$V_{CE} = V_{CC} - I_C R_L = V_{CC} - V_{CC}$$
$$V_{CE} = 0$$
4. As shown in figure 30-3D, the transistor operates as a switch in the ON (CLOSED or SATURATION) position. The collector-to-emitter voltage, V_{CE}, equals zero.

The ideal transistor circuit shown in figure 30-3 dissipates zero power when in the ON or OFF positions. Power is dissipated by the transistor *only* when it is switching between ON and OFF. The power dissipated by the transistor can be calculated from the following equation:

$$P_D = I_C V_{CE}$$

Eq. 30.3

When the transistor is OFF, $I_C = 0$. Therefore,

$$P_D = 0 \times V_{CE} = 0$$

When the transistor is ON, $V_{CE} = 0$, and:

$$P_D = I_C \times 0 = 0$$

Transistors are used in digital logic circuit applications because they are easily and quickly switched from OFF (OPEN or CUTOFF) to ON (CLOSED or SATURATION). When a series of transistors are used in a digital computer and are permitted to operate in those two states (OFF or ON), the computer itself is small, economical to operate, reliable, and extremely fast.

(R30-30) What conditions exist in an ideal transistor switch when it is in the OFF position?

(R30-31) What conditions exist in an ideal transistor switch when it is in the ON position?

INTRODUCTION TO DIGITAL LOGIC GATES

A mathematical system describing two-valued logic, known as *Boolean algebra*, consists of manipulation of digital data which can change discretely from one to the other of two distinct states. Boolean algebra makes it possible to simplify the circuits and connections used in computer logic circuits. Logic algebra, or *Boolean algebra*, deals with circuits whose inputs and outputs can be only TRUE or FALSE, ON or OFF, HIGH or LOW, CLOSED or OPEN. All of these conditions can be represented by the binary digits 1 and 0.

Two voltage levels represent the two binary digits in digital systems. *Positive logic* results when the higher of the two voltages represents a 1 and the lower voltage represents a 0. In *negative logic*, the lower of the two voltages represents a 1 and the higher voltage represents a 0. If a digital system has logic level voltages of +5 V and 0 V, the +5 V may be represented by a 1 and the 0 V may be represented by a 0. The positive and negative logics are then defined as follows:

Positive Logic: High = 1 (+5 V)

Low = 0 (0 V)

Negative Logic: High = 0 (0 V)

Low = 1 (+5 V)

Both positive and negative logic are used in digital systems. However, positive logic is the more common and will be used in this text.

Because logic circuits switch between the 0 and 1 states, they operate as switches that OPEN and CLOSE. Therefore, they are known as *logic gates*, or more simply, gates. Actually, a logic gate is an electronic device that performs an operation in Boolean algebra on one or more inputs to produce an output. The output of the gate occurs when certain conditions are met. A *truth table* shows all of the possible inputs and the resultant outputs of the logic gate circuit. Such a table is used to analyze the operation of the gate.

Logic gate circuits are made using discrete combinations of diodes, transistors and resistors. Digital integrated circuits (ICs) provide a complete circuit which is small in size. Users of digital logic circuits select ICs for both computer and industrial control applications. An integrated circuit (IC) is a single, monolithic chip of semiconductor material in which the electronic circuit elements are fabricated.

An IC chip is about 0.01 inch thick and may vary in size from about 0.03 inch by 0.03 inch to 0.3 inch by 0.3 inch. A typical size is about 0.05 inch by 0.05 inch. The number of gates on one chip may vary from 1 or 2 to many thousands.

ICs are classified according to the number of gates they contain. *Small-scale integration* (SSI) refers to chips containing 12 or fewer gates. *Medium-scale integration* (MSI) refers to chips containing more than 12 gates, but fewer than 100 gates. *Large-scale integration* (LSI) refers to chips containing 100 or more gates.

Standard logic symbols are used to represent the various logic circuits. Three basic logic gates are analyzed in this chapter: the AND gate, the OR gate, and the NOT gate.

(R30-32) Define a logic gate.

(R30-33) Name the two types of logic used in digital systems.

(R30-34) A digital system has logic level voltages of +10 V dc and 0 V dc. Represent these two logic level voltages by the binary digits (1 and 0) in a. positive logic and b. negative logic.

THE AND GATE The basic operation of logic multiplication, commonly known as the AND operation or function, is performed by the AND gate. This gate is composed of two or more inputs and a single output. Figure 30-4A shows the standard logic symbol for an AND gate with two input variables, A and B. The output of this gate is AB or X. The output is expressed in Boolean algebra notation and is read A and B (not as A times B).

The operation of the AND gate is such that the output is HIGH (1) only when *all* of the inputs are HIGH (1). If any of the inputs are LOW (0), the output is LOW (0).

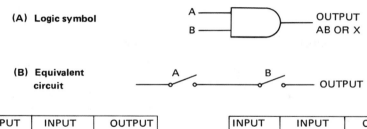

(A) Logic symbol

A — B — OUTPUT AB OR X

(B) Equivalent circuit

A B OUTPUT

INPUT A	INPUT B	OUTPUT AB OR X
0	0	0
0	1	0
1	0	0
1	1	1

(C) Positive logic truth table

INPUT A	INPUT B	OUTPUT AB OR X
LOW	LOW	LOW
LOW	HIGH	LOW
HIGH	LOW	LOW
HIGH	HIGH	HIGH

(D) High-low truth table

Fig. 30-4 Two-input and gate

An AND gate determines when certain conditions are simultaneously true. The AND gate is an all-or-nothing gate because there must be a 1 at all of the inputs to obtain a 1 at the output. The HIGH (1) level is the prime or active output level for the AND gate. An AND gate can have any number of inputs greater than one.

The operation of a gate, in terms of its logic, can be expressed by a truth table that lists all of the input combinations and the corresponding outputs. Figures 30-4C and 30-4D show the truth tables for a two-input AND gate. The truth tables are the same for an AND circuit composed of discrete components and for one consisting of an integrated circuit. Figure 30-4B illustrates an equivalent AND circuit using two switches.

The schematic shown in figure 30-5A illustrates a simple form of AND gate. The voltage truth table for the diode logic (DL) discrete component circuit is shown in figure 30-5B. The operation of this circuit occurs as follows:

1. A 0-V input to terminal A or terminal B is obtained by connecting this terminal to the ground terminal (negative terminal of the power supply).
2. When a 0-V input is applied to terminals A and B simultaneously, diodes D_1 and D_2 conduct. An ideal conducting diode acts like a CLOSED switch (a short circuit). This means that terminal C is at the ground potential. As a result, the output voltage, V_O, is 0 V.
3. When a 0-V input is applied to terminal A and a + 5-V input is applied to terminal B, diode D_1 conducts. Diode D_2 acts like an OPEN switch and does not conduct. Terminal C is at the ground potential. Therefore, the output voltage, V_O, is 0 V.
4. When +5 V is applied to terminal A and 0 V is applied to terminal B, diode D_2 conducts. Diode D_1 acts like an OPEN switch and does not conduct. Terminal C is at the ground potential. Therefore, the output voltage, V_O, is 0 V.

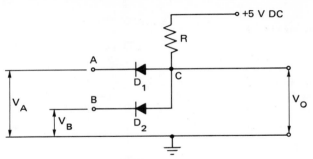

(A) DL logic circuit

INPUT A	INPUT B	OUTPUT V_O
O V	O V	O V
O V	+5 V	O V
+5 V	O V	O V
+5 V	+5 V	+5 V

(B) Voltage truth table

Fig. 30-5 Diode logic (DL) and gate

5. When +5 V is applied to both terminal A and to terminal B, diodes D_1 and D_2 do not conduct. That is, both diodes act like OPEN switches. This means that there is no current through resistor R, terminal C is at +5 V, and the output voltage, V_O, is +5 V.

(R30-35)   Draw the logic symbol for a three-input AND gate.

THE OR GATE   The logic operation performed by the OR gate is logic addition, also known as the OR operation or function. The OR gate has two or more inputs and a single output. Figure 30-6 shows the standard logic symbol for an OR gate with two input variables A and B. The output of this gate is expressed in Boolean algebra notation as A + B, or X. This notation is read as A or B, *not* as A plus B.

The operation of the OR gate is such that a HIGH (1) output is produced when any of the inputs is HIGH (1). The output is LOW (0) only when all of the inputs are LOW (0). The OR gate determines when certain conditions are met simultaneously. The OR gate is an any-or-all gate because a 1 at any input gives rise to a 1 at the output. The HIGH (1) level is the prime or active output level for the OR gate. An OR gate can have any number of inputs greater than one.

Figures 30-6C and 30-6D show the truth tables for a two-input OR gate. The truth tables are the same for an OR circuit composed of discrete components and one consisting of an integrated circuit. Figure 30-6B shows an equivalent OR circuit using two switches.

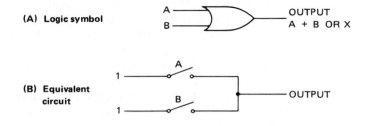

(A) Logic symbol

(B) Equivalent circuit

INPUT A	INPUT B	OUTPUT A + B OR X
O	O	O
1	O	1
O	1	1
1	1	1

(C) Truth table

INPUT A	INPUT B	OUTPUT A + B OR X
LOW	LOW	LOW
HIGH	LOW	HIGH
LOW	HIGH	HIGH
HIGH	HIGH	HIGH

(D) High-low truth table

Fig. 30-6 Two-point OR gate

A simple form of OR gate is shown in the schematic of figure 30-7A. The voltage truth table for this diode logic (DL) discrete component circuit is given in figure 30-7B. The circuit operates in the following manner:

1. A 0-V input is applied to terminal A or terminal B by connecting that terminal to the ground terminal (negative terminal of the power supply).
2. When a 0-V input is applied to terminals A and B simultaneously, diodes D_1 and D_2 act like OPEN switches and do not conduct. Terminal C is at ground potential and the output voltage, V_O, is 0 V.

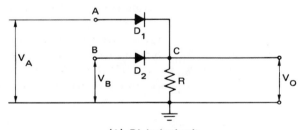

(A) DL logic circuit

INPUT A	INPUT B	OUTPUT V_O
O V	O V	O V
+5 V	O V	+5 V
O V	+5 V	+5 V
+5 V	+5 V	+5 V

(B) Voltage truth table

Fig. 30-7 Diode logic (DL) OR gate

3. When an input of +5 V is applied to terminal A and 0 V is applied to terminal B, diode D_1 conducts, but diode D_2 does not conduct. Terminal C is at +5 V because the conducting diode D_1 acts like a CLOSED switch. The output voltage, V_O, is + 5 V.

4. When 0 V is applied to terminal A and +5 V is applied to terminal B, diode D_1 does not conduct. Diode D_2 conducts because it is forward biased. Terminal C is at +5 V because the conducting diode D_2 acts like a CLOSED switch. The output voltage is + 5 V.

5. When an input of +5 V is applied simultaneously to terminals A and B, diodes D_1 and D_2 both conduct because they are forward biased. Since both conducting diodes act like CLOSED switches, terminal C is at +5 V. The output voltage, V_O, is + 5 V.

(R30-36)   Draw the logic symbol for a three-input OR gate.

THE NOT GATE (INVERTER GATE)   Figure 30-8 gives the logic symbol and truth tables for a NOT or INVERTER gate. This gate has a single input, A. The output is the complement of the input and is expressed as $\overline{A}$. The small circle, or bubble, at the output of the NOT gate is the standard symbol representing inversion or complementation.

An inverter gate changes or converts one logic level to the other logic level. If a HIGH input is applied, the output is LOW. If a LOW input is applied, the output is HIGH.

Figure 30-9A illustrates one method of constructing a NOT circuit. This circuit is called an RTL (resistor-transistor logic) circuit because the resistor R_B is used to couple the input to the transistor. This circuit operates in the following manner:

1. If a HIGH of +5 V is applied to the input of the circuit in figure 30-9B, the base-emitter junction is forward biased. As a result, the transistor is turned ON.

2. The base current, I_B, is made large enough to ensure that the transistor operates in the saturation region. The voltage V_{CE} is zero because the transistor acts like a CLOSED switch. Terminal C is at ground potential and V_O is 0 V. As a result, a HIGH input yields a LOW output.

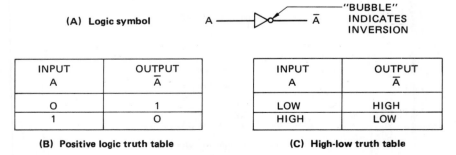

(A) Logic symbol   A ——▷o—— $\overline{A}$ "BUBBLE" INDICATES INVERSION

INPUT A	OUTPUT $\overline{A}$
O	1
1	O

(B) Positive logic truth table

INPUT A	OUTPUT $\overline{A}$
LOW	HIGH
HIGH	LOW

(C) High-low truth table

Fig. 30-8 NOT (INVERTER) gate

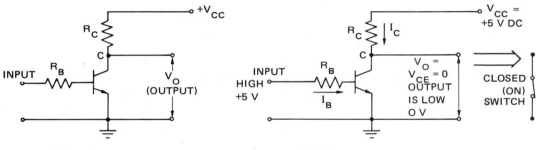

(A) RTL inverter circuit   (B) Transistor operating at saturation

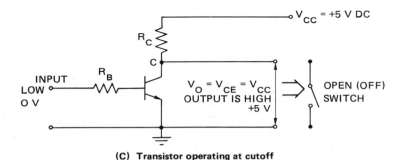

(C) Transistor operating at cutoff

Fig. 30-9 Operation of an RTL inverter

3. If a LOW of 0 V is applied to the input of the circuit in figure 30-9C, the base-emitter junction is reverse biased. The student must remember that applying 0 V means that the negative terminal of the power supply is connected to the input.

4. Since the transistor is operating in the cutoff region, it acts like an OPEN switch. There is no current through resistor R_C. Terminal C is at a potential of +5 V and V_O is +5 V. Therefore, a LOW input results in a HIGH output.

(R30-37) What is the meaning of the small circle, or bubble, at the output of a NOT gate?

(R30-38) What is the purpose of an inverter?

LOGIC GATE COMBINATIONS   Boolean expressions may be implemented by using OR, AND, and NOT gates. The Boolean function $X = (A + B)\overline{C}$ is implemented by the combination of three gates shown in figure 30-10.

(R30-39) Find the output if $A = 1$, $B = 0$, and $C = 1$ in figure 30-10.

(R30-40) Draw the logic gate combinations that will implement the Boolean function $X = A + \overline{A}B$.

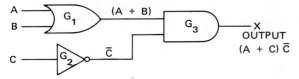

Fig. 30-10 Combinations of logic gates required to implement the function X = (A + B)$\overline{C}$

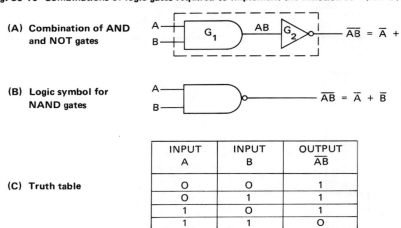

(A) **Combination of AND and NOT gates**

$\overline{AB} = \overline{A} + \overline{B}$

(B) **Logic symbol for NAND gates**

$\overline{AB} = \overline{A} + \overline{B}$

(C) **Truth table**

INPUT A	INPUT B	OUTPUT $\overline{AB}$
O	O	1
O	1	1
1	O	1
1	1	O

Fig. 30-11 NAND gate

THE NAND GATE   A very important Boolean logic function is the NAND operation. The NAND (a contraction of NOT AND) gate has two inputs (A and B) and an output $\overline{AB}$. As shown in figure 30-11A, the NAND gate consists of an AND gate followed by a NOT or INVERTER gate. A single gate, called a NAND gate, is available to perform the important NAND operation. Figure 30-11B gives the standard logic symbol for a two-input NAND gate. The symbol consists of an AND gate followed by a small circle, or bubble, to indicate complementation. The truth table for the NAND gate is shown in figure 30-11C.

NAND gates may have more than two inputs. In the case of more than two inputs, the output is the complement of the logic product of the inputs. For example, for a three-input NAND gate, the output is $\overline{ABC}$. Therefore, the NAND gate has an output of 1 when any or all of the inputs are 0, and an output of 0 only when all of the inputs are 1.

By applying one of the theorems of Boolean algebra, the output of a NAND gate can be expressed as follows:

$$X = \overline{AB} = \overline{A} + \overline{B} \qquad\qquad \text{Eq. 30.4}$$

The function of the NAND gate can be stated in two ways:

1. The output of a NAND gate is equal to the complement of the AND of the input variables, or $\overline{AB}$.
2. The output of a NAND gate is equal to the OR of the input variable complements, or $\overline{A} + \overline{B}$.

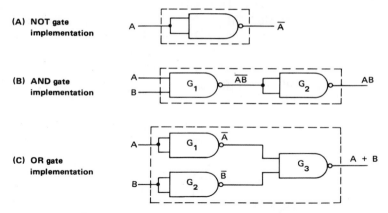

(A) NOT gate
    implementation

(B) AND gate
    implementation

(C) OR gate
    implementation

Fig. 30-12 Universal application of NAND gates

NAND gates can be used to produce any logic function. The NAND gate is known as a universal gate because it can be used to generate the NOT function, the AND function, and the OR function. Figure 30-12A shows how an INVERTER (NOT) gate can be made from a NAND gate by connecting all of the inputs. In effect, a single common input is created. The AND function can be implemented using only NAND gates, as shown in figure 30-12B. An OR function can be implemented using NAND gates as shown in figure 30-12C.

(R30-41) What is the output of a three-input NAND gate with the following input values: $A = 1$, $B = 1$, and $C = 0$?

(R30-42) List the three logic functions which can be implemented with NAND gates.

THE NOR GATE Another very important Boolean logic operation is the NOR operation. The NOR (a contraction of NOT OR) gate has two inputs (A and B) and an output $\overline{A + B}$. As shown in figure 30-13A, the NOR gate consists of an OR gate fol-

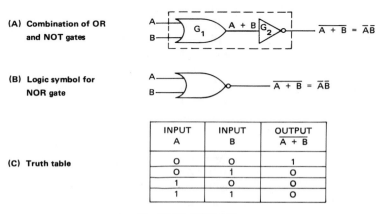

(A) Combination of OR
    and NOT gates

(B) Logic symbol for
    NOR gate

(C) Truth table

INPUT A	INPUT B	OUTPUT $\overline{A + B}$
O	O	1
O	1	O
1	O	O
1	1	O

Fig. 30-13 NOR gate

lowed by a NOT or INVERTER gate. However, as in the case of the NAND operation, a single NOR gate is available. Figure 30-13B shows the standard logic symbol for a two-input NOR gate. The symbol consists of an OR gate followed by a small circle, or bubble, to indicate complementation.

NOR gates may have more than two inputs. For this case, the output is the complement of the logic addition of the inputs. For example, the output of a three-input NOR gate is $\overline{A + B + C}$. The NOR gate has an output of 0 when any or all of the inputs ae 1, and an output of 1 only when all of the inputs are 0.

By applying one of the theorems of Boolean algebra, the output of a NOR gate can be expressed as follows:

$$X = \overline{A + B} = \overline{A}\,\overline{B} \qquad\qquad \text{Eq. 30.5}$$

The function of the NOR gate can be stated in two ways:

1. The output of a NOR gate is equal to the complement of the OR of the input variables, or $\overline{A + B}$.
2. The output of the NOR gate is equal to the AND of the input variable complements, or $\overline{A}\,\overline{B}$.

NOR gates can be used to produce any logic function. The NOR gate is known as a universal gate because it can be used to generate the NOT function, the AND function and the OR function. Figure 30-14A shows how an INVERTER (NOT) gate can be made from a NOR gate by connecting all of the inputs. In effect, a single common input is created. The AND function can be implemented using only NOR gates, as shown in figure 30-14B. An OR function can be implemented using NOR gates as shown in figure 30-14C.

(R30-43) What is the output of a three-input NOR gate with the following input values: A = 1, B = 1, and C = 0?

(R30-44) List the three logic functions which can be implemented with NOR gates.

(A) NOT gate implementation

(B) AND gate implementation

(C) OR gate implementation

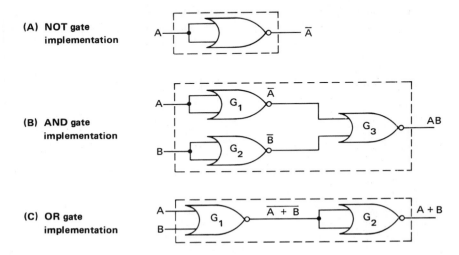

Fig. 30-14 Universal application of NOR gates

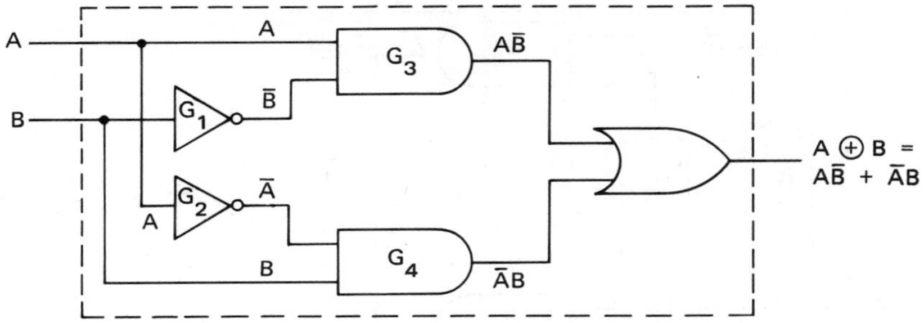

Fig. 30-15 Implementation of the EXCLUSIVE-OR function

Fig. 30-16 Logic symbol for a two-input EXCLUSIVE-OR gate

THE EXCLUSIVE-OR GATE

The normal OR operation applied to two input variables A and B results in an output equal to 1 when either A = 1 or B = 1, or when both A = 1 and B = 1. A very useful Boolean operation, the EXCLUSIVE-OR operation, excludes the case when A = B = 1. The output of the EXCLUSIVE-OR operation is 1 when A = 1 or B = 1, but *not* when *both* inputs are equal to 1. Table 30-10 gives the truth table for the EXCLUSIVE-OR operation, for the case of two variable inputs. The EXCLUSIVE-OR operation is used in digital logic systems to decide if two binary input numbers are equal or not equal.

The expression for this function is:

$$X = A\bar{B} + \bar{A}B \qquad \text{Eq. 30-6}$$

The symbol for the EXCLUSIVE-OR operation is $\oplus$. Equation 30.6 can be rewritten as follows:

$$X = A\bar{B} + \bar{A}B = A \oplus B \qquad \text{Eq. 30.7}$$

One method of implementing the EXCLUSIVE-OR function is to use two AND gates, two INVERTERS, and one OR gate, figure 30-15. Figure 30-16 shows the standard logic symbol for a single gate which can implement this function.

(R30-45) What is the major difference between an OR gate and an EXCLUSIVE-OR gate?

(R30-46) Describe one of the uses of the EXCLUSIVE-OR gate.

TABLE 30-10 Truth Table for EXCLUSIVE-OR Function

Input A	Input B	Output X A $\oplus$ B
0	0	0
0	1	1
1	0	1
1	1	0

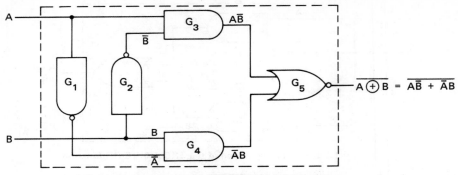

(A) Implementation of the EXCLUSIVE-NOR function

(B) Logic symbol for a two-input
EXCLUSIVE-NOR gate

Fig. 30-17 The EXCLUSIVE-NOR function

THE EXCLUSIVE NOR GATE

When an INVERTER or NOT gate is added to the EXCLUSIVE-OR circuit, the the result is an EXCLUSIVE-NOR function. A method of implementing the EXCLUSIVE-NOR function is to use two NAND gates, two AND gates, and a NOT gate, figure 30-17A. The standard logic symbol for a single gate which can implement the EXCLUSIVE-NOR function is shown in figure 30-17B.

The equation for the EXCLUSIVE-NOR operation is:

$$X = \overline{A\overline{B} + \overline{A}B} = \overline{A \oplus B} \qquad \text{Eq. 30.8}$$

The student should observe that Equation 30.8 is the inversion of Equation 30.7.

(R30-47) Draw a truth table for the EXCLUSIVE-NOR gate.

LABORATORY EXERCISE 30-1 CHARACTERISTICS AND OPERATION OF A DISCRETE DL AND GATE

PURPOSE In completing this exercise, the student will examine the characteristics and operation of a discrete DL AND gate.

EQUIPMENT AND MATERIALS
2 Dc power supplies
1 Multimeter
2 Diodes, 1N914
1 Resistor, 5.1 kΩ, 2 W

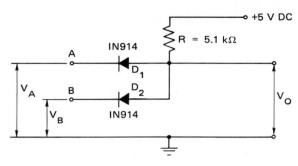

Fig. 30-18

PROCEDURE
1. Construct the circuit shown in figure 30-18.
   a. Apply 0 V to both input A and input B by connecting the ground terminal to the input of each diode, D_1 and D_2.
   b. Measure the output voltage, V_O.
   c. Apply 0 V to input A and +5 V to input B.
   d. Measure the output voltage, V_O.
   e. Apply +5 V to input A and 0 V to input B.
   f. Measure the output voltage, V_O.
   g. Apply +5 V to both input A and input B.
   h. Measure the output voltage, V_O.
2. Answer the following questions.
   a. Draw a voltage truth table using the data and results of steps 1.a. through 1.h.
   b. Compare the truth table prepared in step 2.a. with the truth table given in figure 30-5B. Explain any discrepancies between the two tables.
   c. Explain the operation of the AND gate.

**LABORATORY
EXERCISE 30-2
CHARACTERIS-
TICS AND OPERA-
TION OF A
DISCRETE DL
OR GATE**

PURPOSE In completing this exercise, the student will examine the characteristics and operation of a discrete OR gate.

**EQUIPMENT
AND MATERIALS**
2 Dc power supplies
1 Multimeter
2 Diodes, 1N914
1 Resistor, 5.1 kΩ, 2 W

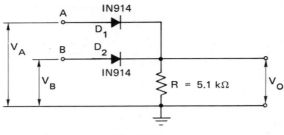

Fig. 30-19

PROCEDURE

1. Construct the circuit shown in figure 30-19.
   a. Apply 0 V to both input A and input B by connecting the ground termi-
      nal to the input of each diode, D_1 and D_2.
   b. Measure the output voltage, V_O.
   c. Apply +5 V to input A and 0 V to input B.
   d. Measure the output voltage, V_O.
   e. Apply 0 V to input A and +5 V to input B.
   f. Measure the output voltage, V_O.
   g. Apply +5 V to both input A and input B.
   h. Measure the output voltage, V_O.
2. Answer the following questions.
   a. Draw a truth table using the data and results of steps 1.a. through 1.h.
   b. Compare the truth table prepared in step 2.a. with the truth table shown
      in figure 30-7B. Explain any discrepancies between the two tables.
   c. Explain the operation of the OR gate.

**LABORATORY
EXERCISE 30-3
CHARACTERIS-
TICS AND OPERA-
TION OF A
DISCRETE DTL
NAND GATE**

PURPOSE

In completing the exercise, the student will examine the characteristics and operation of a discrete NAND gate.

**EQUIPMENT
AND MATERIALS**

2 Dc power supplies
1 Multimeter
3 Diodes, 1N914
1 Resistor, 5.1 kΩ, 2 W
2 Resistors, 1 kΩ, 2 W
1 Transistor, 2N2222

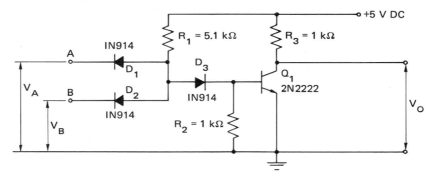

Fig. 30-20

PROCEDURE

1. Construct the circuit shown in figure 30-20.
   a. Apply 0 V to both input A and input B by connecting the ground termi-
      nal to the input of each diode, D_1 and D_2.
   b. Measure the output voltage, V_O.
   c. Apply 0 V to input A and +5 V to input B.
   d. Measure the output voltage, V_O.
   e. Apply +5 V to input A and 0 V to input B.
   f. Measure the output voltage, V_O.
   g. Apply +5 V to both input A and input B.
   h. Measure the output voltage, V_O.
2. Answer the following questions.
   a. Draw a truth table using the data and results of steps 1.a. through 1.h.
   b. Compare the truth table prepared in step 2.a. with the truth table given
      in figure 30-11. Explain any discrepancies between the two truth tables.
   c. Explain the operation of the NAND gate.

LABORATORY
EXERCISE 30-4
CHARACTERIS-
TICS AND OPERA-
TION OF A
DISCRETE DTL
NOR GATE

PURPOSE

In completing this exercise, the student will examine the characteristics and operation of a discrete NOR gate.

EQUIPMENT
AND MATERIALS

2 Dc power supplies
1 Multimeter
2 Diodes, 1N914
1 Resistor, 5.1 kΩ, 2 W

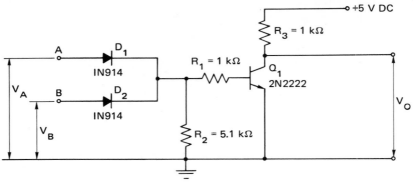

Fig. 30-21

2 Resistors, 1 kΩ, 2 W
1 Transistor, 2N2222

PROCEDURE   1. Construct the circuit shown in figure 30-21.
   a. Apply 0 V to both input A and input B by connecting the ground ter-
      minal to the input of each diode, D_1 and D_2.
   b. Measure the output voltage, V_O.
   c. Apply 0 V to input A and +5 V to input B.
   d. Measure the output voltage, V_O.
   e. Apply +5 V to input A and 0 V to input B.
   f. Measure the output voltage, V_O.
   g. Apply +5 V to both input A and input B.
   h. Measure the output voltage, V_O.
2. Answer the following questions.
   a. Draw a truth table using the data and results of steps 1.a' through 1.h.
   b. Compare the truth table prepared in step 2.a. with the truth table shown
      in figure 30-13C. Explain any discrepancies between the two tables.
   c. Explain the operation of the NOR gate.

EXTENDED   1. Write the decimal number 6045.043_{10} in positional notation.
STUDY TOPICS 2. What are the basic characteristics of the binary number system?
3. Convert the following binary numbers to decimal numbers:
   a. 11011_2
   b. 101110_2
4. Determine the decimal equivalent of the binary number1011100.10101_2.
5. Find the binary equivalent of the decimal number 44_{10}.
6. Convert the decimal number 1453_{10} to its binary equivalent.
7. Add the following binary numbers:
   a. $100101_2 + 000101_2$
   b. $110111_2 + 110111_2$
8. Convert 7765_8 to its decimal equivalent.
9. Convert 791_{10} to octal notation.

10. Convert 5276_8 to binary notation.
11. Convert 111111101111000_2 to octal notation.
12. Convert 11011011001_2 to hexadecimal notation.
13. Convert $5E.D_{16}$ to binary notation.
14. Convert 1472_{10} into the 8421 code.
15. Convert 100001000111_{BCD} to decimal notation.
16. What is the sign and value of the data word 01100000?
17. Write the truth table for a three-input AND gate using 1 and 0 positive logic.
18. Write the truth table for a three-input OR gate using 1 and 0 positive logic.
19. Draw the logic gate combinations required to implement the Boolean function:
$$X = \overline{AB + C}$$
20. What is the output of the logic gate combinations used to implement the function of topic 19 if all of the inputs are 1?

<table>
<tr><td>

Chapter

31

</td><td>

Digital Logic Applications

</td></tr>
</table>

OBJECTIVES   After studying this chapter, the student will be able to

- discuss IC flip-flops.
- analyze discrete multivibrators.
- describe the Schmitt trigger circuit.
- explain the operation of various digital devices.
- discuss the function of the components in a digital computer.
- describe a microprocessor and its applications.
- explain the function of the components in a microcomputer.

INTRODUCTION   The flip-flop belongs to a category of digital logic circuits known as *multivi-*
TO FLIP-FLOPS   *brators*. This chapter describes the three basic types of multivibrators:

1. the bistable multivibrator.
2. the monostable multivibrator.
3. The astable multivibrator.

The *bistable multivibrator* is also called a *flip-flop*.

The output of a combinational logic circuit (as described in Chapter 30), at any instant, is determined solely by its inputs at that instant. This type of circuit does not have the capacity to remember, or store, the effects of inputs at previous instants. Without adding memory devices or storage elements, a combinational logic circuit cannot be used to perform even the simplest function when it depends upon past inputs.

A *sequential circuit* is a digital logic circuit in which the outputs depend upon previous inputs, as well as present inputs. This type of circuit remembers the sequence of inputs in its immediate past history. The digital computer is an example of a device which contains sequential circuits. These circuits enable the computer to perform various arithmetical operations upon previously stored data and store the results for later use.

The flip-flop is a basic storage element of a digital logic circuit. It has two stable states and can store a single bit in either state. The flip-flop is in the HIGH STATE (logic 1) when it stores a 1; it is in the LOW state (logic 0) when it stores a 0. The flip-flop has the capacity to remain in either the one (1) or the zero (0) state until some change in its input causes it to change its state. The change in the input to a flip-flop can cause it to flip from zero (0) to one (1), or to flop from one (1) to zero (0).

The flip-flop is the basic building element for the larger sequential networks that are described later in this Chapter. These sequential networks are fundamental components of digital computers and include counters, shift registers, and memory registers.

Every flip-flop has one or more inputs. The inputs and the present state of the flip-flop determine the state to be assumed next. Flip-flops have two outputs consisting of the stored bit and its complement. Both outputs usually are accessible to the external circuit. The two outputs may be designated as 0 and 1, FALSE and TRUE, LOW and HIGH, or $\bar{Q}$ and Q.

The state of a flip-flop can also be designated by SET (S) and RESET (R). These terms are defined as follows:

SET	RESET
Q output is logic 1	Q output is logic 0
$\bar{Q}$ output is logic 0	$\bar{Q}$ output is logic 1.

Flip-flops are available as complete digital logic units in the form of integrated circuits (ICs). To analyze the operation of a flip-flop IC, it may be described in terms of its discrete logic gates or discrete components.

(R31-1) Name the three basic types of multivibrators.

(R31-2) What are the two characteristic features of the flip-flop?

(R31-3) Define: a. a combinational logic circuit and b. a sequential logic circuit.

(R31-4) What are the logic outputs when a flip-flop is a. RESET and b. SET?

FLIP-FLOP INPUT-OUTPUT SIGNALS Figure 31-1 illustrates a digital pulse signal. This signal changes from logic 0 to logic 1, and then returns to logic 0 after a fixed time, which is usually short. In figure 31-1, the pulses occur during the time intervals t_1 to t_2, t_3 to t_4, and t_5 to t_6. The *leading* (positive) *edge* of a pulse is the edge at which the pulse rises from logic 0 to logic 1. The *trailing* (negative) *edge* of a pulse is the edge where the pulse falls from logic 1 to logic 0.

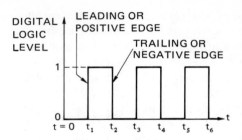

Fig. 31-1 Digital pulse signal

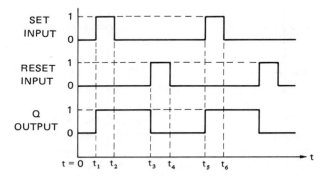

Fig. 31-2 Timing diagram of a flip-flop with SET and RESET control

The timing diagram of a digital circuit shows the sequence of input and output signals in a graphical form. Figure 31-2 shows the timing diagram of a flip-flop with SET and RESET control. This timing diagram can be analyzed in the following manner:

1. At $t = 0$, the Q output is logic 0. Both the SET and RESET inputs are at logic 0. The flip-flop is in the RESET state.
2. At $t = t_1$, the SET input changes to logic 1. The Q output becomes logic 1. In other words, the flip-flop changes to the SET state. Actually, the transition between states is not instantaneous. There is a *propagation delay time* before an input to a flip-flop affects the output. This Chapter, however, treats the circuits as ideal circuits and neglects the propagation delay time.
3. At $t = t_2$, the SET input changes to logic 0. The Q output remains at logic 1. That is, the flip-flop remains in its stable SET state.
4. At $t = t_3$, the RESET input changes to logic 1. The Q output becomes logic 0 and the flip-flop is in the RESET state.
5. At $t = t_4$, the RESET input changes to logic 0. The Q output remains at logic 0 and the flip-flop remains in its stable RESET state.
6. At $t = t_5$, the SET input changes to logic 1. The Q output becomes logic 1. That is, the flip-flop changes from the RESET state to the SET state.

A CLOCK is a logic device that generates a precise pattern of periodically occurring pulses, as shown in figure 31-3. The time for one complete cycle is called the *period*. The frequency of the CLOCK is the reciprocal of the period.

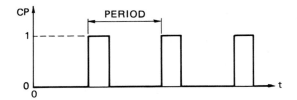

Fig. 31-3 CLOCK pulse signal

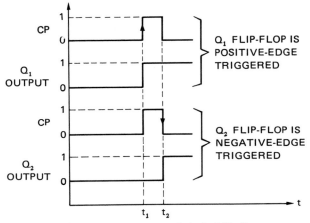

Fig. 31-4 Timing diagrams of clocked flip-flops

CLOCK signals of several megahertz are commonly used in digital logic circuits. The CLOCK may be considered as a control signal that synchronizes the sequence of operation in a sequential circuit.

Flip-flops controlled by CLOCK pulses (CP) are called *clocked* or *gated* flip-flops. Some gated flip-flops are positive-edge triggered. This means that the flip-flops respond to the positive edge of the CLOCK pulse. Other flip-flops are negative-edge triggered and respond to the negative edge of the CLOCK pulse. The timing diagrams of clocked flip-flops with both types of triggering are shown in figure 31-4.

(R31-5) Define: a. the positive edge of a pulse and b. the negative edge of a pulse.

(R31-6) What is the timing diagram for a digital circuit?

(R31-7) What is a CLOCK?

(R31-8) What is a clocked or gated flip-flop?

RS FLIP-FLOP Figure 31-5A shows an RS (RESET-SET) flip-flop logic circuit implemented with two NOR gates connected in series, or cross-coupled. The logic symbol for this circuit is shown in figure 31-5B. The SET input is S, the RESET input is R, and the output, or state, of the flip-flop is Q. The complement of the state, $\overline{Q}$, is

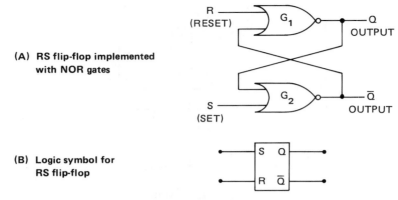

(A) RS flip-flop implemented
with NOR gates

(B) Logic symbol for
RS flip-flop

Fig. 31-5 RS flip-flop

TABLE 31-1 RS Flip-Flop Truth Table

Inputs		Outputs		
S	R	Q	$\bar{Q}$	Comments
1	0	1	0	Set
0	0	1	0	No change
0	1	0	1	Reset
0	0	0	1	No change

also available as an output. The RS flip-flop retains a value of Q = 0 or Q = 1, even after the input which caused these values is removed.

Table 31-1 is the truth table for four consecutive operations of the RS flip-flop. An analysis of the truth table and figure 31-5 are given as follows:

1. When S = 1 and R = 0, the flip-flop is SET. As a result, the output is Q = 1, $\bar{Q}$ = 0. A 1 is stored in the flip-flop.
2. When S = 0 and R = 0, the flip-flop retains or remembers its previous output (Q = 1, $\bar{Q}$ = 0). This means that a 1 is stored even after the SET input is removed.
3. When S = 0 and R = 1, the flip-flop is RESET. Thus, the output is Q = 0, $\bar{Q}$ = 1. A 0 is stored in the flip-flop.
4. When S = 0 and R = 0, the flip-flop retains or remembers its previous output (Q = 0, $\bar{Q}$ = 1). This means that a 0 is stored even after the RESET input is removed.

The timing diagram of figure 31-6 shows the output corresponding to a train of S and R input pulses. The Q output is assumed to be LOW initially.

(R31-9)   Assume Q is logic 1. Logic 0 is applied to the S input and the R input is at logic 0. What is the result of an RS flip-flop?

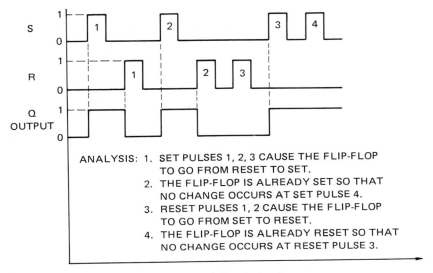

Fig. 31-6 RS flip-flop timing diagram

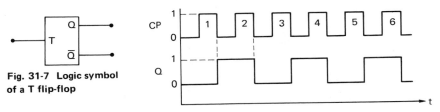

Fig. 31-7 Logic symbol of a T flip-flop

Fig. 31-8 Timing diagram of a T flip-flop

T FLIP-FLOP A T flip-flop is represented by the logic symbol shown in figure 31-7. This circuit is also called a *complementing flip-flop*, a *toggle flip-flop*, a *triggered flip-flop*, or a *clocked flip-flop*. This type of flip-flop changes state with each CLOCK pulse. There are no controlling inputs other than the CLOCK. When a flip-flop changes state with each clock pulse, it is said to be "toggling." This feature gives rise to the name toggle (T) flip-flop. The CLOCK input is indicated with a T, CP, or CL. Toggle operation is illustrated in the timing diagram of figure 31-8. Note that the flip-flop changes state with each negative going edge of the clock pulses.

For figure 31-8, the period of the output waveform is twice that of the input waveform. Therefore, the frequency of the output is one-half the frequency of the input. This means that the T flip-flop is a frequency dividing device because it divides the input (CLOCK) frequency by two. This circuit is used in digital counters and frequency control applications.

(R31-10) What are the two characteristics of the T flip-flop whose timing diagram is given in figure 31-8?

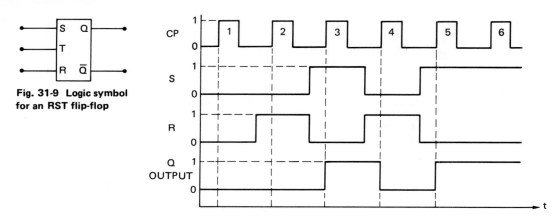

Fig. 31-9 Logic symbol for an RST flip-flop

Fig. 31-10 Timing diagram of an RST flip-flop

RST FLIP-FLOP   The RST flip-flop combines the SET, RESET, and toggle features in a single unit. The logic symbol for the RST flip-flop is shown in figure 31-9. The R input provides RESET operation, the S input provides SET operation, and the T input provides toggle, or clocked, operation. The timing diagram, figure 31-10, is based on the assumption that the flip-flop is initially RESET. An analysis of the timing diagram is as follows:

1. START: Flip-flop is RESET and Q is LOW.
2. CLOCK pulse 1: S is LOW and R is LOW. Q does not change and the flip-flop remains in the RESET state.
3. CLOCK pulse 2: S is LOW and R is HIGH. Q does not change and the flip-flop remains in the RESET state.
4. CLOCK pulse 3: S is HIGH and R is LOW. The flip-flop is SET and Q is HIGH.
5. CLOCK pulse 4: S is LOW and R is HIGH. The flip-flop is RESET and Q is LOW.
6. CLOCK pulse 5: S is HIGH and R is LOW. The flip-flop is SET and Q is HIGH.
7. CLOCK pulse 6: S is HIGH and R is LOW. Q does not change and the flip-flop remains in the SET state.

The RST flip-flop has the indeterminate condition of an RS flip-flop when both the S and R inputs are HIGH at the same time. The flip-flop tries to SET and RESET simultaneously, causing a race condition when the CLOCK pulse is removed. The final state of the flip-flop depends upon the delays through the gates. Therefore, the condition of the flip-flop is indeterminate.

The timing diagram of the RST flip-flop shows that the device is controlled by the logic states of the S and R inputs only when the CLOCK pulse is applied. Note that the flip-flop changes its state when the leading or positive edge of the CLOCK pulse is applied. In other words, the flip-flop is positive-edge triggered. When the CLOCK input is LOW, the S and R inputs can be changed without affecting the state of the flip-flop.

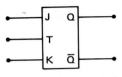

Fig. 31-11 Logic symbol for a JK flip-flop

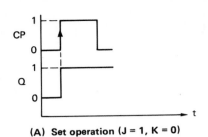

(A) Set operation (J = 1, K = 0)

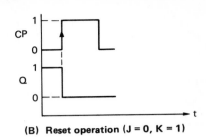

(B) Reset operation (J = 0, K = 1)

Fig. 31-12 SET and RESET operation of a JK flip-flop with positive-edge triggering

(R31-11) What is the common disadvantage of an RS and an RST flip-flop?

(R31-12) What are the two conditions which must exist simultaneously if an RST flip-flop is to change from the RESET to the SET state?

(R31-13) Can the S and R inputs control the state of an RST flip-flop if the CLOCK input is LOW?

(R31-14) Assume an RST flip-flop is RESET. What happens to the state of the flip-flop when a CLOCK pulse is applied simultaneously with a logic 1 input at R?

JK FLIP-FLOP Figure 31-11 shows the logic symbol of a JK flip-flop. The J and K designations for the inputs are arbitrary and have no significance.

The operation of the JK flip-flop is identical to that of the RST flip-flop with one exception. The JK flip-flop does not have an indeterminate state (as was described previously for the RST flip-flop). When both the J and K inputs are HIGH, the JK flip-flop is essentially a T flip-flop. The JK flip-flop combines the features of both the RST and T flip-flops and is commonly used in digital systems as a shift register stage, a toggle stage for counting operations, and a control logic stage.

The JK flip-flop can be designed for either positive-edge triggering or negative-edge triggering. Figure 31-12 shows the operation of a JK flip-flop with positive-edge triggering. The timing diagram shown in figure 31-13 is based on the assumptions that the flip-flop is RESET initially and that it operates with positive-edge triggering. The timing diagram is analyzed in the following manner:

1. START: Flip-flop is RESET and Q is LOW.
2. CLOCK pulse 1: J is LOW, K is LOW, and Q does not change. Thus, the flip-flop remains in the RESET state.
3. CLOCK pulse 2: J is HIGH and K is LOW. The flip-flop is SET and Q is HIGH.
4. CLOCK pulses 3 through 8: A toggle condition exists and Q changes state with each pulse.
5. CLOCK pulse 9: J is LOW and K is HIGH. The flip-flop is RESET and Q is LOW.
6. CLOCK pulse 10: J is LOW and K is LOW. Q does not change and the flip-flop remains in the RESET state.

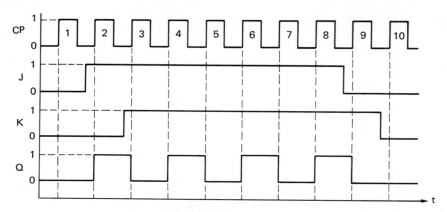

Fig. 31-13 Timing diagram of a JK flip-flop

TABLE 31-2 D Flip-Flop Truth Table

Input D	Output Q	State	Comments
1	1	Set	The Q output as-
0	0	Reset	sumes the state of D at each clock pulse

(R31-15)   What input conditions are required to achieve the toggle operation of a JK flip-flop?

(R31-16)   Compare a JK flip-flop and an RST flip-flop.

D FLIP-FLOP   The logic symbol for a D or DATA flip-flop is given in figure 31-14. This flip-flop has a D (DATA) input, a CP (CLOCK pulse) input, and outputs Q and $\overline{Q}$. The D flip-flop is used in applications where a single data bit (1 or 0) is to be stored. Positive-edge triggering is normally used, but it is possible to design D flip-flops with negative-edge triggering.

The truth table for the D flip-flop is shown in Table 31-2. The timing diagram is shown in figure 31-15. The analysis of the timing diagram is as follows:
1. The D flip-flop has positive-edge triggering.
2. A HIGH (logic 1) on the D input when a CLOCK pulse is applied will SET the flip-flop. The state of the D input (HIGH) is transferred to the Q output on the leading edge of the CLOCK pulse. A logic 1 is stored.
3. A LOW (logic 0) on the D input when a CLOCK pulse is applied will RESET the flip-flop. The state of the D input (LOW) is transferred to the Q output on the leading edge of the CLOCK pulse. In this case, a logic 0 is stored.

(R31-17)   Compare the Q output of a D flip-flop with the state of the D input.

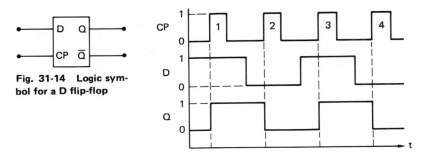

Fig. 31-14   Logic symbol for a D flip-flop

Fig. 31-15   Timing diagram of a D flip-flop

THE COLLECTOR-COUPLED BISTABLE MULTIVIBRATOR   The discrete component, collector-coupled bistable multivibrator, figure 31-16A, has two stable states. In one stable state, Q_1 is ON and Q_2 is OFF. In the other stable state, Q_2 is ON and Q_1 is OFF. Note that the circuit is completely symmetrical. Each transistor is biased from the collector of the other transistor.

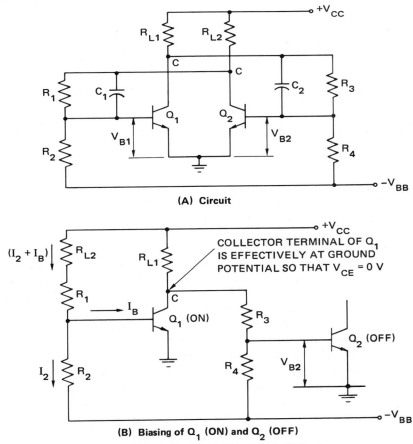

(A) Circuit

(B) Biasing of Q_1 (ON) and Q_2 (OFF)

Fig. 31-16 Collector-coupled bistable multivibrator

The circuit can be triggered from one state to the other only when a triggering input voltage is applied by means of a suitable triggering circuit connected to the base of each transistor. These connecting terminals correspond to the S (SET) and R (RESET) terminals of the integrated circuit (IC) RS flip-flop. The output signals Q and $\overline{Q}$ are obtained from the collector of each transistor.

The circuit of figure 31-16B can be analyzed for the condition when Q_1 is ON and Q_2 is OFF.

1. When Q_2 is OFF, there is no collector current through R_{L2} to Q_2.
2. Resistors R_{L2}, R_1, and R_2 form a voltage divider biasing the base of Q_1 from $+V_{CC}$ and $-V_{BB}$.
3. When Q_1 is ON (operating at saturation), the collector voltage of Q_1 is 0 V (ideally). The collector of Q_1 is at ground potential effectively.
4. Resistors R_{L1}, R_3, and R_4 form a voltage divider biasing the base of Q_2 from $+V_{CC}$ and $-V_{BB}$.
5. Since the collector of Q_1 is at the ground potential (effectively), the voltage V_{B2} biases Q_2 below the ground level. That is, the base of Q_2 is negative. Q_2 remains OFF.
6. The circuit remains in this condition indefinitely (with Q_1 ON and Q_2 OFF.) When Q_1 is triggered OFF, Q_2 switches ON and remains ON. Its base is biased by means of the voltage divider consisting of R_{L1}, R_3, and R_4. The collector of Q_2 is effectively at the ground potential. Therefore, the base of Q_1 is biased negatively, Q_1 remains OFF, and Q_2 remains ON indefinitely.

(R31-18) How can the circuit of figure 31-16A be changed from one state to the other state?

(R31-19) Where are the S (SET) and R (RESET) terminals in figure 31-16A?

(R31-20) Where are the Q and $\overline{Q}$ logic outputs obtained in figure 31-16A?

MONOSTABLE MULTIVIBRATOR (ONE-SHOT)

The *monostable multivibrator,* or the *ONE-SHOT multivibrator,* has only one stable state. This circuit remains in its stable state until it is triggered into its unstable state. The unstable state is temporary and lasts for a length of time determined by the circuit component values. The circuit always returns to the stable state unless it is retriggered.

The logic symbol for a ONE-SHOT is given in figure 31-17. This symbol shows that there is a single trigger input and both the Q and $\overline{Q}$ outputs are available. The IC form of the ONE-SHOT consists of a single chip in an encapsulated package. R and C, which control the pulse width, are connected externally to leads brought out from the appropriate points of the internal circuitry.

An important function of a ONE-SHOT circuit is to generate a time delay output pulse. Often in digital circuits, a pulse must occur after some fixed time interval. This time interval can be used to trigger the monostable circuit which then completes its cycle. At the end of the trigger cycle, the ONE-SHOT gen-

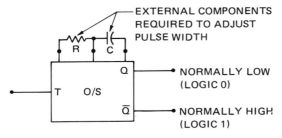

Fig. 31-17 Logic symbol for a ONE-SHOT (monostable multivibrator)

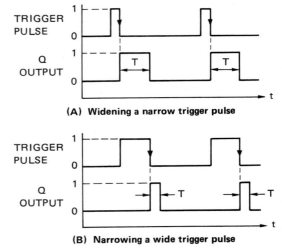

(A) Widening a narrow trigger pulse

(B) Narrowing a wide trigger pulse

Fig. 31-18 Pulse shaping actions of a ONE-SHOT circuit

erates an output pulse having a constant width and amplitude. Figure 31-18 illustrates two pulse shaping actions of a ONE-SHOT circuit. The analysis of this figure is described as follows:

1. The output of the ONE-SHOT is a delayed pulse. The pulse occurs after the negative going edge of the trigger pulse.
2. When the input, or trigger, signal is a train of narrow pulses, the ONE-SHOT can provide a train of wider output pulses having a pulse interval T, as shown in figure 31-18A.
3. When the input, or trigger, signal is a train of wide pulses, the ONE-SHOT can provide a train of narrow output pulses, as shown in figure 31-18B.
4. In both cases, the output pulse is initiated by the negative going edge of the input trigger signal.

(R31-21)   Discuss the two states of a ONE-SHOT circuit.

(R31-22)   What is the purpose of the R and C components in figure 31-17?

(R31-23)   Name two functions of a ONE-SHOT circuit.

(R31-24)   When does the output pulse begin for the ONE-SHOT circuit shown in figure 31-18?

THE DISCRETE COMPONENT ONE-SHOT CIRCUIT   Figure 31-19 shows the basic cross-coupled feedback circuit for a discrete component ONE-SHOT multivibrator. An analysis of the operation of this circuit is as follows:

1. Assume that the ONE-SHOT is in its normal or stable state. The Q output is LOW (logic 0) and the $\overline{Q}$ output is HIGH (logic 1).
2. The capacitor C is fully charged to a value equal to $V_{CC} - V_{BE2}$.

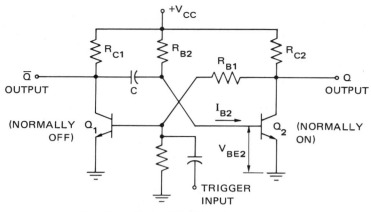

Fig. 31-19 Basic ONE-SHOT discrete circuit

3. The base current, I_{B2}, is supplied to Q_2 through R_{B2}. Transistor Q_2 is ON and its output is LOW. This output Q is the normal or true output of the ONE-SHOT.

4. The collector of Q_2 reverse biases the emitter-base junction of Q_1 so that Q_1 is OFF. Therefore, the output of Q_1, or $\bar{Q}$, is HIGH (logic 1).

5. Steps 1–4 represent the stable condition of the ONE-SHOT.

6. Assume that a short positive trigger pulse is applied to the trigger input.

7. Q_1 is turned ON momentarily so that its output is LOW (logic 0).

8. As Q_1 switches from HIGH to LOW, its collector is coupled through capacitor C to the base of Q_2. As a result, Q_2 is turned OFF.

9. Capacitor C begins to discharge through R_{B2} and Q_1 (which is ON, or at SATURATION). This is the temporary condition which is responsible for the unstable state of the ONE-SHOT circuit.

10. While capacitor C is discharging, Q_2 is OFF and the Q output of the circuit is HIGH. The student must remember that the Q output is the TRUE or normal output. This output has changed from its normal LOW state to a temporary or unstable HIGH state.

11. This unstable, or momentary, condition continues until V_{BE2} reaches a value of +0.7 V. At this point, Q_2 turns ON, Q_1 turns OFF, and the ONE-SHOT returns to its normal stable state.

12. The time constant of R_{B2} and C determines the pulse width of the output of the ONE-SHOT. This time constant corresponds to the length of time that Q_2 is OFF.

(R31-25) What factor(s) determines the pulse width of the Q output of the ONE-SHOT multivibrator shown in figure 31-19?

(R31-26) Which transistor in figure 31-19 is normally ON?

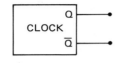

Fig. 31-20 Logic symbol for an astable multivibrator

Fig. 31-21 Clock output waveform

THE ASTABLE MULTIVIBRATOR (CLOCK)

The *astable multivibrator* has no stable state. Both of the states of this device are unstable and the circuit oscillates between them. The two states are cyclical. In other words, the circuit must run through the cycle of one state before it can transfer to the other state. A trigger input is not required because the circuit is unstable. The circuit begins to operate as soon as the proper bias voltages are supplied. The output from either side is a train of pulses. Each pulse is of equal duration and is separated by time intervals of equal length. The output from one side of the circuit is the complement of the output of the other side.

The logic symbol for the astable multivibrator is shown in figure 31-20. This IC circuit provides a CLOCK signal which can be used as a timing train of pulses to operate digital circuits. The CLOCK output waveform is shown in figure 31-21.

(R31-27)   Discuss the two states of an astable multivibrator.

(R31-28)   What use is made of the output of an astable multivibrator?

THE DISCRETE COMPONENT ASTABLE MULTIVIBRATOR CIRCUIT

A discrete component, basic astable multivibrator circuit is shown in figure 31-22. This circuit is a free-running, or oscillator, circuit. It has a cross-coupled feedback arrangement in which both cross-coupled feedback paths are capacitively coupled. The student should compare this circuit with the basic ONE-SHOT circuit shown in figure 31-19. The operation of the CLOCK circuit is analyzed in the following steps:

1. Assume that the analysis begins when Q_1 switches from OFF to ON; that is, from HIGH to LOW.
2. As Q_1 switches from HIGH to LOW, its collector is coupled through capacitor C_2 to the base of Q_2. Therefore, Q_2 is turned OFF.

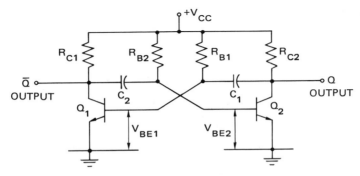

Fig. 31-22 Basic astable multivibrator circuit

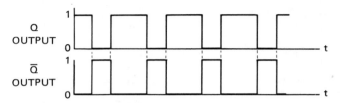

Fig. 31-23 Astable multivibrator output waveforms

3. Q_2 remains OFF until V_{BE2} reaches a value of 0.7 V. The OFF time of Q_2 is determined by the time constant of R_{B2} and C_2.
4. When V_{BE2} reaches a value of 0.7 V, Q_2 turns ON. Its collector is coupled through capacitor C_1 to the base of Q_1. Therefore, Q_1 is turned OFF.
5. Q_1 remains OFF until V_{BE1} reaches a value of 0.7 V. The OFF time of Q_1 is determined by the time constant of R_{B1} and C_1.
6. When Q_1 turns back ON, the cycle repeats itself. The circuit continues to oscillate back and forth without external triggering. The circuit requires only the dc power source.
7. The output Q from the collector of Q_2 is the complement of the output $\overline{Q}$ from the collector of Q_1. This statement is true because when Q_2 is ON, Q_1 is OFF, and vice versa, figure 31-23.

(R31-29)   Compare the Q and $\overline{Q}$ outputs of a CLOCK circuit.

THE SCHMITT TRIGGER CIRCUIT   The basic Schmitt trigger circuit is shown in figure 31-24. This circuit is used in waveshaping applications. It has two operating states which are opposites. Normally, the input signal is not a pulse waveform but a varying ac voltage. The Schmitt trigger circuit is level sensitive and switches the output state at two distinct levels:

1. LTL, or lower trigger level.
2. UTL, or upper trigger level.

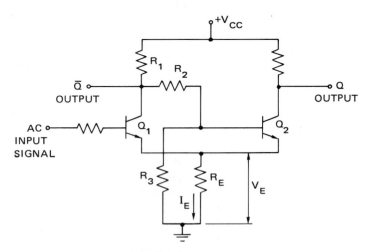

Fig. 31-24 Basic Schmitt trigger circuit

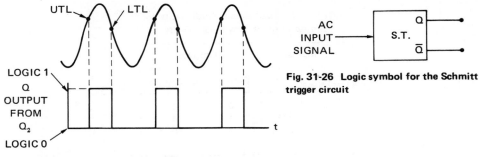

Fig. 31-26 Logic symbol for the Schmitt trigger circuit

Fig. 31-25 Waveforms of the Schmitt trigger circuit

The output of the Schmitt trigger circuit is a digital output having either a logic 0 or logic 1 level, figure 31-25. The logic symbol is shown in figure 31-26.

The operation of the Schmitt trigger circuit shown in figure 31-24 is described as follows:

1. With no input signal, Q_1 is OFF. That is, the collector output $\overline{Q}$ is HIGH.
2. The HIGH on the collector of Q_1 biases Q_2 ON. Therefore, the output Q is LOW.
3. The emitter current of Q_2 produces a voltage drop V_E across R_E. This voltage biases the emitter of Q_1 to a value equal to V_E. This voltage establishes the upper trigger level, UTL.
4. The input signal voltage must exceed V_E to turn Q_1 ON.
5. When the input signal voltage reaches the UTL, Q_1 turns ON and Q_2 turns OFF.
6. The emitter bias voltage V_E is now determined by the emitter current of Q_1, since Q_1 is conducting and Q_2 is OFF. This voltage establishes the LTL.
7. The input signal voltage must drop below the LTL value to permit Q_1 to turn OFF and Q_2 to turn back ON.

(R31-30) What use is made of the Schmitt trigger circuit?

(R31-31) Why is the Schmitt trigger circuit said to be level sensitive?

(R31-32) Define: a. LTL and b. UTL.

BASIC ADDERS The basic rules for binary addition were given in Chapter 30. For convenience, these rules are repeated here:

Rule 1: $0 + 0 = 0$

Rule 2: $0 + 1 = 1$

Rule 3: $1 + 0 = 1$

Rule 4: $1 + 1 = 10$ (One plus one equals a sum of 0 with a carry of 1 to the next bit position)

The half adder (HA) is a logic device which adds two bits of binary data. It performs the binary addition operations shown by the truth table given in Table 31-3. S represents the sum of A and B and C_O represents the carry output. Note that $C_O = 0$ in every case, except the last where the addition of $1 + 1$ results in a carry of 1 to the next bit position.

TABLE 31-3 Half Adder Truth Table for the Addition of Two Binary Digits

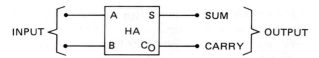

Inputs		Outputs	
A	B	Sum(S)	Carry(C_O)
0	0	0	0
0	1	1	0
1	0	1	0
1	1	0	1

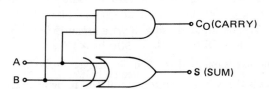

Fig. 31-27 Block diagram symbol for a half adder

The inputs of the half adder accept two binary digits. The half adder produces two binary digits on its outputs (a sum and a carry bit), as shown in the block diagram of figure 31-27. The half adder can be implemented by an EXCLUSIVE-OR gate and an AND gate, figure 31-28.

The full adder accepts three inputs and generates a sum output and a carry output. The full adder can accept input carries as shown in the block diagram symbol given in figure 31-29. The operation of the full adder is illustrated by the truth table given in Table 31-4. The complete logic circuitry for a full adder

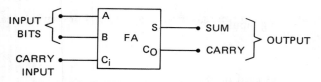

Fig. 31-28 Implementation of a half adder

Fig. 31-29 Block diagram symbol for a full adder

TABLE 31-4 Full Adder Truth Table

Inputs			Outputs	
A	B	C_i	C_O	S
0	0	0	0	0
0	0	1	0	1
0	1	0	0	1
0	1	1	1	0
1	0	0	0	1
1	0	1	1	0
1	1	0	1	0
1	1	1	1	1

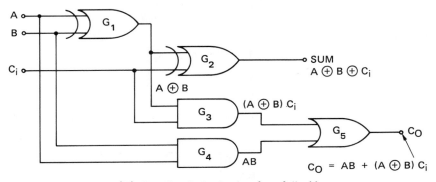

(A) Complete logic circuitry for a full adder

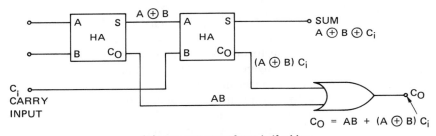

(B) Arrangement of two half adders

Fig. 31-30 Full adder implementation

is shown in figure 31-30A. The arrangement of two half adders to form a full adder is shown in figure 31-30B.

(R31-33) What is the difference between a half adder and a full adder?

(R31-34) A half adder has the following inputs: $A = 0$, $B = 1$. What are the resulting outputs?

(R31-35) A full adder has the following inputs: $A = 0$, $B = 0$, and $C_i = 1$. What are the resulting outputs?

REGISTERS Registers are memory devices used to store and manipulate data. There are two types of registers: memory registers and shift registers.

The method used to enter or remove stored information accounts for the following designations for both types of registers:

1. Serial register: one in which the data is entered or removed one bit at a time.
2. Parallel register: one which accepts or transfers all bits of data simultaneously.

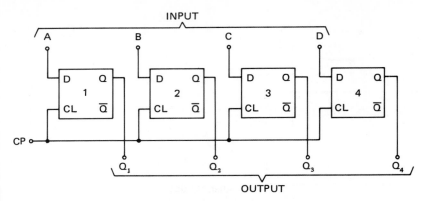

Fig. 31-31 Four-bit memory register

3. Serial-parallel register: one which has a serial input and a parallel output.

4. Parallel-serial register: one which has a parallel input and a serial output.

The functions of a memory or storage register are:

1. to accept information in the form of a binary number,

2. to hold the information after the input providing it is removed, and

3. to make the information available as an output.

Memory registers are constructed using flip-flops. One flip-flop is required for each bit in the word to be stored. Figure 31-31 shows a four-bit memory register using four D flip-flops. This register stores the input ABCD as the output Q_1, Q_2, Q_3, Q_4. The CLOCK signal CP when pulsed is the memory transfer input that transfers A to Q_1, B to Q_2, C to Q_3, and D to Q_4. Once the input ABCD is stored, it remains in the memory register until new data is to be stored by pulsing the memory transfer again.

Although a shift register can store data too, it normally is used to process or move data from one stage of the register to an adjacent stage. The shift is from left to right in a right-shift register; from right to left in a left-shift register; and in both directions in a left-shift, right-shift register.

(R31-36) Name the two types of registers.

(R31-37) What are the four register designations?

(R31-38) What is the primary function of a shift register?

MULTIPLEXERS AND DE-MULTIPLEXERS

A *multiplexer* is a combinational circuit device that selects data from one of two or more input lines and transmits it on a single output line. The device also has control or selection inputs that permit digital data on any one of the inputs to be switched to the single output line. Several messages may be sent over the single output line. This is accomplished by sending in succession the first character of each message. The second character of each message is then sent, followed by the third character, and so on until each character of each message is sent. A block diagram symbol for a two-input multiplexer is shown in figure

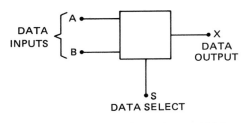

Fig. 31-32 Block diagram of a two-input multiplexer

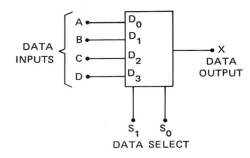

Fig. 31-33 Block diagram of a four-input multiplexer

31-32. This multiplexer transmits data signals A and B over the output line X. The data select signal S determines which of A or B is the output X. When S = 1, the output X = A. When S = 0, the output X = B. The select signal is a pulse signal alternating between 1 and 0 so that first A is the output, then B, then A, and so on. The duration of the logic 1 pulse must be long enough to transmit the code representing each character.

Figure 31-33 gives the block diagram of a four-input multiplexer. The two data selection lines make it possible to select each of the four data input lines. A two-bit binary code on the data select input lines allows the data on the corresponding data input to pass through to the data output. If a binary 0 (S_1 = 0, S_0 = 0) is applied to the data select lines, the data on input D_0 appears on the data output line. Table 31-5 summarizes the operation of the four-input multiplexer.

A demultiplexer receives the multiplexed signal and separates it into the original separate signals. Figure 31-34 shows the block diagram of a two-channel demultiplexer whose input is the multiplexed signal X and whose output is composed of the two separate, or demultiplexed, signals A and B. The data select signal, S, provides alternately A, then B, then A, and so on.

(R31-39)   Define: a. multiplexer and b. demultiplexer.

(R31-40)   The four-input multiplexer shown in figure 31-33 has the following inputs: A = 1, B = 0, C = 1, D = 0, S_0 = 0, and S_1 = 1. Determine the output X.

TABLE 31-5 Operation of a Four-Input Multiplexer

| Data Select Inputs | | | Output |
Binary Number	S_1	S_0	(Input Selected)
0	0	0	D_0 (A)
1	0	1	D_1 (B)
2	1	0	D_2 (C)
3	1	1	D_3 (D)

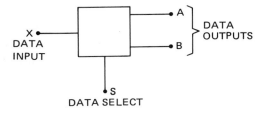

Fig. 31-34 Block diagram of a two-channel demultiplexer

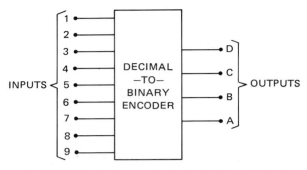

Fig. 31-35 Block diagram of a decimal-to-binary encoder

TABLE 31-6   Truth Table for the Decimal-to-Binary
                          Encoder

Decimal Input	Binary Output			
N	A	B	C	D
0	0	0	0	0
1	0	0	0	1
2	0	0	1	0
3	0	0	1	1
4	0	1	0	0
5	0	1	0	1
6	0	1	1	0
7	0	1	1	1
8	1	0	0	0
9	1	0	0	1

ENCODERS AND DECODERS

Codes are used to transmit numbers or characters over telephone lines, or to place them into a computer. The number or character may be represented by an equivalent binary number. An *encoder* is a device that translates the character into its coded equivalent. A *decoder* is a device that translates the code back into the character it represents. In other words, the encoder converts data into a form that can be interpreted by a digital circuit.

The block diagram of a decimal-to-binary encoder is shown in figure 31-35. The truth table for this encoder is given in Table 31-6. The encoder accepts a decimal digit N as an input from a keyboard and yields its binary equivalent $(ABCD)_2$. The encoder has outputs A, B, C, and D which are actuated (made logic 1), or not actuated (made logic 0), depending upon the input N. Table 31-6 shows that each input activates one or more of the A, B, C, D outputs to result in the binary equivalent. The only exception is the decimal 0 which activates no output lines since its binary equivalent is 0000.

Figure 31-36 shows a block diagram of a binary-to-decimal decoder. This circuit is designed to decode the binary numbers $(ABCD)_2$ listed in Table 31-6. This decoder has ten outputs, one for each of the decimal digits. When the select

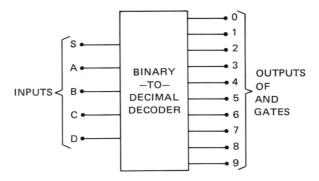

Fig. 31-36 Block diagram of a binary-to-decimal decoder

signal S equals 1 an output appears; when S equals 0, all outputs are 0. The signal S is also called an *enable input*. That is, the signal enables the gate, or allows the gate to be activated, whose output is to be displayed. The select signal S must be an input to all of the AND gates of the decoder shown in figure 31-36. For example, if the input signal SABCD = 10100, the gate labeled 4 is the only gate with an output of 1. Thus, the output of the decoder is the decimal number 4.

(R31-41) Define: a. encoder and b. decoder.

(R31-42) Does the decimal 0 actuate any output lines of a decimal-to-binary encoder?

(R31-43) Explain the meaning of an enable input to a gate.

(R31-44) The input signal, in figure 31-36, is SABCD = 00100. Which gate has an output of 1?

(R31-45) The input signal, in figure 31-36, is SABCD = 10101. What is the decimal output?

COUNTERS A binary counter has the following characteristics:
1. It is a sequential circuit that counts or tallies the number of input pulses it receives.
2. It is a memory device that stores the number of input pulses.
3. The number of input pulses, or the count, can be determined at any time, because the bits are those stored in the flip-flops that make up the counter.

Counters are used in digital computers, timing circuits, and signal generators.

The four-bit binary counter shown in figure 31-37A stores the number of input pulses as a binary number. Four negative-edge triggered flip-flops are used since one flip-flop is required for each bit needed in the count. The input E is the enabling input. The timing diagram is given in figure 31-37B. The operation of the counter is described as follows:
1. Assume that a direct RESET input (not shown) clears the four flip-flops. The enable input E = 1.
2. All four outputs (A, B, C, D) are at logic 0.

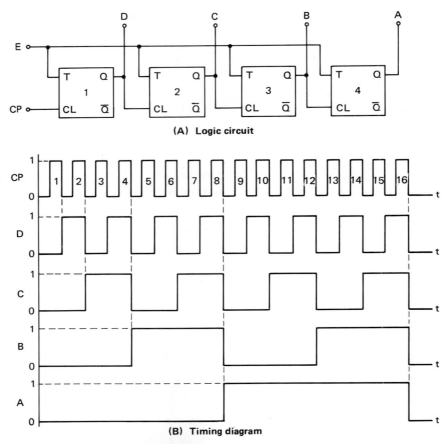

(A) Logic circuit

(B) Timing diagram

Fig. 31-37 Four-bit binary counter

3. When pulse 1 of the CLOCK pulse occurs, the first flip-flop toggles and sets the D bit to logic 1. The other bits, C, B, and A, remain at logic 0 because flip-flops 2, 3, and 4 all have CP = 0.

4. Pulse 2 resets flip-flop 1 to logic 0. D goes from logic 1 to logic 0. Flip-flop 2 toggles because it is negative-edge triggered. Therefore, the C bit is at logic 1.

5. Pulse 3 causes flip-flop 1 to toggle resulting in D = 1 again. C remains at 1.

6. Pulse 4 resets D and C to logic 0. The transition of C from logic 1 to logic 0 sets B to logic 1. The circuit continues to operate as shown in the timing diagram of figure 31-37B.

The outputs of the flip-flops form the stored binary number $(ABCD)_2$. From the timing diagram, it can be seen that the successive values, as the negative edge of the CLOCK pulse occurs, are 0000, 0001, and so on to 1111. These values are the binary equivalents of the decimal numbers 0, 1, and so on to 15. As a result, it is evident that a binary counter of four flip-flops can count 2^4 or 16 decimal integers before starting over at 0. Additional flip-flops are required if there are more decimal integers to be counted.

The circuit shown in figure 31-37A is called an *asynchronous counter* because all of its flip-flops are not set simultaneously by a CLOCK. This counter is also called a *ripple counter* because a certain amount of time is required as each bit changes sequentially. In other words, the effect appears to "ripple" through the counter. The counter is also called an *up counter* because it counts up in the sequence 0, 1, 2, and so on to 15.

(R31-46) How many flip-flops are required for a counter capable of counting 32 decimal integers?

(R31-47) What are three designations for the four-bit binary counter shown in figure 31-37A?

(R31-48) What is the meaning of each of the designations listed in answer to question (R31-47)?

ANALOG-TO-DIGITAL (A/D) AND DIGITAL-TO-ANALOG (D/A) CONVERTERS An analog function is converted to digital form by an analog-to-digital (A/D) converter. This conversion permits the analog function to be processed by a digital system. A digital-to-analog (D/A) converter contains the circuitry necessary to convert the information contained in a binary digital input word to an analog dc output voltage corresponding to the bit pattern in that word.

Figure 31-38 shows D/A and A/D converters as well as the timing diagram of

(A) D/A converter

(B) A/D converter

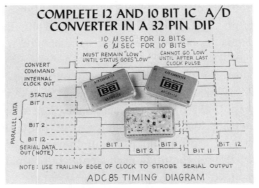

(C) Timing diagram of an A/D converter

Fig. 31-38

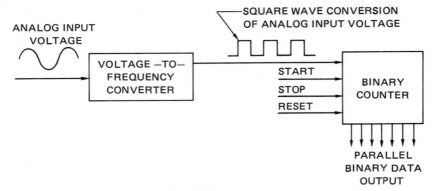

Fig. 31-39 Block diagram of a voltage-to-frequency A/D converter technique

an A/D converter. The block diagram of a simple voltage-to-frequency A/D converter technique is given in figure 31-39. The operation of this converter is analyzed as follows:

1. The technique uses a voltage-to-frequency converter and a binary digital counter.
2. The analog voltage is the input to the voltage-to-frequency converter. The output produced is a square wave.
3. The frequency of the square wave is proportional to the magnitude of the input analog voltage. This conversion actually represents A/D conversion. For practical applications, the binary counter is usually added.
4. The pulses from the voltage-to-frequency converter are counted by the binary counter.
5. The output of the circuit is a parallel digital word whose magnitude is directly proportional to the analog input signal. The primary application of this technique is in digital meters.

A four-bit digital-to-analog converter is shown in the block diagram of the voltage divider D/A converter of figure 31-40. Each input is one of the four digital bits. A binary 1 may be represented on the input by a +15 V level. A binary 0 may be represented by a 0 V level, or ground. The output voltage values for this D/A converter are listed in Table 31-7. If more bits are used in the binary number, the changes or steps in the output voltage are smaller. As a result, the output is a smoother curve which is a closer approximation to the analog function being represented. In the D/A converter of figure 31-40, each step of the output is 1/15 of the voltage level representing a binary 1. For the five-bit converter, each step of the output is 1/31 of the voltage level, and so on.

(R31-49)   Define: a. an A/D converter and b. a D/A converter.

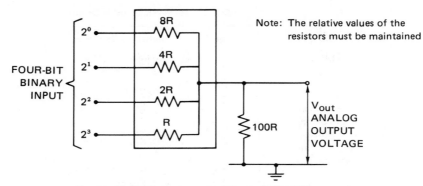

Note: The relative values of the resistors must be maintained

FOUR-BIT BINARY INPUT

2^0 8R

2^1 4R

2^2 2R

2^3 R

100R

V_{out} ANALOG OUTPUT VOLTAGE

Fig. 31-40 Block diagram of a voltage divider D/A converter

TABLE 31-7 Output Voltage Values for the D/A Converter Shown in Figure 31-40

Binary Input				Output Voltage
2^3	2^2	2^1	2^0	V_{out}
0	0	0	0	0
0	0	0	1	1
0	0	1	0	2
0	0	1	1	3
0	1	0	0	4
0	1	0	1	5
0	1	1	0	6
0	1	1	1	7
1	0	0	0	8
1	0	0	1	9
1	0	1	0	10
1	0	1	1	11
1	1	0	0	12
1	1	0	1	13
1	1	1	0	14
1	1	1	1	15

(R31-50) How many steps are there in the output of an eight-bit D/A converter?

READOUTS The output unit in many digital systems is a readout display. For many applications, a readout with several digits is needed. Multidigit displays are available in light-emitting diode (LED) and liquid crystal display (LCD) types.

Figure 31-41 shows a common display format composed of seven light-emitting elements or segments. By lighting certain combinations of these segments, each of the ten decimal digits is produced. The activated segments for each of the ten decimal digits are listed in Table 31-8.

(R31-51) List two multidigit display types.

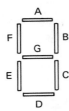

A

F   B
  G
E   C

D

Fig. 31-41 Seven-segment display showing the arrangement of segments

TABLE 31-8 Seven-Segment Display

Digit	Activated Segments
0	A, B, C, D, E, F
1	E, F
2	A, B, G, E, D
3	A, B, C, D, G
4	B, C, F, G
5	A, C, D, F, G
6	C, D, E, F, G
7	A, B, C
8	A, B, C, D, E, F, G
9	A, B, C, F, G

THE DIGITAL COMPUTER

The *digital computer* is an assembly of electronic circuits which perform operations that can be represented mathematically in the binary number system or by Boolean equations. The computer performs the binary and logical operations on binary-coded data very reliably, rapidly, predictably, and in sequence. A computer can do only what it is programmed to do; a computer cannot think for itself.

Programmed tasks are carried out by the digital computer as follows:

1. It interprets various combinations of 1s and 0s as commands, or orders, to perform its operations.
2. It follows a sequential series of commands, unless it is commanded to alter that sequence.
3. It performs tests, and makes programmed decisions based on the results of those tests.
4. Operations are performed in microseconds.
5. It has memory banks that store information for retrieval.
6. It has means for bidirectional communications with the outside world.

Figure 31-42 is a block diagram of a digital computer. The four functional units of the typical digital computer are:

1. The input/output unit (I/O)
2. The arithmetic logic unit (ALU)
3. The timing and control unit
4. The memory unit

The input/output (I/O) unit contains the digital logic necessary to interface the computer to the external equipment. In other words, this unit can generate and verify synchronization between the computer and the external devices to ensure a reliable transfer of data. For this reason, the input/output unit is the main avenue of communication between the computer and the outside world.

The pieces of equipment, or hardware, which are attached by wires to the main functional units are called *peripheral devices.* Some of the input peripheral devices are: punched card, magnetic tape, magnetic disk, typewriter, and punched tape. Some of the output peripheral devices are: high-speed printer, magnetic tape, magnetic disk, typewriter, punched tape, and video display.

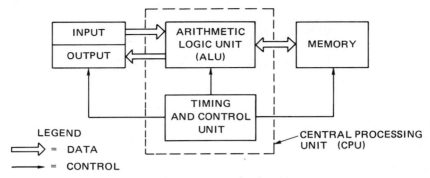

Fig. 31-42 Block diagram of a digital computer

The arithmetic logic unit (ALU) includes all of the hardware used in modifying data to solve equations or to perform other logical, or mathematical, operations on the data. The ALU also includes the logic that indicates the successful completion of each arithmetic operation.

The memory unit stores data for recall and processing as used. Digital information is stored in the form of words. Each word consists of a specified number of binary bits. The *word length* of the computer is the number of binary bits that can be handled by the computer simultaneously in a single operation. The majority of computers used in industrial process control have a word length of 16 bits.

The timing and control logic is the remaining functional unit of the digital computer. This unit synchronizes and controls each internal operation of the computer. The logic receives its basic instructions from the programmer. The timing and control logic interprets each instruction in the computer program. This logic is connected directly to each internal unit of the computer. It controls directly the operation of the computer.

The combination of the arithmetic logic unit and the timing and control unit is known as the *central processing unit* (CPU). This unit performs the arithmetic operations and logic decisions initiated by the program governing the computer.

(R31-52)   Name the four functional units of a digital computer.

THE MICRO-PROCESSOR (μP)   The microprocessor is a single large-scale integrated (LSI) chip that can perform arithmetic and logical functions under program control. It replaces the central processing unit, shown in figure 31-42. Therefore, the microprocessor implements the functions of the arithmetic logic unit and the associated control unit of a conventional digital computer. A μP, or LSI chip, is an assembly that contains about 70% of the computing ability of a very small digital computer. The μP replaces thousands of discrete transistors and diodes used in a conventional digital computer. Figure 31-43 is the data sheet for a μP.

9440 MICROFLAME™
16-BIT BIPOLAR MICROPROCESSOR

GENERAL DESCRIPTION — The 9440 MICROFLAME single-chip 16-bit bipolar processor, packaged in a 40-pin DIP, is implemented using Fairchild's Isoplanar Integrated Injection Logic technology (I3L™). Though structurally different from the CPUs of the Data General NOVA line of minicomputers, the 9440 offers comparable performance and executes the same instruction set.

- EIGHT 16-BIT ON-CHIP REGISTERS
- 64 DIRECTLY ADDRESSABLE I/O DEVICES, EACH WITH THREE BIDIRECTIONAL I/O PORTS
- PRIORITY INTERRUPT HANDLING WITH UP TO 16 PRIORITY LEVELS
- FAST DIRECT MEMORY ACCESS AT MEMORY SPEEDS
- 16-BIT 3-STATE BIDIRECTIONAL INFORMATION BUS
- FLEXIBLE OPERATOR CONSOLE CONTROL USING ONLY FOUR LINES
- POWERFUL, WIDELY USED INSTRUCTION SET
- MULTIFUNCTION INSTRUCTIONS FOR EFFICIENT MEMORY USAGE
- FOUR CLASSES OF INSTRUCTIONS; TOTAL OF 2192 DIFFERENT INSTRUCTIONS
- EIGHT ADDRESSING MODES, 32K 16-BIT WORDS (64K BYTES) ADDRESSING RANGE
- 5 V POWER SUPPLY
- TTL INPUTS AND OUTPUTS
- TYPICAL 1 W POWER DISSIPATION
- FULL MILITARY TEMPERATURE RANGE VERSION
- SINGLE-CLOCK STATIC OPERATION, ON-CHIP OSCILLATOR, DC TO 12 MHz CLOCK RATE
- COMPATIBLE HIGH SPEED MEMORIES AVAILABLE (93481/93483)
- TOTAL SOFTWARE WITH FIRE™ SOFTWARE PACKAGE

CONNECTION DIAGRAM
DIP (TOP VIEW)

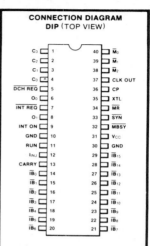

SYSTEM DIAGRAM

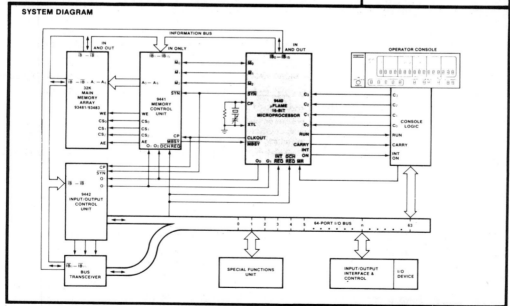

© 1978 Fairchild Camera and Instrument Corporation   Printed in U.S.A.   002-11-0005-058   15M

Fig. 31-43 μP data sheet

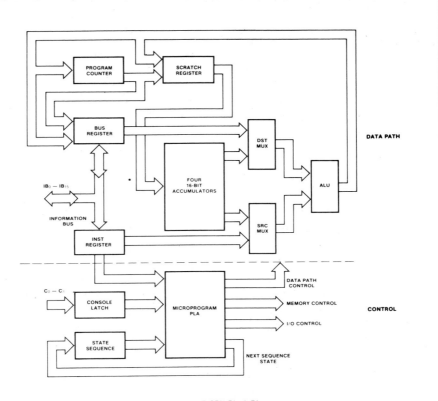

Fig. 1 9440 MICROFLAME CPU Block Diagram

9440 ARCHITECTURE *(Figure 1)*
The 9440 single-chip 16-bit bipolar processor, packaged in a 40-pin DIP, is implemented using Fairchild's Isoplanar Integrated Injection Logic technology (I3L™). Though structurally different from the CPUs of the NOVA line of minicomputers, the 9440, as a microprocessor, offers comparable performance and executes the same instruction set.

The processor is a stored program machine using homogeneous external memory, i.e., instructions and data are stored in the same memory. Although the processor handles 16 bits of information, only 15 bits are used for addressing the memory. Thus, the intrinsic memory capacity of a 9440 system is 32,768 16-bit words.

The 9440 consists of a collection of data paths and all the necessary control circuitry. It governs peripheral I/O equipment, performs the arithmetic, logic and data handling operations and sequences the program.

Data Paths — The data path portion includes a bank of four 16-bit general-purpose registers (accumulators AC0 - AC3), two multiplexers, an ALU, and four 16-bit special registers — scratch register, bus register, instruction register and program counter. Internal data flows between the various registers via 4-bit-wide data paths.

Fig. 31-43 μP data sheet (Cont'd)

(R31-53)   Define a microprocessor.

Some applications of microprocessors include:
Commercial:
autotransaction systems
banking and financial terminals
credit card verification systems
inventory control systems
point-of-sale terminals
security systems
Communications:
error detection
message handling
multiplexers
programmable controllers
remote terminals
switching systems
Consumer:
automotive monitoring and control
educational systems
home microcomputer
educational toys and games
programmable appliances
Data Processing:
auditing and security
communications interface
input/output controllers
office computers
performance monitoring
peripheral processors
programmable calculators
Industrial:
data acquisition systems
environmental monitoring
numerical control
process control
sensor-based systems
Instrumentation:
analytical chemical and medical equipment
automatic test equipment
electronic instruments
Military:
communications
navigation systems
simulators and training equipment

Fig. 31-45 μC module with a μP as a central processing unit

Fig. 31-44   Single-chip microcomputer (8748 LSI μC)

THE MICRO-COMPUTER (μC)   A microcomputer is a microprocessor-based computer. It has complete computer capability, such as memory and timing, and it can communicate with the outside world. Before 1977, a microcomputer required several chips. However, in 1977, the INTEL Corporation introduced a one-chip microcomputer, known as the Model 8748, figure 31-44. The single chip of this microcomputer contains the central processing unit, the program memory, the data memory, the connections between the input and output signals, and the clocks and timers. A microcomputer module with a microprocessor as its CPU is shown in figure 31-45. This single-board microcomputer brings all address, data, and control lines to the board connector.

(R31-54)   Define a microcomputer.

THE CALCULATOR   Figure 31-46 is a block diagram of a calculator, or simple microcomputer. The basic components of this calculator are:
1. The μP or CPU which performs the control and computing functions.
2. The memory which stores the program and data.
3. The input module, or the keyboard used to direct the operation of the calculator.
4. The output module or the light emitting diode (LED) display.
5. The input/output interface which is required to connect the input module and the output module to the microprocessor.

The following steps show how the calculator adds 25.2 and 50.9:
1. The number 25.2 is entered on the keyboard.
2. This number is stored in a register of the microprocessor.
3. The operation "+" is specified by the operator at the keyboard. This operation is memorized in a special microprocessor register until it can be completed.

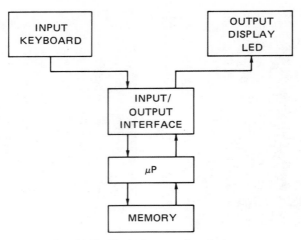

Fig. 31-46 Block diagram of a calculator

4. The second number, 50.9, is entered by the operator at the keyboard.
5. This number is stored, or accumulated, in a register of the microprocessor.
6. The operator indicates the operation "=". This command is the same as ordering the execution of the arithmetic program which was memorized in the special register of Step 3. The addition program stored in the memory is then executed. The result of the addition is deposited in one of the microprocessor registers, called the accumulator.
7. The result, 76.1, is sent from the accumulator to the LED unit to be displayed to the operator.
8. The displayed result is stored in the internal accumulator until it is cleared by pressing the "C" key, or until the operator specifies a new operation.

μP APPLICATIONS   Microprocessors are used in many industrial control applications, such as the one shown in figure 31-47. The automatic welders in a heavy machine assembly area are controlled by the microprocessor. The X, Y, and Z coordinates of both the workpiece and the rod must be known before the automated system can operate properly. The rod must be kept at a precisely controlled distance from the welding surface. If the rod is too close, it sticks to the surface. If the rod is too far from the surface, a poor weld or no weld at all is the result. The μP monitors the current through the arc and the voltage between the rod and workpiece. In this way, precise distance control is maintained.

A microprocessor-controlled waveform generator is shown in figure 31-48. Any waveform that can be drawn can be entered into this instrument and then generated as an output. This waveform generator can easily generate and control the irregular waveforms required in medical, biological, and materials research.

A microprocessor programmable controller is shown in figure 31-49. This controller is used by process industries to control temperature, pressure, flow, and speed.

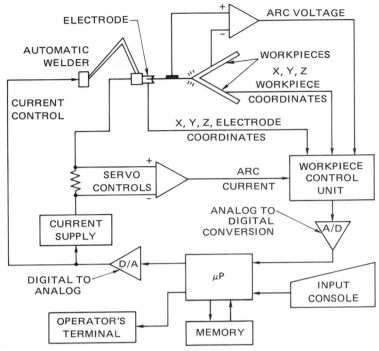

Fig. 31-47 μP control of an automatic welder

Fig. 31-48 μP-controlled arbitrary waveform generator

(A) Microprocessor programmable controller (front view)

(B) Microprocessor programmable controller (internal view)

Fig. 31-49 Microprocessor programmable controller

(A) Processor box showing processor modules

(B) Processor box

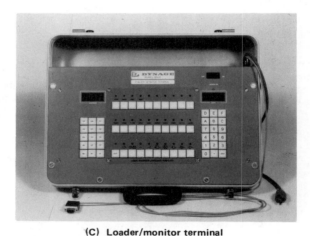

(C) Loader/monitor terminal

Fig. 31-50 μP-based programmable controller

A microprocessor-based, programmable controller is shown in figure 31-50. The group of processor modules and input/output modules provide complete analog and digital sensor-line, input/output control. Users can arrange the components to build systems to satisfy specific needs.

LABORATORY EXERCISE 31-1 CHARACTERISTICS AND OPERATION OF A DISCRETE TRANSISTOR INVERTER

PURPOSE   In completing this exercise, the student will examine the characteristics and operation of a discrete transistor inverter circuit.

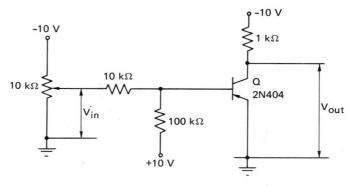

Fig. 31-51

EQUIPMENT **AND MATERIALS**	1 2N404 transistor 1 Potentiometer, 10 kΩ, 5 W 1 Resistor, 1 kΩ, 2 W 1 Resistor, 10 kΩ, 2 W 1 Resistor, 100 kΩ, 2 W 3 Power supplies, dc 1 Multimeter

PROCEDURE

1. Construct the circuit shown in figure 31-51.
   a. Energize the three dc power supplies.
   b. Vary the potentiometer so that V_{in} is 0 V, as indicated on the multimeter.
   c. Measure V_{out}.
   d. Increase V_{in} by 0.1-V steps until V_{out} changes abruptly.
   e. Adjust the potentiometer so that V_{in} = –10 V. Record V_{out}.
2. Answer the following questions:
   a. Explain how the circuit of figure 31-51 operates as an inverter.
   b. Plot a graph of V_{out} versus V_{in}. Indicate the following regions on the graph: 1. saturation, 2. cutoff, and 3. active.
   c. Explain the meaning of each of the regions marked on the graph plotted in step 2.b.

LABORATORY
EXERCISE 31-2
CHARACTERIS-
TICS AND
OPERATION OF A
DISCRETE
BISTABLE
MULTIVIBRATOR

PURPOSE   In completing this exercise, the student will examine the characteristics and operation of a discrete bistable multivibrator.

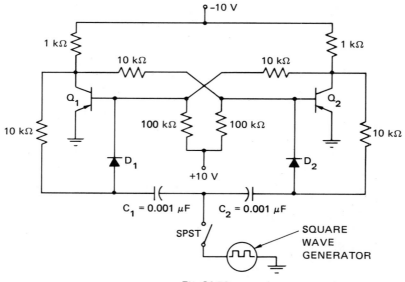

Fig. 31-52

<table>
<tr><td>**EQUIPMENT**
AND MATERIALS</td><td>2 1N914 Diodes
2 2N404 Transistors
2 Resistors, 100 kΩ, 2 W
4 Resistors, 10 kΩ, 2 W
2 Resistors, 1 kΩ, 2 W
2 Capacitors, 0.001 μF
1 Oscilloscope
1 Square wave generator
2 Power supplies, dc
1 Switch, SPST</td></tr>
</table>

PROCEDURE

1. Construct the circuit shown in figure 31-52.
   a. Energize the dc power supplies with the SPST switch OPEN.
   b. Measure the base and collector voltages of Q_1 and Q_2.
   c. Close the SPST switch.
   d. Adjust the frequency of the square wave generator to 100 Hz. Adjust the output voltage to 10 V (peak-to-peak).
   e. Measure the base and collector voltages of Q_1 and Q_2.
   f. Open the SPST switch.
   g. Study the circuit and determine a method of resetting the FLIP-FLOP.
   h. RESET the FLIP-FLOP.
   i. Measure the base and collector voltages of Q_1 and Q_2.
2. Answer the following questions:
   a. Explain how the circuit shown in figure 31-52 operates as a FLIP-FLOP.
   b. Where is the S (SET) terminal of this circuit?
   c. Where is the R (RESET) terminal of this circuit?

LABORATORY EXERCISE 31-3 CHARACTERISTICS AND OPERATION OF A DISCRETE MONOSTABLE MULTIVIBRATOR CIRCUIT

PURPOSE In completing this exercise, the student will examine the characteristics and operation of a discrete monostable multivibrator.

EQUIPMENT AND MATERIALS
1 1N914 Diode
2 2N404 Transistors
3 Resistors, 1 kΩ, 2 W
1 Resistor, 10 kΩ, 2 W
1 Resistor, 47 kΩ, 2 W
1 Resistor, 100 kΩ, 2 W
1 Potentiometer, 100 kΩ, 5 W
2 Capacitors, 0.001 μF
1 Oscilloscope
2 Power supplies, dc
1 Square wave generator
1 Switch, SPST

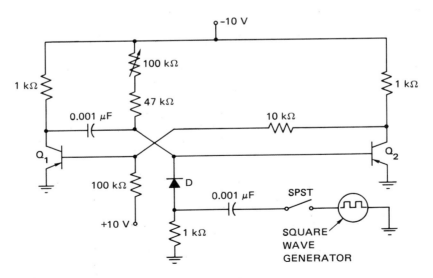

Fig. 31-53

PROCEDURE   1. Construct the circuit shown in figure 31-53.
  a. Energize the dc power supplies with the SPST switch OPEN.
  b. Measure the base and collector voltages of Q_1 and Q_2.
  c. Close the SPST switch.
  d. Adjust the frequency of the square wave generator to 1 000 Hz. Adjust the output voltage to trigger the ONE-SHOT multivibrator.
  e. Measure the base and collector voltages of Q_1 and Q_2.
  f. Increase the frequency in 5 000-Hz steps to 51 000 Hz. Measure the base and collector voltages of Q_1 and Q_2 at each step.
  2. Answer the following questions:
  a. Explain how the circuit shown in figure 31-52 operates as a ONE-SHOT multivibrator.
  b. Explain the waveforms measured during this experiment.
  c. What determines the pulse width limitations of the monostable multivibrator?
  d. What determines the frequency limitations of this circuit?

LABORATORY EXERCISE 31-4 CHARACTERIS-TICS AND OPERATION OF A DISCRETE ASTABLE MULTIVIBRATOR CIRCUIT

PURPOSE   In completing this exercise, the student will examine the characteristics and operation of a discrete astable multivibrator.

EQUIPMENT AND MATERIALS
2 2N404 Transistors
2 Resistors, 1 kΩ, 2 W
1 Resistor, 47 kΩ, 2 W
2 Resistors, 100 kΩ, 2 W
2 Capacitors, 0.1 μF
1 Oscilloscope
1 Power supply, dc

PROCEDURE   1. Construct the circuit shown in figure 31-54.
  a. Energize the dc power supply.
  b. Measure the base and collector voltages of Q_1 and Q_2.
  c. Measure and record the time Q_1 and Q_2 are ON during a complete cycle.
  d. Replace the 100-kΩ resistor connected to the base of Q_2 with a 47-kΩ resistor.
  e. Repeat steps 1.a. through 1.c.

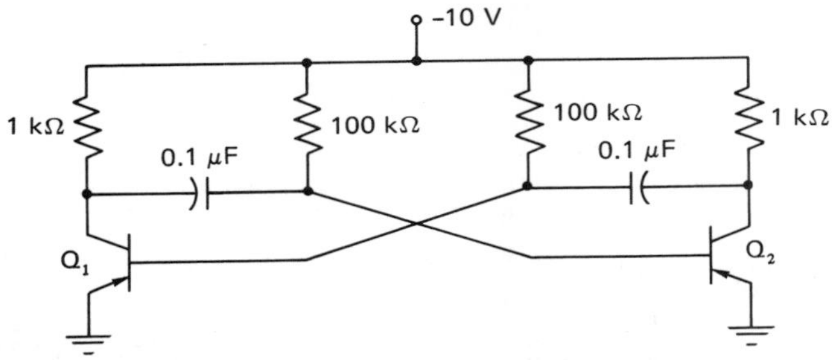

Fig. 31-54

2. Answer the following questions and perform the calculations required:
   a. Explain how the circuit shown in figure 31-54 operates as an astable multivibrator.
   b. Explain the waveforms measured during the experiment.
   c. Calculate the frequency of the multivibrator from the data of step 1.c. and step 1.e.

EXTENDED STUDY TOPICS

1. What is the purpose of a CLOCK?
2. Assume Q is logic 0 in an RS flip-flop circuit. Logic 1 is applied to the R input and the S input is at logic 0. What is the result on the flip-flop?
3. List five IC flip-flops.
4. Compare the two states of a bistable multivibrator, a monostable multivibrator, and an astable multivibrator.
5. Why is an astable multivibrator called a free-running, or oscillator, circuit?
6. For the full adder shown in figure 31-30A, determine the logic state (1 or 0) at each point in the circuit for the following inputs:
   a. $A = 1, B = 1, C_i = 1$
   b. $A = 0, B = 1, C_i = 1$
7. The four-input multiplexer shown in figure 31-33 has the following inputs: $A = 0, B = 1, C = 1, D = 0, S_0 = 1, S_1 = 0$. Determine the output X.
8. The input signal, in figure 31-36, is SABCD = 11001. What is the decimal output?
9. What are the three characteristics of a binary counter?
10. What is a synchronous counter?
11. Why are A/D and D/A converters used?
12. What is the difference in the operation of a μC and a calculator?
13. What are the minimum requirements for a digital computer?
14. What are the main advantages of a μC as compared to a conventional computer system?
15. List the general categories of digital instruction operations.
16. What is the essential difference between a μP and a μC?
17. What is the difference between data and instruction in a μP?

Chapter 32	Audio Systems

OBJECTIVES   After studying this chapter, the student will be able to

- explain audio and high-fidelity concepts.

- discuss stereo and quadraphonic sound.

- describe the characteristics of audio system components.

- list the applications and advantages of the Dolby noise reduction system.

- discuss CB radio principles.

AUDIO CONCEPTS   Sound consists of vibrations. When a violin is played, the instrument vibrates. These vibrations create motion in the air surrounding the violin. The motion spreads outward from the source like the waves in a pond when a stone is thrown into the water. Even a speaker's voice is an instrument that sets up waves or pressure variations in air. Therefore, sound can be defined as a wave motion in air which produces an auditory sensation in the ear due to a change in air pressure at the ear.

Sound waves travel through the air at a speed of about 1 127 feet per second. An electrical signal is generated when these sound waves reach a microphone. The signal may be stored as a pattern of magnetism on a tape, a series of wiggles in a record groove, or used to modulate a radio carrier wave. The electrical signal is then amplified and played through a loudspeaker which recreates the original air vibrations, and therefore, the original sound.

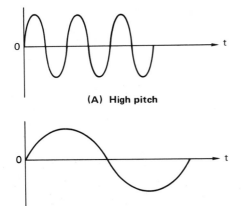

(A) High pitch

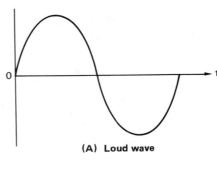

(A) Loud wave

(B) Low pitch

Fig. 32-1 Pitch of a musical note

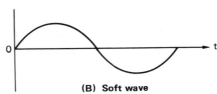

(B) Soft wave

Fig. 32-2 Loudness of a sound wave

The audible range, or the range of human hearing, consists of vibrations in the frequency range from 20 Hz to 20 000 Hz. Generally, teenagers and children can hear the higher frequencies. Adults usually have an upper hearing limit of 16 000 Hz. This limit is reduced to 10 000 Hz or less with increasing age. The human ear does not hear all frequencies with the same intensity. The greatest sensitivity is for frequencies between 500 Hz and 6 000 Hz.

Pitch is a property of a musical tone that is determined by its frequency. The pitch increases as the frequency increases. Figure 32-1 shows the difference between musical notes with a high pitch and those with a low pitch. All musical instruments manufactured in the United States are standardized to a frequency of 440 Hz. The tone of this frequency corresponds to that of the 49th key of a standard 88-note piano. Some sounds contain a random mixture of frequencies. As a result, the vibrations do not have a regular rate of recurrence. Sounds without a definite pitch include jangling metal, rushing air, and splashing water.

Every sound has a certain loudness or volume. The amplitude of the wave, which marks its energy content, indicates the loudness of the wave. As shown in figure 32-2, the amplitude of a sound wave determines whether the sound is loud or soft.

(R32-1) Define sound.

(R32-2) List three uses of the electrical signal output of a microphone.

(R32-3) What is the audible frequency range?

(R32-4) Define pitch.

(R32-5) What effect does the amplitude of a sound wave have upon its volume?

HIGH-FIDELITY CONCEPTS

High fidelity (hi-fi) is the accurate and faithful reproduction of a musical score. A hi-fi system must fulfill the following requirements:

1. The hearer must feel that he or she is listening to real performers.
2. The voices must sound real.
3. It must be possible to separate and identify each musical instrument.
4. The instruments must sound real.

Every note on a musical scale has a fixed frequency called the *pitch*, or *fundamental frequency*. Middle C_4 on a piano has a fundamental frequency of 261.23 Hz. The note A_4, above middle C_4, has a fundamental frequency of 440 Hz. In addition to producing notes at the fundamental frequencies, each musical instrument delivers a whole series of other frequencies known as *overtones* or *harmonics*. These harmonics consist of multiples of each fundamental frequency. In other words, the second harmonic is two times the pitch frequency.

Each instrument has its own manner of emphasizing the various harmonics. The relative strength or amplitude of these harmonics accounts for the characteristic tone quality, or timbre, of different musical instruments. Also, the way in which an instrument treats the harmonics is responsible for the lifelike character of the instrument. *Timbre* is the characteristic of a musical instrument that distinguishes it from another instrument. If the harmonics are removed by filters, all instruments playing the same fundamental note will sound the same.

High-fidelity reproduction requires that a system reproduce all of the musical frequencies from the deepest bass to the highest treble. All frequencies must be reproduced with equal efficiency or loudness. All of the overtones must be delivered in precisely the same proportion as they are produced in the concert hall or studio. Low, middle, and high notes must maintain their amplitude and time relationship to one another. The system must not emphasize some frequencies more than others.

The flat frequency response curve, figure 32-3, shows the intensity of each reproduced test tone as a function of its frequency. The test tones all have the same intensity. Therefore, a good system must reproduce the tones with the same intensity. Manufacturers specify audio amplifier frequency response as 20 Hz to 20 000 Hz ±1 dB. This means that at no point in the range from 20 Hz to 20 000 Hz does the response deviate by more than 1 dB from true flat-

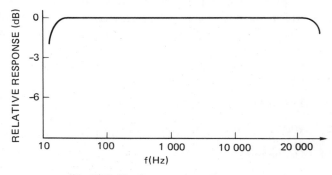

Fig. 32-3 Flat frequency response

ness. (Chapter 26 stated that the minimum change in acoustical power must be at least 1 dB to be perceptible.)

The system must also have excellent *transient response*. This response is the ability of the system to follow rapidly changing signals accurately. The term transient refers to sounds that appear suddenly and cease just as suddenly, such as the scraping of the bow on a violin, the pluck of the pick on a guitar, the rush of breath on a flute, drum beats, and crashing piano chords.

Low distortion is another important quality of hi-fi system reproduction. The system must not add harmonics that are not present in the original signal. Such added harmonics cause harmonic distortion. The system must not allow two simultaneous sounds to interfere with each other and thereby create other sounds. This interference is called *intermodulation distortion*, or *IM distortion.* These distortion factors should be less than 1% at the rated power of the system for true high-fidelity reproduction.

A block diagram showing the components of a simplified single-channel high-fidelity system is given in figure 32-4. The system operates as follows:

1. The *phonograph cartridge* transforms mechanical energy into electrical energy. The cartridge is also known as a *pickup*, or *transducer*.
2. The output of the transducer is amplified by the preamplifier and the power amplifier.
3. The output of the power amplifier is fed into the loudspeaker system. This system consists of three speakers and a crossover filter network. The network allows each speaker to respond to its own band of frequencies.
4. The loudspeaker is the electromechanical system which changes electrical energy to the mechanical energy needed to produce the audible sound waves.

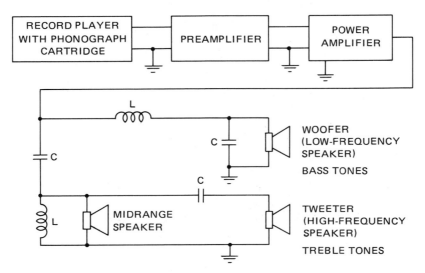

Fig. 32-4 Single-channel high-fidelity system

5. The crossover filter network directs the low frequencies (20 Hz to 300 Hz) to the low-frequency speaker, or the *woofer*. The output tones are the bass tones.
6. The frequencies from 300 Hz to 3 000 Hz are directed to the midrange speaker.
7. The high frequencies (3 000 Hz to 20 000 Hz) are directed to the high-frequency speaker, or the *tweeter*. The output tones are the treble tones.
8. When a system is designed for two-speaker operation, the woofer is designed to respond to a frequency range of 20 Hz to 3 000 Hz. The tweeter is designed to respond to a frequency range of 3 000 Hz to 20 000 Hz.

(R32-6)   Define high fidelity.

(R32-7)   For the piano note A_4: a. what is the fundamental frequency, b. what is the second harmonic, and c. what is the third harmonic?

(R32-8)   Define overtone.

(R32-9)   What is timbre?

(R32-10)   Why is the response curve of figure 32-3 flat?

(R32-11)   What is the characteristic of a hi-fi system with excellent transient response?

(R32-12)   Describe harmonic distortion.

(R32-13)   Define IM distortion.

STEREO PRINCIPLES   A stereophonic sound system uses two microphones. Each microphone channel has a separate amplifier and loudspeaker. This type of system is known as an *auditory perspective system* because the sound of an orchestra can be reproduced closer to its proper perspective. This means that the origin of each sound is precisely located in space. The acoustic aura of the original performance is reproduced in its entirety in the listener's environment because the stereo sound equipment retains the space factors of the recorded music. A block diagram of a two-channel stereo system is given in figure 32-5.

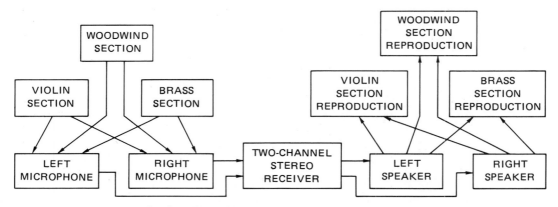

Fig. 32-5 Block diagram of a two-channel stereo system

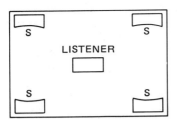

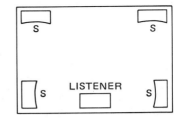

(A) Optimum arrangement   (B) Practical arrangement

Fig. 32-6 Speaker arrangement for quadraphonic reproduction

Stereo reproduction is not concerned with the basic quality of the sound. Stereo reproduction merely indicates two-channel reproduction which may or may not be high-fidelity reproduction. The following factors define the relationship between hi-fi and stereo:

1. The tone quality depends upon the hi-fi reproduction.
2. Stereo reproduction must meet hi-fi standards if it is to be acceptable musically.
3. High fidelity is the important feature of a sound reproduction sytem. The addition of stereo to a hi-fi system is an advantage to the listener.

(R32-14) Define stereo.

(R32-15) What is the advantage of a stereo system?

(R32-16) Compare stereo and high-fidelity reproduction.

QUADRAPHONIC SOUND Quadraphonic sound, or quad, is four-channel sound reproduction. Figure 32-6 shows two possible speaker arrangements for quad listening. A quad system requires four speakers and four separate amplifier channels to feed separate signals to each speaker. For a true high-fidelity system, each speaker consists of low-, midrange, and high-frequency units which can be packaged as separate units or combined in one speaker enclosure.

Four-channel reproduction enhances the ambience of sound. *Ambience* refers to the total pattern of the sound reverberations and reflections in the studio or concert hall. Quad distributes the sound energy reflections from the rear and side walls as the music is recorded. The reverberations are duplicated in the listener's environment just as they originally existed.

Quad is no substitute for quality. The amplifier system must meet the high-fidelity requirements of frequency response and freedom from distortion.

(R32-17) Define quadraphonic sound.

(R32-18) What is ambience?

TURNTABLES A turntable consists of two basic parts:

1. The platter which supports the record and rotates it at a constant speed of 33 1/3 or 45 r/min.

2. The tonearm which holds the cartridge and allows the stylus (needle) to move freely across the record as it tracks the record groove.

The quality of the reproduced sound depends upon the ability of the platter and the tonearm to perform their functions with a high degree of efficiency. The ear is very sensitive to minute fluctuations in speed. For critical listening, the platter must have very low wow and flutter (0.1% or less). *Wow* is a relatively slow wavering of pitch due to minute variations in the speed of the turntable or the tape recorder. *Flutter* is a relatively rapid variation in pitch due to minute variations in the speed of the turntable or the tape recorder.

The drive motor and bearings for the platter must be very quiet in operation or the listener will hear rumble from the speakers. *Rumble* is a low-frequency sound or electronic noise due to turntable vibrations. Rumble is reproduced by the amplifying system. The rumble specification of a good turntable should be –65 dB or more below the signal sound level.

Currently, there are three classes of turntables:

1. Belt drive
2. Direct drive
3. Linear drive

The belt-drive system uses an ac or dc motor with a high r/min rating to turn the platter by means of an elastic belt. The belt helps to absorb the vibration and noise of the motor. This system was the professional standard before the introduction of more sophisticated drive systems. It is still used in less expensive turntables. A belt-driven turntable is shown in figure 32-7. This turntable achieves precise speed control with a dc servo-controlled motor, a two-pound cast platter, and a vibration-absorbing belt drive. Inaudible levels are maintained for wow and flutter (within 0.05%) and rumble (–60 dB).

A direct drive system is used by most of the more expensive turntables. A 12- or 16-pole motor is attached directly to the turntable shaft. The motor turns only as fast as the turntable platter, minimizing audible rumble. The rotor of the motor is divided into 12 or 16 sections (poles). Therefore, it tends to rotate in a series of jumps rather than in a smooth continuous motion. This

Fig. 32-7 Belt-driven turntable, Model MT6211

Fig. 32-8 Linear direct drive turntable, Model 6225AC

ratcheting, or cogging, action is reduced by the mass of a heavy platter, but it is never eliminated entirely.

Another direct drive system is linear drive. The platter itself is the rotor of the drive motor. Rather than the customary 12 or 16 poles, there are 120 magnetic poles evenly spaced around the circumference of the platter. Three stationary drive coils produce an overlapping series of magnetic push/pull pulses. The coils are controlled by a three-phase electronic drive system. The large number of poles and the overlapping drive pulses produce nearly constant torque, with no ratcheting action. Figure 32-8 shows a direct drive turntable with a wow and flutter within 0.03% and a rumble of −70 dB.

(R32-19) What are the two basic parts of a turntable?

(R32-20) List the three classes of turntables.

TUNER The tuner is a device which tunes in radio stations and plays their programs through a high-fidelity system. The tuner reproduces the broadcasts with a sound quality far superior to that of ordinary radios. Most FM stations broadcast in stereo by a multiplex process. Because of this process, a single FM transmitter can send out both stereo channels. *Multiplex* (MPX) is the transmission of two or more channels on a single carrier so that they can be recovered independently at the receiver. Multiplex in FM stereo transmission consists of the transmission of a left plus right (sum) signal on the main carrier and a left minus right (difference) signal on the subcarrier. The multiplex decoder in the receiver recovers and separates the independent left and right stereo channels from the multiplexed signal.

Noise is any unwanted disturbance superimposed upon a useful signal. Noise interferes with the information contained in the signal. That is, the greater the noise, the less information is received. The following two examples illustrate the effect of noise interference.

1. The noise in television receivers produces small black or white spots on the picture. Therefore, very severe noise can eliminate the picture.

2. The noise in radio receivers produces crackling and hissing. This noise may be great enough to mask completely the voice or music output of the radio receiver.

Noise is independent of the signal because it exists even when there is no signal. Noise contains sinusoidal components of all frequencies. It is an unwanted signal not derived from or related to the input signal. Noise comes from a variety of sources and is classified as man-made interference or naturally occurring noise. Man-made interference is due to sources such as electric motors, neon signs, power lines, ignition systems, 60-cycle hum, and power supply ripple. Sources of naturally occurring noise include atmospheric disturbances, extraterrestrial radiation, and circuit noise. An example of circuit noise is thermal noise which is caused by the random motion of electrons. This electron motion is due to the temperature of the device and results in the production of an unwanted noise voltage. The noise voltage is uniformly distributed through the practical frequency range. The voltage does not reach a break frequency until 10^{12} Hz.

Sensitivity is the ability of a tuner to pull in weak and distant stations. The sensitivity of a tuner is stated in relation to quieting. *Quieting* is the ability of a tuner to reject the noise from a radio signal to yield a clear, undisturbed audio signal. This signal appears at the output of the tuner and can be fed into the amplifier. Quieting is specified in decibels (dB). Assume that a specification states that the sensitivity is 3 μV. This value means that an incoming signal must have a strength of 3 μV at the antenna terminals so that the noise can be quieted to a level 30 dB below the music level.

The muting control automatically turns the sound volume down or off. FM muting keeps the receiver quiet as the tuner is being tuned across blank spots on the FM dial.

Figure 32-9 illustrates an AM/FM stereo tuner with meters for signal strength and center channel tuning. A third meter reads the FM signal deviation directly in kHz. This tuner has an FM sensitivity of 1.7 μV and a signal-to-noise (S/N) ratio of 70 dB.

Fig. 32-9 AM/FM stereo tuner, Model FM2310

Fig. 32-10 Stereo power amplifier, Model CA2310

(R32-21) Define a tuner.

(R32-22) What is FM multiplex transmission?

(R32-23) Define noise.

(R32-24) Define the sensitivity of a tuner. How is it stated?

(R32-25) What is the purpose of FM muting?

STEREO POWER AMPLIFIER

The power rating of a stereo power amplifier tells how much rms power the amplifier can deliver to the speakers. A rating of 20 to 25 watts per channel is adequate for most home situations. A margin of power is needed to ensure the clear reproduction of certain musical events, such as crashing fortissimos, loud chords struck on a piano, or the deep rolling notes of the low bass.

The stereo power amplifier shown in figure 32-10 has excellent high-fidelity reproduction. It has a power rating of 70 W per channel continuous rms power into an 8-Ω loudspeaker from 20 Hz to 20 000 Hz. The total harmonic distortion for this amplifier is no more than 0.1%.

(R32-26) A manufacturer states that a stereo power amplifier has an rms power rating of 50 W. What does this rating signify?

AM/FM STEREO RECEIVER

The AM/FM stereo receiver, figure 32-11, combines the functions of separate components, such as the tuner, preamplifier, and power amplifier. This receiver has a rating of 10 W per channel continuous rms power into an 8-Ω loudspeaker from 60 Hz to 20 000 Hz. The total harmonic distortion of this receiver is no more than 1% and it has an FM sensitivity of 2.8 μV.

A complete sound system can be made by adding record-playing equipment and speakers to the AM/FM receiver. Formerly, the construction of a system from separate components offered a performance advantage as compared to the quality of AM/FM stereo receiver reproduction. Now, many users claim that this performance edge has been eliminated by the present level of stereo receiver technology.

Fig. 32-11 AM/FM stereo receiver, Model MC2100

(R32-27)   What functions are performed by an AM/FM stereo receiver?

SPEAKERS   A speaker reproduces sound tones and should deliver these tones with clarity and low distortion. A flat frequency response from speakers is essential. Any sharp deviation from flat response is called a *response peak* if certain tones are emphasized, and a *response dip* if certain tones are suppressed. Response peaks should not exceed 5 dB to ensure that the output of the speaker remains smooth and pleasant to the ear.

The power rating of the speaker is an indication of how many watts of amplifier power it can absorb or handle. An amplifier that can deliver 50 watts per channel should not be connected to a speaker with a power rating of 25 watts. The speaker may be damaged if the amplifier volume is turned to its full output.

A speaker can produce full volume using less power than its power rating. Depending upon the efficiency of the speaker, it may require from 5 watts to 25 watts.

The speaker system shown in figure 32-12 can deliver a sound level of up to 112 dB in a typical living room. (Chapter 26 stated that the sound noise in a subway is approximately 100 dB.) This system can be used with amplifiers rated as high as 130 watts. However, the high efficiency of the system gives outstanding performance with a 25-watt amplifier. For this reason, its power rating is specified as 25-W minimum/130-W maximum. Its frequency response is 40 Hz to 20 000 Hz. The top center speaker is a three-inch horn tweeter. The two five-inch speakers in the center are the midrange drivers. The bottom 15-inch speaker is the woofer. The crossover frequencies of the crossover network are 1 kHz/5 kHz.

(R32-28)   What is the primary function of a speaker?

(R32-29)   What is a. a response peak and b. a response dip?

(R32-30)   For the speaker system of figure 32-12, what is the frequency range of: a. the woofer, b. the midrange drivers, and c. the horn tweeter?

Fig. 32-12 Speaker system, Model ST460

THE DOLBY NOISE REDUCTION SYSTEM   The Dolby noise reduction system can be used with high-quality audio recording, transmission channels, or sound movies. A special signal component is derived from four band-splitting filters and low-level compressors. This component is combined with the incoming signal during recording or sending. During reproduction, the additional component is removed and any noises acquired in the channel are reduced, or attenuated, in the process. This noise reduction system can give high-quality performance with regard to noise reduction and signal quality. Figure 32-13 shows some of the professional applications of the Dolby noise reduction system. Figure 32-14 shows a Dolby A-type professional noise reduction module. The Dolby B-type consumer noise reduction circuit is illus-

Professional Recording and Transmission Applications

360

The Dolby 360 is a basic single-channel A-type noise reduction unit for encoding or decoding. This unit is normally used in a fixed mode such as in disc cutting or landline sending or receiving; the operating mode is manually selected.

361

The Dolby 361 is similar to the 360, providing a single channel of A-type noise reduction, but with relay switching of operating mode and tape recorder connections. The changeover can be controlled automatically by the recorder.

M-Series

The Dolby M16H A-type unit is designed specifically for professional multi-track recording, and incorporates 16 channels of noise reduction in a compact chassis only 10½ inches high. The similar M8H is an 8-track version, and the M8XH allows simple extension of the M16H for 24-track use.

Noise Reduction Module

Cat. No. 22

The Dolby noise reduction module, Cat. No. 22 is the basic functional unit employed in all A-type equipment. The Cat. No. 22 is available as a spare or in quantity to OEM users for factory installation. A half-speed version of the module (Cat. No. 40) is also available.

Dolby Laboratories Inc

'Dolby', Dolbyized and the double-D symbol are trademarks of Dolby Laboratories

Motion Picture Industry

364

The Dolby 364 Cinema Noise Reduction Unit is intended primarily for use with Dolby A-type encoded optical sound-tracks. The 364 also includes a standard 'Academy' filter for conventional tracks, and provision for playback of magnetic sound-tracks with or without Dolby system encoding.

E2

The Dolby E2 Cinema Equalizer is a companion unit to the 364, and has been specifically designed to solve the response equalization problems of cinemas. Used with the 364 and Dolbyized optical sound-tracks, the E2 enables most cinemas to achieve modern sound reproduction standards without replacement of existing equipment.

CP100

The Dolby CP100 Cinema Processor is designed for the reproduction of all current and presently foreseeable film sound-track formats including conventional optical and magnetic tracks, Dolby encoded monaural optical tracks, Dolby encoded magnetic sound-tracks and the new stereo optical release prints. Up to three noise reduction modules can be incorporated. Typically, three channels of theatre equalization, as in the E2, will be incorporated, but facilities exist for five channels of equalization and the connection of an external quadraphonic decoder.

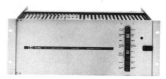

CP50

The new model CP50 is intended for the reproduction of all optical sound-track formats, Dolby encoded and conventional, mono and stereo. The unit is designed to interface with an existing fader and magnetic stereo installation; a wide range of accessories is available.

731 Sansome Street
San Francisco CA 94111
Telephone (415) 392-0300
Telex 34409
Cable Dolbylabs

Professional Encoders for Consumer Media

330

The Dolby 330 Tape Duplication Unit is a professional quality unit with B-type (consumer) noise reduction characteristics. The unit is used for encoding duplicating master tapes in the high-speed duplication of Dolbyized cassettes, cartridges, and open-reel tapes. The 330 is a two-channel unit.

334

The 334 FM Broadcast Unit allows broadcast stations to encode stereo FM broadcasts with the Dolby B-type characteristic. The unit also provides for a reduction of high frequency pre-emphasis to 25 microseconds; this reduces the need for high frequency limiting, thus allowing a significant additional improvement in reception quality.

Test Set (A-type)

Cat. No. 35

The Dolby NRM Test Set, Cat. No. 35, permits rapid verification of performance of Cat. No. 22 Noise Reduction Modules without the need for additional test equipment.

Noise Weighting Filter

Cat. No. 98 A

Noise weighting filter to CCIR/ARM characteristic (recommended by Dolby Laboratories). Filter is used with average responding meter (ordinary millivoltmeter) allowing noise measurements to be made on tape recorders, tapes, FM tuners,etc., with results which correlate closely with the subjective effect of the noise. Filter can be used for the testing of professional and consumer equipment.

346 Clapham Road
London SW9
Telephone 01-720 1111
Telex 919109
Cable Dolbylabs London

S77/272/357

Fig. 32-13 Professional applications of Dolby noise reduction units

Fig. 32-14A Dolby A-type noise reduction module

Fig. 32-14B Internal view of Dolby A-type noise reduction module

Fig. 32-15 Dolby B-type consumer noise reduction circuit

trated in figure 32-15. Figure 32-16A shows a studio standard three-head cassette deck with Dolby system. This equipment has a signal-to-noise ratio of 62 dB. A studio standard stereo cassette deck with Dolby and wireless remote electronic editing is shown in figure 32-16B. The deck has a signal-to-noise ratio of 50 dB without Dolby and 56 dB with Dolby. Figure 32-16C shows a studio standard cassette/8-track recording deck with Dolby and wireless remote cassette editing. It has a S/N ratio with Dolby of 56 dB (cassette) and 52 dB (8-track).

Listeners have had to endure background noise from the earliest days of recording. At first, most of the noise was due to the phonograph needle rubbing the walls of the record groove. Lately, the noise consists of the hiss produced by recording tape. Scientists and engineers have learned that noise is detected by the ear and brain of a listener under certain conditions. The design of the Dolby system is based upon research into electronics and hearing. The patented Dolby circuit is programmed with information about the way background noise is heard by the listener. By putting the music through the Dolby circuit before it

Fig. 32-16A Studio standard three-head cassette deck with Dolby, Model CR5120

Fig. 32-16B   Studio standard stereo cassette deck with Dolby and remote electronic editing, Model CR4025

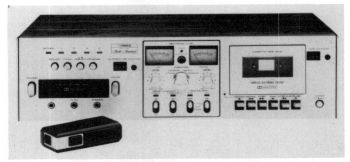

Fig. 32-16C Studio standard cassette/8-track recording deck with Dolby and wireless remote cassette editing, Model ER-8155

is recorded or broadcast, and again when the program is played back, the noise introduced by recording or broadcasting is reduced considerably. Figure 32-17 explains the difference between making an ordinary recording and a Dolbyized recording.

Dolby stereo is a high-fidelity stereo sound system which is widely used in theaters. It is a step-by-step process that affects sound recording on the movie set; looping, mixing, and dubbing as the film is edited; the sound tracks of release prints; and theater sound systems. The Dolby system produces the sound realism which people enjoy from their home stereos, but could not experience in movie theaters until recently.

(R32-31)   What is the Dolby system?

CITIZEN'S BAND (CB) RADIO   Citizen's band radio, or CB, is a two-way radio service licensed by the Federal Communications Commission (FCC). It is intended for short distance (under 150 miles) personal and business radiocommunications. The addition of a CB two-way radio in an automobile is almost like having a telephone in the car. CB makes it possible to communicate with people in places where telephones normally are not installed. The CB radio can be installed in the home, office,

Music

Music consists of sounds of different pitch and loudness, separated by intervals of silence. Loud and soft sounds are shown here as long and short lines. The 'music' represented in the diagram at left starts very loud and gradually becomes very quiet.

What the Dolby system does first

Before a recording is made, the Dolby system 'listens' to the music to find the places where tape noise might later be heard when the tape is played. This happens mainly during the quietest parts of the music. When it finds such a place, the system automatically increases the volume so that the music is recorded at a higher level than normal. An important feature which enables the Dolby system to do this effectively is its ability to distinguish sounds of different pitch as well as sounds of different loudness.

Noise

Even the best recording tape makes a constant hissing sound when it is played, whether or not any music has been recorded on it. The best professional tape recordings contain this noise, just as home cassette recordings do. However, the problem is much more serious with cassettes because of the slow speed of the tape and the narrow width of the recording track. These cause tape noise to be even more annoying, because of the drastically reduced level of music which can be recorded on the tape.

The recording

In a Dolbyized recording, those parts of the music which have been made louder stand out clearly from the noise. This is what makes Dolbyized recordings sound unusually brilliant even when played without the special Dolby circuit. Listeners without Dolby equipment use Dolbyized cassettes, for example, just like any other cassettes.

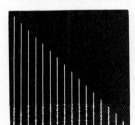

Music and noise

When a tape recording is played, the noise of the tape obliterates silent passages, conceals the quietest musical sounds, and interferes with many sounds, even some louder than the noise itself. Bass tones need to be extremely loud for tape noise not to be heard during the music.

What the Dolby system does during playback

When Dolbyized recordings are played back on a recorder equipped with the Dolby circuit, the volume is automatically reduced in all of the places where it was increased during recording. This makes the music sound exactly right, because the loudness of every note is just the same as it was at the start. At the same time the noise of the tape, which is now mixed with the music, is reduced in all of the same places – just the places where the noise would otherwise have been heard.

Fig. 32-17A Making an ordinary recording

Fig. 32-17B Making a Dolbyized recording

car, truck, recreational vehicle, boat, golf cart, airplane, tractor, mobile home, or it can be carried by the operator. Table 32-1 lists the frequencies of each of the 40 CB channels. Any channel can be used except Channel 9 which is reserved for emergency personnel and motorist assistance only.

The equipment needed for CB operation consists of a two-way radio, a microphone or handset, and an antenna. The two-way radio, or *transceiver*, is a combination transmitter and receiver. The transmitter section is used to transmit or send a radio signal. The receiver section receives radio signals from other CB radios. The transceivers installed in cars, trucks, boats, planes, and other vehicles are known as *mobile transceivers*. *Base station transceivers* are those installed in homes, offices, or other fixed locations. Another type of CB transceiver is hand-held and portable and is known as a walkie-talkie.

Figure 32-18 illustrates a 40-channel CB walkie-talkie. The CB transceiver shown in figure 32-19 has a squelch control. The squelch control eliminates background noise when listening for calls or monitoring the CB radio. Normally, this control is adjusted to the point where it just silences the receiver. The 40-channel mobile CB transceiver shown in figure 32-20 has priority switching

TABLE 32-1 CB Channel Frequencies

Channel Number	Frequency (MHz)	Channel Number	Frequency (MHz)
1	26.965	21	27.215
2	26.975	22	27.225
3	26.985	23	27.255
4	27.005	24	27.235
5	27.015	25	27.245
6	27.025	26	27.265
7	27.035	27	27.275
8	27.055	28	27.285
9	27.065	29	27.295
10	27.075	30	27.305
11	27.085	31	27.315
12	27.105	32	27.325
13	27.115	33	27.335
14	27.125	34	27.345
15	27.135	35	27.355
16	27.155	36	27.365
17	27.165	37	27.375
18	27.175	38	27.385
19	27.185	39	27.395
20	27.205	40	27.405

Fig. 32-18 40-channel CB walkie-talkie

Fig. 32-19 40-channel CB transceiver with squelch control

Fig. 32-20 40-channel CB transceiver with priority switching circuit

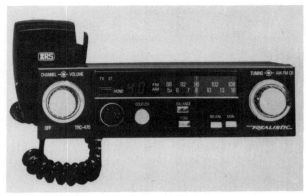

Fig. 32-21 40-channel CB/AM/FM stereo radio

which enables the operator to change instantly from any channel to Channel 9 or 19. Channel 19 is widely used by truckers and operators of all moving vehicles. The 40-channel CB/AM/FM stereo radio shown in figure 32-21 is a complete mobile entertainment and communications center which mounts in or under the dashboard. Sensitive AM and FM stereo circuits provide immediate access to news and music. A special monitor switch allows the operator to receive CB calls while listening to AM or FM.

(R32-32) What is CB?

(R32-33) List the basic equipment needed for CB operation.

(R32-34) What is a transceiver?

EXTENDED STUDY TOPICS

1. Two tones of the same frequency and intensity sound simultaneously. Express the resulting sound intensity, as compared to the sound intensity of a single tone, in dB units.
2. What is the speed of a. a sound wave in air and b. a radio wave in air?
3. What are the general functions performed by a high-fidelity system?
4. What factors determine the characteristic tone quality or timbre of a musical instrument?
5. List the requirements for a system that can achieve high-fidelity reproduction.
6. What are the crossover frequencies for the system shown in figure 32-4?
7. Define: a. a woofer and b. a tweeter.
8. What is a crossover filter network?
9. Is the high-fidelity system shown in figure 32-4 a stereo system?
10. Define: a. wow, b. flutter, and c. rumble.

<table>
<tr><td>Chapter
33</td><td>Instrumentation Fundamentals
and Devices</td></tr>
</table>

OBJECTIVES After studying this chapter, the student will be able to

- explain the principles of operation of negative feedback circuits.
- describe process control systems and their components.
- discuss sensors and transducers.
- explain the principles of measuring pressure, flow, and temperature.
- use process controllers and transmitters.

INTRODUCTION An instrumentation system is an integrated combination of unlike, but interacting, elements that function to achieve an objective. Instruments and instrumentation systems are used to compute, control, detect, display, measure, communicate, and observe physical quantities. A flow of information in the form of signals occurs in any instrumentation system. Figure 33-1 is the block diagram of a single-channel data acquisition system. This figure shows each device, or subsystem, as a block. Each block is labeled to indicate the function that is to be performed without indicating how it is performed.

(R33-1) Define an instrumentation system.

NEGATIVE FEEDBACK CONCEPTS An important feature of amplifier design is the negative feedback loop. This means that part of the output is returned to the input to improve the amplifier performance. When the amplitude and phase of the signal fed back are such that the overall gain is less than the forward gain of the amplifier, then negative feed-

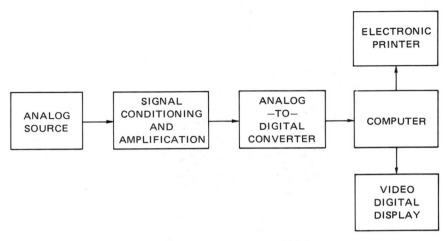

Fig. 33-1 A single-channel data acquisition system

back exists. The returning signal is out of phase with the input signal. The advantages gained by negative feedback outweigh the loss in overall gain. It is a simple matter to achieve more voltage or current gain by adding stages of amplification. The improvements in the performance of a negative feedback amplifier are described as follows:

1. The gain is stabilized and can be made nearly independent of device parameters. The gain can be made insensitive to temperature changes, aging of components, and variations from typical values.

2. The gain can be made almost independent of reactive elements. This means that the gain will be insensitive to frequency. No one amplifier can amplify equally well at all frequencies. At the upper and lower cutoff frequencies, the voltage gain is only 70.7% of the midfrequency value. Figure 33-2 shows that the bandwidth with negative feedback is increased by sacrificing gain.

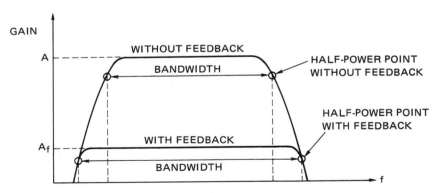

Fig. 33-2 Effect of negative feedback on bandwidth and gain of an amplifier

3. As a result of the wider bandwidth, the time constant of a circuit component is modified to obtain faster response. A limiting factor in the performance of control systems is the response speed of the control elements. For example, the fluid level in a chemical tank drops a given amount in a finite time, and an actuating motor with inertia comes up to a given speed in a finite time. The speed of response is greatly increased with negative feedback, and so the time constant of the system is reduced.

4. The gain can be made selective to discriminate against noise, distortion, or system disturbances. Negative feedback cannot improve a signal that is distorted before it enters the amplifier. Distortion generated internally can be reduced with negative feedback. Examples of such distortion include gain variations, power supply ripple, frequency distortion, phase distortion, amplitude or harmonic distortion, and intermodulation distortion.

5. The input and output impedance of an amplifier can be improved greatly. A high input impedance minimizes the loading effect of the amplifier on a signal source. A low output impedance provides efficient power transfer to the load.

(R33-2) How does negative feedback affect the following amplifier characteristics: a. magnitude of gain, b. stability of gain, c. bandwidth, d. time constant, e. internally generated distortion, f. input impedance, and g. output impedance?

NEGATIVE FEEDBACK CIRCUIT ANALYSIS

Figure 33-3 shows a negative feedback amplifier. For this amplifier,

A = voltage gain of the internal amplifier

$$A = \frac{V_{out}}{V_{error}}$$

Eq. 33.1

β = voltage gain of the feedback circuit

$$\beta = \frac{V_f}{V_{out}}$$

Eq. 33.2

V_f = feedback voltage = βV_{out}

$V_{error} = V_{in} - \beta V_{out} = V_{in} - V_f$ = input to the internal amplifier. This error volt-

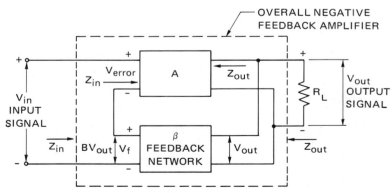

Fig. 33-3 Negative feedback amplifier

age is the difference between the input signal and the feedback signal.

A_f = voltage gain of the overall feedback amplifier.

$$A_f = \frac{A}{1 - \beta A}$$ (See Chapter 27 for the derivation of A_f)

$$A = \frac{V_{out}}{V_{in}}$$ Eq. 33.3

If the numerator and denominator of Equation 33.3 are divided by A, then

$$A_f = \frac{A/A}{1/A - \beta A/A}$$

$$= \frac{1}{1/A - \beta}$$

If A is very large (10 000 or greater), then

$$A_f \cong -\frac{1}{\beta}$$ Eq. 33.4

Equation 33.4 shows that if the gain *without* feedback is very large, the gain *with* feedback depends only upon the feedback element, and not upon the amplifier itself. A high-gain amplifier with negative feedback is a practical way to obtain a specific, accurate, and constant value of gain. Such amplifiers are widely used in instrumentation systems where the output of a device must be amplified greatly to be useful.

Problem 1   The negative feedback amplifier shown in figure 33-3 operates at the following conditions: V_{out} = 10 V, V_{error} = 1 mV, and V_f = 250 mV. Calculate: a. A, b. β, c. V_{in}, and d. A_f.

Solution   a. $A = \dfrac{V_{out}}{V_{error}}$

$$= \frac{10 \text{ V}}{1 \times 10^{-3} \text{ V}}$$

= 10 000 Open loop gain of the internal amplifier.

b. $\beta = \dfrac{V_f}{V_{out}}$

$$= \frac{(250 \times 10^{-3}) \text{ V}}{10 \text{ V}}$$

= 0.025

In a practical negative feedback circuit, β is always between 0 and 1.

c. $V_{in} = V_{error} + V_f$

$$= 1 \text{ mV} + 250 \text{ mV}$$

$$= 251 \text{ mV}$$

d. $A_f = \dfrac{V_{out}}{V_{in}}$

$$= \frac{10 \text{ V}}{(251 \times 10^{-3}) \text{ V}}$$

= 39.8

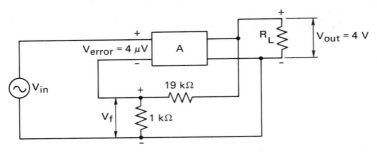

Fig. 33-4 Voltage divider feedback network of a negative feedback amplifier

(R33-3)   How does negative feedback affect the gain of the amplifier described in Problem 1?

The feedback network of a negative feedback amplifier can be a voltage divider. Figure 33-4 shows this type of feedback network.

(R33-4)   Using figure 33-4, calculate: a. V_f, b. β, c. A, and d. A_f.

(R33-5)   Using Equation 33.4, calculate A_f for Problem 1.

PROCESS CONTROL CONCEPTS   Process control is concerned with the control of a manufacturing process by automatic equipment. A human being is not a necessary or integral part of the regulation of the process. A *dynamic variable* is any physical parameter which can change naturally or because of external influences. Process control deals with dynamic variables requiring regulation in industrial applications. Dynamic variables include flow rate, force, speed, humidity, level, light intensity, pressure, and temperature.

The main objective of process control is to hold the dynamic variable at or near some desired specific value. Some corrective action must be provided constantly to achieve this goal. Regulation concerns the maintenance of the dynamic variable at the desired value. Thus, process control *regulates* a dynamic variable.

(R33-6)   Define process control.

(R33-7)   What is a dynamic variable?

(R33-8)   What does regulation accomplish?

(R33-9)   What is the objective of process control?

PROCESS CONTROL ANALYSIS   This section investigates the factors to be considered in automating a process. Later in this chapter, the hardware used for process control systems is described.

Figure 33-5 is the block diagram of an automatic feedback process control system. The following components are part of this system:

1. The *sensor* is a piece of hardware that measures system variables. Normally, the sensor changes, or transduces, the measured quantity into another form of energy. It is the first piece of hardware in any process control system. Sensors provide the feedback signals by detecting the

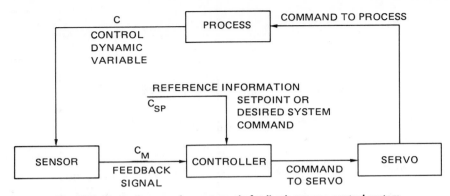

Fig. 33-5 Block diagram of an automatic feedback process control system

present condition, state, or value of a process variable. The output of the sensor, C_M, is a feedback signal that is a measured representation of the dynamic variable being controlled.

2. Another part of the hardware is the *controller* which accepts both the information from the system sensors and the reference information. (This reference information is the setpoint or desired system command.) The controller calculates the correct response that serves as the input command resulting in the automatic control of the variable.

3. The controller contains the prescribed set of rules or equations, known as an *algorithm*, by which the solution to the automatic control problem is calculated. The controller implements these rules or equations in the process of calculating the signals that are sent to the process actuators.

4. The setpoint, C_{SP}, is the desired value at which the controlled dynamic process variable, C, is to be controlled.

5. Servomechanisms, or servos, are hardware devices which accept control system commands. The servos convert, or transduce, these commands into mechanical motion. Servomechanisms are the devices which actually cause the system variables to change value according to the computations issued by the control system controller.

Negative feedback *decreases* the difference between the desired effect and the actual effect. The difference between the desired effect (controlled setpoint C_{SP}), and the actual effect (feedback signal C_M), or the actuating error, is the input to the controller, figure 33-6. The *summing point*, S, is also called a *subtractor*, or a *comparator*.

Figure 33-7 shows the physical diagram of a process control flow system and its representation in block diagram form.

(R33-10)   What is the output of a sensor?

(R33-11)   Name the first piece of hardware in any process control system.

(R33-12)   Which two signals are presented to the controller of figure 33-5?

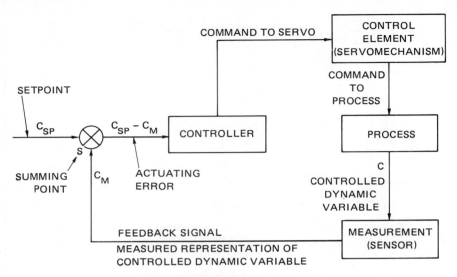

Fig. 33-6 Process control loop block diagram

(R33-13)   Define setpoint.

(R33-14)   What is a servomechanism?

(R33-15)   What is an actuating error?

SENSORS AND TRANSDUCERS

A *transducer* is a physical piece of hardware that can transform, or convert, a physical variable present in one energy system to a proportional value in a more convenient energy system. For example, the gasoline tank level device in an automobile converts the liquid level information into an electrical signal that is proportional to the gasoline level in the tank. This mechanism is actually a transducer that changes a liquid level value to an electrical signal. As another example, liquid contained in a glass thermometer is also a transducer because it converts temperature information to a liquid level height in the thermometer tube.

As stated previously, a sensor is a transducer that senses, or measures, the condition of a process control system variable. The sensor then provides an input concerning this variable to the rest of the process control system.

(R33-16)   Are all sensors also transducers?

(R33-17)   Are all transducers also sensors?

(R33-18)   Explain why the gasoline gauge in an automobile is called an electromechanical transducer?

PRESSURE MEASUREMENT

Pressure is defined as a force spread over an area. The dimensions of pressure are those of a force over those of the area: pounds per square inch (psi), or grams per square centimeter (g/cm^2).

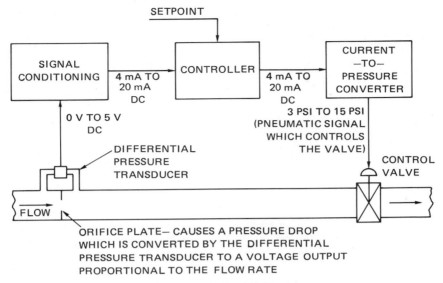

SETPOINT

| SIGNAL CONDITIONING | 4 mA TO 20 mA DC | CONTROLLER | 4 mA TO 20 mA DC | CURRENT —TO— PRESSURE CONVERTER |

0 V TO 5 V DC

DIFFERENTIAL PRESSURE TRANSDUCER

3 PSI TO 15 PSI (PNEUMATIC SIGNAL WHICH CONTROLS THE VALVE)

CONTROL VALVE

FLOW

ORIFICE PLATE— CAUSES A PRESSURE DROP WHICH IS CONVERTED BY THE DIFFERENTIAL PRESSURE TRANSDUCER TO A VOLTAGE OUTPUT PROPORTIONAL TO THE FLOW RATE

(A) Physical diagram of the system

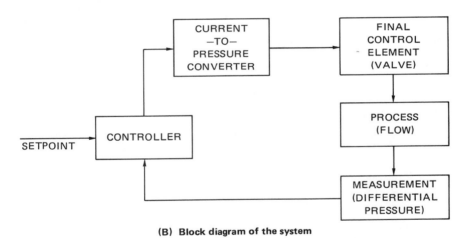

SETPOINT → CONTROLLER → CURRENT —TO— PRESSURE CONVERTER → FINAL CONTROL ELEMENT (VALVE) → PROCESS (FLOW) → MEASUREMENT (DIFFERENTIAL PRESSURE)

(B) Block diagram of the system

Fig. 33-7 A process control flow system

One method of measuring low pressure makes use of the ability of pressure to push up a column of liquid against the effects of gravity. The manometers shown in figure 33-8 measure pressure in this manner. An unknown pressure is brought to one of the connections of the manometer. The other end of the manometer is open to the ambient pressure (atmospheric or room pressure). The liquid in the manometer (mercury or water) moves in response to the difference between the pressures in the two tubes. Therefore, the manometer is a *differential* measuring device because one pressure is compared and measured

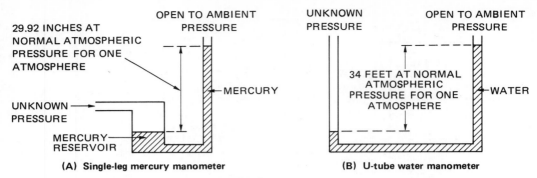

29.92 INCHES AT NORMAL ATMOSPHERIC PRESSURE FOR ONE ATMOSPHERE

OPEN TO AMBIENT PRESSURE

MERCURY

UNKNOWN PRESSURE

MERCURY RESERVOIR

(A) Single-leg mercury manometer

UNKNOWN PRESSURE

OPEN TO AMBIENT PRESSURE

34 FEET AT NORMAL ATMOSPHERIC PRESSURE FOR ONE ATMOSPHERE

WATER

(B) U-tube water manometer

Fig. 33-8 Manometers

with respect to another pressure. Whenever an unknown pressure is compared to the ambient pressure, the resulting differential pressure measurement is called the *gauge pressure*. Normal atmospheric pressure equals 14.7 psi. This pressure will raise a column of mercury 29.92 inches, or a column of water 34 feet.

The panel-mounted mercury manometer shown in figure 33-9 is a precision instrument. Applications of this manometer include chemical process control, infrared spectroscopy, aneroid instrument calibration, and other areas of research, development, and production. A U-tube mercury column sonar manometer is shown in figure 33-10. This device uses ultrasonic pulses to measure the column height. The digital, solid-state electronic system of this manometer is highly responsive to changes in the mercury level. Digital readings are dis-

Fig. 33-9 Panel-mounted mercury manometer

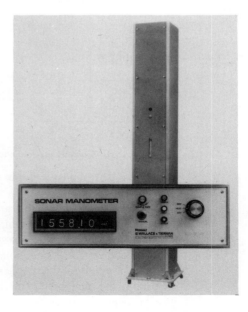

Fig. 33-10 U-tube sonar manometer

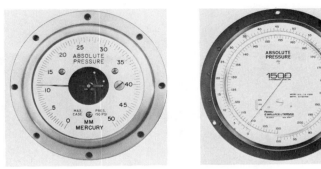

Fig. 33-11 Absolute pressure gauges

played instantaneously. The output is an accurate binary coded digital (BCD) signal which can actuate electronic data processing equipment and an electronic display. Applications of the sonar manometer include production line calibration of pressure transducers; continuous monitoring of pressure in wind tunnels, engine test stands, and vacuum chambers; and use as a barometer for automatic weather stations.

When a pressure measurement is compared to the pressure of outer space, the measurement is called *absolute pressure*. (The pressure of outer space is accepted as absolute zero.) Normal atmospheric pressure is 14.7 psi absolute. Absolute pressure gauges are shown in figure 33-11.

A common mechanical pressure sensor is the bourdon tube. This device is made of a thin, springy metal formed into a flattened tube. One end of the tube is sealed and the other end of the tube is open. Figure 33-12 shows the internal components of a bourdon tube pressure gauge.

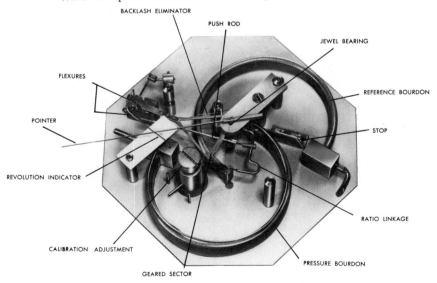

Fig. 33-12 Internal components of a bourdon tube pressure gauge

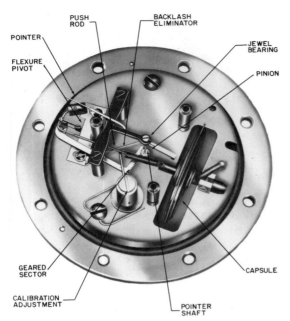

Fig. 33-14   Differential pressure transducer

Fig. 33-13 Capsule-type mechanism of a pressure gauge

Another common pressure-to-mechanical motion transducer uses a bellows. The bellows is made from a springy material which is formed into a thin-walled tube. This tube is then worked so that deep convolutions, or ridges, are formed in it. The bellows is an excellent pressure-sensing device. It is usually more sensitive than a bourdon tube. As a result, it can sense lower pressures. If a spring is added to prevent excessive elongation, or lengthening, the bellows can be used to sense higher pressures. Figure 33-13 shows the internal components of a capsule-type mechanism whose operation is similar to that of the bellows.

A differential pressure transducer is shown in figure 33-14. This transducer is designed to measure the differential pressure of liquids or gases, and has an output of 5 V (full scale). The pressure transducer shown in figure 33-15A measures the pressure of corrosive liquids or gases from 0 psig to 500 psig (pounds per square inch gauge pressure). Figure 33-15B shows a pressure transducer which can be used in systems where the absolute pressures range from 0 psia to 10 000 psia (pounds per square inch absolute).

(R33-19)   Define: a. gauge pressure, b. absolute pressure, and c. differential pressure.

(R33-20)   What is the gauge pressure measurement at normal atmospheric conditions?

(R33-21)   What is the absolute pressure measurement at normal atmospheric conditions?

(R33-22)   Is a differential pressure measurement a gauge measurement, an absolute pressure measurement, or neither one?

(R33-23)   List two common devices used for pressure measurements.

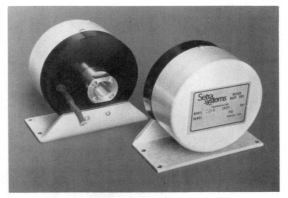

Fig. 33-15A Gauge pressure transducer

Fig. 33-15B Absolute pressure transducer

(R33-24)   What is a manometer?

FLOW MEASUREMENT   The measurement of flow is an important operation in a process industry. The rate of flow of fluids, which are conducted through pipelines, must be determined. The flow at a particular moment can be measured by the direct view flowmeter shown in figure 33-16A. The flowmeter shown in figure 33-16B can display the percentage of maximum flow.

A recording flowmeter, figure 33-17, is used to monitor flow in the following applications: sewer flow measurement, industrial waste measurement, waste-water treatment plant flow measurement, natural stream flow measurement, and irrigation flow measurement. The flowmeter can be used in any application where a known relationship exists between the liquid level and the flow rate. The liquid level is measured using an ISCO bubbler system, figure 33-17. A small inside diameter tube is inserted in the primary device. Air is supplied by

Fig. 33-16A Direct view flowmeter

Fig. 33-16B Flowmeter which measures the percentage of maximum flow

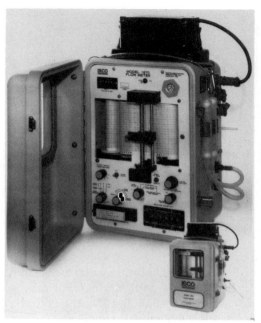

Fig. 33-17 Recording flowmeter

an internal compressor. The air bubbles out of the tube at a constant rate, resulting in a tube pressure which is proportional to the liquid level in the primary device. A transducer measures this pressure and converts it into a digital electronic signal which is proportional to the liquid level. A module stores, in digital form, the level-flow rate relationship of the primary device. This module accurately converts the liquid level signal to the flow rate. In other words, the microprocessor-based circuitry of the flowmeter uses the stored level-flow information to convert the level signal from the transducer to a signal which is proportional to the flow rate. The flow rate, or level, is recorded on a rugged, built-in strip chart recorder. The total flow is displayed on a six-digit resettable totalizer. The module stores level-flow rate information on four different primary devices. The device to be used is selected by a front panel switch.

The clamp-on ultrasonic flowmeter, figures 33-18A and 33-18B, is commonly used in industry because it is easy to install and has a variety of applications in the measurement, control, survey, and totalization of fluid flow. It is economical and reliable. It is clamped to the outside of the user's pipe. This means that it is not necessary to shut down the operation to install or remove the flowmeter. The use of this type of flowmeter does not create the problems which normally occur when a flow sensor is inserted into the flow stream. Fluids that can be handled by such a flowmeter include: No. 6 fuel oil, raw sewage, water, seawater, demineralized water, latex, gasoline, and cooling

Fig. 33-18A Clamp-on ultrasonic flowmeter in a fixed installation on a large water line

Fig. 33-18B Clamp-on ultrasonic flowmeter used as a survey tool

Fig. 33-18C Control room installation of three rack-mounted flowmeters

water. Figure 33-18C shows the control room installation of three rack-mounted flowmeters.

The ultrasonic flowmeter shown in figure 33-19 can measure fluid flow with an accuracy of 99.5%. This flowmeter contains solid-state electronics with advanced microcircuitry, housings designed for process environments, transducers, flow sections, and interconnecting cable. The transducers are secured in bosses welded to the flow section. The transducers are in direct contact with the fluid to ensure the best sonic performance, but they do not project into the line of flow. The flowmeter can be used to monitor fluids that will propagate sound pulses. This type of flowmeter has been used successfully to measure the flow of the following fluids: alcohol, ammonia, aromatic amines, brine, chlorine, crude oil, diesel fuel, hydrochloric acid, jet fuel, liquid natural gas, liquid nitrogen, nitric acid, No. 6 fuel oil, phosphorus, phosphorus trichloride, sulfur, tetraethyl lead, titanium tetrachloride, turbine fuel, waste water, and potable water.

Another method of flow measurement involves measuring the pressure drop due to a restriction in the flow path. This technique is called the differential pressure, or head, method of measurement. The differential pressure is pro-

Fig. 33-19 Ultrasonic flowmeter

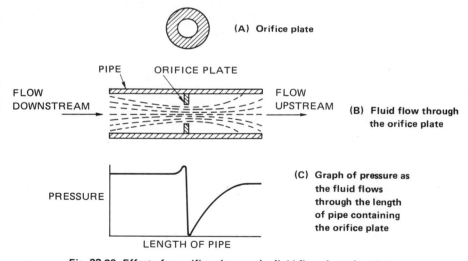

(A) Orifice plate

PIPE   ORIFICE PLATE

FLOW
DOWNSTREAM

FLOW
UPSTREAM

(B) Fluid flow through the orifice plate

PRESSURE

LENGTH OF PIPE

(C) Graph of pressure as the fluid flows through the length of pipe containing the orifice plate

Fig. 33-20 Effect of an orifice plate on the fluid flow through a pipe

duced when an orifice plate is inserted in a pipe (refer to figure 33-7A). Figure 33-20 shows the effect of this orifice plate on the fluid flow. The orifice plate causes a flow restriction. However, the length of the flow path through this restriction is negligible when compared to the diameter of the pipe. The restriction causes an increase in the flow velocity, and a decrease in the pressure. The orifice contracts the flow to a smaller cross-sectional area at a point slightly behind the orifice, in the downstream direction.

When a liquid flows through a pipe containing an orifice plate, the upstream pressure and flow velocity are constant until just before the fluid enters the orifice. Figure 33-20C shows this point where the pressure increases with a corresponding slight decrease in the flow velocity. As the flow enters the orifice, there is a sudden increase in the flow velocity, and a rapid decrease in the fluid pressure. The point of greatest flow contraction occurs behind the orifice in the downstream direction. At this point, the flow velocity has its greatest value and the corresponding pressure is at a minimum. Figure 33-20C shows that after the flow passes the point of greatest contraction, the downstream fluid velocity decreases gradually and the pressure increases. The pressure does not rise to the value of the pressure ahead of the orifice because of friction losses.

(R33-25) What is the principle governing the effect of an orifice plate?

(R33-26) When referring to pressure, what does the term "head" mean?

(R33-27) List two effects that an orifice plate has upon the fluid flow at the point of the greatest flow contraction.

TEMPERATURE MEASUREMENT

Heat is a form of pure energy. Temperature is the amount of heat, or lack of heat, contained by a substance. Temperature is also a measure of the driving force that causes heat to flow between two bodies with different temperatures. A thermocouple is a device that consists of two wires, made of different metals, which are electrically joined at two points, or junctions. When these two junctions are at different temperatures, an electromotive force (emf) is generated. This emf is proportional to the temperature difference between the two junctions of the thermocouple.

Some industrial thermocouples are shown in figure 33-21. The encased thermocouples, figure 33-22, are fast response devices which sense critical temperatures in motors, generators, turbines, pumps, and compressors. These thermocouples are used to sense the air temperature, the surface temperature, and the temperature of the stator windings and bearings. The thermocouples shown in figure 33-23 have a thermocouple junction laminated into a flat configuration between thin layers of flexible insulation. These devices are fast and accurate and can sense air, gas, liquid, or surface temperatures in a range from −328°F to +500°F. They are used to measure, record, or control temperatures in many commercial, industrial, and laboratory applications, including: baths, ovens, motors, avionics, computer memory planes, process pipelines, medical instruments, refrigeration, electrical components, and electrical equipment.

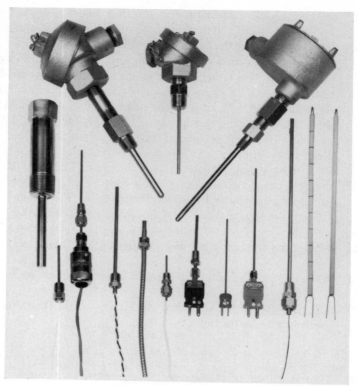

Fig. 33-21 Industrial thermocouples

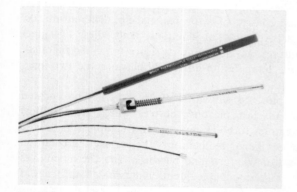

Fig. 33-22 Thermocouples used with electrical machinery

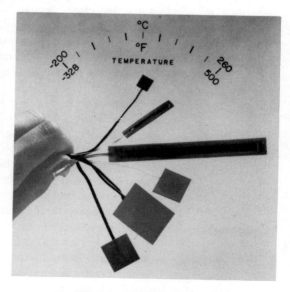

Fig. 33-23 Laminated thermocouples

Fig. 33-24 Digital thermocouple test set

The thermocouple test set shown in figure 33-24 provides a direct digital temperature readout from thermocouples. This instrument is ideal for industrial field use and laboratory applications because it is simple to operate, easy to read, very accurate, portable, and rugged.

(R33-28) Define temperature.

(R33-29) What is a thermocouple?

PROCESS CONTROLLERS

A process controller is a pneumatic, electronic, and/or mechanical piece of hardware which applies the theory of automatic control systems to industrial process control problems. The controller maintains a process control variable at a predetermined value by comparing its existing value to the desired value. It also controls an actuator in response to the actuating error, to reduce that error to the smallest value possible consistent with the design of the controller.

The controller can control one variable in a process, or it can control hundreds of process variables at the same time. The controller may be either analog or digital, or it can be a combination of both. A temperature controller is shown in figure 33-25A and a temperature recorder is shown in figure 33-25B. Figure 33-26 shows a recording controller. Figure 33-27 shows a pneumatic receiver recording controller with its chart and chart plate removed.

Fig. 33-25A Temperature controller

Fig. 33-25B Temperature recorder

Fig. 33-26 Recording controller

Fig. 33-27 Pneumatic receiver recording controller

The control display station shown in figure 33-28 is mounted directly in a control room panel. A change in a variable is noted by the measurement pointer as it moves in relation to the setpoint pointer. Push buttons in the manual output portion of the display station enable a smooth, balanceless transfer to be made from automatic to manual operation or from manual to automatic operation. Automatic or manual status is shown by the internal lighting of the correct push button.

A central monitoring and control operation is very important in process control. The control or supervisory center shown in figure 33-29 contains four video screens. Three of the screens are operator stations which are normally

Fig. 33-28 Control display station

Fig. 33-29 Supervisory center

Fig. 33-30 Process control panel

used for overview, group (or utility), and alarm functions. The fourth video screen is a supervisory station. A process control panel is shown in figure 33-30.

TRANSMITTERS   The signals from the process variables may go directly to the controller with no further signal handling. This is the case if the controller is located close to the process. However, many controllers are in physical locations remote from the process. In this situation, transmitters are needed to interface with the process. The transmitters receive a variable signal from the process transducer, or sensor and provide enough power to the signal to ensure that it can be transmitted over the required distance. The transmitted signal must not contain additional noise. In addition, the accuracy and sensitivity of the signal must not be reduced.

The Instrument Society of America (ISA) specifies several standard signal levels. When the signal is to be transmitted in pneumatic form, the standard levels are 3 psi to 15 psi. The *maximum* pressure of 15 psi represents the full-scale signal deviation, or 100% signal amplitude. The *minimum* pressure of 3 psi represents the minimum possible signal amplitude, or 0% signal amplitude. The pneumatic pressure from the transmitter varies between 3 psi and 15 psi as the process variable changes in an analog fashion from 0% to 100%.

The ISA primary standard for electrical signal transmission is 4 mA to 20 mA of current. The 4-mA value represents the *minimum* level of the process variable. The 20-mA value represents the *maximum* level of the process variable. Transmitters for flow, speed, buoyancy, pressure, temperature, and liquid level control are shown in figure 33-31. Electrical and pneumatic flow transmitters are shown in figure 33-32. The pneumatic calibrators shown in figure 33-33 are portable and accurate. They are used for onsite calibration of pneumatic controls and instrumentation. Figure 33-34 shows an assortment of instrumentation control devices.

Fig. 33-31A Flow transmitters

Fig. 33-31B Speed transmitter

Fig. 33-31C Buoyancy transmitter

Fig. 33-31D Pressure transmitter

Fig. 33-31E Temperature transmitter

Fig. 33-31F Liquid level transmitters

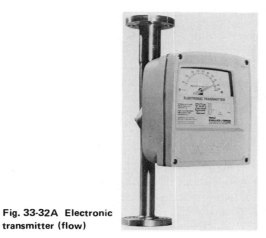

Fig. 33-32A Electronic transmitter (flow)

Fig. 33-32B Pneumatic transmitter (flow)

Fig. 33-33 Portable pneumatic calibrators

Fig. 33-34 Assortment of instrumentation control devices

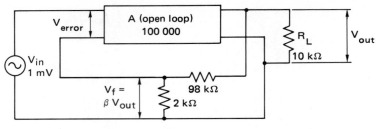

Fig. 33-35

EXTENDED STUDY TOPICS

1. An amplifier consists of three stages. Each stage has a voltage gain equal to −10. Find: a. the overall gain without feedback, b. the overall gain with feedback when 0.05 of the output is returned to the input.

2. List five benefits of negative feedback.

3. Using the negative feedback amplifier shown in figure 33-35, calculate: a. the gain of the voltage divider, b. the closed loop gain, c. the output voltage, and d. the error voltage.

4. List seven dynamic variables which require regulation in industrial process control.

5. What is the primary objective of process control?

6. What is an algorithm?

7. Name three functions of a process controller.

8. What is a control element?

9. Is the gasoline tank level transducer in the automobile also a sensor?

10. The process variable of a system varies between a minimum value of 50 and a maximum value of 100. What is the output of a pneumatic transmitter when the process variable has an output of 75?

11. What is the output of an electrical transmitter under the operating conditions of topic 11?

Chapter 34	Testing and Measuring Instruments I

OBJECTIVES

After studying this chapter, the student will be able to

- demonstrate current, voltage, and resistance measurement techniques.
- analyze dc and ac meter movements.
- calculate the results of ammeter and voltmeter loading.
- discuss dc and ac ammeters and voltmeters according to types, applications and limitations.

CURRENT MEASUREMENT

Direct current or alternating current can be measured using a deflection-type instrument called an *ammeter*. Figure 34-1 shows the internal components that form the deflection system. Three panel ammeters are shown in figure 34-2.

 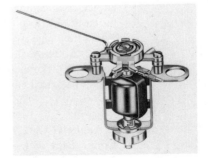

Fig. 34-1 Internal components which form the deflection system of an ammeter

Fig. 34-2 Panel ammeters

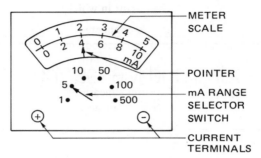

Fig. 34-3 Multirange dc milliammeter

In addition to its internal components, the multirange dc ammeter shown in figure 34-3 consists of:

1. A calibrated scale; a pointer is deflected over this scale to indicate the value of the current being measured.
2. Two input terminals identified as + and −.
3. A range switch to select the current range for each current being measured.

For the milliammeter shown in figure 34-3, the scales are read as follows:

1. When the range switch is set to the 100-mA position, the full-scale deflection is 100 mA. The bottom numbers of the meter scale (0 mA to 10 mA) are read using a multiplying factor of 10. For example, if the pointer indicates a reading of 6 on the lower scale, the actual current being measured is: 6 mA × 10, or 60 mA.
2. When the range switch is set to the 500-mA position, the full-scale deflection is 500 mA. In this case, the top scale is read using a multiplying factor of 100. For example, if the pointer indicates a reading of 3 on the top scale, the actual current being measured is: 3 mA × 100, or 300 mA.

Rules for Ammeter Use. There are three basic rules for using ammeters.

1. Connect the ammeter in series with the load or circuit in which the current is to be measured.

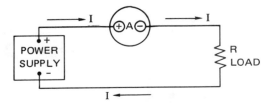

Fig. 34-4 Connecting a dc ammeter with correct polarities in a circuit

2. Use an ammeter with a full-scale deflection rating greater than the maximum expected current.

3. Use an ammeter having an internal resistance that is much lower than the resistance of the circuit in which the ammeter is inserted.

When a dc ammeter is connected, the polarities must be as shown in figure 34-4. If the ammeter is connected with opposite polarities, the current enters the minus (–) terminal and leaves the plus (+) terminal. In this case, the pointer deflects to the left of the 0 mark on the scale, rather than moving upscale to the right. For alternating-current measurement, the terminal polarity is not important because the pointer of an ac ammeter deflects upscale regardless of the polarity.

(R34-1) What is an ammeter?

(R34-2) What is the actual current being measured by the ammeter shown in figure 34-3 when the range switch is set to 1 mA and the pointer is at 4 on the lower scale?

(R34-3) What is the actual current being measured by the ammeter shown in figure 34-3 when the range switch is set to 50 mA and the pointer is at 3 on the upper scale?

(R34-4) Is terminal polarity important when using: a. dc ammeters and b. ac ammeters?

VOLTAGE MEASUREMENT

The potential difference, or electrical pressure, between two points is measured by means of a *voltmeter.* The terminal polarity must be observed when using a dc voltmeter. For an ac voltmeter, an upscale deflection of the pointer is obtained regardless of the terminal polarities.

Figure 34-5 shows one dc voltmeter being used to measure the terminal voltage of a battery and one dc voltmeter being used to measure the voltage across

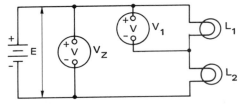

Fig. 34-5 Measuring dc voltages

Fig. 34-6A   Single range ac voltmeter

Fig. 34-6B   Multirange ac voltmeter

the lamp, L_1. A single range ac voltmeter is shown in figure 34-6A, and a multirange ac voltmeter is shown in figure 34-6B.

Rules for Voltmeter Use. There are four basic rules for using voltmeters.

1. Voltmeters are always connected in parallel with the voltage being measured.
2. A dc voltmeter must be connected using the correct terminal polarity to obtain an upscale deflection.
3. The voltmeter should have a full-scale deflection rating which is greater than the maximum expected voltage.
4. The internal resistance of the voltmeter should be high when compared with the resistance of the component whose voltage is being measured. (The reason for this rule is given later in this chapter.)

(R34-5)   What is a voltmeter?

(R34-6)   Is the terminal polarity important when using a. dc voltmeters, and b. ac voltmeters?

(R34-7)   What is the difference in the way an ammeter and a voltmeter are connected in a circuit?

RESISTANCE MEASUREMENT   Resistance is measured by an instrument called an *ohmmeter*. The ohmmeter can be a separate instrument, or it can be a part of a multimeter, figure 34-7. The ohmmeter contains a battery or an internal power supply so that it can pass a current through the resistance to be measured. Figure 34-8 shows the method of connecting a resistor to the terminals of the multimeter in figure 34-7.

Ohmmeters usually have a range switch and a pointer that moves over a calibrated scale. However, the ohmmeter scale, figure 34-7, differs from the scales of the ammeter and the voltmeter. The left-hand side of the scale is the 0-ohms

Fig. 34-7 Multimeter

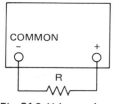

Fig. 34-8 Using an ohm-
meter (VOM multime-
ter) to measure a resis-
tance value

position. The right-hand side of the scale is marked ∞, or infinity. It can be seen that the ohmmeter scale is nonlinear, becoming more and more cramped from left to right. (Some multimeters have the 0-ohms position on the right-hand side of the scale and ∞ is on the left-hand side.)

Rules for Ohmmeter Use. There are six basic rules for using an ohmmeter.
1. An ohmmeter must not be connected to a circuit while the power supply is switched ON.
2. The ohmmeter terminals must be connected in parallel with the resistance to be measured.
3. The polarity of the ohmmeter connection is not important.
4. If the component whose resistance is to be measured forms part of a circuit, the power supply must be switched OFF. In addition, one terminal of the component must be disconnected from the circuit before the ohmmeter is connected.
5. The full-scale deflection rating of the ohmmeter should be greater than the maximum expected resistance.
6. The most accurate resistance measurements are made when the pointer of the ohmmeter is positioned near center scale.

(R34-8) What are the resistance multipliers for the multimeter shown in figure 34-7?

(R34-9) What is an ohmmeter?

(R34-10) Name a characteristic of the ohmmeter scale.

DC METER MOVEMENTS The two most common dc meter movement configurations are the D'Arsonval design and the taut band design. Both movements are examples of a permanent magnet moving coil (PMMC) deflection meter. The movements operate on the same basic principle as the dc motor. The PMMC device can be connected to act as a moving coil ammeter, a moving coil voltmeter, or an ohmmeter.

Basically, the PMMC instrument consists of a permanent magnet to provide a magnetic field and a small lightweight coil which pivots in the magnetic field.

Fig. 34-9 Armature and element assembly of a PMMC instrument

A torque is exerted on the coil by the interaction of the magnetic field and the field set up by the current passed through the winding of the coil. The resulting deflection of the coil is indicated by a pointer moving over a calibrated scale. Figure 34-9 shows the armature and element assembly of a PMMC instrument. A side view of the D'Arsonval meter movement, without the permanent magnet, is shown in figure 34-10A. A front view of the movement, including the permanent magnet, is shown in figure 34-10B.

A controlling force is needed in addition to the deflecting force provided by the interaction of the coil current and the magnetic field from the permanent magnet.

1. The controlling force returns the coil and pointer to the zero position when there is no current through the coil.
2. The controlling force also serves to balance the deflecting force. As a result, the pointer remains stationary for any constant value of current through the coil.

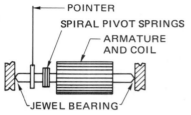

Fig. 34-10A Side view of the D'Arsonval meter movement (less the permanent magnet)

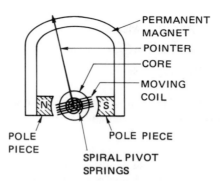

Fig. 34-10B Front view of the D'Arsonval meter movement (including the permanent magnet)

3. The spiral springs shown in figures 34-10A and 34-10B supply the controlling force. These springs are also used as the connections by which current is conducted through the coil.

When a current flows in the coil, the armature assembly deflects in a clockwise direction. The amount of deflection is proportional to the strength of the current. The scale is marked in units of current. For a PMMC instrument, the scale is linear because the scale divisions for a given change of current are equal at all points along the scale. The movement of the pointer is limited by mechanical stops at the high and low ends of the scale. The stops are located just past the zero and full-scale markings.

The taut band suspension (TBS) movement is nearly the same as the D'Arsonval movement. It differs only in the way in which the armature coil is mounted. This system is superior to all types of pivot and jewel mechanisms for dc and ac instruments. Bearing and control springs are not required in the taut band suspension system. Each end of the moving element is supported by a short hairlike band made from a special high-strength metal alloy. The material is drawn to a rectangular cross section of approximately 0.005 inch in width and 0.0005 inch thick. The dimensions are controlled to an accuracy of less than five millionths of an inch. The bands are permanently anchored to the moving element of the instrument and to U-shaped springs. These springs maintain the proper band tension and help to isolate the movement from shock and vibration. Additional cushioning against shock is provided by small stops which prevent too much axial and radial movement. The rectangular taut bands provide the restoring torque and also carry current to the moving coil. The components of the taut band suspension system are shown in figure 34-11. The taut band mechanism is shown in figure 34-12. The taut band meter movement has two main advantages over the D'Arsonval design:

1. Greater sensitivity. Sensitivity is the amount of current required for a full-scale deflection (FSD). A high-quality D'Arsonval movement may

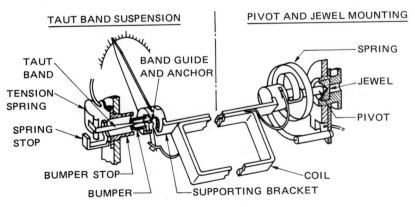

Fig. 34-11 Taut band suspension system

Fig. 34-12 Taut band mechanism

have a full-scale sensitivity of 50 μA. Taut band movements are available with a full-scale sensitivity of 2 μA.

2. Greater durability. The jewel bearing construction of D'Arsonval movements means that they are more easily damaged than taut band movements.

(R34-11)   What are the two common dc meter movement configurations?

(R34-12)   Name the three dc instruments that use the PMMC device.

(R34-13)   The controlling force is supplied by which part of a D'Arsonval PMMC device?

(R34-14)   Is polarity important when using a PMMC ammeter or voltmeter?

(R34-15)   List the two advantages of the taut band meter movement over the D'Arsonval design.

DC AMMETERS   The basic dc ammeter movement has a single current scale, such as 0 to 1 mA, or 0 to 100 μA. To measure larger currents, a shunt resistor is connected in parallel with the meter, figure 34-13. The value of the shunt resistor determines the amount of current entering the parallel combination of the shunt and the meter

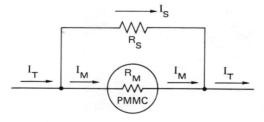

Fig. 34-13 Shunt arrangement for a PMMC meter used as a dc ammeter

coil. Using the current division relationship in a parallel circuit, the current through the meter is expressed by the following equation:

$$I_M = I_T \left(\frac{R_S}{R_M + R_S} \right)$$ Eq. 34.1

where,

I_M = current through the meter coil
I_T = the total current being measured
R_S = the shunt resistance
R_M = the meter coil resistance

Problem 1   A PMMC instrument has a coil resistance of 200 Ω. A full-scale deflection occurs at a current of 100 μA. If the instrument is to be used as an ammeter with a full-scale deflection of 1 A, determine the value of the shunt resistance.

Solution   At the full-scale deflection, I_M = 100 μA (meter coil current)

$$V_M \text{ (meter voltage)} = I_M R_M$$
$$= (100 \times 10^{-6} \text{ A})(200 \ \Omega)$$
$$= 20 \text{ mV}$$

Meter voltage = shunt voltage = 20 mV

$I_T = I_S + I_M$
$I_S = I_T - I_M$
$\quad = 1 \text{ A} - 100 \ \mu\text{A}$
$\quad = 1 \text{ A} - 0.000 \ 1 \text{ A}$
$\quad = 0.999 \ 9 \text{ A (current through the shunt)}$

$$R_S = \frac{V_M}{I_S}$$
$$= \frac{20 \text{ mV}}{0.999 \ 9 \text{ A}}$$
$$\cong 0.020 \ \Omega$$

Manufacturers usually specify shunts in relation to the full-scale deflection of the ammeter, not the meter coil, and the shunt voltage drop. Shunts can be installed within the meter case or on the outside. Figure 34-14 shows a 1 000-A

Fig. 34-14 1 000-A dc ammeter used with a 1 000-A, 50-mV external shunt

Fig. 34-15 75-A, 100-mV shunt

dc ammeter which is used with an external 1 000-A, 50-mV shunt. A 75-A, 100-mV shunt is shown in figure 34-15. When I_T is much larger than the meter coil current, I_M, the shunt current rating is specified as the full-scale deflection rating of the ammeter, or I_T.

(R34-16) Calculate the value of the shunt resistor needed to change the range of the ammeter described in Problem 1 to 10 A.

DC AMMETER LOADING An ammeter must be connected in series with the load whose current is being measured. This means that the ammeter must have a very low resistance. When the load resistance is only slightly larger than the ammeter resistance, the addition of the ammeter to the circuit can result in a large change in the load current. The effect, which is known as *ammeter loading*, is examined in Problem 2.

Problem 2 An ammeter with a coil resistance of 100 Ω is to measure the current supplied to a 400-Ω resistor from a 100-V source. Calculate: a. the current through the resistor before the ammeter is connected, and b. the current through the resistor after the ammeter is added to the circuit. Refer to figure 34-16.

Solution a. $I = \dfrac{E}{R_L}$

$= \dfrac{100 \text{ V}}{400 \text{ }\Omega}$

$= 250 \text{ mA}$

b. $I = \dfrac{E}{R_M + R_L}$

$= \dfrac{100 \text{ V}}{100 \text{ }\Omega + 400 \text{ }\Omega}$

$= \dfrac{100 \text{ V}}{500 \text{ }\Omega}$

$= 200 \text{ mA}$

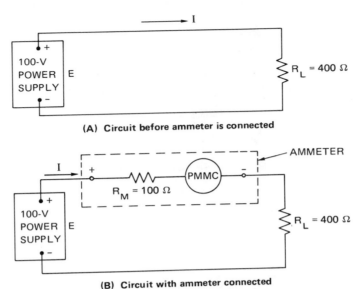

(A) Circuit before ammeter is connected

(B) Circuit with ammeter connected

Fig. 34-16 Effect of ammeter loading on the circuit current

(R34-17)   What is ammeter loading?

MULTIRANGE DC AMMETER   A multirange dc ammeter can be assembled using several shunt resistors and a rotary switch to select the desired range, figure 34-17A.

The ammeter is connected in series with the circuit in which the current is to be measured. Therefore, the rotary switch must not cause an open circuit in the shunt. If all of the shunts are open circuited, the full-load current will flow through the PMMC. This amount of current may destroy the moving coil.

The make-before-break rotary switch, figure 34-17B, protects the shunts of the multirange ammeter from becoming open circuited. The wide-ended moving contact always moves to the next terminal and connects to it before it loses contact with the previous terminal. During the switching time, two shunts are in parallel with the moving coil. This condition ensures that no shunt is open circuited.

(R34-18)   Why is a make-before-break type of range selector switch required for a multirange ammeter?

DC VOLTMETERS   A PMMC instrument cannot be used as a voltmeter without changes because the coil resistance is too low. The voltmeter must be connected in parallel with the component whose voltage is being measured. For this reason, a resistor is connected in series with the instrument to determine the range of the voltmeter.

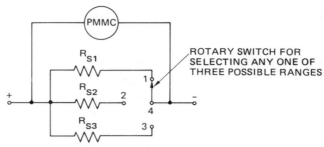

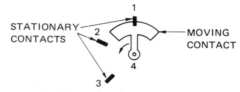

(A) Multirange DC ammeter circuit

(B) Make-before-break rotary switch

(C) Switching from position 1 to position 2

Fig. 34-17 Multirange dc ammeter circuit and rotary switch for range changing

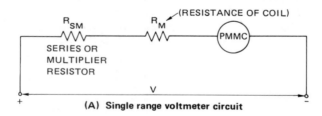

(A) Single range voltmeter circuit

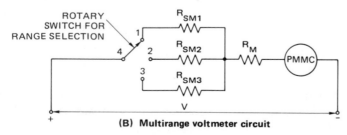

(B) Multirange voltmeter circuit

Fig. 34-18 Dc voltmeter circuits

Fig. 34-19 Self-contained dc voltmeter with a single range

This resistor is known as a multiplier resistor. Figure 34-18 shows single-range and multirange voltmeter circuits. A self-contained single range dc voltmeter is shown in figure 34-19.

The multirange dc voltmeter differs from the multirange dc ammeter in the following ways:

1. A break-before-make rotary selector switch is used with the voltmeter. That is, the moving contact disconnects from the previous terminal before it connects to the next terminal.
2. The voltmeter must have a high resistance because it is connected in parallel with a component.

The total voltmeter resistance is found from the following equation:

$$R_V = R_{SM} + R_M \qquad \text{Eq. 34.2}$$

where,

R_V = total voltmeter resistance
R_{SM} = resistance of the series, or multiplier, resistor
R_M = meter coil resistance

However, $R_{SM} \gg R_M$, and

$$R_V \cong R_{SM} \qquad \text{Eq. 34.3}$$

Problem 3   A PMMC meter having a full-scale deflection current of 50 μA is to be used as a voltmeter with a full-scale deflection of 100 V. Calculate the value of the multiplier resistor. Neglect the value of the coil resistance.

Solution   I (for FSD) = 50 μA

$$R_{SM} = \frac{V}{I}$$

$$= \frac{100 \text{ V}}{50 \mu\text{A}}$$

$$= 2 \text{ M}\Omega$$

(R34-19)   Calculate the value of the multiplier resistor needed to convert the PMMC meter described in Problem 3 to a voltmeter with a full-scale deflection of 10 V.

(R34-20) How does a multirange dc voltmeter differ from a multirange dc ammeter?

VOLTMETER SENSITIVITY Voltmeter sensitivity, or the resistance per volt, is an important characteristic of a voltmeter.

$$\text{Sensitivity (ohms per volt)} = \frac{1 \text{ V}}{I_{FSD}} \qquad \text{Eq. 34.4}$$

where I_{FSD} = the current required for full-scale deflection.

Problem 4 Calculate the sensitivity of a voltmeter with a full-scale deflection current of 50 μA.

Solution
$$\begin{aligned}
\text{Sensitivity (ohms per volt)} &= \frac{1 \text{ V}}{I_{FSD}} \\
&= \frac{1 \text{ V}}{50 \text{ } \mu A} \\
&= 20\ 000 \text{ } \Omega/\text{V}
\end{aligned}$$

The total voltmeter resistance is determined by multiplying the sensitivity (ohms per volt) by the voltmeter range:

$$R_V = (\text{sensitivity})(\text{range}) \qquad \text{Eq. 34.5}$$

Problem 5 Find the total voltmeter resistance of a voltmeter having a sensitivity of 20 000 Ω/V on a 50-V range.

Solution
$$\begin{aligned}
R_V &= (\text{sensitivity})(\text{range}) \\
&= (20\ 000 \text{ } \Omega/\text{V})(50 \text{ V}) \\
&= 1 \text{ M}\Omega
\end{aligned}$$

(R34-21) Calculate the sensitivity of a voltmeter with a full-scale deflection current of 1 mA.

(R34-22) What is the total voltmeter resistance of the voltmeter in question (R34-21) on a 100-V range?

DC VOLTMETER LOADING A voltmeter must draw a very small current so that it does not change the circuit conditions. A voltmeter with a low resistance can have a large effect on a circuit. This condition, which is called *voltmeter loading*, is shown in the following problem.

Problem 6 For the circuit shown in figure 34-20, calculate the voltage across resistor R_2, a. when the voltmeter *is not* in the circuit, and b. when the voltmeter *is* in the circuit.

Solution
$$\begin{aligned}
\text{a. } V_2 &= (E)\frac{R_2}{R_1 + R_2} \\
&= \frac{(100 \text{ V})(1 \text{ k}\Omega)}{100 \text{ k}\Omega + 1 \text{ k}\Omega} \\
&\cong 0.99 \text{ V}
\end{aligned}$$

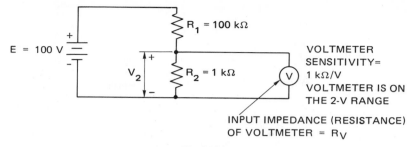

Fig. 34-20

b. R_V = (sensitivity)(range)

   = $(1 \text{ k}\Omega/\text{V})(2 \text{ V})$

   = $2 \text{ k}\Omega$

$R_V \parallel R_2 = \dfrac{(2 \text{ k}\Omega)(1 \text{ k}\Omega)}{2 \text{ k}\Omega + 1 \text{ k}\Omega}$

   = $666.7 \ \Omega$

$V_2 = 100 \text{ V} \left(\dfrac{666.7 \ \Omega}{100 \text{ k}\Omega + 666.7 \ \Omega} \right)$

   $\cong 0.66 \text{ V}$

(R34-23)   What is voltmeter loading?

(R34-24)   What is the input impedance (resistance) of the voltmeter described in Problem 6?

(R34-25)   If the voltmeter in Problem 6 is changed to one which has a sensitivity of 20 000 Ω/V, what is its reading? (Assume the voltmeter is on the 2-V range.)

(R34-26)   What conclusion can be reached by analyzing the answer to question (R34-25)?

DC DIGITAL VOLTMETER   The digital voltmeter (DVM) displays a voltage value in the form of a number, rather than by the deviation of a pointer. More significant figures can be read with the DVM. This advantage is very important when a high accuracy is required. The digital voltmeter makes an analog-to-digital conversion of the incoming analog voltage signal.

   The DVM shown in figure 34-21 has an accuracy of 0.1% and an input impedance of 10 MΩ.

Fig. 34-21 Digital voltmeter (DVM)

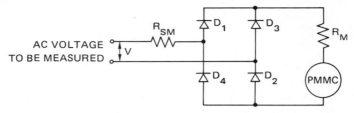

Fig. 34-22 Ac rectifier-type voltmeter

AC AMMETERS AND VOLTMETERS

PMMC ammeters and voltmeters read zero on alternating current because they respond to the average value. For an alternating-current, or voltage, wave the average level is zero.

An ac voltmeter may use a PMMC element in addition to a full-wave bridge rectifier circuit, as shown in the ac rectifier-type voltmeter of figure 34-22. The actual current that flows through the moving coil is a rectified current. The pointer of the meter reads $0.637 V_m$. However, the rms value of the voltage is the desired ac measurement. There is a direct relationship between the rms value ($0.707 V_m$) and the average value ($0.637 V_m$) for pure sine waves. Therefore, the instrument is designed so that its scale can be marked to indicate rms volts. This voltmeter is accurate only for pure sinusoidal waves.

A rectifier-type ac ammeter is shown in figure 34-23. The current transformer in the circuit gives the ammeter a low terminal resistance and a low voltage drop. In addition, the transformer provides the voltage needed to operate the PMMC device. The range of the ammeter may be changed by switching to different taps on the primary winding of the current transformer. Make-before-break rotary switches must be used for range switching. Figure 34-24 shows a multirange ac ammeter.

The electrodynamometer is similar to the PMMC dc movement. However, a pair of stationary coils connected in series is used instead of the permanent magnet, figure 34-25. As shown in figure 34-26, this instrument has a positive deflection for current flowing in either direction. In figure 34-26A, the fixed and moving coils set up fluxes so that like poles are adjacent to each other. The like poles repel and cause the moving coil to be deflected clockwise. As a result, the

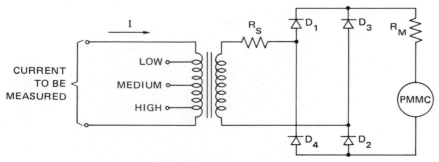

Fig. 34-23 Ac rectifier-type ammeter

Fig. 34-24 Multirange ac ammeter

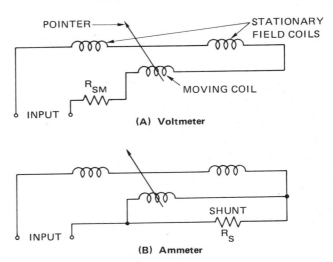

(A) Voltmeter

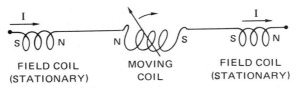

(B) Ammeter

Fig. 34-25 Electrodynamometer movement

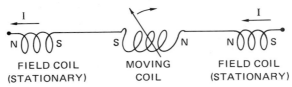

(A) Current flowing from left to right
produces clockwise (positive) deflection

(B) Current flowing from right to left
produces clockwise (positive) deflection

Fig. 34-26 Deflection of an electrodynamometer instrument
showing the positive deflection caused by the repelling action
of magnetic poles with like polarity

pointer moves upscale from left to right. When the current reverses, as shown in figure 34-26B, like poles are again adjacent to each other and the pointer moves upscale from left to right.

The electrodynamometer is not a polarized device and its terminals are not marked + and −. This instrument reads direct-current values and the rms value of alternating current.

The iron vane movement consists of two iron vanes and a coil of wire which carries the current being measured. One vane is movable and the other vane is

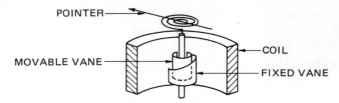

POINTER

COIL

MOVABLE VANE

FIXED VANE

Fig. 34-27 Basic construction of an iron vane meter movement

stationary. The pointer is attached to the movable vane. Figure 34-27 shows the basic construction of an iron vane meter movement. Current flowing through the coil magnetizes the two vanes with the same polarity. Like poles repel each other and the movable vane deflects by an amount which is proportional to the value of the current. This meter can be used to measure both direct and alternating current.

The iron vane meters are the simplest type of ac meters. They are very popular for use in industrial ac measuring applications because they cost less and are sturdy.

The characteristics of ammeters and voltmeters are summarized in Table 34-1.

(R34-27)   A voltage wave has the following equation: e = 170 sin 377t. When a PMMC voltmeter is connected to this voltage source, what is its reading?

(R34-28)   Why does the rectifier-type ac voltmeter, figure 34-22, read $0.637 V_m$?

(R34-29)   What is the mathematical relationship between the rms and average values of a pure sine wave?

(R34-30)   Name four types of ac ammeters and voltmeters.

EXTENDED STUDY TOPICS

1. What are the basic elements of a multirange dc ammeter?
2. What are the three basic rules for using ammeters?
3. For the multimeter shown in figure 34-7, the most accurate resistance measurements are obtained when the pointer is in what position? Explain the answer to this question.
4. What is the purpose of the controlling force in a PMMC ammeter?
5. A PMMC instrument has a coil resistance of 100 Ω and gives a full-scale deflection for a current of 1 mA. If the instrument is to be used as an ammeter with a full-scale deflection of 1 A, calculate the value of the shunt resistance required.
6. An ammeter having a coil resistance of 50 Ω is to be used to measure the current supplied to a 200-Ω load resistor from a 20-V source. Calculate: a. the current through the load resistor (ammeter unconnected), and b. the current through the load resistor (ammeter connected in the circuit).
7. A PMMC meter having a full-scale deflection current of 2 μA is to be used as a voltmeter having a full-scale deflection of 100 V. Calculate the value of the multiplier resistance. (Note: neglect the value of the coil resistance.)

TABLE 34-1 Characteristics of Ammeters and Voltmeters

Type	Use	Frequency Ranges	Voltage Ranges	Sensitivity	Current Ranges	Accuracy
D'Arsonval (PMMC)	Dc only	Dc only	10 mV - 10 kV	1 000 Ω/V to 20 000 Ω/V	1 μA - 1 000 A	0.1% to 2%
Iron Vane	Dc or ac	25–125 Hz	1 V - 750 V	100 Ω/V	10 mA - 50 A	0.5% to 2%
Rectifier with PMMC	Ac only	0–20 000 Hz	2.5 V - 1 500 V	20 000 Ω/V (dc) 5 000 Ω/V (ac)	1 mA - 1.5 A	1% to 3%
Electrodyna- mometer	Dc or ac	0–200 Hz	1 V - 300 V	30 Ω/V	1 A - 50 A	0.1% to 1%
Digital Electronic	Dc or ac	0–1 MHz	1 μV - 10 kV	> 10 MΩ	10^{-13} A - 1 A	0.005 - 0.1% (dc) 0.1% - 2% (ac)

8. Two resistors are connected in series across a 250-V dc supply. R_1 = 330 kΩ and R_2 = 220 kΩ. Calculate the voltage across R_2, a. when there is no voltmeter in the circuit, and b. when the circuit contains a PMMC voltmeter having a sensitivity of 10 kΩ/V on the 250-V range.

9. Which type of voltmeter has the least effect on a circuit? Refer to Table 34-1. Why is this true?

10. Which type of movement is widely used for ac measurements in the power industry? Refer to Table 34-1.

<table>
<tr><td>

Chapter

35

</td><td>

Testing and Measuring Instruments II

</td></tr>
</table>

OBJECTIVES   After studying this chapter, the student will be able to

- explain the principles of operation and uses of multimeters, bridges, and signal generators.
- discuss the uses of logic analyzers.
- list the applications of spectrum analyzers.
- explain the basic circuits in an oscilloscope.
- use an oscilloscope to measure voltage, current, frequency, and phase shift.
- describe the use of oscilloscope probes.

MULTIMETERS   A multimeter is an instrument which can measure voltage, current, and resistance. An analog multimeter, or VOM (abbreviation of voltmeter-ohmmeter-milliammeter), is shown in figure 35-1.

The VOM contains all of the circuits needed to measure voltage, current, and resistance. A schematic diagram of the internal components of a VOM is given in figure 35-2. (Refer to the instruction manual provided with the VOM.)

The VOM uses a D'Arsonval movement for direct-current measurements. For alternating-current measurements, a rectifier is used in addition to the PMMC movement. Resistance measurements are made by an ohmmeter circuit which applies the voltage from a battery across a series connection consisting of a known internal resistance and an unknown external resistance. The D'Arsonval movement determines the value of the unknown resistance by measuring the re-

Fig. 35-1 Analog VOM Model 260

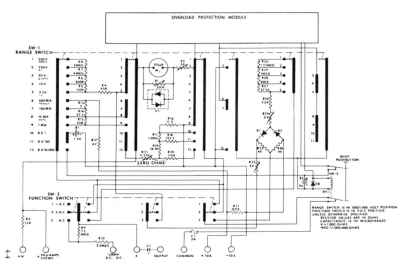

Fig. 35-2 Schematic diagram of the internal components of an analog VOM

sulting current. The output voltage of the battery must be checked regularly to ensure that the circuit will operate properly.

The main controls of the VOM are a function switch and a range switch. The *function switch* sets the instrument to the quantity to be measured:

1. Dc current
2. Dc voltage
3. Ac voltage
4. Resistance

The *range switch* selects the instrument range to be used. If the range switch is set at 250 V, and the function switch is set at +dc, figure 35-1, the meter can measure up to 250 V dc at the full-scale deflection. The test leads are connected to the – (COMMON) and + jacks.

When the VOM is used as a dc voltmeter, it has a sensitivity of 20 000 Ω/V and full-scale voltage ranges of 250 mV to 1 000 V. When the VOM is used as an ac voltmeter, it has a sensitivity of 5 000 Ω/V and full-scale voltage ranges of 2.5 V to 1 000 V. As a dc ammeter, the VOM has full-scale current ranges of 50 μA to 10 A. Resistance values can be measured from 0 Ω to 20 MΩ. The center of the scale is then equal to 120 000 Ω. The dc voltage ranges have an accuracy of 2% of the full-scale deflection. The ac voltage ranges have an accuracy of 3% of the full-scale deflection.

One of the disadvantages of the analog VOM is its input impedance which is considered low for some applications. The use of electronic multimeters can overcome this problem. This type of multimeter is available in two types:

1. Analog electronic instruments — vacuum-tube voltmeters (VTVM) and transistor voltmeters (TVM)
2. Digital electronic instruments

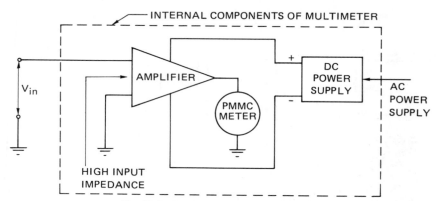

INTERNAL COMPONENTS OF MULTIMETER

Fig. 35-3 Analog electronic multimeter

The VTVM and TVM (analog instruments) use a circuit similar to that shown in figure 35-3. The amplifier presents a high-input resistance, or impedance, to the voltage source being measured. The input impedance of the solid-state FET VOM, figure 35-4, is 12 MΩ dc and 10 MΩ ac on all scales. The input impedance of the VTVM shown in figure 35-5 is 16 MΩ dc and 1 MΩ ac on all scales.

The amplifier shown in figure 35-3 can provide enough power to drive the PMMC deflection instrument. The multimeter contains a dc power supply which provides the power to operate the amplifier and the deflection instrument. The dc power supply is energized by an external ac power supply. Therefore, power is not drawn from the source being measured.

The digital multimeter (DMM) has a direct numerical readout, thus reducing human error in reading the instrument. Figure 35-6 shows two digital multimeters. The DMM meter in figure 35-6A has an input resistance of 10 MΩ on all dc scales, and an input impedance of 10 MΩ in parallel with a 75-pF capacitor on all ac scales.

Fig. 35-4 Solid-state FET VOM

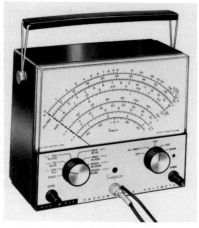

Fig. 35-5 Vacuum tube voltmeter

Fig. 35-6 Digital multimeters

(R35-1)   What is a multimeter?

(R35-2)   List the two main VOM controls.

(R35-3)   What is the input resistance of the Model 260 VOM on the 10-V dc scale?

(R35-4)   Name the two types of electronic multimeters.

(R35-5)   What is the source of the power needed to operate electronic multimeters?

BRIDGES   Bridge circuits are used to measure component values such as resistance, inductance, and capacitance. Bridge circuit measurements are very accurate because the circuit compares the value of an unknown component to that of an accurately known, or standard, component. The measurement accuracy is directly related to the accuracy of the bridge components and not to the accuracy of the null, or balance, indicator.

The Wheatstone bridge circuit, figures 35-7A and 35-7B, is used to make precise measurements of resistances ranging from a fraction of an ohm to several megohms. The resistors R_1 and R_2 are precision resistors and are known as the *ratio arms*. The variable resistor R_3 is called the *standard arm* of the bridge. The null detector (galvanometer) is a PMMC instrument. Its pointer indicates zero at the center of the scale. The null detector is a very sensitive instrument. Very small currents

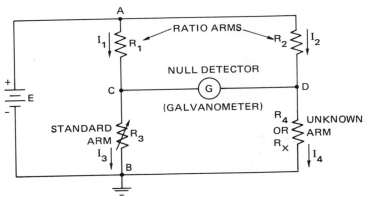

Fig. 35-7A Wheatstone bridge circuit

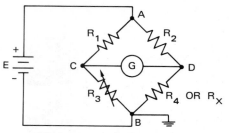

Fig. 35-7B Usual representation of a Wheatstone bridge circuit

through the coil of the galvanometer cause it to deflect to the left or to the right of zero. The galvanometer is used to detect the zero current condition, which is why it is called a null detector.

In Chapter 7, it was shown that when the galvanometer indicates a null condition, the voltages on each side of the instrument must be equal. That is, $V_C = V_D$, and the unknown R_X can be determined by the following equation:

$$R_X = R_3\left(\frac{R_2}{R_1}\right)$$ Eq. 35.1

When $R_2/R_1 = 1$, then $R_X = R_3$ so that direct reading of the value of R_3 is an indication of the value of the unknown R_X. A commercial dc resistance bridge is shown in figure 35-8.

The ac bridge, figure 35-9, consists of four bridge arms, an oscillator, and a null detector. The null detector responds to ac unbalance currents. An inexpensive null detector consists of a pair of headphones. Another type of null detector consists of an ac amplifier with either an output meter, or an electron ray tube, or tuning eye, indicator.

The general equation for bridge balance is obtained by using complex notation for the impedances of the bridge circuit. When $V_C = V_D$:

$$Z_1 Z_4 \underline{/(\theta_1 + \theta_4)} = Z_2 Z_3 \underline{/\theta_2 + \theta_3}$$ Eq. 35.2

Fig. 35-8 Dc resistance bridge

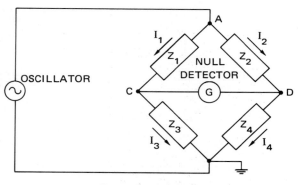

Fig. 35-9 Ac bridge

Fig. 35-10 Universal measuring bridge

Equation 35.2 shows that an ac bridge can be balanced only when two conditions are met simultaneously.

1. Condition 1: the products of the magnitudes of the opposite arms must be equal.

$$Z_1 Z_4 = Z_2 Z_3 \qquad \text{Eq. 35.3}$$

2. Condition 2: the sum of the phase angles of the opposite arms must be equal.

$$\theta_1 + \theta_4 = \theta_2 + \theta_3 \qquad \text{Eq. 35.4}$$

The universal measuring bridge, figure 35-10, is a very useful instrument. It is used to measure resistance, inductance, and capacitance in production checking, faultfinding, and development engineering. It is a self-contained and portable 1% bridge. It has a high sensitivity which means that the full range of dc resistance measurement can be used, without turning to external voltage supplies. An internal oscillator permits the measurement of inductance, capacitance, and ac resistance over a wide range at a frequency of 1 000 Hz. An external signal operating between 20 Hz and 30 kHz may be connected to a jack socket on the instrument. The measured value can be displayed in digital form.

(R35-6) What component values can be measured by bridge circuits?

(R35-7) Why does a bridge circuit have a very high measuring accuracy?

(R35-8) Define null balance.

(R35-9) To balance an ac bridge, which two conditions must be met simultaneously?

SIGNAL GENERATORS   Two types of electrical signals are used in electronics: 1. sinusoidal waveforms, and 2. nonsinusoidal waveforms, such as pulses, square waves, and triangular waves.

Signal generators are the devices which supply the signals for testing electronic circuits and systems.

(A)   **Fig. 35-11 Function generators**   (B)

Function Generator. The *function generator* is an important signal source. It provides a variety of wave shapes, includung triangular waves, square waves, sine waves, and pulses. *Sinusoidal* waveforms are used to determine the full frequency response of various devices. *Triangular* waveforms, when used with oscilloscopes, can determine the overload, or clipping, point of amplifiers. They also provide time bases for oscilloscopes and paper recorders. *Square* waveforms are used to determine the low-frequency, high-frequency, and transient response of amplifiers. Pulses and square waves are used as clock and signal sources in logic circuitry.

The function generator shown in figure 35-11A can be used on all audio circuits from the microphone to the speaker. This instrument supplies sine, square, and triangular waveforms from 1 Hz to 500 kHz. It has a maximum output of 8.5 V rms. The device also has a dc level control for adding an output dc level on all waveforms. This control is continuously variable from –6 V to +6 V.

The function generator, figure 35-11B, supplies sine, square, triangular, pulse, and ramp signals. It has a frequency range of 0.001 Hz to 10 MHz, and a maximum peak-to-peak output of 40 V.

Figure 35-12A shows a low-distortion oscillator. It can supply output sinusoidal waveforms ranging from 10 Hz to 110 kHz, with an output level up to 10 V rms. The ultra low-distortion oscillator, figure 35-12B, is a general purpose

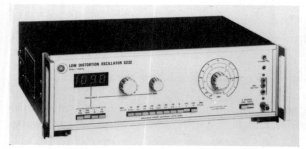

Fig. 35-12A Low-distortion oscillator

Fig. 35-12B Ultra low-distortion oscillator

laboratory and service instrument. It features very low distortion and a flat frequency response. The oscillator operates over a frequency range of 10 Hz to 110 kHz with less than –100 dB distortion. Its output level is up to 3 V rms.

Pulse Generator. The *pulse generator* is an electronic instrument that supplies an output in the form of a series of pulses (pulse train). The square wave is a series of pulses which have equal ON times and OFF times. Applications of the pulse generator include the following:

1. Transient analysis of a linear system.
2. Radar testing when the generator is used to simulate the video of a radar return.
3. Component testing.
4. Communication system testing.
5. Integrated circuit testing.

The pulse generator shown in figure 35-13 has an output ranging from 80 mV to 10 V. The pulse repetition rate of this device ranges from 1 Hz to 100 MHz.

The CB signal generator shown in figure 35-14 has a keyboard entry mode which allows the instant selection of all 40 CB channels. This device has an LED (light emitting diode) channel display, and an output level ranging from 0.3 μV to 100 mV.

(R35-10)   List two types of electrical signals.

(R35-11)   Define a square wave.

Fig. 35-13 Pulse generator

Fig. 35-14 CB signal generator

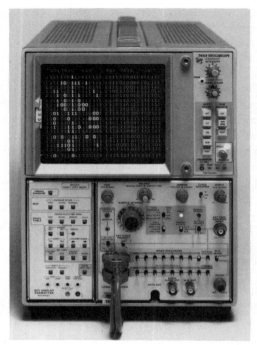

Fig. 35-15   Logic analyzer with data displayed in the form of a binary state diagram

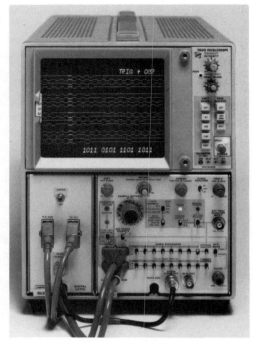

Fig. 35-16   Logic analyzer with data displayed in the form of a timing diagram with 16 channels

LOGIC ANALYSIS   The *logic analyzer* is an instrument of fairly recent origin. It is an essential instrument for logic design and troubleshooting. This instrument can perform the following functions:

1. Acquire a number of channels simultaneously.
2. Sample data synchronously, or asynchronously.
3. Store data in memory.
4. Trigger on a selected word.
5. Display data that occurs before a fault condition.

The logic analyzer presents information in one of two possible states: 1. the logic state, figure 35-15; and 2. the logic timing state, figure 35-16.

Ten-pin Harmonica Connector. The 10-pin harmonica connector, figure 35-17, is an efficient method for hooking up data lines. It places test points in the logic circuitry within easy reach. The P6451 probe can be used with the logic analyzers shown in figures 35-15 and 35-16 to eliminate the time-consuming process of locating each test point in the circuitry, and then attaching each connector. When a 10-pin harmonica connector is used, the data lines are attached quickly by plugging the harmonica terminal into the circuit board. This connector ensures easy access to data during all stages of development, manufacturing, and service.

(R35-12)   List the two states of information which can be obtained using a logic analyzer.

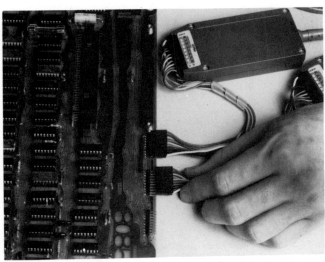

Fig. 35-17 Ten-pin harmonica connector for the P6451 logic analyzer probe

SPECTRUM ANALYZERS Spectrum analyzers are used to evaluate the frequency spectra of signals such as microwave signals, audio signals, radio frequency signals, biomedical waveforms, and mechanical vibration waveforms. A spectrum shows in graph form the relationship between the absolute, or relative value of a parameter and frequency. Every physical parameter of a system, including electromagnetic, mechanical, thermal, and hydraulic systems, has a unique spectrum.

Electronics deals with these parameters in the form of signals which can be fixed or changing electrical quantities, such as voltage, current, or power. Any signal can be expressed as the sum of a dc component and a number of sinusoidal components ranging from one to infinity. The relative phases and frequencies of the components, at every instant, add algebraically to equal the magnitude and polarity of the original signal at that instant. A constant amplitude signal does not vary with time. As a result, its spectrum contains only a dc value, or a zero frequency component. In the spectrum of a varying signal, the dc component corresponds to the average value of the time domain signal. For some signals, the dc component is zero. This system of analysis was formulated by the French mathematician Fourier. Therefore, the spectrum is called the *Fourier spectrum.*

Figure 35-18 compares two signals with their Fourier spectra. Figure 35-18A shows a pure sine wave having a constant amplitude in the time domain. Figure 35-18B shows the frequency domain plot of this wave. This plot is a single vertical line positioned at the sine-wave frequency. Its amplitude corresponds to the sine-wave amplitude. A distorted signal is represented in the time domain plot shown in figure 35-18C. The following equation expresses this waveform:

$$e = 10 \sin (2\pi ft) + 5 \sin (6\pi ft) - 3 \sin (10\pi ft)$$

The time domain waveform is produced by adding three sine waves algebraically: (1) 10 V at 100 Hz, (2) 5 V at 300 Hz, and (3) 3 V at 500 Hz. The phase relation-

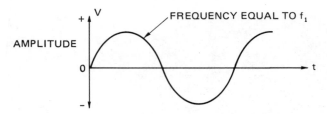

Fig. 35-18A Pure sine wave (constant repetitive amplitude) in the time domain

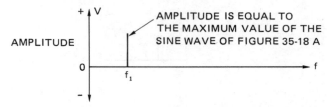

Fig. 35-18B Frequency domain plot of the sine wave of figure 35-18A

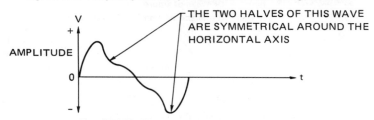

Fig. 35-18C Distorted wave in the time domain

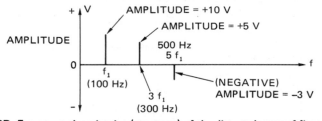

Fig. 35-18D Frequency domain plot (spectrum) of the distorted wave of figure 35-18C

ship of these three waves is such that all cross the zero voltage axis every 1/200th of a second. Figure 35-18D shows the frequency domain plot for this wave. The spectrum analysis results in three vertical lines at different frequencies and with different amplitudes. The 500-Hz component is drawn as a *negative* going line, because it *subtracts* from the other two components in the time domain summation equation. The distorted time domain waveform shown in figure 35-18C is symmetrical around the zero voltage axis. Therefore, its average value is zero, and its spectrum, figure 35-18D, has no dc component, or zero-frequency component. Because the waveform shown in figure 35-18C is symmetrical around the zero voltage axis, it contains only odd harmonics. It also has the characteristic of half-wave symmetry. *Half-wave symmetry* means that the negative half-cycle of a wave is an exact duplicate of the positive half-cycle.

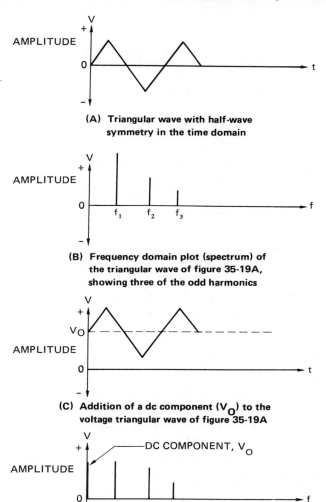

(A) Triangular wave with half-wave
     symmetry in the time domain

(B) Frequency domain plot (spectrum) of
     the triangular wave of figure 35-19A,
     showing three of the odd harmonics

(C) Addition of a dc component (V_O) to the
     voltage triangular wave of figure 35-19A

(D) Frequency domain plot (spectrum) of
     the wave of figure 35-19C, showing the
     dc spectrum and three of the odd
     harmonics

Fig. 35-19 Effect of a dc component on the frequency spectrum

Figure 35-19 shows how a dc component affects the frequency spectrum of a
wave. The triangular wave, figure 35-19A, has half-wave symmetry. Therefore,
its spectrum contains odd harmonics only. Figure 35-19B shows three of the
odd harmonics of this wave. When a dc component, V_O, is added to this wave,
the resulting wave moves up, as shown in figure 35-19C. The spectrum of figure
35-19D is the same as the one shown in figure 35-19B. However, a line equal in

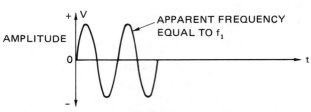

(A) An apparent pure sine wave in the time domain, as shown on an oscilloscope

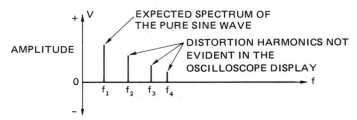

(B) Spectrum analysis of the wave of figure 35-20A, as displayed on a spectrum analyzer

Fig. 35-20 Analyzing an apparent pure sine wave on the oscilloscope and the spectrum analyzer

amplitude to V_O is drawn at the zero frequency value. It can be seen that adding a dc component to a waveform does not affect the harmonics. The only change in the spectrum is the addition of a line at the zero frequency.

The oscilloscope is a time domain instrument in which the periodic signal is viewed as a function of time. The vertical axis represents voltage and the horizontal axis represents time. This means that the oscilloscope displays the instantaneous value, v, of a periodic wave.

The spectrum analyzer displays the amplitude and frequency of each component of a wave. Therefore, the spectrum analyzer displays the peak values of the spectral components versus frequency. This display gives information that often is not available from a conventional oscilloscope.

The oscilloscope display shown in figure 35-20A appears to be a perfect sine wave. However, the spectrum analyzer display in figure 35-20B shows that the wave has three unexpected harmonics. Modulated signals can be measured on the oscilloscope, figure 35-21A. The spectrum analyzer accurately measures the modulated wave and indicates the frequency and possible distortion of the sidebands, figure 35-21B. Spectrum analyzers are shown in figure 35-22.

(R35-13) Define a spectrum.

(R35-14) What is the dc component of a pure sinusoidal waveform?

(R35-15) What is the effect of adding a dc component to a periodic wave having half-wave symmetry?

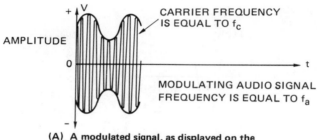

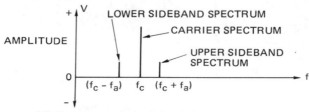

(A) A modulated signal, as displayed on the oscilloscope

(B) Spectrum analysis of the modulated wave of figure 35-21A, as displayed on a spectrum analyzer

Fig. 35-21 Analyzing a modulated wave on the oscilloscope and the spectrum analyzer

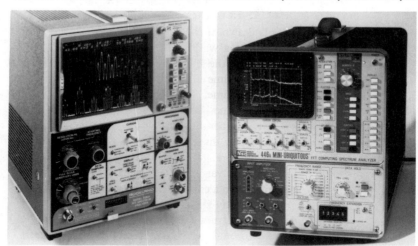

Fig. 35-22 Spectrum analyzers

THE OSCILLOSCOPE

The oscilloscope, also known as a CRO or scope, is a universal measuring instrument. It can measure a variety of rapidly changing electrical signals having one or more of the following characteristics:

1. The signals may be repetitive.
2. They may occur only once.
3. They may last a fraction of a microsecond.

Fig. 35-23   Oscilloscope Model 7104
with a 1-GHz bandwidth

Fig. 35-24 Internal view of a storage oscilloscope

The ocsilloscope display shows the changes in the signal amplitude on its vertical axis and the amount of time the event lasts on its horizontal axis. Using the oscilloscope, the user can determine the polarity of the signal (positive or negative), its amplitude, and its duration. Some general purpose scopes can view one signal only, at 100 Hz to 500 kHz. Other scopes can view one or more signals, from dc to 1 GHz (10^9 Hz). The oscilloscope shown in figure 35-23 has a 1-GHz bandwidth.

Storage scopes can store the signal trace until it is needed. Figure 35-24 shows the internal view of a storage scope. A block diagram of an oscilloscope is shown in figure 35-25.

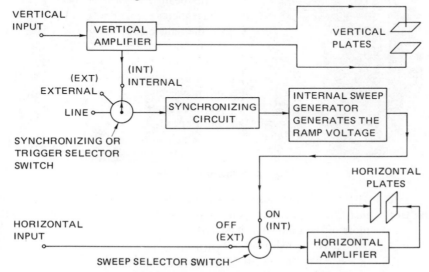

Fig. 35-25 Basic block diagram of an oscilloscope

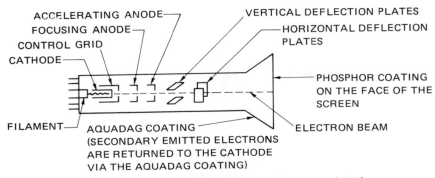

Fig. 35-26 Internal components of a cathode ray tube (CRT)

Cathode Ray Tube. The cathode ray tube (CRT) is the heart of the oscilloscope. It produces the electron beam and focuses and accelerates the beam toward a phosphor screen. The electron beam converts its kinetic energy to light energy when it strikes the phosphor coating on the face of the screen. Figure 35-26 shows the internal components of a cathode ray tube. When zero voltage is applied to the deflection plates, the electron beam strikes the center of the CRT to cause a small luminous spot. The electron beam can be deflected horizontally and vertically by applying a voltage to the deflection plates. The deflection is directly proportional to the voltage at the specific plates. When a dc voltage is applied to the deflection plates, the beam can be positioned horizontally and vertically.

Figure 35-27 shows the deflection of a beam when a *sawtooth*, or *ramp*, sweep waveform is applied to the horizontal plates and the voltage on the vertical plates is zero. Initially, the beam is positioned at the extreme left of the CRT screen. As the ramp voltage increases, figure 35-27A, the luminous spot on the screen is deflected toward the right. If the CRT has a six-inch diameter, and requires 100 V across the horizontal plates for a deflection of one inch, then the beam deflects six inches toward the right when a 600-V sawtooth signal is applied. The spot moves from left to right at the rate of one inch per second. On a six-inch CRT screen, the beam sweeps across the face of the screen in six seconds. At the end of this time, the ramp voltage suddenly drops to 0 V and the beam retraces to its original position, as shown in figure 35-27D.

When a sinusoidal signal, having the same frequency as the ramp voltage at the horizontal plates, is applied simultaneously to the vertical plates, the beam traces an exact representation of the sinusoidal signal on the screen. If the frequency of the sinusoidal wave is doubled, two cycles of the sinusoid appear on the screen. By regulating the ratio of the sweep, or ramp, frequency to the input signal frequency, portions of the input signal or many cycles can be displayed.

To present a stationary pattern on the CRT screen, the sweep voltage frequency must equal a fixed multiple of the frequency of the vertical input signal being viewed. The display must start each trace at the same point on the verti-

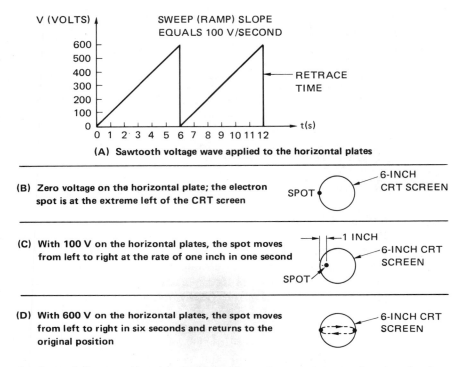

(A) Sawtooth voltage wave applied to the horizontal plates

(B) Zero voltage on the horizontal plate; the electron spot is at the extreme left of the CRT screen

(C) With 100 V on the horizontal plates, the spot moves from left to right at the rate of one inch in one second

(D) With 600 V on the horizontal plates, the spot moves from left to right in six seconds and returns to the original position

Fig. 35-27 Deflection of a beam with zero voltage on the vertical plates and a sawtooth voltage wave applied to the horizontal plates

cal input signal. This requirement is handled by the trigger generator circuit which synchronizes the sweep circuit, as shown in the block diagram of figure 35-28.

Note in the figure that the trigger sources are labeled internal (INT), external (EXT), and LINE. When the trigger mode switch is set to INT, the trigger generator signal comes from the signal being measured. In addition, the sweep generator may be synchronized to an external signal (trigger mode switch on EXT), or to the 60-Hz line frequency (trigger mode switch on LINE).

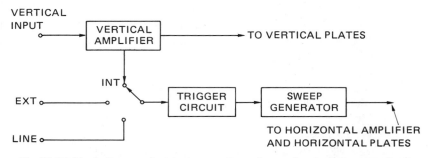

Fig. 35-28 Block diagram of a trigger generator and a synchronization sweep circuit

Fig. 35-29 Dual trace oscilloscopes

The horizontal sweep generator produces the ramp voltage needed to deflect the electron beam linearly across the CRT screen. The sweep generator output voltage cannot drive the horizontal plates directly. This voltage must be amplified by the horizontal amplifier, as shown in figure 35-25. The sweep selector switch has two positions:

1. When the sweep selector switch is in the ON position, the sweep generator is connected directly to the internal sweep generator to produce a sweep across the CRT screen.

2. When the sweep selector switch is in the OFF, or EXT position, the horizontal amplifier input is connected to the horizontal input terminal. The beam remains stationary in this position. Thus, the scope acts like an X-Y recorder when an external signal is fed into the horizontal input terminal and another signal is fed into the vertical input terminal. This type of operation is used to measure phase and frequency.

The oscilloscopes shown in figure 35-29 have dual trace displays. This feature allows the technician to perform time and amplitude comparisons be-

THE ELECTRONIC SWITCH ALTERNATES BETWEEN EACH
PREAMPLIFIER FOR A SWEEP PERIOD IN THE *ALTERNATE
MODE;* THE ELECTRONIC SWITCH ALTERNATES BETWEEN
EACH PREAMPLIFIER FOR A FIXED INTERVAL (5–10 µs),
INDEPENDENTLY OF THE SWEEP PERIOD, IN THE *CHOP
MODE*

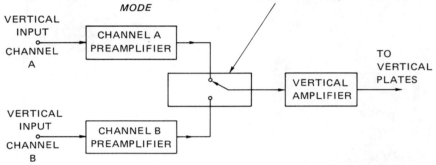

**Fig. 35-30 Block diagram of the method for generating a dual trace display using an electronic
switch in the alternate or chop mode**

tween two waveforms. This is accomplished with one sweep generator and two
vertical amplifiers operating in the alternate, or *chop*, mode, figure 35-30. The
alternate mode displays one vertical channel for a full sweep. On the next
sweep the other channel is displayed. In other words, the output of each vertical channel is displayed alternately. The alternate mode of operation is used to
view high-frequency signals when the sweep speeds are much faster than the
phosphor decay time of the CRT screen. In the alternate mode, the electronic
switch shown in figure 35-30 alternates between each amplifier for a sweep
period. In the chop mode, the electronic switch alternates between each amplifier for a fixed interval of time, approximately 5 µs – 10 µs. This interval is
independent of the sweep period. In the chop mode, a small time segment of
the sweep is given to one vertical channel, and the next time segment is given to
the other vertical channel. The chop mode is very useful when low-frequency
signals are to be viewed. For such signals, the alternate mode can cause a severe
flicker of the display.

When voltage measurements are to be made, the peak-to-peak voltage is measured on the scope. The number of divisions of the peak-to-peak voltage displayed is substituted in the following equation:

$$E_{p-p} = (\text{Number of Divisions})(\text{vertical sensitivity}) \qquad \text{Eq. 35.5}$$

The peak voltage is:

$$E_{peak} = \frac{E_{p-p}}{2} \qquad \text{Eq. 35.6}$$

For a pure sine wave, the rms voltage is:

$$E = \frac{E_{p-p}}{2\sqrt{2}} \qquad \text{Eq. 35.7}$$

Problem 1 A sinusoidal signal has a peak-to-peak display of five divisions on a scope. The vertical sensitivity is set at 1 volt per division. Calculate: a. E_{p-p}, b. E_p, and c. $E(rms)$.

Solution a. E_{p-p} = (5 divisions)(1 V/division)

$= 5$ V

b. $E_p = \dfrac{E_{p-p}}{2}$

$= \dfrac{5 \text{ V}}{2}$

$= 2.5$ V

c. $E(rms) = \dfrac{5 \text{ V}}{2\sqrt{2}}$

$= 1.77$ V

The CRO cannot measure current directly because of its high input impedance. To measure current indirectly, a resistor of known value is connected in series with the circuit whose current is to be measured. The resistance value must not be large enough to affect the actual current in the circuit. The voltage drop across the resistor is measured by the scope. The current is determined using Ohm's Law:

$$I = \frac{E}{R}$$ Eq. 35.8

The frequency of a sinusoidal wave can be measured on the scope using the sweep method. The waveform to be measured is fed into the vertical input. The scope is adjusted to give a stable waveform with at least one complete cycle covering a large section of the CRO screen. The number of horizontal divisions for a complete cycle is measured. The period is found as follows:

T = (number of horizontal divisions for one cycle)(sweep speed)   Eq. 35.9

where,

T = period, in seconds

sweep speed = seconds per division

Then,

$$f = \frac{1}{T}$$ Eq. 35.10

Problem 2 A sinusoidal signal has a display of eight horizontal divisions on a scope for one cycle. The sweep speed is set at 10 μs per division. Calculate the frequency of the signal.

Solution T = (8 divisions)(10 μs/division)

$= 80 \ \mu$s

$f = \dfrac{1}{T}$

$= \dfrac{1}{80 \times 10^{-6} \text{s}}$

$= 12\ 500$ Hz

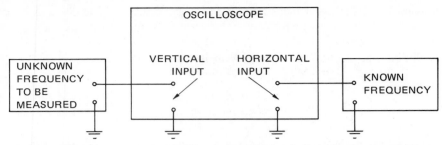

Fig. 35-31 Lissajous pattern method of measuring frequency (horizontal sweep must be in the OFF or EXT position)

Lissajous Pattern. A more accurate way of measuring frequency is the frequency comparator method. This method yields a pattern called the *Lissajous* pattern. Figure 35-31 illustrates this method of measuring frequency. An unknown signal is fed into the vertical input, and a known frequency is fed into the horizontal input. Then the horizontal sweep generator is set to OFF. If both input signals are at the same frequency and amplitude, a circular pattern is displayed, figure 35-32D. If the frequency being measured is higher than the known frequency, the Lissajous patterns shown in figures 35-32A, 35-32B, and 35-32C are displayed. The frequency of the unknown signal can be found from the following equation:

$$f_v = \frac{L_h \, f_h}{L_v}$$ Eq. 35.11

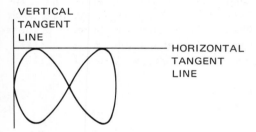

(A) 2 to 1 ratio (two loops touch the horizontal tangent line and one loop touches the vertical tangent line)

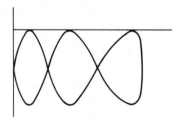

(B) 3 to 1 ratio

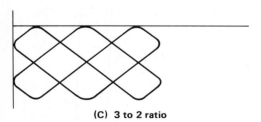

(C) 3 to 2 ratio

THIS CIRCULAR FIGURE IS DISPLAYED ONLY WHEN BOTH INPUT SIGNALS HAVE THE SAME AMPLITUDE, OTHERWISE AN ELLIPSE IS DISPLAYED

(D) 1 to 1 ratio

Fig. 35-32 Lissajous patterns for determining the frequency of an unknown signal

where,

f_v = frequency of vertical input signal
f_h = frequency of horizontal input signal
L_h = number of loops touching the horizontal tangent line.
L_v = number of loops touching the vertical tangent line.

Problem 3   The Lissajous pattern shown in figure 35-32C is displayed on a CRO. The frequency of the horizontal input wave is 60 Hz. What is the frequency of the unknown wave?

Solution f_h = 60 Hz
Horizontal tangent loops = L_h = 3
Vertical tangent loops   = L_v = 2

$$f_v = \frac{L_h\, f_h}{L_v}$$

$$= \frac{(3)(60 \text{ Hz})}{2}$$

$$= 90 \text{ Hz}$$

The phase relationship between two sinusoidal waves of the same frequency may be measured directly by simultaneously viewing both waveforms on a dual trace scope. The two waveforms appear in the proper time relationship as shown in figure 35-33. The phase shift angle can be found from the following equation:

$$\theta = \frac{\text{No. of divisions of phase shift}}{\text{No. of divisions for one full cycle}} \times 360° \qquad \text{Eq. 35.12}$$

Problem 4   Calculate the phase angle, or the angle of lag, of wave B, figure 35-33.

Solution $\theta = \dfrac{1}{12} \times 360°$

$= 30°$

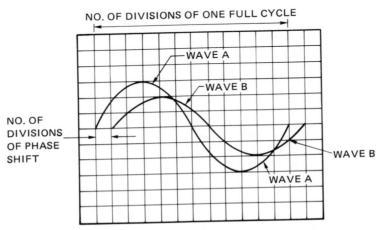

Fig. 35-33 Phase shift measurement

A second method for determining the phase relationship between two sinusoidal waves of the same frequency is to feed one wave into the vertical input, and the other wave into the horizontal input. The internal sweep generator of the scope must be in the OFF position. Figure 35-34 shows several typical Lissajous phase shift patterns resulting from this method. The phase shift (θ), in degrees, can be found from the following equation:

$$\theta = \text{arc sin}\left(\frac{A}{B}\right)$$

Eq. 35.13

where A is the intercept on the vertical axis, and B is the overall height of the pattern displayed.

(R35-16) What is an oscilloscope?

(R35-17) List three types of signals which can be displayed on the scope.

(R35-18) To which plates is the sawtooth deflection signal applied?

(R35-19) To which input should a sinusoidal waveform be applied so that it can be observed on a scope?

(R35-20) Name three trigger modes.

(R35-21) What is the function of the oscilloscope internal horizontal sweep generator?

(R35-22) When should the alternate mode of operation of a dual trace scope be used?

(R35-23) When is it necessary to use the chop mode of operation of a dual trace scope?

(R35-24) How can current be measured using a scope?

(R35-25) What are the two methods used to measure frequency?

(R35-26) What are the two methods used to measure phase shift?

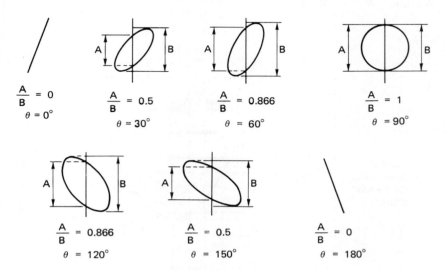

Fig. 35-34 Lissajous phase shift patterns

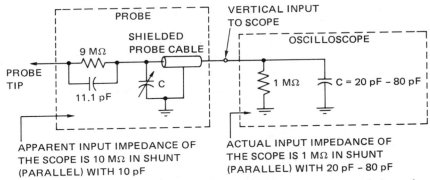

Fig. 35-35 Use of a compensated probe to present a higher input resistance and a lower shunt capacitance while attenuating all frequencies equally

OSCILLOSCOPE PROBES

The input impedance of a general purpose scope is 1 MΩ in parallel with a 10 pF to 80 pF capacitance. The 60 Hz signal, or other stray signals, can be picked up by any wire connected from the vertical input terminal of the scope to the test circuit. A coaxial cable may be used to shield the probe from these stray signals. However, the coaxial cable increases the input shunt capacitance and loads the circuit. A compensated, or *attenuator*, probe overcomes this undesirable characteristic. Typical probes have attenuation ratios of 10 to 1, 50 to 1, and 100 to 1.

The compensated probe performs the following functions:

1. It offers a higher input resistance.
2. It offers a lower shunt (parallel) capacitance.
3. It attenuates all frequencies equally.

The circuit of a 10 to 1 compensated probe is shown in figure 35-35. This probe increases the input resistance to 10 MΩ and lowers the shunt capacitance to 10 pF. A probe and its accessories are shown in figure 35-36.

(R35-27) List the three functions performed by an oscilloscope probe.

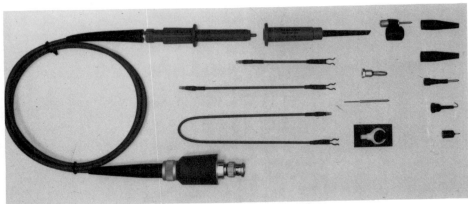

Fig. 35-36 Oscilloscope probe with accessories

1. List the four measurable quantities which can be selected by the function switch of the VOM shown in figure 35-1.

2. What is the input impedance of the Model 260 VOM on the 10-V ac scale?

3. A Wheatstone bridge circuit, figure 35-7A, has the following values for the ratio arms: $R_1 = 5 \text{ k}\Omega$ and $R_2 = 10 \text{ k}\Omega$. The null condition is indicated on the galvanometer when $R_3 = 2 \text{ k}\Omega$. Determine the value of the unknown resistor R_X.

4. The impedances of the ac bridge shown in figure 35-9 are:
   $Z_1 = 100 \ \Omega \underline{/80^\circ}$
   $Z_2 = 250 \ \Omega \underline{/0^\circ}$
   $Z_3 = 400 \ \Omega \underline{/30^\circ}$
   Express Z_4 in polar notation.

5. What is a signal generator?

6. List the five functions a logic analyzer can perform.

7. Name the two characteristics of a wave with half-wave symmetry.

8. What is the effect of regulating the sweep frequency upon the display of an oscilloscope?

9. To obtain a stationary pattern on a CRO, what condition must be met?

10. How is the condition required by topic 9 accomplished?

ANSWERS TO REVIEW QUESTIONS

CHAPTER 1

R1-1 10^{10} or 1.0×10^{10}
R1-2 10^{-12} or 1.0×10^{-12}
R1-3 2.9×10^4
R1-4 1.0×10^{12}
R1-5   1 000 000
R1-6   14 580 000
R1-7 2.22×10^{-3}
R1-8 5.0×10^{-11}
R1-9   0.000 009 22
R1-10   0.000 01
R1-11 9.7×10^{-1}
R1-12 -2.25×10^2
R1-13 -5.0×10^{-5}
R1-14 2.0×10^3
R1-15 5.0×10^2
R1-16   f = 710 000 Hz
R1-17   P = 1 000 W
R1-18   900 pC
R1-19   The mile is larger
R1-20 30.2×10^{-6} H
R1-21   3 720 μH
R1-22   55 μH
R1-23   (a)   56.4
            (b)   0.222
            (c)   255
R1-24   10 080 minutes
R1-25   1.333
R1-26   2.0
R1-27 $0°$ to $90°$
R1-28 θ is an angle between $0°$ and $360°$. It can be equal to $0°$, but it is less than $360°$.
R1-29 θ is an angle between $0°$ and $360°$. It is greater than $0°$, but it can be equal to $360°$

CHAPTER 2

R2-1   Chemical energy to heat energy
        Heat energy to mechanical energy
        Mechanical energy to electrical energy
        Electrical energy to magnetic energy
        Magnetic energy to electrical energy
        Electrical energy to heat energy, light energy, sound energy, mechanical energy.

R2-2   Heat energy — electric heater
        Light energy — incandescent lamp
        Mechanical energy — oil burner motor
        Sound energy — radio

R2-3   No
R2-4   Field force
R2-5   Zero
R2-6   No
R2-7   Less
R2-8   Field force or electrostatic force
R2-9   Repel
R2-10   Like positive charges repel each other.
R2-11   The rods will attract each other.
R2-12   Repulsion
R2-13   Attraction
R2-14   Force of attraction doubles
R2-15   Protons and neutrons have approximately the same mass. Protons have a positive electrical charge. Neutrons have no electrical charge.
R2-16   Protons
R2-17   Two
R2-18   92
R2-19   5
R2-20   32
R2-21 1.4×10^{24}
R2-22   Zero
R2-23   4
R2-24   Tungsten

CHAPTER 3

R3-1   Negative
R3-2   Battery, load or energy consuming element, connecting wires or conductors
R3-3   Negative
R3-4   Positive
R3-5   BCDA
R3-6   1.0 C
R3-7 $1\ 123.2 \times 10^{18}$ electrons
R3-8 $2\ 246.4 \times 10^{18}$ electrons

R3-9 A is (+) and B is (−) following the conventional direction of current flow.

R3-10 $I = 0$ A

R3-11 $I = 100$ mA

R3-12 I 500 C

R3-13 Battery and generator (or alternator)

R3-14 0.5 V

R3-15 Pointer will swing against the resting stop as it moves counterclockwise from zero.

R3-16

R3-17

CHAPTER 4

R4-1 2 kΩ

R4-2 Resistance is the same in both cases. A linear resistor has a constant resistance value.

R4-3 R is bilateral
Diode is unilateral

R4-4 Current varies directly with voltage.

R4-5 1 624.3 CM

R4-6 6.385 Ω

R4-7 4.006 Ω

R4-8 B and D

R4-9 100 V

R4-10 0 V

R4-11 Equal to 50 V

R4-12 ±5% to ±20%

R4-13 4 230 Ω to 5 170 Ω

R4-14 8.9 Ω ±10%

CHAPTER 5

R5-1 1 lb = 4.45 N

R5-2 The heat energy of the flashlight lamp

R5-3 The voltage doubles. $V = 12.6$ V

R5-4 $P = 3.0$ kW

R5-5 833 mA

R5-6 No

R5-7 (a) volts (b) watts

R5-8 (a) line drop $V = IR$

  (b) line loss $P = I^2 R$ or $\dfrac{V^2}{R}$

R5-9 45.9%

R5-10 24 kWh

R5-11 $R_{int} + R = R_L$

R5-12 (a) 20 Ω (b) 20 Ω (c) 100 V
  (d) 200 V (e) 500 W (f) 1 000 W

CHAPTER 6

R6-1 2 A

R6-2 0 A

R6-3 Open circuit prevents current flow.

R6-4 I_3 (through R_3) = $I_T = I_1 = I_2$

R6-5 Current enters the − terminal of the power supply and leaves the + terminal.

R6-6 E is an active element. R_1, R_2, and R_3 are passive elements.

R6-7 No. Switch S is open.

R6-8 100 V

R6-9 (a) $I_T = 10$ A (b) $R_{eq} = 12\ \Omega$

R6-10 (a) $G_{eq} = 0.083$ S (b) $I_T = 10$ A

R6-11 $I_3 = 10$ A, $I_5 = 8$ A

R6-12 $E = 240$ V

R6-13 $R_{eq} = 26.7\ \Omega$

R6-14 I_1 remains at 6 A
  I_2 remains at 4 A

R6-15 $R_{eq} = 12\ \Omega$

R6-16 (a) R_{eq} increases (b) Current in other branches does not change (c) I_T decreases

R6-17 Not changed.

R6-18 1.5 kΩ

R6-19 60 V

CHAPTER 7

R7-1 R_2 and R_3

R7-2 R_1 and R_4

R7-3 Same current flows through R_1 and R_4

R7-4 8.2 V

R7-5 82 V

R7-6 100 V = 8.2 V + 9.84 V + 82 V
  100 V = 100 V

R7-7 $I_T = I_1 = I_4 = I_9 = 1.15$ A
  $I_2 = 0.767$ A
  $I_3 = 0.383$ A
  $V_1 = 11.5$ V

$V_2 = V_3 = 15.32$ V
$V_4 = 58.65$ V
$V_5 = V_6 = V_7 = V_8 = 0$ V
$V_9 = 34.5$ V

R7-8 $I_1 = 5$ A, $I_2 = 4$ A, $I_3 = 1$ A
R7-9 $V_{AE} = 20$ V
R7-10 $R_{AB} = 37.5$ Ω
R7-11   Must draw same total current.
        Must dissipate same total power.
R7-12   +120 V
R7-13   –120 V
R7-14 $I_1 = 10$ A, $I_2 = 5$ A, $I_N = 5$ A
R7-15   The magnitude of the neutral current
        remains at 10 A, but I_N flows in the
        direction of I_1.
R7-16   10 A
R7-17   Load R_3 does not have any effect on the
        neutral current since it is connected across
        the two outside line conductors.
R7-18 $V_D = V_{DB}$ = voltage across R_4
R7-19   + (positive with respect to reference or
        ground)

CHAPTER 8
R8-1   Positive
R8-2   Negative
R8-3   +50 V
R8-4   The insulating characteristic of the
        dielectric.
R8-5   The energy stored by the capacitor in the
        electrostatic field.
R8-6   50 μF
R8-7   (a) 737.5 pF   (b) 0.442 5 μC
R8-8   0.2 mil
R8-9   0.17 C
R8-10   3 μF
R8-11   250 mC
R8-12   1 s
R8-13   63.2 V
R8-14   5 s
R8-15   28.53 V

CHAPTER 9
R9-1   South pole
R9-2   The magnets will move apart or repel each
        other.

R9-3   (a) South pole to north pole
        (b) North pole to south pole
R9-4   e = 0 V
R9-5   e = 6 V
R9-6   The ignition system would be inoperative.
R9-7   16 H
R9-8 $E_L = 1\ 000$ V
R9-9 $L_{eq} = 2$ H
R9-10   2 s
R9-11 $v_L = 36.8$ V
R9-12 $v_R = 63.2$ V
R9-13   10 s
R9-14   A is –; B is +

CHAPTER 10
R10-1   (a) V = +50 V
        (b) V = +50 V
        (c) V = +50 V
R10-2   (a) V = –50 V
        (b) V = –50 V
        (c) V = –50 V
R10-3   B to A
R10-4   Induced voltage doubles
R10-5   Twice the speed causes the rate of change
        of flux linkages, $\dfrac{\Delta\phi}{\Delta t}$ to double.
R10-6   90° and 270°
R10-7 $e_1 = +60$ V, $e_2 = –60$ V
R10-8 $E_{P-P} = (2)(100$ V$) = 200$ V
R10-9   One = +100 V, the other = –100V
R10-10 360°
R10-11 1 1/2 cycles
R10-12 No
R10-13 The current varies from 0 to its maximum
        value I_m and then back to 0. It follows
        the sinusoidal Equation 10.5.
R10-14 I = 10.6 mA
R10-15 $E_{rms} = 141.4$ V
R10-16 (a) $\dfrac{\pi}{6}$ radians (b) 90°
R10-17 e = 170 V sin 157t
R10-18 90° lagging
R10-19 90° leading
R10-20 I lags E by 30°
R10-21 No
R10-22 25 mA

CHAPTER 11

R11-1 Circuit 1:
$$E = 100 \text{ V}, I = 5 \text{ A}, \theta = 0°$$
Circuit 2:
$$E = 150 \text{ V}, I = 10 \text{ A}, \theta = 45°$$
Circuit 3:
$$E = 200 \text{ V}, I = 20 \text{ A}, \theta = -45°$$

R11-2 Circuit 2 has a leading current. Circuit 3 has a lagging current. The current and voltage are in phase in Circuit 1.

R11-3 $56°$ (I leads E)

R11-4 $V_2 = -3 - j10$

R11-5 $I = 107.7 \text{ A} \underline{/-111.8°}$

R11-6 $I = 40 \text{ A} + j60 \text{ A}$

R11-7 $V_T = 20.9 \underline{/142.6°}$

R11-8 $11.05 \underline{/95.2°}$

R11-9 $5 - j2$

R11-10 $-1 + j11 \text{ A}$

R11-11 $11.05 \underline{/95.2°}$

R11-12 $30 \underline{/-70°}$

R11-13 $22 - j59$

R11-14 $2 \underline{/20°}$

R11-15 $-0.4 + j2.2$

R11-16 $i = 7.07 \sin(377 t + 100°)$

CHAPTER 12

R12-1 (a) $I_m = 10 \text{ A}, I = 7.07 \text{ A}$
(b) $i = 10 \text{ A} \sin(157t - 30°)$
(c) $E = 141.4 \text{ V} \underline{/-30°}$
$I = 7.07 \text{ A} \underline{/-30°}$

R12-2 (a) $X_L = 35.35 \text{ Ω}$
(b) $i = 2.828 \text{ A} \sin(\omega t - 40°)$
(c) $E = 70.7 \text{ V} \underline{/50°}$
$I = 2 \text{ A} \underline{/-40°}$

R12-3 (a) $f = 318.3 \text{ Hz}$ (b) $X_L = 0 \text{ Ω}$

R12-4 In figure 12-9A: electrons flow from A to B.
In figure 12-9B: electrons flow from B to A.

R12-5 (a) $X_c = 17.675 \text{ Ω}$
(b) $i = 5.656 \text{ A} \sin(\omega t + 30°)$
(c) $E = 70.7 \text{ V} \underline{/-60°}$
$I = 4 \text{ A} \underline{/30°}$

R12-6 (a) $I = 1.5 \text{ A}$
(b) $I = 0$

R12-7 Lagging

R12-8

R12-9 They are numerically equal.

R12-10 The impedance angle is always positive in a series RL circuit as shown in figure 12-4. The circuit phase angle is negative since the current lags the voltage $E = 120 \text{ V} \underline{/0°}$ by $62°$.

R12-11 An RL series circuit has a lagging phase angle. An RC series circuit has a leading phase angle.

R12-12 $\theta = -78°$

R12-13 $I = 0.899 \text{ A}$

R12-14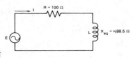

R12-15 $L = 0.234 7 \text{ H}$

CHAPTER 13

R13-1 Less than $90°$

R13-2 $I = 0.707 \text{ A} \underline{/-45°}$

R13-3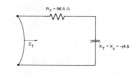

R13-4 $V_R = 60 \text{ V} \underline{/53°}$

R13-5 $Z_T = 20 \text{ Ω} \underline{/90°}$

R13-6 A coil having no resistance and whose inductive reactance $X_L = 20 \text{ Ω}$

R13-7 $I_2 = 12 \text{ A} \underline{/-53°}$

R13-8

R13-9 I_2 lags I_1 by $90°$

R13-10 $R = 10 \text{ Ω}$

R13-11 $R_S = 70.7 \text{ Ω}, X_S = j70.7 \text{ Ω}$

R13-12 $R_S = 70.7 \text{ Ω}, X_S = -j70.7 \text{ Ω}$

R13-13

R13-14 $R_{eq} = 82.6 \text{ Ω}, C_{eq} = 68 \text{ μF}$

R13-15

R13-16 $+25.2°$, leading circuit

R13-17 $Z_{eq} = 55.4 \text{ Ω} \underline{/18°}$

R13-18 $-22.4°$

CHAPTER 14

R14-1 Loop 1 (ABCA): $E_1 - I_1 R_1 - I_3 R_3 = 0$

Loop 2 (ADCA): $E_2 - I_2 R_2 - I_3 R_3 = 0$
At Node C: $I_1 + I_2 = I_3$

R14-2 $I_3 = I_1 + I_2$

R14-3 $I_1 = 3.01$ mA, $I_2 = 5.265$ mA, $I_3 = 8.275$ mA

R14-4 $I_{R_2} = 359$ mA

R14-5 $I_{R_3} = 296$ mA

R14-6 $I_L = 0.1$ A

CHAPTER 15

R15-1 $V = 100$ V

R15-2 100 V

R15-3 100 V

R15-4 $P_m = 400$ W

R15-5 From $0°$ to $180°$: e is +, i is +, and p = ei is +.
From $180°$ to $360°$: e is –, i is –, and p = ei is +.

R15-6 $V_L = 754$ V

R15-7 $V_L = 754$ V

R15-8 $V_L = 754$ V

R15-9 $V_C = 530$ V

R15-10 $V_C = 530$ V

R15-11 $V_C = 530$ V

R15-12 I lags E by $45°$

R15-13 $V_R = V_L$, V_R lags V_L by $90°$

R15-14 500 Ω

R15-15 $I_T = 50$ A

R15-16 $I_T = 9.899$ A

R15-17 $\theta = 45°$

R15-18 $P = 72$ W

R15-19 PF = 0.394

R15-20 $P_1 = 72$ W

R15-21 The two answers are identical since the only power dissipated in the circuit is that power dissipated in the resistor R.

R15-22 $I_C = 10.2$ A

R15-23 10.2 A

R15-24

R15-25 $I_T = 10$ A $\angle\,0°$

R15-26 $1\,715$ VA

R15-27 $1\,200$ VA

CHAPTER 16

R16-1 $f_r = 796$ kHz

R16-2 $I = \infty$

R16-3 $X_L = 2\,000$ Ω, $X_c = 2\,000$ Ω, $V_R = 100$ V, $V_L = 200$ V, $V_C = 200$ V

R16-4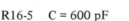

R16-5 C = 600 pF

R16-6 L = 100 mH

R16-7 C = 0.025 μF

R16-8 Q = 20

R16-9 P = 4 W

R16-10 P = 2 W

R16-11 $I_L = 10$ mA

R16-12 P = 20 mW

R16-13 For frequencies outside the pass band, the parallel resonant circuits offer low-impedance paths and the series resonant circuit offers a high-impedance path. Therefore, these frequencies are attenuated or rejected.

R16-14 For frequencies outside the stop band, the series resonant circuits offer high-impedance paths and the parallel resonant circuit offers a low-impedance path. Therefore, these frequencies pass from input to output with little opposition.

CHAPTER 17

R17-1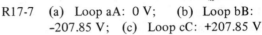

R17-2 Loop aA: 240 V
Loop bB: 0 V

R17-3 In at a, out on A

R17-4 $I_N = 29.2$ A $\angle\,-7.49°$

R17-5 $P_T = 6\,952$ W

R17-6

R17-7 (a) Loop aA: 0 V;   (b) Loop bB: -207.85 V;   (c) Loop cC: +207.85 V

R17-8 0 V

R17-9 0 A

R17-10 0 A

R17-11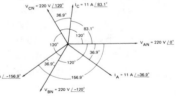

R17-12 $P_T = 5\ 806$ W

R17-13 $I_A = 17.32$ A $\underline{/\ 30°}$

R17-14 $I_A + I_B + I_C = 0$ A

The phasor sum of the three line currents in a delta-connected system must equal 0 A.

R17-15 (a) $I_{AC} = 50$ A $\underline{/\ 0°}$

  $I_{BA} = 50$ A $\underline{/\ 120°}$

  $I_{CB} = 50$ A $\underline{/\ -120°}$

(b) $I_A = 86.6$ A $\underline{/\ -30°}$

  $I_B = 86.6$ A $\underline{/\ 90°}$

  $I_C = 86.6$ A $\underline{/\ -150°}$

R17-16 $P_T = 3$ kW

R17-17 $E_{AB} = 208$ V $\underline{/\ -150°}$

R17-18 $P_T = 749$ W

R17-19 Figure 17-9A: ABC; Figure 17-9B: CBA
Figure 17-9C: ABC

CHAPTER 18

R18-1 Primary voltage; impedance of primary winding

R18-2 TR = 3

R18-3 (a) $V_2 = 60$ V   (b) $0.12\ \dfrac{V}{T}$

  (c) $0.12\ \dfrac{V}{T}$

R18-4 $I_2 = \dfrac{V_2}{Z_L}$

R18-5 TR = 10

R18-6 (a) $V_1 = 400$ V   (b) $V_1 I_1 = 1\ 600$ VA
  (c) $V_2 I_2 = 1\ 600$ VA

R18-7 The real part of the impedance seen by the source must be equal to the real part of the internal impedance of the source. The imaginary parts must be conjugate.

R18-8 TR = 25 to 1

R18-9 Positive

R18-10 Negative

R18-11 (a) Connect terminals 2 and 3. The 480-V primary supply voltage should be connected to terminals 1 and 4.

  (b) Connect terminals 1 and 3
  Connect terminals 2 and 4

The 240-V primary supply voltage should be connected to terminals 1 and 4.

R18-12 $R_1 = 1\ 600\ \Omega$

R18-13 No electrical isolation between primary and secondary windings.

CHAPTER 19

R19-1 The wind

R19-2 Because the output voltage of an ac generator is an alternating sinusoidal waveform.

R19-3 Field coils

R19-4 Practically zero output voltage

R19-5 1 500 r/min

R19-6 Decrease speed of generator

R19-7 Until electric clocks have regained correct time

R19-8 No. A change in alternator speed affects the frequency.

R19-9 VR = +4.35%

R19-10 When the induced emf is zero and when it is ready to change its polarity.

R19-11 To vary the strength of the magnetic field and thereby control the output voltage of the generator.

R19-12 No. Without a load the series circuit is incomplete.

R19-13 No. The current supplied to the electromagnetic field does not flow through the external load. This allows the generator to be operated up to its rated speed and the induced emf is reached before the external load is connected.

R19-14 No

CHAPTER 20

R20-1 A motor converts electrical energy into mechanical energy. A generator converts mechanical energy into electrical energy.

R20-2 There must be a magnetic field.
There must be a current-carrying conductor in that magnetic field.

R20-3   The counter emf is zero at the instant of starting. Therefore, the starting current is limited only by the resistance of the motor.

R20-4   Constant speed, low starting torque.

R20-5   The motor will reverse direction.

R20-6   Low starting torque

R20-7   A dc shunt motor has excellent speed regulation. It is practically a constant speed motor. A dc series motor has poor speed regulation. It has very high speed at low loads and low speed at high loads.

R20-8   The short-shunt compound motor has the shunt field connected directly across the armature. The long-shunt compound motor has the shunt field connected across the series combination of the series field and the armature.

R20-9   Constant speed

R20-10   Speed of the rotating magnetic field.

R20-11   No

R20-12   The field of the copper shading ring lags behind the main field.

R20-13   To create a rotating magnetic field and thereby supply the starting torque for motor operation.

R20-14   The field of winding 1 lags $90°$ (approximately) behind the field of winding 2.

R20-15   Both motors start as repulsion motors. At rated speed, a device short circuits all of the commutator bars of a repulsion-induction motor so that it continues to run as an induction motor.

R20-16   Approximately zero

R20-17   No effect

CHAPTER 21

R21-1   By an increase in the light intensity

R21-2   Peak power output = 24 W

R21-3   (a) A photocell is a two-terminal device which converts light energy into electrical energy.
         (b) A solar cell is a photovoltaic cell which produces voltage when exposed to sunlight.

(c) A solar module is a combination of solar cells connected in series, or in parallel, or in series-parallel.

(d) A solar array is a combination of modules designed to meet the requirements of a particular application.

R21-4   To allow the silicon surface to be accessible to the incident light.

R21-5   640 cells arranged in 20 parallel groups consisting of 32 cells per group.

R21-6   The fuel cell power plant is a one-step electrochemical system. The conventional power plant requires three steps. (See Chapter 2, figure 2-1.)

R21-7   Primary battery cannot be recharged. Secondary battery can be recharged.

R21-8   Secondary battery since it can be recharged.

R21-9   From the positive copper electrode to the negative zinc electrode.

R21-10   The zinc electrode slowly dissolves into the electrolyte.

R21-11 $R_L = 2.7 \, \Omega$

R21-12 $I_L = 20 \, A$

CHAPTER 22

R22-1   Electrons, protons, neutrons

R22-2   Electrically neutral

R22-3   (a) A positive ion is an atom that has lost one or more electrons.
         (b) A negative ion is an atom that has gained one or more electrons.

R22-4   K shell: 2 electrons
         L shell: 8 electrons
         M shell: $14 - (2 + 8) = 4$ electrons

R22-5   They both have four electrons in their outermost shell.

R22-6   Silicon and germanium atoms are electrically neutral because the total number of orbital electrons equals the total number of protons in each nucleus.

R22-7   The holes have no electrical charge.

R22-8   Conduction band, valence band.

R22-9   The forbidden band is the band between the conduction band and the valence band in which no electrons can exist.

R22-10   The core represents the nucleus and the inner complete shells of each atom. The core has four more positive charges (from its protons) than it has negative charges from the electrons in the inner complete shells of each atom.

R22-11   Neutral

R22-12   A pure semiconductor

R22-13 2.5×10^{13}

R22-14 1.5×10^{10}

R22-15   Valence electrons and holes flow in opposite directions.

R22-16   Only negatively charged electrons

R22-17   (a) Mobility is the movement of free electrons and holes in a semiconductor.
         (b) Recombination occurs when a free electron falls into a hole so that the hole disappears.
         (c) Lifetime is the time interval between the creation and disappearance of an electron-hole pair.

R22-18   (a) Doping is the addition of impurity atoms to an intrinsic semiconductor.
         (b) A tetravalent atom is one which has four electrons in its valence shell.
         (c) An extrinsic semiconductor is a pure crystal to which impurity atoms have been added.
         (d) Bulk resistance is the resistance of a doped semiconductor.

R22-19   No. The n-type semiconductor is electrically neutral since the impurity atom has five valence electrons which are balanced by the +5 charge in the core.

R22-20   Electrons (negative charges)

R22-21   Increase in temperature increases the number of minority carriers.

R22-22   No. The p-type semiconductor is electrically neutral since the impurity atom has three valence electrons which are balanced by the +3 charge in the core.

R22-23   Electrons (negative charges)

R22-24   Negative acceptor ions, positively charged holes, electron-hole pairs created by thermal agitation.

R22-25   Positive donor ions, negatively charged electrons, electron-hole pairs created by thermal agitation.

R22-26   (a) Barrier potential is an electrostatic field force offered by the p-n junction to the flow of majority carriers.
         (b) Potential hill is another term for barrier potential. It equals 0.7 V for silicon and 0.3 V for germanium.
         (c) The depletion region is an area on each side of a p-n junction which is depleted of majority charge carriers.

CHAPTER 23

R23-1   The positive side of the dc voltage source must be connected to the p-type section. The negative side of the dc voltage source must be connected to the n-type section.

R23-2   (a) 0.7 V   (b) 0.3 V

R23-3   (a) Conduction is by majority carriers.
         (b) I_F is very low until V_F is greater than the barrier potential.
         (c) Resistance of diode is very low in forward direction.
         (d) Diode has almost linear response above the knee of the characteristic curve.

R23-4   The positive side of the dc voltage source must be connected to the n-type section. The negative side of the dc voltage source must be connected to the p-type section.

R23-5   Reverse saturation current is the current flow in a reverse biased diode due to the flow of minority carriers across the junction. The reverse biased diode is forward biased for minority carriers.

R23-6   At room temperature, electron-hole pairs are created due to thermal agitation. Therefore, the p-section has some electrons and the n-section has some holes.

R23-7   Charge carriers increase in numbers in a reverse biased diode caused by minority carriers striking atoms in the depletion region and freeing valence electrons. An avalanche of carriers and a resulting abrupt increase in current follows.

R23-8   Very little current flows with reverse bias. It is in the order of microamperes. A very large reverse current flows when the reverse breakdown voltage is reached. This current can destroy the diode. Large current in the order of milliamperes flows with forward bias.

R23-9   Hole flow

R23-10 Green

R23-11 (a) V_{RM} = 75 V
        (b) V_{BR} = 100 V
        (c) I_F = 75 mA
        (d) $I_{FM \, (surge)}$ = 500 mA
        (e) I_R = 25 nA
        (f) V_F = 1 V
        (g) P = 250 mW

R23-12 (a) I_m = 170 mA   (b) I_{dc} = 54.1 mA

R23-13 PIV = 170 V

R23-14 (a) I_m = 170 mA   (b) I_{dc} = 108.2 mA

R23-15 With the same input signal:
        I_m is the same for a full-wave rectifier circuit and a half-wave rectifier circuit.
        I_{dc} for a full-wave rectifier circuit is twice I_{dc} for a half-wave rectifier circuit.

R23-16 PIV = 340 V

R23-17 V_{dc} is the same for a full-wave rectifier circuit with a center-tapped transformer and a bridge rectifier circuit.

R23-18 PIV = 240 V

R23-19 V_m

R23-20 The ripple voltage will increase.

CHAPTER 24

R24-1   A bipolar junction transitor is a semiconductor which has been doped to obtain a single NPN crystal or a single PNP crystal.

R24-2   One crystal

R24-3   Three

R24-4   Emitter, base, collector.

R24-5   Reverse saturation current

R24-6   Doubles for each $10°C$ rise in temperature

R24-7   Emitter − base junction: forward biased
        Collector − base junction: reverse biased

R24-8 $I_E = I_B + I_C$

R24-9 $I_C = \alpha I_E + I_{CBO}$

R24-10 Less current

R24-11 (a) I_B 105 μA   (b) I_C = 5.145 mA

R24-12 -0.7 V

R24-13 +0.7 V

R24-14 Common-base (CB), common-emitter (CE), common-collector (CC).

R24-15 Region between saturation and cutoff regions on the output characteristic curves.

R24-16 Region between I_C axis and output characteristic curves.

R24-17 Region between V_{CB} axis and the characteristic curve for I_E = 0.

R24-18 $I_{C \, (sat)}$ = 2 mA

R24-19 Transistor has much higher current gain in CE mode.

R24-20 β = 49

R24-21 Positive

R24-22 The output voltage is equal to the input voltage minus the voltage drop between the base and emitter. The output voltage is taken between the emitter and the common collector.

R24-23 (a) 75 V   (b) 0.010 μA   (c) 20
        (d) 1.3 V   (e) 1.5 V

R24-24 (a) I_{BQ} = 40 μA
        (b) V_{CEQ} = 7.4 V
        (c) I_{CQ} = 1.25 mA

CHAPTER 25

R25-1   (a) $I_{C \, max}$ = 3.5 A
        (b) $V_{CE \, max}$ = 16 V
        (c) P_D = 60 W

R25-2   Gain = 2 B

R25-3   G = 10 000

R25-4   Gain = -6 dB

R25-5 dB_T = 55 dB

R25-6   Gain = 20 dB

R25-7   Gain = –3 dB$_m$

R25-8   Output voltage is 180° out of phase with input voltage.

R25-9   G = 1 000

R25-10 Gain = 30 dB

R25-11 A_V = 1 000

R25-12 A_i = 50

R25-13 I_C = 200 mA

R25-14 G' = 14 400

CHAPTER 26

R26-1   Infinity

R26-2   Source, gate, drain.

R26-3   Since it requires only majority carriers in order to function.

R26-4   A BJT is a current-operated device. A JFET is a voltage-operated device.

R26-5   BJT Polarities
    Base-emitter: forward biased
    Collector-base: reverse biased
    JFET Polarities
    Gate-source: reverse biased
    Drain-source: reverse biased

R26-6   Due to the field effect created by the depletion layers around each *p-n* junction.

R26-7   Increase

R26-8   (a) Depletion layers almost touch.
    (b) $I_D = I_{DSS}$

R26-9 I_{DSS}

R26-10 $V_P < V_{DS} < V_{DS\,(max)}$ (Eq. 26.1)

R26-11 (a) 4 V to 21 V
    (b) –4 V to 0 V; 0 mA to 9 mA

R26-12 I_D = 1 mA

R26-13 r_d = 10 k Ω

R26-14 (a) V_{DSO} = 40 V
    (b) 400 mW
    (c) g_m or Y_{fs} = 1 000 μS (for typical operation)
    (d) 5 mA
    (e) 50 pA
    (f) 4.5 V
    (g) –4.5 V
    (h) r_d = 100 kΩ

R26-15 Excellent voltage amplification and high input impedance.

R26-16 180° phase reversal between the output and input voltages

R26-17 A_V = –28

R26-18 Less than 1

R26-19 Same except Equation 26.8 has no minus sign.

R26-20 A_V = 27.2

R26-21 Insulated gate FET

R26-22 Less I_D with operation in depletion mode.

CHAPTER 27

R27-1   (a) Output is not fed back to input.
    (b) Portion of output is fed back to input.

R27-2   (a) $\beta A < 0$ and A_f is less than A.
    (b) $0 < \beta A < 1$ and A_f is greater than A.

R27-3   Negative feedback amplifier

R27-4   Frequency determining network, amplifier, feedback network.

R27-5   Noise voltages

R27-6   Because it has no external input

R27-7   Feedback is provided by the inductive voltage divider. The voltage across L_2 is fed back to the input.

R27-8   Feedback is provided by the capacitive voltage divider. The voltage across C_2 is fed back to the input.

R27-9   50%

R27-10 40 W

R27-11 P_C = 106.7 W; $P_{lsb} = P_{usb}$ = 6.65 W

R27-12 The most desirable modulation is 100%. The maximum useful radiated power is equal to one-third of the total radiated power with 100% modulation.

R27-13 186 000 miles per second

R27-14 Industrial communication band

R27-15 Filters IF carrier, filters dc component of the AM wave.

R27-16 Supplies a sinusoidal output from the local oscillator at a frequency 455 kHz higher than the station which is tuned.

Mixes the incoming signal with the output of the local oscillator and gives a difference frequency of 455 kHz.

R27-17   The amplitude of the FM signal does not vary.

The carrier frequency deviates (is modulated) according to the amplitude of the audio signal.

R27-18   (a)   88 MHz to 108 MHz
(b)   10.7 MHz

R27-19   Ensures that no amplitude variation occurs.

R27-20   (a)   AM   (b)   FM

R27-21   Electron gun, yoke, phosphor-coated screen.

R27-22   (a)   550 lines   (b)   262.5 lines

CHAPTER 28

R28-1   20

R28-2   Ground potential is the zero voltage reference level for all input, output, and power supply voltages.

R28-3 $+150°$

R28-4 $-30°$

R28-5   Yes

R28-6   No

R28-7   Differential amplifier

R28-8   Yes

R28-9   Zero

R28-10 $V_0 = 2$ V

R28-11 V_0 will vary as Z_L varies

R28-12 $V_0 = -15$ V

R28-13   The practically infinite input impedance of the op amp

R28-14   Negative feedback

R28-15   At the noninverting terminal

R28-16   At the inverting terminal

R28-17 $I_1 = 200 \, \mu A$

R28-18 $V_0 = 2$ V $\underline{/\,225°}$

R28-19   No

R28-20 $A = -5$

R28-21   (a) $V_0 = +4$ V   (b) $R_F = 200$ kΩ

R28-22   (a) $A = 2$ (b) $V_0 = 2$ V
(c)   No effect

CHAPTER 29

R29-1   Zener diodes are designed to operate in the breakdown region with reverse bias. Rectifier junction diodes are designed to operate with forward bias.

R29-2   300 000 volts per centimeter

R29-3   Zener effect, avalanche effect.

R29-4   (a)   30 V   (b)   ±5%   (c)   8.5 mA
(d)   40 Ω (e)   1 000 Ω (f)   0.25 mA
(g)   31 mA   (h)   1 W

R29-5   An inverter is an uninterruptible power system which supplies ac power to a load from a dc source of power.

R29-6   Transformer, rectifier, filter, regulator.

R29-7   The transformer

R29-8   To convert the ac input voltage to the required dc voltage level.
To hold the output voltage constant even though the load current changes.

R29-9   Voltage regulation is a measure of the change in load voltage with load current

$$\% \text{ regulation} = \frac{V_{OC} - V_L}{V_L} \times 100\%$$

R29-10   Regulation improves; percent regulation decreases

R29-11   Zero

R29-12 $R_0 = 8.34 \, \Omega$

R29-13 $I_{L \, (max)} = I_S - I_{Z \, (min)}$

R29-14 $I_{L \, (min)} = I_S - I_{Z \, (max)}$

R29-15   (a) $R_{L \, (max)} = \dfrac{V_Z}{I_{L \, (min)}}$

(b) $R_{L \, (min)} = \dfrac{V_Z}{I_{L \, (max)}}$

R29-16 $I_{Z \, (min)} = 8$ mA; $I_{Z \, (max)} = 72$ mA

R29-17 $I_S = 72$ mA

R29-18 $R_{L \, (max)} = \infty$

R29-19   (a) I_L varies from 0 mA to 76 mA
(b) I_Z varies from 9.5 mA to 85.5 mA
(c) I_S remains constant at 85.5 mA
(d) V_0 remains constant at 10 V
(e) R_L can vary from $R_{L \, (min)} = 132 \, \Omega$ to $R_{L \, (max)} = \infty$
(f) R_S is constant at 117 Ω

R29-20   The power rating of the Zener diode must not be exceeded.

With the lowest value of R_L (with I_L at a maximum value), the Zener diode must still draw a minimum current $I_{Z (min)}$ equal to 0.1 I_{ZM}.
$I_{Z (max)}$ must be limited to 0.9 I_{ZM} when R_L is at its highest value (I_L is then at its minimum value).

R29-21 Limited power handling capability. Limited regulation and ripple reduction since Z_{ZT} is not zero.

R29-22 The Zener diode is the reference element; the series transistor is the control element.

R29-23 Regulation is achieved by forcing V_{CE1} to decrease by the same amount as the unregulated input voltage V_{in}.

R29-24 A source of power
A load to be supplied with power
A trigger circuit

R29-25 The trigger circuit supplies a pulse of sufficient magnitude to the gate to establish the gate current (I_G) needed to turn the SCR on.

R29-26 No

R29-27 No

R29-28 Yes. See figure 29-19.

R29-29 Phase control

R29-30 Time (in degrees) during which the SCR conducts.

R29-31 50% of available input power

R29-32 100% of available input power

R29-33 100% of available input power

R29-34 A p-n junction diode is a unidirectional device. A DIAC is a bidirectional device.

R29-35 A five-layer device called a bidirectional diode thyristor.
A three-layer device called a bidirectional trigger diode.

R29-36 Conduction angle decreases.

R29-37 Varying the resistor R_1 controls the conduction angle.

CHAPTER 30

R30-1 Analog devices: liquid-filled thermometer, slide rule

Digital devices: adding machine, calculator

R30-2 Analog device measures; digital device counts

R30-3 Ten

R30-4 Ten

R30-5 $385.23_{10} = (3 \times 10^2) + (8 \times 10^1) + (5 \times 10^0) + (2 \times 10^{-1}) + (3 \times 10^{-2})$

R30-6 A bit is a digit in a binary number. It is a contraction of the words "binary digit."

R30-7 MSB = 8; LSB = 0

R30-8 10_{10}

R30-9 $1011_2 = 11_{10}$

R30-10 $101010_2 = 42_{10}$

R30-11 $24_{10} = 11000_2$

R30-12 1100011_2

R30-13 8

R30-14 8

R30-15 $1024_8 = 532_{10}$

R30-16 $435_{10} = 663_8$

R30-17 $170_8 = 001111000_2$

R30-18 $10110000011_2 = 2603_8$

R30-19 16

R30-20 16

R30-21 $1010101101_2 = 2AD_{16}$

R30-22 $F3.2A_{16} = 11110011.00101010_2$

R30-23 $150_{10} = 000101010000_{BCD}$

R30-24 $100101110100_{BCD} = 974_{10}$

R30-25 (a) A word is a prearranged, fixed length group of bits.
(b) A byte is an 8-bit group.
(c) A nibble is a 4-bit group.

R30-26 (a) 16 bits
(b) 2 bytes
(c) 4 nibbles

R30-27 Sign is negative (–).
Value is –21

R30-28 65 536

R30-29 Zero voltage drop when forward biased.
Zero current when reverse biased.

R30-30 Input voltage V_{BE} = 0 or transistor input is reverse biased.
I_B = 0

$I_C = 0$

$V_{CE} = V_{CC}$

$P_D = 0$

Transistor is operating in cutoff position.

R30-31 Input voltage V_{BE} is positive; transistor input is forward biased.

I_B is large enough so that $I_C R_L = V_{CC}$

$V_{CE} = 0$

$P_D = 0$

Transistor is operating in the saturation position.

R30-32 A logic gate is an electronic device that performs a Boolean algebra operation on one or more inputs to produce an output.

R30-33 Positive logic, negative logic.

R30-34 (a) Positive Logic

High = 1 (+10 V dc)

Low = 0 (0 V dc)

(b) Negative Logic

High = 0 (0 V dc)

Low = 1 (+10 V dc)

R30-35

A
B
C
OUTPUT
ABC or X

R30-36

A
B
C
OUTPUT
A + B + C or X

R30-37 It is the standard designation for inversion or complementation.

R30-38 To change or convert one logic level to the other logic level.

R30-39 X = 0

R30-40

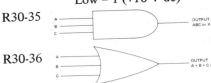

R30-41 1

R30-42 NOT, AND, OR.

R30-43 0

R30-44 NOT, AND, OR.

R30-45 An OR gate has an output of 1 if both A = 1 and/or B = 1. An EXCLUSIVE OR gate excludes the case when A = B = 1; in other words, it has an output = 0 when A = B = 1.

R30-46 It is used in digital logic systems whenever it is necessary to decide whether two binary input numbers are equal or not equal.

R30-47

TRUTH TABLE EXCLUSIVE-NOR GATE		
INPUT A	INPUT B	OUTPUT X A + B
0	0	1
0	1	0
1	0	0
1	1	1

CHAPTER 31

R31-1 Bistable, monostable, astable

R31-2 Has two stable states; two outputs which are logically inverse signals.

R31-3 (a) Output determined by input at that instant.

(b) Output determined by previous inputs as well as present inputs.

R31-4 (a) RESET: Q output is logic 0

$\overline{Q}$ output is logic 1

(b) SET: Q output is logic 1

$\overline{Q}$ output is logic 0

R31-5 (a) Edge at which pulse rises from logic 1 to logic 0

(b) Edge at which pulse falls from logic 1 to logic 0.

R31-6 A graphical display which shows the sequence of input and output signals.

R31-7 A logic device that generates a precise pattern of periodically occurring pulses.

R31-8 Flip-flops controlled by CLOCK pulses.

R31-9 No change occurs. Flip-flop remains SET.

R31-10 The flip-flop is negative-edge triggered.

The frequency of the output is one-half the frequency of the input.

R31-11 Both flip-flops have the same indeterminate state which occurs when both the S and R inputs are HIGH at the same time.

R31-12 S input must be at logic 1. CLOCK pulse must be applied simultaneously.

R31-13 No

R31-14 No change

R31-15 J = 1 and K = 1

R31-16 The operation of a JK flip-flop is identical to that of an RST flip-flop except that the JK flip-flop does not have an indeterminant state.

R31-17 The Q output assumes the state of the D input at each CLOCK pulse.

R31-18 By the application of a suitable triggering circuit connected to the base of each transistor.

R31-19 The base terminals of each transistor

R31-20 From the collector of each transistor.

R31-21 The ONE-SHOT circuit has one stable state. It remains in this state until triggered into its unstable state. The unstable state is temporary. The circuit returns to the stable state unless it is retriggered.

R31-22 To set the duration that the monostable remains in its unstable state. This in turn determines the output pulse width.

R31-23 Generation of a time-delay output pulse.
Pulse shaping, in other words, widening or narrowing an input pulse.

R31-24 Output begins with negative going edge of the input trigger signal.

R31-25 The time constant of R_{B2} and C. This in the main determines the length of time that Q_2 is off.

R31-26 Q_2

R31-27 It has no stable state. The circuit continuously oscillates between both states.

R31-28 The output, which is a CLOCK signal, is used as a timing train of pulses to operate digital circuits.

R31-29 The Q output is the complement of the $\overline{Q}$ output.

R31-30 Waveshaping purpose. When the input is an ac sinusoidal waveform, the output is a digital output at the logic 0 or logic 1 level.

R31-31 Because its output state switches at two distinct levels of the ac input signal.

R31-32 (a) LTL is the lower trigger level.
(b) UTL is the upper trigger level.

R31-33 A half adder adds two bits of binary data. A full adder adds three bits of binary data. It can accept two input bits and a carry in (C_i).

R31-34 $S = 1; C_0 = 0$

R31-35 $S = 1; C_0 = 0$

R31-36 Memory registers, shift registers.

R31-37 Serial, parallel, serial-parallel, parallel-serial.

R31-38 To process or move data from one stage of the register to an adjacent stage.

R31-39 (a) A multiplexer is a combination circuit device that selects data from one of two or more input lines and transmits it on a single output line.
(b) A demultiplexer is a device that receives the multiplexed signal and separates it into the original separate signals.

R31-40 (See Table 31-5)

R31-41 (a) An encoder is a device that converts data into its coded equivalent such as a decimal-to-binary encoder.
(b) A decoder is a device that translates the code back into the original data such as a binary-to-decimal decoder.

R31-42 No

R31-43 An enable input allows a gate to be activated.

R31-44 None

R31-45 5

R31-46 Five

R31-47 Asynchronous counter, ripple counter, up counter.

R31-48 All of its flip-flops are not set simultaneously by a CLOCK in an asynchronous counter. A certain amount of time is required as each bit changes sequentially in a ripple counter. An up counter counts up in the sequence 0, 1, 2, and so on.

R31-49 (a) An A/D converter changes an analog function to digital form.
(b) A D/A converter changes a binary digital input word to an analog dc voltage corresponding to the bit pattern in that word.

R31-50 256

R31-51 Light-emitting diode (LED)
Liquid crystal display (LCD)

R31-52 The input/output unit (I/O)
The arithmetic logic unit (ALU)
The timing and control unit
The memory unit

R31-53 A microprocessor is a single large-scale integrated (LSI) chip that can perform arithmetic and logic functions under program control. It replaces the central processing unit of the conventional digital computer (see figure 31-42).

R31-54 A microcomputer is a microprocessor-based computer with a microprocessor as its CPU.

CHAPTER 32

R32-1 Sound is a wave motion in air which produces an auditory sensation in the ear by a change in air pressure at the ear.

R32-2 Signal may be stored as a pattern of magnetism on a tape.
Signal may be stored as a series of wiggles in a record groove.
Signal may be used to modulate a radio carrier wave.

R32-3 20 Hz to 20 000 Hz

R32-4 Pitch is a property of a musical note determined by its frequency.

R32-5 Determines whether sound wave is loud or soft.

R32-6 The accurate and faithful reproduction of an original music score.

R32-7 (a) 440 Hz   (b) 880 Hz   (c) 1 320 Hz

R32-8 Overtone is a harmonic of the fundamental frequency.

R32-9 Timbre is the characteristic of a musical instrument which permits it to be distinguished from another.

R32-10 All test tones have the same intensity and a good amplifier system reproduces equal test tones with the same intensity.

R32-11 The ability of a hi-fi system to follow rapidly changing signals accurately.

R32-12 Harmonic distortion is the addition of harmonics that were not present in the original signal.

R32-13 IM distortion is intermodulation distortion caused by two simultaneous sounds interfering with each other and producing tones not present in the input signal. These tones are not necessarily harmonically related to the input tones.

R32-14 Stereo is two-channel reproduction.

R32-15 The origin of each sound is precisely located in space. The exact position of each player or singer can be sensed. Stereo delivers the sound in its natural acoustic environment.

R32-16 Stereo is two-channel reproduction. High-fidelity is a faithful reproduction of the original sound. Stereo may or may not be hi-fi. Stereo has no connection with the basic quality of sound.

R32-17 Quad is four-channel reproduction.

R32-18 Ambience describes the acoustic surroundings in which the music is played. It refers to the total pattern of the sound reverberations and reflections in the studio or concert hall.

R32-19 Platter, tonearm.

R32-20 Belt drive, direct drive, linear drive.

R32-21 A device for tuning in radio stations and playing their programs through a high-fidelity system.

R32-22 The transmission of two or more channels on a single carrier so that they can be recovered independently at the receiver.

R32-23 Noise is an unwanted disturbance superimposed upon the useful signal. It is independent of the signal.

R32-24 Sensitivity of a tuner is its ability to pull in weak or distant stations. It is stated in μV at the antenna terminals necessary in order that the noise may be quieted to a level 30 dB below the music.

R32-25 FM muting keeps the receiver quiet while the tuner is being tuned across blank spots on the FM dial.

R32-26 The 50-W rating signifies that each stereo channel is capable of delivering 25 W to the speakers.

R32-27   Contains the functions of a tuner, pre-amplifier, and power amplifier.

R32-28   A speaker is a reproducer of sound tones which should be delivered with clarity and low distortion independent of volume over a specific range.

R32-29   (a)   Deviation from flat response, if certain tones are emphasized.
         (b)   Deviation from flat response, if certain tones are suppressed.

R32-30   (a)   40 Hz to 1 kHz
         (b)   1 kHz to 5 kHz
         (c)   5 kHz to 20 kHz

R32-31   The Dolby system is an audio noise reduction system which is suitable for use with high quality audio recording, transmission channels, or sound movies.

R32-32   CB is a two-way radio service licensed by the FCC. It is intended for short distance (under 150 miles) personal and business radio communications.

R32-33   Two-way radio, microphone or handset, and an antenna.

R32-34   Combination transmitter and receiver.

CHAPTER 33

R33-1   An instrumentation system is an integrated combination of unlike, but interacting, elements that function to achieve an objective.

R33-2   (a)   Reduced
         (b)   Stabilized
         (c)   Increased
         (d)   Reduced
         (e)   Reduced
         (f)   Increased
         (g)   Decreased

R33-3   Negative feedback reduced the open loop gain (without feedback) from 10 000 to 39.8, the closed loop gain (with feedback)

R33-4   (a) $V_f = 0.2$ V   (b) $\beta = 0.05$
         (c) $A = 1\ 000\ 000$ (d) $A_f = 20$

R33-5 $A_f \cong 20$

R33-6   Process control is the control of a manufacturing process by automatic equipment.

R33-7   A dynamic variable is any physical parameter which can change naturally or because of external influences

R33-8   Regulation maintains the effects of the variable within a desired range.

R33-9   Process control regulates a dynamic variable.

R33-10   Feedback signal (C_M) which is a measured representation of the controlled variable.

R33-11   Sensor

R33-12   Feedback signal (C_M) and setpoint signal (C_{SP}).

R33-13   The setpoint, C_{SP}, is the desired value at which the controlled dynamic process variable, C, is to be controlled.

R33-14   Servomechanisms are hardware devices which accept control system commands and convert or transduce these commands into mechanical motion.

R33-15   Difference between setpoint and feedback signal $(C_{SP} - C_M)$

R33-16   Yes

R33-17   No

R33-18   The gasoline gauge converts gas level, which may be supplied to it as an electrical signal, into the mechanical rotation of a pointer.

R33-19   (a)   Gauge pressure is the measurement of an unknown pressure with respect to ambient pressure.
         (b)   Pressure relative to an absolute standard.
         (c)   Differential pressure is the measurement of one pressure with respect to another pressure. It is the difference between two pressures.

R33-20   Zero

R33-21   14.7 psia

R33-22   Neither

R33-23   Bourdon tube, bellows or capsule.

R33-24   A manometer is a differentail pressure measuring device.

R33-25 Measuring the pressure drop caused by a restriction in the flow path.

R33-26 Differential pressure

R33-27 Flow velocity is greatest, corresponding pressure is the least.

R33-28 Temperature is a measure of the degree of hotness or coldness of a substance. It is also a measure of the driving force which has the ability to cause heat to flow between two bodies at different temperatures.

R33-29 A thermocouple is a device consisting of two wires, made of different metals, which are electrically joined at two joints. When these two junctions are at different temperatures, an emf is thermoelectrically generated and is proportional to the difference in temperatures.

CHAPTER 34

R34-1 An instrument used to measure current.

R34-2 0.4 mA

R34-3 30 mA

R34-4 (a) Yes   (b) No

R34-5 An instrument used to measure the potential difference, or electrical pressure, between two points in a circuit.

R34-6 (a) Yes   (b) No

R34-7 An ammeter must be connected in series with the element whose current is being measured. A voltmeter must be connected in parallel with the element whose voltage is being measured.

R34-8 $R \times 1$; $R \times 10$; $R \times 100$; $R \times 1$ kΩ; $R \times 10$ kΩ; $R \times 100$ kΩ; $R \times 1$ MΩ

R34-9 An instrument used to measure resistance.

R34-10 Scale is nonlinear.

R34-11 D'Arsonval design, taut band design.

R34-12 Moving coil ammeter, moving coil voltmeter, ohmmeter.

R34-13 The spiral springs

R34-14 Yes

R34-15 Greater sensitivity, greater durability.

R34-16 $R_S \cong 0.002$ Ω

R34-17 Ammeter loading refers to the effect that the resistance of an ammeter has upon the circuit current.

R34-18 To prevent full load current flowing through the moving coil and possibly causing its destruction.

R34-19 $R_{SM} = 200$ kΩ

R34-20 Requires a break-before-make rotary selector switch; requires a high series resistance

R34-21 Sensitivity = 1 000 Ω/V

R34-22 $R_V = 100$ kΩ

R34-23 Voltmeter loading is the effect of the resistance of a voltmeter on the circuit.

R34-24 $R_V = 2$ kΩ

R34-25 0.99 V

R34-26 A highly sensitive voltmeter has negligible effect on circuit conditions.

R34-27 Zero

R34-28 This meter reads the average value of voltage in this circuit.

R34-29 1.11

R34-30 Iron vane, rectifier with PMMC, electrodynamometer, digital electronic.

CHAPTER 35

R35-1 A multimeter is a multifunction instrument which has the capability of measuring voltage, current, and resistance.

R35-2 Function switch, range switch.

R35-3 200 000 Ω

R35-4 Analog electronic instruments (VTVM or TVM)
Digital electronic instruments

R35-5 An external ac power supply

R35-6 Resistance, inductance, capacitance.

R35-7 Since it measures the deviation of an unknown component from that of an accurately known or standard component.

R35-8 A condition of balance in the bridge circuit which results in zero current.

R35-9 The products of the magnitudes of the opposite arms must be equal.
The sum of the phase angles of the opposite arms must be equal.

R35-10 Sinusoidal waveforms
Nonsinusoidal waveforms such as pulses, square waves, triangular waves.

R35-11 A square wave is a pulse or series of pulses which have equal ON times and OFF times.

R35-12 Logic state; logic timing state

R35-13 A spectrum displays the relative amplitude of the sinusoidal components (comprising the wave) versus frequency.

R35-14 Zero

R35-15 Has no effect on harmonics
The only spectral change is a new line at zero frequency.

R35-16 A universal measuring instrument capable of displaying (versus time) and measuring a wide variety of rapidly changing electrical signals.

R35-17 Repetitive
Only occurs once
Lasting a fraction of a microsecond

R35-18 Horizontal plates

R35-19 Vertical input

R35-20 Internal, external, line.

R35-21 To produce a sawtooth (ramp) voltage needed to deflect the electron beam across the CRT screen.

R35-22 For viewing high-frequency signals in cases where sweep speeds are much faster than the phosphor decay time of the CRT screen.

R35-23 When viewing low-frequency signals where the alternate mode might cause a severe flicker of the display.

R35-24 By connecting a resistor, whose value is known, but has negligible effect, in series with the circuit whose current is to be measured. The voltage across the resistor is measured on the scope. The current is found from Ohm's Law:
$$I = \frac{E}{R}$$

R35-25 Sweep method
Frequency comparator method which gives Lissajous pattern

R35-26 Alternate sweep method
Using Lissajous pattern analysis

R35-27 It offers a higher input inpedance.
It offers a lower shunt (parallel) impedance.
It attenuates all frequencies equally.

Acknowledgments

Sponsoring Editor, Technical Division: William W. Sprague
Senior Editor: Marjorie A. Bruce
Editorial Assistant: Mary V. Miller
Consulting Editor, Electronics Technology: Richard L. Castellucis
Southern Technical Institute
Marietta, Georgia

The author and editors would like to express their appreciation to the reviewers listed below who contributed many long hours to this text. Their extensive technical reviews and suggestions were invaluable in developing the material in the text.

Dr. Margaret R. Taber
Department of Electronic Technology
Purdue University
(Formerly of Cuyahoga Community College, Cleveland, Ohio)

J. Michael Jacob
Electrical Engineering Technology
Purdue University
(Formerly of Florence-Darlington Technical College, Florence, S.C.)

David L Favin
Bell Telephone Laboratories

The following companies contributed generously of illustrations and technical data:

Ad-Vance Magnetics, Inc.: 9-13, 9-14
AMP Incorporated: 28-3
Allen-Bradley Co.: 4-8, 4-12B, 4-17A, 4-17B, 4-19I
Allis-Chalmers Corporation: 20-15A, 20-15B
Beckman Instruments, Helipot Division: 4-17C
Bell Industries, J. W. Miller Division: 9-21A, 9-21B, 16-19, 16-20, 27-4
Bell Laboratories: 21-4
Beman Manufacturing, Inc.: 4-19E, 4-19F
James G. Biddle Co.: 4-12A, 10-9A, 10-9B, 17-10, 18-3, 33-24, 35-10
Bodine Electric Co.: 20-25
Bourns, Inc., Magnetics Division: 9-16
Burr-Brown Research Corp.: 31-38A, 31-38B, 31-38C
Caddock Electronics, Inc.: 4-19B, 4-19C, 4-19D
Christie Electric Corporation: 23-20A, 23-20B, 29-7A, 29-7B
Clifton Precision, Litton Systems, Inc.: 20-3
Colt Industries, Central Moloney Transformer Division: 18-1, 18-16
Controlotron Corp.: 33-18A, 33-18B, 33-18C
Conver Corp.: 29-17
Cutler-Hammer, Shallcross Resistor Products: 4-19H
Dale Electronics, Inc.: 9-17, 9-20, 29-24
Digitec-United Systems Corp.: 4-26
Dolby Laboratories, Inc.: 32-13, 32-14A, 32-14B, 32-15, 32-17A, 32-17B
Dormeyer Industries, Div. of A. F. Dormeyer Manufacturing Co., Inc.: 9-6, 9-7
Dranetz Engineering Laboratories, Inc.: 10-11, 15-5
Duncan Electric Co., Inc.: 18-19, 18-22
E. I. DuPont de Nemours & Co. (Inc.) Instrument Products Division: 33-19
Dynage, Inc.: 31-50A, 31-50B, 31-50C

Educational Products, Division of Stock Sales Company: 21-3
Elamex, S. A.: 9-8
Electronic Navigation Industries, Inc.: 18-6, 25-4A, 25-4B
Electric Power Research Institute: 21-14, 21-15, 21-16, 21-17
ERDA (Energy Research and Development Administration): 21-18, 21-19
Essex Engineering Company: 31-49A, 31-49B
Fairchild Camera and Instrument Corporation: 28-4, 28-5, 31-43A, 31-43B
Fasco Industries, Inc.: 20-17A, 20-17B, 20-17C
Fenwal Electronics: 4-24, 4-25
Fisher Corporation: 32-7, 32-8, 32-9, 32-10, 32-11, 32-12, 32-16A, 32-16B, 32-16C
The Foxboro Company: 33-26, 33-27, 33-28, 33-30, 33-31A, 33-31B, 33-31C, 33-31D,
    33-31E, 33-31F, 33-34
Charles T. Gamble Industries: 4-2, 4-19A
Gen Rad: 18-12A, 18-12B, 35-8
General Electric Company: 21-25B, 34-9, 34-15
Guardian Electric Manufacturing Co.: 9-10A, 9-10B
Gulton Industries, Inc.: 29-25A, 29-25B, 29-26A, 29-26B
Hevi-Duty Electric Co.: 18-8
Hewlett-Packard Co.: 10-19, 10-23, 27-24B, 27-24C
Hickok Electric Instruments Co.: 35-11A, 35-14
Honeywell, Inc.: 33-29
Hy-Cal Engineering: 33-21
Intel Corporation: 31-44
ISCO (Instrumentation Specialties Co.): 33-17
E. F. Johnson Co.: 8-1
Keithley Instruments, Inc.: 24-7, 24-9
Kepco, Inc.: 3-17A, 3-17B
Leeds & Northrup Co.: 4-11
Marconi Instruments: 35-12A, 35-12B
McGraw-Edison Company, Power Systems Division: 15-11A, 15-11B, 21-25D
Micro Power Systems Incorporated: 24-2, 26-4, 26-12A, 26-12B
Minco Products, Inc.: 33-22, 33-23
Motorola Semiconductor Products, Inc.: 25-1
NASA — Langley Research Center: 2-2, 2-3, 2-4
NASA — Lewis Research Center 21-8, 21-9, 21-10, 21-11, 21-12
Nicolet Scientific Corporation: 35-22B
North American Philips Controls Corporation: 20-2
Ohio Semitronics, Inc.: 5-13
Panasonic: 21-24, 21-25A, 21-27, 21-28, 21-29, 21-30, 21-31A
Pennwalt — Wallace & Tiernan Division: 33-9, 33-10, 33-11A, 33-11B, 33-12, 33-13, 33-16A,
    33-16B, 33-32A, 33-32B, 33-33A, 33-33B
Philips Test and Measuring Instruments, Inc.: 35-11B, 35-13, 35-29C
H. K. Porter Co., Inc. Electrical Division: 18-18
Power Tech, Inc.: 25-3A, 25-3B
Radio Shack ®, A Division of Tandy Corporation: 22-21B, 24-3A, 24-3B, 26-5, 32-18,
    32-19, 32-20, 32-21
Ray-O-Vac Division, ESB Incorporated: 21-20A, 21-20B, 21-20C, 21-20D, 21-20E, 21-25C,
    21-31B
RCA: 22-21D, 26-21, 28-2A, 29-16A, 29-16B
RFL Industries, Inc.: 9-3A, 9-3B, 9-3C, 9-3D, 27-13, 27-14, 27-15
Rockland Systems Corporation: 16-21, 16-22, 35-22A
S & C Electric Company: 18-24
Sangamo Capacitor Division of Sangamo Weston, Inc.: 8-2
Semtech Corporation: 22-21A, 23-11, 23-16A, 23-16B, 23-16C

Sencore: 27-24A
Setra Systems Inc.: 33-14, 33-15A, 33-15B
Siemens Corp.: 27-5, 29-3
Siemens-Allis, Inc.: 19-2, 19-3, 20-15C
Simpson Electric Co.: 2-20, 3-24, 3-25, 33-25A, 33-25B, 34-1A, 34-1B, 34-2, 34-6A, 34-7, 34-12, 34-21, 35-1, 35-2, 35-4, 35-5, 35-6A, 35-6B, 35-29B
Solar Power Corporation, Affiliate of Exxon Enterprises, Inc.: 21-13
Solid State Devices, Inc.: 22-21C
Sorenson (A Raytheon Co.): 29-6A, 29-6B, 29-6C, 29-6D
Sprague Electric Company: 20-20, 23-21, 28-2B
Square D Co., Electromagnetic Industries: 18-21, 18-23
Standard Power Inc.: 29-8A, 29-8B
Struthers-Dunn, Inc.: 9-11
The Superior Electric Company: 20-11, 20-12, 20-13, 20-14
Tektronix, Inc.: 24-29A, 35-15, 35-16, 35-17, 35-23, 35-24, 35-36
Texas Instruments Incorporated: 23-8A, 23-8B, 24-17A, 24-17B, 31-45
Topaz Electronics: 29-9A, 29-9B
Transmission and Distribution, Cleworth Publishing Co., Inc.: 15-12A, 15-12B
Union Carbide Corporation, Battery Products Division: 3-15A, 3-15B, 3-15C, 3-15D, 21-26
Victoreen Instrument Division, Sheller-Globe Corp.: 4-19G
O. S. Walker Co., Inc.: 9-9A, 9-9B
Wavetek: 10-17, 31-48
Western Electro-Mechanical Co., Inc.: 18-20
Westinghouse Electric Corporation: 5-11, 5-12, 15-10, 18-4, 18-11, 18-14, 18-15, 19-4, 19-8, 19-10, 20-22, 34-6B, 34-11, 34-14, 34-19, 34-24
Weston Instruments, Inc.: 3-10, 3-12, 3-19A, 3-19B, 5-7
Winco, Div. of Dyna Technology, Inc.: 19-1A, 19-1B, 19-1C

Index